AF580335

Surviving With The Biosphere

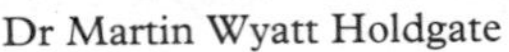

Dr Martin Wyatt Holdgate

Dr Mostafa Kamal Tolba

To
two outstanding leaders of the environmental movement,
Martin Wyatt Holdgate
and
Mostafa Kamal Tolba
and their successors in their respective
world-spanning key positions,
this book is dedicated
with admiration and affection

Surviving With The Biosphere

Proceedings of the Fourth International Conference on Environmental Future (4th ICEF), held in Budapest, Hungary, during 22–27 April 1990

Edited by
NICHOLAS POLUNIN
and
JOHN BURNETT

EDINBURGH UNIVERSITY PRESS

Edinburgh University Press Ltd
22 George Square, Edinburgh

Set in Linotron Plantin
by Koinonia Ltd, Bury, and
printed in Great Britain by
The Alden Press Ltd,
London and Oxford

A CIP record for this book
is available from the British Library

ISBN 0 7486 0314 X

Preface

The first International Conference on Environmental Future, or ICEF, took place in Jyväskylä, Finland, in 1971, and was a major input to the United Nations Conference on the Human Environment, which was held in Stockholm, Sweden, in 1972. Indeed, practically all of the threats which have now emerged publicly as menacing our world were discussed, or at least adumbrated, in the first ICEF. The 2nd ICEF took place in 1977 in Reykjavik, Iceland, and the 3rd was held in Edinburgh, Scotland, in 1987.

The present volume represents the Proceedings of the 4th ICEF, which took place in Budapest, Hungary, in late April 1990. Although not indicated in the titles of the first three ICEFS – respectively *The Environmental Future, Growth Without Ecodisasters?*, and *Maintenance of The Biosphere* – a major objective of them has been to lead up to the effective consideration of Mankind's and Nature's prospects of 'Surviving With The Biosphere', which was the theme and title of this fourth Conference in the series. Indeed, without a reasonably healthy Biosphere comprising practically our entire life-support, we humans cannot possibly survive at all equably on this Planet Earth, and the chances of doing so elsewhere now seem remote. Consequently, The Biosphere – ranging as it does from the highest levels of Earth's atmosphere to the deepest depths in the lithosphere *at which any form of life exists naturally* – must be enabled to survive. Indeed to ensure The Biosphere's survival is probably the most fundamentally important consideration for us all.

NICHOLAS POLUNIN & JOHN BURNETT

Acknowledgements

We express our grateful thanks to the following, each of whom participated in at least one meeting of the International Steering Committee, in Switzerland or England, from September 1988 onwards: Professor Lynton K. Caldwell, Professor Andrew S. Goudie, Dr Martin W. Holdgate, Dr Erik Jensen, Dr Bent E. Juel-Jensen, Dr Donald F. McMichael, Dr Norman Myers, Dr Ivan Polunin, Dr N. V. C. Polunin, Professor M. E. D. Poore, Dr Arthur H. Purcell, David Shreeve, Dr John R. Vallentyne, Professor Gabór Vida, and Dr Arthur H. Westing.

Between them, the above made numerous helpful suggestions, assisted with key decisions regarding the conference, and well satisfied our set stipulation of having representatives from not fewer than six countries involving at least four continents.

The conference could not have been carried through without the help and support of numerous friends and well-wishers, and in this we were again extremely fortunate – first of all in our hosts and joint sponsors, the Hungarian Academy of Sciences, and then in our generous co-sponsors, the United Nations Population Fund (UNFPA), the Japan Shipbuilding Industry Foundation, and an anonymous donor on behalf of the World Conservation Union (IUCN). Additional generous supporters were the United Nations Environment Programme (UNEP), the Government of Hungary through their Ministry of Environment and Water Resources, and the World Wide Fund for Nature (WWF International).

We would like to convey our special thanks and deep appreciation for their help to Professor István Láng, General Secretary of the Hungarian Academy of Sciences; Professor Gabór Vida (our co-Organizer & Local Agent in Budapest) and his ever-attentive understudy Dr Andréas Demeter; Mrs Georgina Pechlof for invariably finding remedies; Mrs Helen E. Polunin for organizing and leading the accompanying persons' excursions; the Secretary-General's Personal Assistant, Mrs Lynn M. Curme, for her help in many and often vital ways; and Dr Richard G. Miller, of the Foresta Institute for Ocean and Mountain Studies.

Contents

Prologue and Opening of the Conference

INTRODUCTION
by the Secretary-General as Chairman

Your Excellencies, distinguished – all-invited – participants, and friends; ladies and gentlemen. Welcome to the opening of the Fourth International Conference on Environmental Future, on the background and auspices of which you will find a micro-history in your satchels. I have been asked by our co-hosts, the Hungarian Academy of Sciences, to introduce to you our splendid Patron, His Excellency Professor Dr F. Brunó Straub. This I consider it a unique honour to do – to introduce a national leader who has occupied also the Presidency of the World's scientific summit – to an eminent gathering of globally-minded authorities on our [supremely] vital theme and in his own country! For if The Biosphere does not survive, our civilization assuredly will not.

Professor Straub has occupied an extraordinarily wide range of high offices in his continuingly active life. Primarily a biochemist, he has been President of the Hungarian National Council for Environment and Nature Protection, Vice-President of the Hungarian Academy of Sciences – whose current President, Professor Iván T. Berend, is also gracing us with his presence here today – and, on the world scene, President of the International Council of Scientific Unions (ICSU) [which is indeed the World's scientific summit]. Until recently he was President of the Presidential Council of the People's Republic of Hungary and so, effectively, head-of-state of this very dear country.

Ladies and Gentlemen, I present to you His Excellency Professor Dr Brunó Straub.

RESPONSE
by Professor Dr F. Brunó Straub, Patron of the Conference

Mr President, Ladies and Gentlemen: I am very glad to greet you here on this occasion in Budapest and I see that eminently a rainy day was provided to keep the people in the audience and not straying away because the view is really beautiful from here if the sun is shining. Now you know that your Conference is gathering in a time which [sees] big changes in Hungary. We had the second round of the elections on the 8th and now the Government is being formed which means that we have completed a two-years' procedure of changing from our [former] one-party system to our [new] multi-party democratic system [and] we hope from our state centrally-oriented economy to a market-oriented economy. And in the autumn, I think in September, we shall have elections at the municipal level, which will provide for the people in the country, in the villages, and in the towns, to decide for themselves what they [want] in any respect.

Now those factors mean that the chances of doing work for environmental protection and for the future of the environment are great. The chances [are there], but it's very important to use these chances in order to do something, because young people mostly gave demonstrations during the past years and that was [all] very nice; they came together and demanded things but nothing happened. Now the time is coming when something has to be done, and the problem according to my mind is [that] we have to decide the priorities. There are so many things to do in environmental protection, but what are the priorities? One has to be careful not to spend money on something which can be done at a later time; we have to do first things first.

As an example I would say that I had a discussion for a long time with no success with people in the water authority about what is the first thing to do: let the people have tap-water in the smallest villages, or to have that only in some and provide for the sewage? Now what has been done for decades was to increase the flow of water to the villages and to neglect sewage. In our country which gets all its water from rain or by surface water from abroad, we get 90% [of the river water] from the Danube and 10% from the Tisza and other small rivers coming [into] the eastern part of Hungary.

In this country the water supply is critical. I was very glad to see in the Programme that there will be a discussion about this tonight. Now [we need] a decision [as to] what to do first: to supply more drinking-water through pipelines to every village, or to supply drinking-water and at the same time to provide for sewage. This is an example where I believe knowledge must be applied, [yet] in environmental actions it is nowadays still difficult to come to decisions because we are at a stage where feelings are running high and knowledge is still rather low. Therefore sometimes we are misled by feelings and do not in fact do what has to be done.

[Even] when it comes to experts there is [still] a difficulty. I do remember our discussion upon the disposal of high-level waste from atomic power-stations, where I've participated in two groups. One was purely scientific, so to speak, of physicists and technologists and so on, [while] the other [was] where the people were already engaged by the atomic industry, and the people within the industry were very strongly opposed to the other group, [feeling] that they are just interfering – they don't know what it's about because the only people who understand the problem are [our]selves. This is the difficulty, then, to decide: do these people have a vested interest in fathering something on which [we] are convinced they are wrong, or are these absolutely honest people who are telling you the right things whereas the outsiders do not know what they are talking about?

Similar situations arise in forestry I think [and] water management everywhere. Therefore sound knowledge [surely] is one of the most important commodities [which] we need nowadays. And I do hope the present discussions will contribute to it. I mentioned the sewage and water problem: we know that international agreements are already binding us to do something about acid rain and its causes, and whatever we can do about the climate; but I think if we spread our resources too thin then we [will not be] doing what's expected from us in the future.

So I believe [that] not isolated actions but coherent action in a logical order is needed, and we expect to learn in that line something from your discussions. I hope that you will also have the chance to look around in our country and see that it is not [everywhere] spoiled but that there are some fine places still. I wish you good discussion. Thank you.

Professor Straub was followed by three others who were among those seated on the platform for the Conference's opening. First came *Professor Iván T. Berend, President of the Hungarian Academy of Sciences*:

Mr President, Ladies and Gentlemen: On behalf of the Hungarian Academy of Sciences, I warmly welcome you in Budapest on the occasion of the 4th International Conference on Environmental Future: Surviving With The Biosphere. It is symbolic, and it gives great emphasis to *this* occasion that the Opening Ceremony of the Conference is starting exactly a day after Earth Day [was] celebrated yesterday, which means that a lot of interest and great attention is paid to these environmental issues.

Your regular meetings and conferences make a fantastic job of revelations of environmental issues. You started almost 20 years ago in Finland, [when already] you raised most of those basic questions which are really in the forefront of environmental problems. Of course, meantime in this almost 20 years, the world learned [much], partly from your work and partly from others' work, [so that] peoples and governments are [now] much better informed about environmental problems than before. They are much better prepared to carry out proper actions to save the environment, to save the future, and we are really glad and happy with all these [advances]. However, of course the quality of life on Earth is not improving but actually the reverse.

The bulk of the work is before us: it still belongs to the future, [as is particularly] visible and understandable in the relatively poor countries [such as] the East European countries, including Hungary. This country has had an economic strategy in recent decades – a kind of forced industrialization – when ecological consequences were not really [under] consideration, and all these negative side-effects of forced rapid industrialization were partly neglected [until] environmental issues started to arise and attracted interest in the [latest] years or decade. It is also quite symbolic that the 1980s became a decade of environmental debate in Hungary as well. The basic reason was the very well-known case of the Danube Dam.

The Danube Dam issue [was raised by] a governmental decision, together with the Czechoslovak Government and an investment project, [and would involve such] a major change in the environment [that it] caused a tremendous debate inside the country. I am rather proud that the Hungarian Academy of Sciences belonged to those in this country [who] from the beginning opposed this project, and already [in] the early 1980s and mid-1980s worked out ... alternative proposals for the Government. That started a long debate, a long fight, and in a way it showed [up], and is a symbol of, an environmental issue which became a major political problem of the country, and played a very important role in the development of the opposition groups against the Government – because it was not only the environment but [also] the political environment which came under scrutiny.

[The Dam issue showed up] as a negative effect or consequence of an uncontrolled governmental action, [initiated] without consideration of warnings of scholars and other people. [We should all realize] that this period [of your] 4th Conference is a period of major change in this country and all the neighbouring countries. It is a period when we are working out a new economic strategy for the future, and I hope very much that this new economic strategy will be [an] environment-oriented new policy which will not follow a one-sided effort of industrialization but will consider, [and] will care for, the environment. [It must] care [for] all the future and [not] destroy the future in [the furtherance of] a short-sighted economic concept.

I think that we need all the scholarly work [possible] for these countries and those parts of the world which are poor and relatively backward, and where environmental issues are still [widely] neglected. We are waiting for the new results of these kinds of talks and discussions, and I wish [you] again, on behalf of the Hungarian Academy of Sciences, a good Conference, a very, very fruitful debate, good lectures, and a very pleasant stay in this country. Thank you very much.

Next came *Dr János Zákonyi, Chief Ministerial Counsellor of the Hungarian Ministry for Environment and Water Management*, in the absence as yet of a Minister of the reconstituted Ministry:

Ladies and Gentlemen, Mr President: On behalf of the Hungarian Ministry for Environment and Water Management, I have the privilege to welcome you and to express also our pleasure that Hungary can host this meeting of prominent representatives of international environmental science and international organizations. Ladies and Gentlemen, I think that a lot of very valuable ideas [have been] given already by excellencies [and others] here. I do not want to repeat, and I have the instruction to be very [brief], but still I would like to point out just one issue.

As has already been mentioned, in Hungary we have now a period of enormous changes. We environmentalists – official and unofficial environmentalists – are convinced that this period has to be used for the restructuring process together with improvement of the environment. We have now the historical chance to influence more the economic processes [and] political processes [than has been possible hitherto].

Just two more remarks: I am very sorry, Ladies and Gentlemen, that [although] the Hungarian Meteorological Service belongs to our Ministry, [even together we] could not manage for you to have better weather! The [other] remark I'd like to make is that the environmental Ministers of the COMECON countries have their meeting at this time in Hungary. With the agreement of Professor Polunin, we [are hoping] to organize a meeting for you with them tomorrow – perhaps to hear together the lecture of Dr Mostafa Tolba and then to have a possibility for [participatory dialogue or at least] a little conversation. I think science and practical state administration have to work together in a better way [than we have been able to] accomplish until now. To finish my speech, Ladies and Gentlemen, I wish you a very successful Conference, a pleasant stay, and good impressions of our country. Thank you.

Last came *Mrs Judit Vásárhelyi, representing Hungarian Independent [Nongovernmental] Organizations*:

Mr President, Ladies and Gentlemen: On behalf of the Hungarian environmental groups, circles, associations, foundations, etc., namely the Hungarian nongovernmental organizations, it is an honour for me to welcome you here in Budapest, one day after Earth Day, and on this Earth Weekday. I won't repeat commonplaces about the environment. All these commonplaces are true. On one hand, Frau Brundtland's report put an emphasis upon the need for increased participation of nongovernmental agencies and organizations in the field of protection of the environment. On the other hand, the young Hungarian environmental movements and groups need professionalization, better organization, and more expertise. For these two reasons I highly appreciate the opportunity to listen to your lectures, to your work, your approach, opinion, and results. To be in opposition with the Ministry, I think, on behalf of nongovernmental organizations, that it *is* very good weather today, for the sake of the Earth, because we have [had] a very dry period, so this [rain] is a blessed sort of relief! Just to finish, I wish you very effective work for the sake of the common future. Thank you.

Response
of Chairman

We thank you, Your Excellency and eminent Lady and Gentlemen, for your wise, and wherever possible, encouraging, words, and signify that we have also received impressive and often heartwarming messages from several other special sources. These include Finland where these Conferences began, various interested parts of the United Nations, the Chairman of our joint Co-sponsors the Japan Shipbuilding Industry Foundation, and the Executive Director of the United Nations Population Fund (UNFPA), (the Director-General of IUCN is with us here), and from the former Governor-General of Australia in his unique capacity of Australia's Ambassador for the [world] Environment. There being no time to read even a chosen selection of these messages now, I ask you to allow Sir John Burnett and myself, as Editors of the Proceedings volume, to handle this matter by publishing herein a few of these messages.

This leaves me with one immediate task to perform. Ladies and Gentlemen, a Conference that does not conclude with an agreed statement will at least miss a chance, and so we are hoping to agree upon a Budapest Imperative on Survival With The Biosphere.

A Selection of Messages received from Co-Sponsors and other Supporters and Well-wishers

Message from *Dr Nafis Sadik, Executive Director, United Nations Population Fund (UNFPA):*

[Professor Polunin], distinguished representatives, colleagues and friends. On behalf of the United Nations Population Fund, I should like to say how pleased we are to be able to serve as a co-sponsor of this most important Fourth International Conference on Environmental Future. One cannot help but feel that this Conference will make an important contribution to the well-being of future generations, especially in view of the crucial environmental issues to be discussed, the expertise that participants bring to this Conference, and the effectiveness of previous [such] Conferences to forge consensus and influence policy.

We are particularly pleased to see that population problems figure prominently on the Conference agenda. Why should we be concerned about this?

The trends and patterns of population growth and distribution, resources, environment, and development, are inextricably linked. At present, the human [species] numbers 'only' 5.25 [thousand millions]. Yet, already our industries pour acid rain on the temperate forests, rivers, and lakes; they damage the essential layer of ozone between the Earth and the Sun's ultraviolet rays; we have degraded millions of hectares of agricultural land; we are clearing away the tropical rain-forests and the thousands of species [that are] found in them; and now it seems that we are changing the climate itself – warming the global atmosphere in a process whose full consequences cannot yet be calculated.

Looking ahead, the world's population is likely to grow by an average of 100 millions per year during the 1990s. And over the longer term, progress in reducing fertility rates has been [so limited] that the eventual total world population cannot [be hoped] to stabilize at 10 [thousand millions] – as projected by the United Nations in 1984 – but rather is likely to reach 11 [thousand millions]. In many of the poorest [of the less-developed] countries, population momentum is such that the demands of human numbers for food, health services, schooling, and employment, [is projected to] double in a mere 20 to 30 years.

How will the planet cope when 90% of future growth is expected to [occur in] the less-developed countries? Increasing numbers of scientists, development officials, and non-governmental organizations, are concerned that rapid population growth will have negative consequences for global carbon-dioxide emissions, waste generation, and water resources. This applies particularly to countries where poverty is rampant, appropriate technologies are lacking, and imbalances in the distribution of population growth are adding to urban congestion. Nor has any assurance been provided that the transfer of non-polluting technologies and related financial assistance between more- and less-developed countries will be sufficient to offset implications of rapid population growth.

Perhaps the most troublesome, if not politically immense, question concerning population growth, environment, and sustainable development, is to ask: Is it possible that contributions to conserve the environment in more developed countries – those with the means to invent and adopt non-polluting technologies and clean up their wastes – might be cancelled out by the huge, anticipated population effects in less-developed countries? These are the kinds of questions [which] we hope this Conference will help us [to] answer and [in time] resolve.

The challenges are, of course, daunting. Yet today as never before, a common humanity is emerging in the quest to protect a mutual friend – our very homeland, Planet Earth. Representation here by so many distinguished people is testimony to that fact. We, at the United Nations Population Fund, join you in the search for a common solution to Safeguarding Our Future.

NAFIS SADIK, *Executive Director*
United Nations Population Fund (UNFPA)
220 East 42nd Street
New York
NY 10017, USA

Message from *Mr Ryoichi Sasakawa, Chairman of the Japan Shipbuilding Foundation,* received in Geneva, Switzerland, with accompanying letter addressed to the Conference's Secretary-General:

It is my great honour to present my views on such an important subject as the protection of the human environment to this distinguished audience of political leaders, scientists, and administrators, as well as to those representatives of the conscience of global citizens. The diverse issues of environmental protection have indeed become the most important task for Mankind to confront. I have a strong belief in the harmony [of] people who have their own characters [yet feel] 'The World is One Family: All Mankind are Brothers and Sisters'. This faith of mine applies to the questions of environmental protection in its truest form.

A sweeping change has been taking place in the world, in particular in [your] part of the world. The code-words of this change are: freedom and participation, human dignity and quality of life, and globalization and interdependence. All of these values are essential requirements for the protection of [our] environment. Aspirations of global citizens for the improvement of human environment and the spirit of our time enhance the critical importance of each [to the] other.

Perhaps the most salient factor of the dramatic changes in the world is the re-emergence of Middle Europe. Budapest, of course, is the centre of Middle Europe. It is symbolic for the world to hear the messages on the most critical issues of environmental protection from the heart of re-emerging Middle Europe. These messages will provide the hopes as well as the tasks for Mankind at this important crossroad of history. It is indeed a great honour and pleasure for the Japan Shipbuilding Industry Foundation to contribute, albeit in a small way, to the efforts of this [occasion of] historic significance.

I wish you great success in your deliberations. Thank you.

RYOICHI SASAKAWA, *Chairman*
Japan Shipbuilding Industry Foundation
Senpaku Shinko Building
1-15-16 Toranomon
Minato-ku
Tokyo 105, Japan

Dr Martin W. Holdgate conveyed the good wishes of the World Wide Fund for Nature, with apologies for his absence from the Director-General, Mr Charles de Haes, as follows:

WWF is pleased to be associated with the Conference, and hopes for results that will improve global understanding of environmental priorities. For its part, the WWF attaches high priority to arresting the current unprecedented loss of global biological diversity and the destruction of tropical forests and wetlands. However, WWF accepts that, as the World Conservation Strategy stated, such damage can only be halted by the universal establishment of procedures which [also] meet human needs and curtail runaway population-growth, catering for human requirements through the sustainable use of natural resources.

WWF works in especially close partnership with UNEP and IUCN, among the wider groups of organizations concerned with the world environment, and sees this meeting as another demonstration of this partnership. At a time when WWF is expanding its work in Eastern Europe, the holding of the 4th ICEF in Hungary is particularly welcome. The WWF congratulates Professor Nicholas Polunin for his tireless efforts in organizing the Conferences, thanks the generous Hungarian hosts, and [feels] confident that the world community [will] benefit greatly from the outcome.

CHARLES DE HAES, *Director-General*
WWF International
Avenue du Mont-Blanc
1196 Gland, Switzerland

Message from *Professor Dr Victor A. Kovda, Former President of SCOPE and the International Society of Soil Science*, and key advocate of the Vernadsky International Centre for Biosphere Studies, addressed to 'Dear Mr President, Professor Polunin; Dear respected colleagues,':

I very much and most sincerely regret that I could not participate in this historic meeting of internationally recognized stars involved in the solution of the problems of preservation of The Biosphere.

My life [has been] dedicated to the study of these problems in my country and other continents of the globe. I am happy that [two] brilliant representatives of my Institute [the Director and Assistant Director] are participants in this important event. I hope this will [mark] the beginning of our more close and regular contacts [continued last] in March 1989 in Moscow–Pushchino. The ideas and suggestions of the participants of the Symposium 'Man and Biosphere', to establish in Pushchino, The Vernadsky International Centre for Biosphere Studies, have been fully realized. In addition, the scientific discussion-club 'Biosphere' has been created as a body of the above Centre and is actively working now. Scientists from all over the world are [being] invited to present lectures which will be published in Pushchino, [where we look forward to seeing you again soon].

[Signed] VICTOR A. KOVDA
Vernadsky International Centre
for Biosphere Studies
PO Box No. 156
142292 Pushchino, USSR

Message from *participants at the 1st ICEF:*

The first International Conference on Environmental Future, [held] in Finland in 1971, was of the utmost importance. It was held at a time when the global scientific community had just observed that the life-supporting machinery of Nature had been damaged by the actions of Man. [For] Man had taken the ingredients of his affluence from Nature by [applying] the force of communicated knowledge accumulated in a long chain of generations – something totally external to the DNA-system. We understood that the damage observed in the life-supporting machinery was deepening rapidly, and we understood that the conclusion of the development [which we were witnessing] would be an ecocatastrophe – a holocaust which had only recently got that name from Paul Ehrlich [but which we have since preferred to call an ecodisaster].

The 1st ICEF was one of those key meetings which gave rise to the development of a new branch of science – environmental science. No other branch of science has grown with such speed and power as environmental science. (During recent years *Pollution Abstracts* has already listed about 10,000 scientific papers annually – and that is, of course, only a small, though clearly-defined, fraction of the total production of this branch of science.)

[Your] gathered biological [especially ecological] knowledge already outlines the means by which we can change our present catastrophic route to a route [at least] assuring us [of] the continuation of life. In spite of this, the deterioration of Nature is continuously increasing. So, we have not yet solved the problem of how to restrict the human inclination to take more ingredients of affluence from Nature than [should be] allowed by Nature. It is necessary to change the behaviour of Man to conform to the true spirit [and edicts of holistic ecology].

SIRKKA-LIISA & PEKKA NUORTEVA
Department of Environmental Conservation
University of Helsinki
Vikki 00710, Finland

Celebration of Earth Day

At the instance of the organizers of Earth Day '90, based on Stanford, California, and in accordance with a decision of the Foundation for Environmental Conservation at its fifteenth Annual General Meeting, the Secretary-General and his Wife gave a Conference Foregathering Drinks Party and Buffet Supper on the top floor of the Conference venue, Hotel Agro, during which he addressed those present as follows:

Your Excellencies, Ladies, and Gentlemen: A very warm welcome, in prospect, to you all for the Fourth International Conference on Environmental Future which we start early tomorrow.

It is my honour, and gives me deep personal satisfaction, to introduce to those of you who may not know him, one of our favourite leaders in the environmental/conservational movement. Concerning another I'll be having a similar pleasure before you on Tuesday.

This one has most generously given up the whole of our Conference week to be with us here in Budapest, though he must be among the busiest executives in the world – as well as one of the most important, at least from our viewpoint.

By now you will have divined that I refer to Dr Martin Wyatt Holdgate, Commander of the Bath, Director-General of the World Conservation Union (IUCN), and a leading member of the International Board of Sponsors of

EARTH DAY

which is being celebrated this very day, April 22nd, practically throughout the world. Earth Day – now often referred to as International Earth Day – has, for some 20 years past, served its vitally important purpose of alerting people everywhere, of every sort, creed, and calling, to caring for our Planet Earth as one unique entity that we humans, by our unprecedented numbers, pandominance, and general profligacy, are threatening more and more grievously.

Dr Holdgate has a slogan, which should lead to a way of life, that I am anxious that he – whose idea it is – should be the first to give out on this multiply special occasion, and, through media here represented, pronounce to the concerned world.

Your Excellencies, Ladies, and Gentlemen, pray silence for *Dr Martin W. Holdgate.*

To which Dr Holdgate responded as follows:

The original Earth Day took place some 20 years ago. It was one of the first major events that gathered people together in their thousands to express their concern about the environment. But it was not truly a worldwide occasion, being concentrated in the United States; and it had a touch of protest about it.

Twenty years ago there were very few organizations concerned with the environment. Now there are literally thousands. [Following the 1st ICEF in 1971 and the UN Conference on the Human Environment in Sweden in 1972: we have been given many warnings of environmental degradation], and now every year seems to bring on more environmental meetings and discussions, [while] national Departments of the Environment, environmental protection laws, and allied agencies, have mushroomed.

Yet despite all this [and very much more], the world environment has continued to deteriorate. Why? One of my colleagues summed it up very well recently when he said 'Thunderstorms of rhetoric and monsoons of conferences and declarations have been followed by droughts of inaction.' The fact is that humanity has been great on talking but weak on doing, and now we face even more urgent environmental problems, on a greater scale, than we did 20 years ago.

We shall only solve these problems if there are fundamental changes in human attitude and behaviour, affecting all of us at the individual, national, and global, levels. Nations need to be convinced that the environment is not just some green fringe, to receive its modest share of resources in order that national parks and Nature reserves can be established alongside factories, motorways, and power stations, but in a very real sense is the natural capital, the natural wealth, of [every] nation, without which that nation cannot expect to prosper.

At the global level, we have to recognize that the current, uneven use of the world's limited environmental resources is in itself an impediment to any escape from poverty and the population explosion by the developing countries. Debts and the structure of world commodity markets, are still causing a net flow of resources from the poor countries to the rich ones, and I am told that last year this totalled some 25,000 million dollars. How can we possibly build a common future for humanity with such distortions?

But above all, it is at the individual level that we need to start. We have to convince people that their own actions can do something to make their own environments better and influence the world as a whole.

One Earth Day every 20 years, or even every year, is clearly not enough. *Every day must be Earth Day*; for without due care of the Earth, future days will be dark indeed.

I give you the toast: 'The Earth, and Its Future' [to which was added by the host] 'Every Day An Earth Day'.

1. Overview: Our Threatened World

MARTIN W. HOLDGATE
Director-General, The World Conservation Union (IUCN),
Rue Mauverney 28, 1196 Gland, Switzerland

INTRODUCTION

An opening address should provide a point of departure. The literature is full of prophecies of the impending doom of global ecology and human institutions. Such generalizations have been valuable in forcing both statesmen and scientists to face the implications of current trends, but more critical analysis is needed if priorities for action are to be judged correctly. This paper is intended as a signpost to such analysis.

It is important to be clear about the meaning of my title. Glib and superficial statements about 'the end of the world' somehow seem to imply that the Earth as a planet is now physically threatened by the excesses of humanity. That is obviously nonsense: so far as we can judge, the Earth will endure for several thousand of millions of years more, until the Sun emerges from its present more or less stable phase to engulf or vaporize the inner planets. The threat we face today is to the world of humanity, and to the living systems within the thin envelope of The Biosphere. This threat arises because of the escalation of human populations, human impacts, and human demands for the renewable and non-renewable resources of the Earth. These trends – which take different forms in developed and developing countries but are present in them all – have brought humanity on to a collision course with the planetary systems that have sustained our species, *Homo sapiens*, throughout the millions of years of our evolution, and are bringing components of our society into conflict with their neighbours in space and time.

A major purpose of this Conference is to discuss strategies for avoiding the worst of these collisions. Such strategies will have to be directed primarily at human targets. Science will be needed to give us insights into the structure and dynamics of environmental systems and their likely response to human actions; but the main instruments for averting threats will be those that modify human perceptions and behavioural patterns.

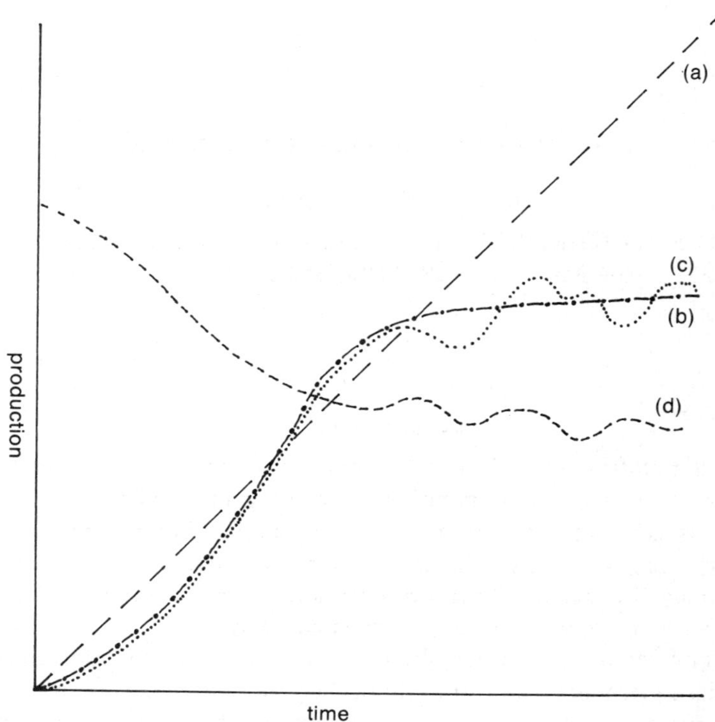

Figure 1.1. Patterns of development. Curve (a) represents the unwarranted expectation of steadily increasing growth for an indefinite period. Curve (b) is the expression of theoretical expectation that development will increase yield to a sustainable steady state. Curve (c) is the more realistic expectation that actual yield will fluctuate according to environmental circumstances. Curve (d) traces actual experince with many mismanaged systems: falling production has first to be halted and may then be restored to a new, sustainable level.

Development of the Threat

The present situation is the cumulative consequence of innumerable small and localized impacts over millennia, superimposed on ecological patterns which are both geographically diverse and highly dynamic. These human impacts have, however, one thing in common despite their regional and cultural diversity: they are the consequence of a set of processes that we call 'development'.

Development is the progressive modification, by humanity, of the environment and ecosystems of the Earth so as to increase the perceived usefulness of that environment to the human species. It changes the pattern of use from low-intensity (as in a forest, where the whole area may be used for hunting and over 50% of the tree species may be used for products – other than firewood) to high-intensity, based on a greater abundance of a reduced number of species. The replacement of forest by pasture and cropland, the alteration of hydrological systems for irrigation, the creation of fish-ponds, the quarrying of stone, and the mining of ores, are all manifestations of development, and have permitted the expansion of humanity – which in turn led to the demand for yet more development. Today we face questions of the sustainability of the development process – not for the first time, but for the first time on the present scale (World Commission, 1987).

Figure 1.1 represents in a generalized manner the range of possible patterns of development and we can illustrate the expansion of human impact with more precision by reference to three 'snapshots' of the world 2,000, 100, and 10, years ago. During the first period, the Roman Empire was at its height and there were major agricultural and urban civilizations in the Mediterranean basin, Ethiopia, Sudan (Imru, this volume), and the Middle East. Some of the cities undoubtedly experienced problems of local water pollution and water-borne disease (Romans complained of the stench in the Tiber in summer, near the outfall of the great drain, the Cloaca Maxima). Plato had already complained about the denudation of the hills of Attica, as erosion followed the excessive cutting of forests (Thirgood, 1981). In Britain, forest clearance, partly at least by fire, had truncated the soil profiles and almost certainly diminished productivity (Dimbleby, 1978). On the southern coasts of the Mediterranean, the grain-fields around great cities such as Leptis were very probably already showing signs of degeneration in the process that, today, we all-too-glibly call 'desertification'. The civilizations of Egypt and Mesopotamia were in a final phase of decaying glory, partly brought about by the non-sustainability of their irrigation systems (Jacobson & Adams, 1958). There were great civilizations also in the Indus basin, and in China, and there were substantial cultures in Central and Andean South America. In some central and southern parts of Africa there were probably the precursors of the civilization that flourished later in such places as Great Zimbabwe. But Australia, most of South America, and large tracts of North America, northern Europe, Africa, and Asia, were still tied to a hunter–gatherer, pastoral, or small-scale cultivation, pattern. Certain large mammals, such as mammoths (*Mammothus* spp.), mastodons (*Mastodon* spp.) and the ground-sloth, (*Mylodon* sp.), had been eliminated, partly at least because of over-hunting (Martin, 1967). Moas (species of *Emeus* and other genera) still thrived in New Zealand only because people had not yet arrived or, anyway, settled there. Even so, to us the world of 2,000 years ago would have seemed highly natural, and the sea would presumably have borne no mark from the impact of contemporary fisheries.

A hundred years ago the world was at the height of the colonial era. Just as Rome two millenia earlier had taken much of the Mediterranean basin and western Europe into its grasp, so Europe in turn dominated large tracts of Africa, South America, and the Indian Sub-continent, and had destroyed some advanced indigenous civilizations in South and Central America. The development of the resources of the colonized regions, with considerable transfer of materials for the benefit of the colonizers, was concentrating great wealth in Europe. Some former colonial regions – notably the United States and the new republics of South America – had established their political independence and were beginning to pursue their own paths of industrial development. In the developed regions, medicine was enhancing human survival, and the combination of wealth, high fertility, and diminishing mortality, allowed populations to grow rapidly. Some of the benefits of the medical advances had been transferred to the colonial territories, aiding population growth there. Industrial air-pollution had become an acute local problem, and the first legislation to curb it, in the United Kingdom, had been on the statute-book for nearly 20 years (Ashby & Anderson, 1981). Water pollution and water-borne diseases were also causing severe problems in many parts of Europe, and people were beginning to question whether dirt and wealth really did go hand-in-hand. The first major technological revolution had borne much fruit, and the ecosystems of considerable parts of the world were substantially changed.

Ten years ago the post-colonial world was well embarked on the period of environmental concern, stimulated particularly by the United Nations Conference on the Human Environment, held in Stockholm in 1972 (McCormick, 1989). Medicine had markedly enhanced survival. Most of Europe had moved through the 'demographic transition', in which birth-control had reduced fertility more or less to match mortality, giving near-stable populations, and North America, Australasia, and Japan, were showing signs of following along the same path. The divergence between the developed and developing countries had become very great, with high wealth and disproportionate consumer consumption in the former and a dependence in the latter on the basic production of natural resources, exported to gain the money required to purchase the instruments of industrialization. Both commodity prices and the costs of industrial technology were largely controlled by the developed nations. Debt, accumulated in the developing countries as a result of misplaced loans, was already beginning to impede the process of development, because too much of their export earnings had to go to service the debt burden. Pollution had expanded from the local to the regional or even global, with the recognition of acid rain as a continental phenomenon in Europe and parts of North America, of marine pollution as an issue in land-locked seas such as the Mediterranean, North Sea, and Baltic, of stratospheric ozone depletion as sufficiently probable to justify the creation of an international committee to coordinate action, and of the 'greenhouse' effect as a more-than-plausible assertion.

Again the North–South divide was evident because virtually all of these major pollution problems originated in the developed countries but were superimposed upon the other problems of the developing regions. Even more seriously, a 'perception curtain' had descended, the developing countries seeing the affluent, over-consuming North's demands for environmental conservation and pollution control, as, at best, an expression of the conscience of the rich and, at worst, a plot to maintain the imbalances of the world. Meanwhile the developed countries saw the unrestricted population growth of the poorer nations, the instability and unpredictability of their governments, and the failure of their economies to industrialize, as severe weaknesses threatening the global future of humanity. That 'perception curtain', with some bitterness and suspicion, persists to this day and constitutes part of the context for our present Conference. It poses a real threat to the world of humanity.

Nature of the Threat

There are four main components to the threat which is now evident around us. They stem from:

(a) The magnitude of the changes which have already been brought about in The Biosphere.
(b) The fundamental nature of the processes that have been modified.
(c) The impetus of current trends.
(d) The lack of mechanisms by which these trends may be halted before severe damage is done.

The Magnitude of Change

The scale of the transformations is evident from the following facts:

(i) The content of carbon in the standing crop of world vegetation has been reduced from a pre-agricultural total of between 800 and 1,100 x 10^{15} g to 680–700 x 10^{15} g a century ago and to 560–590 x 10^{15} g today (Bolin *et al.*, 1979; Olson *et al.*, 1983);
(ii) Some 40% of plant productivity on land is now stated to be utilized directly or indirectly by humanity (Vitousek *et al.*, 1986);
(iii) Some 40% of the original tropical moist forest has been converted to other formations. Temperate-zone deforestation, which began earlier, has removed a significantly higher proportion of the original vegetation. Between 25% and 50% of the original wetlands have been modified. Some 35% of the Earth's land-surface – 75% of the world's rangelands – is affected by various kinds of degradation (World Resources Institute, 1988). One-third or more of that land has lost more than 25% of its productive potential (Postel, 1989). In some countries (*e.g.* Pakistan), well over half the irrigated land is affected by salt accumulation.
(iv) Atmospheric pollution by chlorofluorocarbons has depleted stratospheric ozone over Antarctica in springtime by up to 80% at certain altitudes and for brief periods, and globally this ozone screen, which excludes UV-B radiation that is potentially damaging to primary production as well as to human health, is likely to have been depleted by several percentage points; and
(v) The accumulation of 'greenhouse' gases (notably CO_2, CH_4, N_2O, CFCS, and tropospheric ozone) is calculated to be likely to raise global mean temperatures by 1–2°C by 2030 AD, while associated sea-level rise (due to thermal expansion and the melting of mountain glaciers) will be likely to average 10–20 cm globally (Commonwealth, 1989).

The Processes Affected

The above changes affect fundamental biogeochemical and ecological processes. The impacts of 'greenhouse' gas and CFCs (chlorofluorocarbons) on the atmosphere will modify radiation penetration and retention, and hence the dynamics of the whole system. Changes in global temperature and precipitation patterns will inevitably have profound consequences for ecosystems and human utilization patterns (Commonwealth, 1989). The transformation of forest and other ecosystems, and the degradation of pastures and soils, is already curbing agricultural expansion in areas whose human populations are in desperate need (IUCN, 1989). The transformation processes are commonly negative in their long-term implications for human development (Reid *et al.*, 1988).

These impacts, put simply, mean that, in many parts of the world, human populations are overdrawing their 'natural capital'. That capital consists of the resources of soil, air, water, and living systems, from which come all the essential needs for human life. It is being reduced by bad development which reduces the productive base – either by over-using it and breaking down its inherent fertility and productivity, or by damaging it through the impairments resulting from local or global pollution.

The Impetus of Current Trends

The core of our present dilemma – the real threat to the human future – lies in the fact that current trends have such an impetus that it will not be possible to stop even the damaging ones on any short time-scale. The world, indeed, is clearly going to need accelerated and intensified development if it is to accommodate some 8 to 10 thousand million people a century from now as against the 5.2 thousand millions who are present today. It is easy to be glib, and draw the obvious conclusion that such development must be 'sustainable' (World Commission, 1987). But we have to be clear as to just what we are demanding by such an assertion. There are ambiguities in the way in which the term 'sustainable development' is used, not least because it has become a universal cliché, accepted by an enormous diversity of people of whom few have looked back at the original definition.

Sustainable development is *not* the continuous projection of a curve of growth in the yield from a particular environmental resource, although many politicians and economists act as if it were. Sustainable development is about expanding the productivity of a particular system for human welfare up to the point at which a particular yield can be maintained indefinitely into the future (see Figure 1.1[a]). In all cases it is likely to mean accepting a fluctuating yield from year to year (as every farmer does), because *inter alia* of fluctuations in environmental conditions. In some cases it may involve *reducing* yields below present levels in order to maintain optimal production – for example by reducing the dependence of agricultural systems on nutrient inputs which in turn depend on the industrial use of fossil-fuel energy. It is essentially about *not* overdrawing the natural capital. The concept is easy to articulate. The problems come with turning it into practice. At the technical level, this clearly demands understanding of the environmental systems involved, the provision of some margin of resilience, and the insertion of management feedback loops so as to allow the system to be adjusted in the face of climate change, or of other new factors that affect the asymptote on the production-time curve. This in turn demands human understanding – and a lack of social and economic pressures which may force misuse because of short-term imperatives.

The Impracticability of Control

Projecting into the future, the threats with which this Conference is concerned are aggravated by the impetus of four major processes (each with many component variables), as follows.

1. Human Population Growth;
2. Escalating and Unsustainable Human Demands for Resources;
3. Degenerative Processes in The Biosphere Resulting from Human Pressures, and including:
   (a) Deforestation;
   (b) Desertification;
   (c) Loss of Biological Diversity;
   (d) Global Pollution;
   (e) Climate Change; and
4. Increasing Social Strife caused by Inequalities in the Global Sociopolitical System.

These require brief review.

1. Human Population Growth

The human population of the world rose very slowly from its primitive beginnings to reach approximately one thousand millions by 1800AD, doubled by 1900, and has now more than doubled again to reach 5.3 thousand millions. The period of rapid growth in Europe appears to be past; but elsewhere, notably in Africa and Asia, population growth is still proceeding at between 2% and 4% per annum. It is this, coupled with the fact that some 50% of the population of many developing countries is still under the age of 16, that makes a further near-doubling of world population seem almost inevitable. Even if the types of policy that have been attempted in China were suddenly imposed, and if all people now below reproductive age produced two children per couple, the demographic momentum would carry us globally towards the 8 thousand million mark.

It is commonly assumed that the experience of Europe will be repeated elsewhere; namely that a combination of economic growth, availability of consumer goods, higher-quality employment reducing the demand for manual labour, greater medical and social security so that children are not perceived as a necessity for the support of parents in old age, the availability of birth-control facilities and the removal of taboos on their use, and a broad change in social perception in favour of smaller families will, of itself, bring about a 'demographic transition' towards sustainability. This is no more than hypothesis, essentially based on *post hoc* rationalization of what appears to have happened in western Europe (where, however, there has never been an overt or deliberate population policy). The suggestion that the world's human population will level off at between 8 and 10 thousand millions is no more than a further hypothesis, based on extrapolation. It does not imply that the world can support that number, and it certainly does not imply that all countries everywhere will be able to sustain the numbers born into them before economic growth achieves a demographic transition (if it ever will do).

The real probability is that, in a number of areas, resources will be inadequate (even despite projects to aid development), populations will overshoot them, economic growth will not take place, and there will be social collapses followed by migration. This is already happening on a small scale in many parts of the 'developing world' (IUCN, 1990), and the accretion of such local failures to national level could pose a real threat to the stability of nations. Environmental refugees were said in 1989 to have already become 'the single largest class of displaced persons in the world' (Jacobson, 1989). The more we investigate these difficult interactions, the less certain we can be about the ultimate global demographic situation and the way of alleviating the threats which it poses.

2. Escalating and Unsustainable Human Demands for Resources

It is commonly (and truly) stated that human impact on the world is a function not of numbers alone but of the product of numbers and resource consumption. In many parts of the world, extra human beings make small demand on resources (the *per caput* consumption of energy in Bangladesh, for example, is two gigajoules compared with 278 gigajoules – 139 times as great – in the United States).

The conundrum which we face is a double one. The economic system – including the employment pattern – of the developed countries is still geared to market expansion and increased consumption. While pressure for environmentally benign products, waste avoidance, and recycling, may reduce resource consumption without undermining the system, at the present time the vision of a 'consumption transition', matching the necessary demographic transition and creating stability in energy- and materials-use (preferably well below present levels), linked to greatly diminished releases of pollutants such as 'greenhouse' gases, appears to be no more than an environmentalist's dream. Industry has begun to respond to that challenge, but it will clearly take decades for real change to occur.

Meanwhile the industrializing less-developed nations have thousands of millions of people weighed down by demeaning poverty. The hypothesis of the economic base for the demographic transition rests on the assumption that industrial growth and expansion in goods for consumers will be conducive to population stability. Given the experience of the developed world, however, that will not mean resource consumption stability. If 75% of the world's population even raise their *per caput* resources' demand to half that in the United States or western Europe, this would mean a five- to tenfold increase in energy and materials use, and with present technology this would be intolerably disruptive of The Biosphere. The challenge, linked to that of demography, is immense, and the prospect for millions of escape from the poverty trap is not good.

3. Degenerative Processes in The Biosphere Resulting from Human Pressures

(a) Deforestation

Human impact has already removed much of the forest that was the natural cover of temperate-region Europe, the densely-peopled parts of the Indian Sub-continent, and northern Africa. Today, deforestation is proceeding rapidly in virtually the whole tropical belt, whose rain-forests house much of the richness of the Earth's life. Until recently the best estimate was that losses were of the order of 11 million hectares a year, but new studies imply a much higher figure of 15 or 16, or perhaps even 20, million hectares annually (World Resources Institute, 1990). The statistics are complicated by the fact that not all forest is converted to non-forest land, and some reverts to other kinds of woodland. It is this deforestation that is largely responsible for the calculated drop in the amount of carbon locked up in the standing crop of plants from a 'natural' 1,000 x 10^{15} g to around 560–590 x 10^{15} g today (Olson *et al.*, 1983).

This process is unstoppable because of the rapid human population growth and the need for cropland in many tropical countries. Moreover, not all tropical forest soils are unsuitable for cultivation. For example, between 10 and 15% of the Amazon basin is suitable for agriculture or agro-forestry (J. A. Sayer, pers. comm.). The conversion of such lands is highly probable: the imperative is that the development be sustainable. That in turn depends on how effectively governments in various countries control the conversion of forest to other forms of land-use, how far the demonstration of economic benefits leads to the sustainable use of other forest systems (calculated to deliver more return than conversion under many circumstances: Myers, 1988), and the effectiveness of intensive agricultural development in adjacent areas in taking pressure off the woodlands. These are all socio-political issues requiring the determined attention of the politician as well as the scientist.

(b) Land Degradation (Desertification)

'Desertification', in the sense of severe and often near-irreversible diminution of fertility (or anyway productivity) of soil arising through erosion of the upper layers, salt contamination (often resulting from bad irrigation practice), alkaline contamination (for the same reasons), and irreversible aridification (often as a result of the removal of layers of woody vegetation), has already affected or threatens some 35% of the Earth's land surface (4.5 thousand million hectares) (Postel, 1989). Some 40 million irrigated hectares, about half of them in India and Pakistan, have been affected by salt accumulation. Erosive losses are said to have permanently removed a total area of cultivable land equivalent to one-third of the cropland of China. In the Sahel in the mid-1980s, some 2 million people in Burkina Faso, Chad, Mali, Mauritania, and Niger – between 3% and 16% of the total populations – were displaced as a consequence of drought (Jacobson, 1989). Current population pressures in these countries are increasing faster than the production from agriculture, even though the latter has increased in recent years (IUCN, 1989). In the western, central, and eastern, Sahel, *per caput* food production has declined from self-sufficiency (or even slight surplus) 25 years ago to only between 90 and 82% of self-sufficiency in 1989.

The scientific principles for halting desertification are well established and in some countries, such as Pakistan, remedial measures are being put in hand. But the overall trend in land degradation is adverse, and the application of available knowledge is hampered by lack of resources. Climate change, predicted to increase the aridity of many arid zones, is likely to exacerbate these problems, even though most of the damaged areas are capable of restoration and sustainable use. The key – as with deforestation – lies in recognizing the problem as a social one and in investing adequate resources to deal with it (World Resources Institute, 1989, 1990).

(c) Loss of Biological Diversity

Many extreme statements are being made about the loss of the Earth's biological diversity, and they need careful examination. We start from the difficulty that we do not know how many species the Earth currently supports. Science has described a total of the order of 1.5 millions. Sampling from the canopy of tropical rain-forests has revealed an astonishing diversity of associated insect species, many of them previously undescribed, and apparently with highly restricted distributions. Extrapolation from these figures implies a global total of about 20 million species, while the evidence of restricted distribution indicates that many will be lost when forests are cleared. Such calculations form the basis of many projections of impending extinction.

These estimates can be challenged, but they are the best we have. They lead on to assertions that 15–33% of all species may become extinct in 10–20 years – including 15% of the world's vascular plants and 12% of bird species. Whatever the precise details, there seems little doubt that continued forest clearance and continued development for agriculture and agro-forestry, as well as the increasing expansion of mariculture and other encroachments in the coastal zones, must threaten the overall richness of genetic entities and of ecosystems on Earth.

The significance of these losses can be debated. The functioning of ecosystems as recyclers of elements or maintainers of environmental conditions does not seem to depend on the totality of species that have carved ecological niches into increasingly narrow slices. A functional ecosystem clearly demands a reasonable balance of primary producers, herbivores, carnivores, and decomposers; but an impoverished super-ecocomplex, such as that of the British Isles, with only a few thousands of species, seems to be able to function in these respects as well as a much more diverse system. This is no argument for permitting the impoverishment of all global ecocomplexes on such a scale – largely because we simply do not know either how much redundancy resides in them as functioning entities, or which species are key; but it does suggest that some reduction of diversity may be acceptable without ecosystem collapse on a wide scale. Yet deciding how much, and in what form is obviously a challenge to science.

There are nonetheless reasons of both principle and practice for being concerned about the projected loss of biological diversity. The issue of principle arises from the acceptance in (for example) the World Charter for Nature that species other than *Homo sapiens* have a right to exist. The issue of practicality arises first because we know that ecological diversity is in some way related to resilience, and secondly because the genetic diversity of the world's biota will inevitably be drawn upon if, as seems likely, the planet experiences a more rapid and dramatic climatic modification in the next 50 years than in any comparable period of the known past. Crop breeders continually return to the wild for genetic material with which to increase the productivity, versatility, or disease-resistance of cultivated species. They are likely to have even more need of wild genotypes in order to improve crop strains to mitigate climatic change. Wild species are also a direct source of a wide range of foodstuffs – and the source of pharmaceuticals, some 40% of which are said to originate in wild plants even if they are subsequently synthesised by industry.

Some loss of planetary biological diversity is clearly inevitable if forests and other vegetation-types are to continue to be replaced by agriculture and agro-forestry. Again, the evident need is to control these changes, and minimize the losses and threats. This, as with land degradation, is a social problem: it requires a practical strategy, backed by sufficient investment, to concentrate agriculture and other intensive forms of land-use on areas where the soil and water supplies make this most practicable, and to minimize encroachment on areas of vulnerable habitat.

(d) Global Pollution

Global pollution will be an inescapable threat for the next 40 or so years, for one simple reason. Many of the most damaging substances are also highly persistent in The Biosphere, and are present today in concentrations which mean that they will continue to exert an effect for decades to come – even if fresh inputs are prevented. A good example is provided by the chlorofluorocarbons and other ozone-depleting chemicals. These have a residence-time in the atmosphere of between 80 and 110 years; and even if international action succeeds in stopping their production and use by 2000 AD, as many governments wish, the Antarctic 'ozone hole', and the less-marked but general stratospheric ozone depletion, will remain for the greater part of the next century. Very similarly, the polychlorinated biphenyls and other persistent chlorinated organic compounds, that are present in ecological systems and food-chains, will persist for a long time even if further inputs are halted.

As for 'greenhouse' gases, their residence times in The Biosphere are rather less prolonged, but carbon dioxide concentrations are unlikely to decline significantly for some considerable period after inputs are slowed. The elevated levels of heavy-metals in some aquatic systems, the nitrate that has passed into subterranean limestone aquifers in many densely-farmed areas of northern Europe, and the phosphates that are slowly leaching from intensively-used agricultural areas into the sea, will all take many years to pass out of those systems, even if inputs are reduced dramatically. All that can be done is to moderate the further increment of input of such substances, thereby minimizing the period over which they exert an unacceptable impact.

(e) Climate Change

Climate change is predicted to come about especially as a consequence of the accumulation in the atmosphere of certain 'greenhouse' gases (carbon dioxide, methane, nitrous oxide, chlorofluorocarbons and, under some circumstances, tropospheric ozone and water vapour – which last two, however, do not accumulate). These substances are collectively expected to bring about a warming of global climate by an average of between 1°C by 2030 AD, and an associated rise in sea-level of some 10–30 centimetres on the same time-scale. The warming is unlikely to be latitudinally uniform, the greatest effects – several times the global average – coming in high latitudes in winter, and the least in the tropics. However, the warming on its own is less likely to be a significant biological factor than redistribution of precipitation (generally speaking, the models suggest that the wet areas will get wetter and the dry areas drier), with increases in the intensity, and possibly in the frequency, of storms. Taken altogether, these changes have considerable implications for both biogeography and human development patterns. Although the models are imperfect, and projections of change highly sensitive to varying assumptions within them (Hare, this volume; Bryson, this volume), there is consensus of opinion that it would be sensible to consider now what social adjustments the changes which they indicate would require (Commonwealth, 1989).

An increase of 1°C can roughly be offset by a movement of between 120 and 150 kilometres towards the Poles, or by a vertical ascent of 150 metres. However, such changes over so short a period as 40 years would inevitably pose stresses on the redistributive capacity of plant and animal species, not all of which would have suitable habitats available into which to spread. The adequacy of the world's protected-area network in safeguarding biological diversity needs urgent reappraisal in the face of the possibility that many of these areas will no longer be suitable refuges for the systems for which they were created. Limits of crop tolerance are also likely to be redistributed, placing strains on the energy of the crop breeder and also calling for changes in the dietary habits of some human communities. Where increasing temperature, increasing evaporative loss, and declining rainfall, all combine, as they may in some of the areas of Africa where water is already a limiting resource (Falkenmark, 1990), the stresses on the development process will clearly be very grave indeed, and if these are also areas of high population growth, the result may well be either mass human mortality, or migration, or both.

4. Increasing Social Strife Caused by Inequalities in the Global Socio-political System

All the processes which we have just described result from human activities, and the threats that they pose can only be averted by the establishment of some kind of stable balance between people and environment. The issue is, what kind of balance, and by what means? In biological terms, the mechanisms for limiting the numbers of individuals of a dominant species involve competition for territory or resources, and mortality or non-reproduction of surplus individuals. The human species, after several millennia of increase and expansion, has largely escaped from the social and behavioural systems that have maintained a balance with Nature among certain indigenous groups, and has not established a substitute in its new societies. This is the core of the present dilemma and future threat.

Population increase can, self-evidently, only be reversed in two ways: the conscious decision and concomitant practice of individuals to have fewer children, or a return to mortality-limited populations, with a collapse in medical services, massive environmental degradation, and human migration on a vast scale. Deforestation, desertification, and losses of biodiversity, are all intimately associated with the demands of people for land to cultivate and occupy, and the inefficient ways in which those needs have been, and are being, met. They will only cease when the pressures of human population-growth ease, and more-sustainable modes of development follow. Pollution and climate change are manifestations of expansion in various kinds of industrial activity without adequate forethought as to the consequences. They have been, especially, the product of the developed world whose insatiable resource-consumption and dominance of world commodity and financial markets themselves place burdens on, and inhibit the development of, other regions of the world. Again, fundamental social changes are needed if they are to be halted.

Such patterns have clear roots in history. By nature, the various regions of the world are not uniformly endowed with environmental riches: nations can never be environmentally equal. Human history has been, however, a process of competitive social advance which has, to a considerable degree, aggravated such inequalities. The trends today continue to exacerbate the problems. In recent years, despite international aid, the net flow of resources continues to be from the poorer to the richer countries: in 1989 this flow is said to have been valued at the equivalent of 25 thousand million United States dollars.

The prospects for the future are of increasing strife from such causes, unless national and international behaviour patterns are modified. Even if population growth *can* be halted as a consequence of economic development bringing better employment, enhanced health-care, education, and other facilities, and even if land is developed sustainably, minimizing degradation, deforestation, and losses of biological diversity, this will not happen without massive economic investment. Even if pollution can be curbed, 'greenhouse' gas emissions reduced by energy conservation and other means, CFCs replaced, and human life-styles adapted to less wasteful modes, this will require enormous investment in industrial restructuring – and the transfer of the new technology and approaches to the developing world. In practical terms, while the need for all this is admitted and widely obvious, investment is lagging and the gaps are widening.

Perhaps the greatest threat to The Biosphere arises because of the lack of institutional machinery to bring people into balance with Nature by promoting – and securing – the acceptance of population policies, wiser environmental management, reduced demands for resources, and the other essential steps towards stability. And this is not simply or largely a matter of 'top down' reform of international machinery such as the United Nations or national top-tier governmental structures. It means remedying weaknesses in the structure and organization of human communities, improving the processes of decision-taking, and leading people to modify their behaviour in the light of lessons supposedly learned.

Trends towards better government and wiser administration have not kept up with the above demands stemming from human needs. In many parts of the world, governments are inadequately structured and resource-endowed to deal with environmental problems, and political priorities do not take proper account of the fundamental role of the environment as a basis for development. Few Finance Ministers seem ever to have heard of the concept of 'natural capital'. As pressures intensify, there is a considerable risk that political weakness will exacerbate the impact of environmental deterioration – or render the adjustments that are really essential still impracticable even if the need of making them is obvious. Indeed, environmental pressures in some regions seem more likely to destroy the essential social and administrative structure than to reinforce it.

Ingredients of a Solution

The solutions to these problems must therefore lie in the modification of human behavioural patterns. There are several ingredients in this modification:

(a) Adequate knowledge, including proper evaluation of environmental resources;
(b) Sound communication, leading to public understanding and individual commitment;
(c) Willing response at the group level;
(d) Governmental commitment and action; and
(e) Adequate international cooperation.

(a) An Adequate Knowledge-base

The volume of data about the environment has grown vastly over the past 50, and especially 20, years. Satellite observations have brought precision to our knowledge of the shape of the geoid, its detailed topography, sea-surface configuration, and the geology of large tracts of land. Remote sensing has also demonstrated the distribution of plant biomass and production, and recorded the details of atmospheric circulation. Models have begun to link temperature and circulation patterns in one area to phenomena elsewhere. Compendia of environmental data have proliferated (*e.g.* UNEP, 1987; World Resources Institute, 1989). In other respects, knowledge has lagged behind need, and an analysis undertaken in 1982 for UNEP (Holdgate *et al.*, 1982) reveals that the data-base for pollution is patchy and inadequate.

Global or national data alone are, however, inadequate for detailed environmental management. It is, moreover, unlikely that there will ever be enough knowledge in advance of a problem to be totally prepared for resolving it. The needs are for (a) broad global, national, and subnational, monitoring which defines the overall picture, (b) national environmental resource surveys which (for example) define land-use capability or the centres of biological diversity, and (c) a scientific capacity to investigate the situation in a particular location when a problem arises (for example, undertaking detailed land-management planning or environmental impact assessment). There is also a need for modelling and interpretational capability that will allow particular problems and systems to be dissected, and their response to management or change predicted.

Such scientific surveys, monitoring, and assessment, need to be augmented by economic evaluations which characterize national or local 'natural capital'. The methodology of such assessments is crucial. Economists are commonly blamed for having 'got their sums wrong' so far as the environment is concerned, and it is clear that many extremely important parameters have often been left out of the economic equations. This is true of the role of environmental systems in regenerating atmospheric oxygen, recycling nutrients, and regulating the hydrological cycle. Genetic diversity as represented in the wild relatives of crop plants, and the untapped pharmaceutical products that may or may not lurk in the glades of the forest, are commonly not regarded as a part of national wealth until they are incorporated in the sector that is used or managed by society. New techniques have now indicated the genuine economic value of natural systems and of biological diversity (McNeely, 1988). Other techniques are being developed to correct the poor inter-temporal comparison provided by much economic methodology, and which tends to write down environmental resources by applying to them discounting methods which effectively imply that they have no real value 30 years ahead.

There even remain those who doubt whether economics as a discipline can ever be adequately applied to the environment. Yet so long as economic evaluations are a fundamental basis for decision-taking, it is surely reasonable to correct the valuation of 'natural capital' as far as proves practicable. Indeed, there is a strong case for conducting 'environmental budgets' on an annual basis, in which the impact on the national environmental wealth of alternative development strategies is critically assessed, and the plans of governments are judged accordingly. Bad evaluation and bad economics certainly pose a real threat to the sustainability of development, and need to be remedied as far as possible.

(b) Sound Communication, Public Understanding, and Individual Commitment

It is clear from the preceding analysis that the major threats to The Biosphere arise from the actions of thousands of millions of human individuals. People will not take the action needed to correct this situation unless they understand *why* they are being asked to change their behaviour. They need, first, to have access to the findings of scientific monitoring and scientific analysis, expressed in comprehensible language. The last 20 years have in fact been marked by an immense and welcome growth in accurate and well-presented popular science in developed countries, linked to education, and this needs to be extended globally and augmented by environmental education and training for semi-literate communities – especially those interacting with fragile environments. In India, education *via* village television screens, and simple flip-chart displays taken on bicycles around villages, have had partial success. In Africa, schools' education systems (*e.g.* IUCN's 'Walia' project in Mali) are making some ground. They have demonstrated that communication, to be effective, needs to be based on a deep understanding of the groups to be influenced, and hence to grow within the community.

Such information needs to be action-orientated. People do not just want to know that there are threats: they want guidance on what action to take. In developed countries, one new and growing communications network is linking the environmentalist and the consumer, leading to the rise of 'green consumerism', 'soft tourism', and other forms of self-chosen, environmentally responsible behaviour. Rather similarly, public reaction to scientific accounts, for example of acid rain and oxidant damage to lakes and trees, leads to demands for action such as curbs on power-station emissions to reduce long-range transboundary air pollution, or fitting catalytic converters to cars. All of these impose costs on the public; but they have been made possible by the public's perception of the environmental damage that has to be prevented. Similar demands will need to be created if 'greenhouse' gas emissions are to be cut, and wasteful use of resources curtailed. There is a huge amount to do in these fields.

Individual commitment may prove to be the key to the whole future. In many communities, the most dominant influence on social behaviour is what may be loosely called 'conviction and belief'. Religions feature in this area, although they are not its exclusive component. The fact is that people commonly do things because they believe that they are right, abstaining from things that offend them, and so are generally grouped in their minds as 'wrong'. Taboos imposed under the ancient religions have endured for millennia, while the constraints of legislation tend to have a much shorter term.

For this reason, the world has a great need of environmental prophets. There are such around today, and a few of their names are almost household words in some countries. Inspiring leaders who touch the consciences of the world and make people demand a more equitable, more beautiful, and more uplifting frame for their lives, may be the paramount resource to resolve the threats which we see around us.

(c) Willing Response at the Group Level

People respond within a social context. Our world is threatened because many groups – villagers, foresters, industrial concerns, local authorities or governments – either do not understand the way in which their actions threaten their environment and the futures of their children, or feel that the imperatives of today leave them no choice.

Group decisions will certainly be a key to the aversion of many current threats. The question is likely to be, how far can groups be provided with the information and resources that will allow them to adopt policies which are environmentally sound and socially positive. Some decisions may involve moving away from previous social beliefs (this will certainly be so if the number of children that couples have is to fall to replacement level). Others largely involve investment to do what communities want to do.

Much experience has been gained in promoting sustainable development at the group level among rural communities in the developing world (*cf.* Panos, 1987; IUCN, 1990). It is clear that the ingredients are:

(i) open discussion of needs and opportunities with the whole group (or sub-groups if, for example, men and women need to debate separately);
(ii) letting the group frame their agenda and priorities;
(iii) involving the whole group in the action;
(iv) injecting the resources essential to commence action, *but*;

(v) sustaining momentum by setting aside some economic benefits to fuel further action, so that the initiative develops its own momentum.

There are ground rules, based on experience, for the guidance of group responses in rural and urban communities in many countries. The need is to escalate and broaden the response, always by a 'base up' rather than an imposed 'top down' approach.

(d) Governmental Action to Reinforce and/or Direct Behavioural Change

Governments have played only a limited role in bringing about new perceptions of environment. Indeed, governments are commonly perceived as having slowed progress to a rate that is less than most people would have wanted. Opinion polls commonly reveal that many people think that more should be done about the environment: however, this often translates into demand for government to coerce a third party (such as industry) rather than impose costs on individuals by higher taxes. There is ambivalence in this field, but it seems true that, where public behaviour has altered, governments have not done much towards bringing this alteration about.

The role of governments in this respect is, in fact, limited. They can prohibit certain kinds of action, and require certain other kinds of activity, by use of laws and regulations. Governments can set standards for products, and for emissions to the environment from factories or cars, and they can make it mandatory for people to do certain things such as having their vehicles checked annually, or using smokeless fuel in particular control zones.

Governments can also provide financial incentives, for example making it cheaper to buy unleaded petrol, and they can, through use of land-use planning machinery, direct developments into areas which they judge are most appropriate. All this is, however, to work at the general level and in a top-down way. Governments have not shown themselves to be good at moulding the personal behaviour of individuals, and in this sphere the media have been more successful. In many countries governments have, in fact, often encouraged environmentally unwise or even damaging actions, compounding rather than easing the threats to The Biosphere. There is a huge void of uncertainty over how best to correct this situation, which poses a major challenge to the forthcoming 1992 United Nations Conference on Environment and Development. Governments may demand action, but their activating mechanisms need considerable review if they are to deliver it effectively.

One problem arises from the failure of most governments to recognize that the environment is not just another sectoral concern, to be dealt with by a sectoral department alongside those for industry, transport, agriculture, or public health. Virtually no government in the world treats its environment as the ultimate national wealth, guarding it as jealously as a Finance Minister guards Man-made wealth. The need to evaluate the national 'natural capacity', and assess the wealth-creating or wealth-dissipating nature of national plans, may be accepted, and more than 40 countries have now prepared national Conservation Strategies which are partial means to that end; but the philosophy rarely feeds through into appropriate action. It is difficult to resist the view that some fundamental new thinking is needed urgently by most governments.

(e) International Cooperation

Cooperation between people, groups, and nations, is clearly crucial. The world is uneven environmentally, and will inevitably continue to be so. There is no way in which nations can be made equal, or individuals around the world automatically be given access to the same share of climatic, biological, or geological, bounty. Only mutual assistance, geared to equitable trading patterns and to the redistribution of wealth among nations, is likely to achieve a future in which humanity everywhere has decent living-conditions, and in which human populations are stabilized under economic and social circumstances that provide for human dignity.

The institutional mechanism that is represented by the United Nations and other organizations has a part to play in these developments. However, the institutional machinery, like the governmental, will almost certainly have to follow behind the demands of people, and is unlikely to deal with the situation unless it is first believed in and demanded.

People in developed countries may need, for example, to insist that their governments regard assistance to the poorer countries not as some kind of charity, held at the wholly inadequate target of 0.7% of Gross National Product (GNP) as is suggested currently, but instead as a real investment in a stable future world, and probably requiring assignment of something nearer to 5% of GNP. Conversely, people in developing countries have a right to demand that their national environmental future be a top priority of their respective governments, to be managed with efficiency, integrity, and in continuity. The world has a long way to go in all of these respects.

Conclusions

The central thesis of this overview is simple, namely that the world has already been transformed by the nature and scale of human actions to the point where humanity is threatening its own future. The chief threats stem from pressures on scarce and finite natural resources, both living and non-living; from the waste of those resources through over-consumption, degradation, and competitive exploitation, in which individual works against individual or group against group, motivated by self-seeking. Such an attitude focuses on the present, and undervalues the need to invest in the future.

Human population growth is a major factor in this situation. Ultimately it will stop: that is an inescapable lesson from biology. The question is how and when it will stop. In some countries the demographic transition has come harmoniously, in association with economic growth: but how far is this possible generally? The alternative is misery, famine, social disintegration, and environmental disruption. In some countries this looks like being the most probable outcome.

The problem is aggravated by the perversity of the human–environment interactive systems. Good system-design would be expected to provide regulatory devices involving warning of impending threats, social adjustment, and need for limitations to behaviour, before the overall complex interactive mechanism became imperilled. It is a paradoxical condemnation of humanity that the knowledge which put Men on the moon has failed, not only to engineer such a regulatory mechanism, but even to convince most people of the need for it.

The solutions lie in changing human perceptions and, through changing perceptions, also modifying human behaviour. People can be stimulated by various tools which lie at the disposal of communities today. All the ingredients of a system necessary to bring humanity into harmony with Nature exist. The problem is that they are not being used – or not with sufficient urgency – on an adequate or even significant scale.

The lesson for this Conference is that, while science is essential as a basis for the understanding and action which humanity needs, our first concern should be to analyse how we can alter people's perceptions of the world and then change their behaviour so that they can serve it more wisely. That is a difficult challenge for scientists to face up to, because it takes the natural scientist into dependence upon the political scientist and the sociologist. But such a transition of roles is likely to be a necessary pre-condition for ensuring that our societies are not destroyed by the commonly self-engendered threats which confront them.

References

Ashby, E. & Anderson, M. (1981). *The Politics of Clean Air.* Clarendon Press, Oxford, England, UK: viii + 178 pp.

Bolin, B., Degens, E. T., Kempe, S. & Ketner, P. (Eds.) (1979). *The Global Carbon Cycle.* (SCOPE 13.) J. Wiley & Sons, Chichester–New York–Brisbane–Toronto: xxxv + 491 pp., illustr.

Bryson, Reid A. (1992). This volume, Chapter 22, p. 535 *et seq.*

Commonwealth (Comm – 89) (1989). *Climate Change: Meeting the Challenge: Report of a Commonwealth Group of Experts.* Commonwealth Secretariat, Marlborough House, Pall Mall, London, England, UK: ix + 31 pp., illustr.

Dimbleby, G. W. (1978). Changes in ecosystems through forest clearance. Pp. 3–16 in *Conservation and Agriculture* (Ed. J. G. Hawkes). Duckworth, London, England, UK: xvi + 284 pp., illustr.

Falkenmark, M. (1990). Water, the finite resource. *IUCN Bulletin*, 21(1), March 1990, pp. 14–5.

Hare, F. K. (1992). This volume, Chapter 2, p. 31 *et seq.*

Holdgate, M. W., Kassas, M., and White, G. F. (Eds.) (1982). *The World Environment, 1972–1982: A Report by the United Nations Environment Programme.* Tycooly International Publishing Ltd., Dublin, Ireland: xxxii + 637 pp., illustr.

Imru, Mikael (1992). This volume, *Annexe 5.1*, p. 327 *et seq.*

IUCN (1989). *The IUCN Sahel Studies, 1989.* International Union for Conservation of Nature and Natural Resources (IUCN), Gland, Switzerland: xxi + 154 pp., illustr.

IUCN (1990). Special Report on 'Population and Resources; Managing to Survive'. *IUCN Bulletin*, 21 (1), March 1990, pp. 12–26, illustr.

Jacobsen, T. & Adams, R. M. (1958). Salt and silt in ancient Mesopotamian agriculture. *Science*, 128, pp. 125–8, illustr.
Jacobson, J. (1989). Abandoning Homelands. Pp. 59–76 in *State of the World, (1989): A Worldwatch Institute Report on Progress towards a Sustainable Society* (Ed. Lester R. Brown). W. W. Norton and Company, New York and London: xvi + 256 pp.
McCormick, J. (1989). *The Global Environmental Movement: Reclaiming Paradise.* Belhaven, London, England, UK: xv + 259 pp.
McNeely, J. A. (1988). *Economics and Biological Diversity: Developing and Using Economic Incentives to Conserve Biological Resources.* International Union for Conservation of Nature and Natural Resources (IUCN), Gland, Switzerland: xiv + 236 pp.
Martin, P. S. (1967). Prehistoric overkill. Pp. 75–120 in *Pleistocene Extinctions: The Search for a Cause* (Eds. P. S. Martin & H. E. Wright). Yale University Press, New Haven, Connecticut, USA: 453 pp, illustr.
Myers, N. (1988). Tropical forests: much more than stocks of wood. *J. Tropical Ecology*, 4, pp. 209–21.
Olson, J. S., Watts, J. A. & Allison, L. J. (1983). *Carbon in Live Vegetation of Major World Ecosystems.* (Oak Ridge National Laboratory Environmental Sciences Division publication no. 1997.) Oak Ridge National Laboratory, Oak Ridge, Tennessee, USA: xiii + 164 pp. + map.
Panos (1987). *Towards Sustainable Development.* The Panos Institute, London, England, UK: xvi + 280 pp.
Postel, S. (1989). Halting land degradation. Pp. 21–40 in *State of the World, 1989: A Worldwatch Institute Report on Progress Towards a Sustainable Society* (Ed. Lester R. Brown). W. W. Norton and Company, New York and London: xvi + 256 pp.
Reid, W. V., Barnes, J. N. & Blackwelder, B. (1988). *Bankrolling Successes: A Portfolio of Sustainable Development Projects.* Environmental Policy Institute and National Wildlife Federation, Washington DC, USA. 45 pp., illustr.
Thirgood, J. V. (1981). *Man and the Mediterranean Forest.* Academic Press, London, England, UK: x + 194 pp., illustr.
UNEP (1987). *Environmental Data Report: Prepared for UNEP by the Monitoring and Assessment Research Centre, London, UK.* Basil Blackwell, Oxford, England, UK: vii + 352 pp., illustr.
Vitousek, P. M., Ehrlich, P. R., Ehrlich, A. H. & Matson, P. A. (1986). Human appropriation of the products of photosynthesis. *BioScience*, 36, pp. 368–73.
World Commission (1987). *Our Common Future: Report of the World Commission on Environment and Development.* Oxford University Press, Oxford & New York: xv + 383 pp.
World Resources Institute (1989). *World Resources, 1988–1989: An Assessment of the Resource Base that Supports the Global Economy.* (A Report by the World Resources Institute and the International Institute for Environment and Development in collaboration with the United Nations Environment Programme). Basic Books, New York, NY, USA: xii + 372 pp., illustr.
World Resources Institute (1990). *World Resources 1990–91: A Guide to the Global Environment.* (A Report by the World Resources Institute in collaboration with the United Nations Environment Programme and the United Nations Development Programme.) Oxford University Press, New York, NY, USA: xiv + 383 pp., illustr.

Commentary on Chapter 1

CHAIRMAN: Professor Nicholas Polunin
PANELLISTS AND OTHER CONTRIBUTORS:

Stumm, Láng, Hare, Harun ur Rashid, Oza, Goldberg, Juel-Jensen, Kuenen, Trompf, Holdgate, Westing

Stumm wished to take up one of the many issues raised by **Holdgate**, namely, that of 'anthropogenic' chemicals and their accelerated cycling, global and regional, due to the influence of Man, through the hydrogeochemical cycles. There were approximately 70,000 synthetic organic chemicals in daily use and each one potentially capable of making an impact on millions of organisms and with potential effects on biological diversity. The ordinary man or woman in the street expected scientists to have tested every one of these chemicals for their possible effects on living organisms before they were released into the environment and, thereafter, to monitor them. Scientists, he believed, should have the courage to tell the people, through the media, that this task was utterly impossible. He wished, therefore, to make a short plea for a more conceptual approach and less testing and monitoring, since he was concerned by the present overemphasis on monitoring and toxicity testing versus exploratory research. He accepted that there was a great deal of intelligent monitoring; some programmes, such as those in Canada, were extremely fascinating, but too many other programmes lacked a sound conceptual basis. Monitoring and testing should not, in his view, have priority over exploratory research on:

a. Comparative toxicology;
b. The cycles of energy and elements and their interaction with, and ecological effects on, communities; and
c. The impact of chemicals and other stress-inducing factors on the structure, distribution, and frequency of species.

He gave as an example to illustrate his thesis the situation which arose some years ago in Switzerland when there was a major chemical catastrophe due to a few tons of pesticides being released into the Rhine by the Sandos company. This had a serious, instantaneous impact on the riparian communities and most animals were killed over some hundreds of kilometres of the Rhine.

Two points needed to be noted. Firstly, in fact, all the monitoring installations along the Rhine were of no use in detecting this accident, and were useless in preventing it. Moreover, the predictions of potential toxicity, based on testing the LD_{50}, were wrong by at least one order of magnitude, as demonstrated by the effects on the fish, eels and macro-invertebrates. On the other hand, the course of recovery from the accident was relatively fast – much faster than was generally expected. One reason for this was that, owing to the constant chronic pollution of the Rhine prior to the accident, it did not contain many sensitive species!

Secondly, it could be concluded that chemical accidents, such as that at Sandos, were, from an ecological point of view, of less concern than was persistent, chronic pollution which irreversibly alters the biocoenoses and reduces the diversity of organisms.

These conclusions bear on the perpetual dilemma of what should be regarded as most important, the health of Mankind, or the health of the ecosystems? He considered that at present in government circles, most especially those in the USA, there was an undue emphasis on human health over 'ecological health'. This attitude failed to recognize that the maintenance of resilient ecosystems, capable of resisting structural and functional displacement, was a prerequisite for the well-being and health of Humankind.

Finally, he made a pragmatic suggestion which he considered to be both economically and technologically feasible. He proposed that all the commonly used, mass-produced substances which, because of their use in the household, in agriculture, in trade and manufacturing processes (such as detergents, solvents, agrochemicals, fluorocarbons, etc.), become dispersed throughout the environment, should all, in some way, be satisfactorily degradable. They might be biodegradable, chemically degradable, or photochemically degradable, but only by such action could these synthetic substances be prevented from accumulating in the environment.

Láng drew attention to the fact that there were ongoing activities, and others were planned, which had a bearing on the issues raised both by the Brundtland Commission and by **Holdgate**. For instance a meeting was planned for May 1991 of the Ministers of the Environment, to be preceded by a meeting of concerned scientists, and the major UN Conference on Environment and Development in Brazil in 1992. He hoped that something positive would happen, globally and nationally, as a result of these meetings. However, he wished to focus on two specific environmental problems in Hungary, namely, the energy policy and that for agricultural production, as examples of national problems which needed to be resolved with urgency.

Hungary was using four to five times more energy per $1 of GNP than was either Austria or Finland. That was no way to continue and, without a radical change in government policy, there was little chance of even reaching the average developmental level for Europe, let alone bettering it. However, such a change could not be achieved by Hungary alone. It would require technical assistance from the countries of the group of 24 over the next 10 years.

Agriculture was a different problem. Hungarian agriculture was very good and some 30% of production was exported; even more production was possible. However, that would not make good sense because the market in Europe for agricultural products was not good, and the nature of the present methods of production placed a very great pressure on the environment. So, quality, and hence higher value, not mass production should be the objective in the future. Moreover, that change should be accompanied by a decrease in total agricultural production while maintaining self-sufficiency, economic compensation being achieved by better quality, higher-value products.

Finally, he wished to draw attention to one further aspect of particular importance to the implementation of the Brundtland Report, not only to Hungary but all over the World. Three groups in society were concerned for, and with, environmental issues – the State administration, the scientific community and the variety of environmental movements – the so-called environmentalists. Obviously each group had, and should have, its own approach and its own policies, and should exert pressure on the other groups. But, if they could not find a common language, and if they did not make common cause over many essential activities, then nothing serious would be achieved in the promotion of environmental progress. That was a real problem in Hungary and he believed for many other countries. It would only be solved through changes in the organization and structure of discussions within, and interactions between, all groups. He hoped that the arrangements both preceding, and at, the 1991 Bergen Conference of Ministers might make such a change possible. Major recommendations arising from discussions at the preceding scientific meetings and those of non-governmental organizations would be submitted to

Ministers. He thought it would be wise for them, at their meeting, to adopt the main recommendations. That, he thought, could make things much better; he sincerely hoped so.

Hare remarked that although no-one could express the essence of the situation more clearly than **Holdgate**, some attitudes, such as how and who was to lead governments, needed further elaboration. There had been and still was a tendency for governments to suppose that they lead, but, in fact, they are led! He believed that there was still a widespread attitude, even in Cabinets (and he had served as a member of a Cabinet!), to suppose that environmental issues were a lot of nonsense and that they would blow over. But, because the public did not know of this attitude, cabinets accepted that something would have to be done without damaging the economy (which, of course, had actually produced the environmental damage in the first place) if the Cabinet was to stay in office! This widespread reluctance to act, based on the belief that government was in complete control, should not be underestimated, but there were clear signs that public opinion was having a real effect, for example over cigarette smoking in public buildings, and he believed that that force for change was vastly stronger than any governmental reluctance. He saw no need to be pessimistic.

Harun ur Rashid commented on two points. Firstly, he wished to stress the important distinction, from the point of view of the environment, of the so-called North-South divide – the industrialized as opposed to the developing countries – including his own, Bangladesh. It was a fact that in the Third World 500 million children would go to bed hungry every night: that each of these countries had an enormous population problem, and that each was attempting to provide a minimal standard of living for its people. (It should be recalled that today's 5.2 billions could become 14 billions within 40 years, on some predictions, and the bulk of the increase would be in developing countries.) To change those conditions would need massive government assistance and that, in turn, would require economic development to provide the necessary funding. But economic development, of the kind that had been followed by industrialized countries, and environmental protection were seen to be in competition.

Secondly, the resolution of environmental issues in developing countries was often made an additional condition of foreign assistance. But what was the impact of such conditions on governments' attempts to raise standards of living, and how was it perceived by the people? Naturally, the questions were asked, why was it that industrialized nations, having themselves achieved a satisfactory standard of living now, in effect, told developing countries not to develop, and in what ways were they supposed to develop in order to be able to improve their standard of living? Foreign aid, including technology transfer, improved education, medical progress, and indeed, almost everything that distinguished industrialized countries from developing countries, was always conditional! These dilemmas dominated the effectiveness of actions designed to reduce the North–South divide so perceptively addressed by **Holdgate**. He asked participants to recognize the immense problems the Third World had to grapple with simultaneously – population increase, unemployment, illiteracy, poor economies – even though they recognized that environmental problems were universal, that their solution required a global approach, and they wished to play their part in that global outlook.

(*See* also Annexe 1, where environmental problems in Bangladesh are described as an example of the issues in Third World countries, Eds.)

Oza supported **Harun ur Rashid**'s comments on the Third World. If he were to describe the SE Asian region, including the Indian subcontinent, from a biological point of view, he would rate it as one of the richest biomes in the world. Paradoxically, it was in the grip of hunger and included the poorest of the poor, from the point of view of human suffering and its alleviation. That human condition resulted in increasing land hunger with an inevitable loss of biological variability and diversity. Somehow a balance had to be struck between hungry people, cultivable land and the richness of the natural wildlife.

Goldberg, while supporting and acknowledging **Holdgate**'s thesis, was concerned that the resolution of many environmental problems needed substantial data and he queried how reliable some were. For example, how accurately did we know the population numbers for different parts of the world? We needed much better statistical data, especially because of the phenomenal changes in population numbers and, in addition, an understanding of the causes of such changes. He doubted the competence of many demographers either to document, or to account for, these changes. Migration, for example, was occurring all over the world – even in his own small city in Southern California they were experiencing a mass migration from Guatemala in an attempt to provide a better life for the migrants and their children. But, how were data on migrations accumulated? He was well aware from his own experience with the United Nations of the inadequacies and difficulties in gathering data of substance that could be trusted. These were matters that should be addressed with urgency.

A second, not unrelated, point concerned environmental research programmes. These, in his opinion, could no longer be conducted on a year-to-year basis funded by national or international agencies. Attention would have to be directed to projects lasting for decades at least, perhaps for many generations, in order to provide the data to solve the terribly important problems raised by **Holdgate**. That would require novel methods of global support and funding to ensure continuity.

Juel-Jensen considered that the most important problem raised by **Holdgate** was that of the population explosion: no-one could have any doubts that there were too many people! Knowledge and education, getting the message across, was vital because unless population growth was stopped we would really end up in a mess. There was a further reason for an increased effort in education, of which he had direct personal knowledge through contacts with his outstations in Africa, Thailand and Brazil, namely that, in many countries, in Africa for instance, a tenth of the whole population now carries the HIV or AIDS virus. Its spread would go hand-in-hand with population growth. It was terrifying in the extreme to contemplate the prospect of reducing the population by the spread of AIDS and, if the West had no other obligation, it was to bring home to everyone that the epidemic must be stopped. If it was not, we could end up with a scenario not unlike that painted so vividly by Mary Shelley in *The Last Man* wherein a pathogen eliminated everybody bar one!

Kuenen observed that the natural sciences had a history of several hundred years and had succeeded in producing materials which we could use and, more or less, knew how to handle, although that claim could be debatable. What was needed now was a stronger input from the social sciences, including economics. However, they differed profoundly from the natural sciences. He could see no way of relating their theories to reality, nor was there any hope that, if an adequate theoretical base was developed, anyone would apply it to real situations! He was, therefore, profoundly unhappy at the notion of having to rely absolutely on the further development of the social sciences to save the environment. He did not believe that there was any hope that the social sciences would succeed in time. Equally, he considered that it was not essential for science to continue to refine its information, such as determining the exact amount of atmospheric ozone. What was absolutely essential was to change Man's way of living.

Trompf asked whether **Holdgate** thought that one of the implications of his views was that there was now a necessity to impose 'base-up' community environment units on human populations, paradoxically from the top down? By that he meant whether through the United Nations and, subsequently, at the national level, people should be required to form local community units which were both environmentally aware and prepared to monitor the environment? In the Third World many community units already existed as clans or tribes and he thought it would be easy to graft environmental concerns and duties on to them; but in the more irresponsible West it would certainly be a matter of imposing 'base-up' arrangements from the top.

Holdgate responded by making five points. Firstly, he did not think that **Hare** had posed a very fundamental issue when he questioned who was to lead governments. If governments were in a responsive mood in an essentially unstructured system, then they would be blown hither and thither by every blast of false doctrine rather than by those with the best data, or the best underlying logic. It was, therefore, extremely important to develop within a community a series of articulate, intellectually respectable groups (however strongly self-motivated), who operated with some degree of common social ethos and common basic values, even if that basis was no greater than, or no less than, a respect for the truth. If a whole series of people could interact with governments on a basis of respect for truth and who would present it in a way that seemed right to them, then governments might respond in a better way than at present.

Secondly, a consideration of how governments respond led him to answer **Trompf**'s question. He disliked the notion of imposing a 'base-up' solution from top down. He suggested that an evolutionary model might be more helpful. If governments were to create environments in which such 'base-up' processes could develop, for example, by showing that they have a listening ear, by giving encouragement to groups whose objective was to improve perception, by providing resources down the line to local communities to do just that, and by making it clear that lines are open from the village to the cabinet room, then the evolution of such groups could take place. People would respond more effectively if they thought that action in response to their needs would come from government.

Thirdly, **Goldberg**'s comments on the importance of accurate data were relevant in the context of these considerations; they were the basis of an informed, true view. Good data also bore on **Kuenan**'s *cri de coeur* about the apparent, perceived differences in intellectual rigour between the natural and the social sciences. He acknowledged the truth of **Kuenan**'s view of the perception of the activities of social scientists by natural scientists and noted that most of the former claimed that it was a total misconception. He believed that there should be more dialogue between them. Could the 'hard systems' of the physical scientists be applied more effectively to the intellectual conundrums with which economists had to grapple? He suspected that faced with this, most physicists would retreat to physics with sighs of relief at being required to deal only with measurable parameters, verifiable by experiment!

Fourthly, if progress was to be made then more dialogue was essential because we had to believe in the methodologies of economics. Moreover, better environmental economics were coming forward which needed to be explained to, and studied by, environmentalists.

Fifth and finally, he commented on the views of **Harun ur Rashid** and **Oza**. There was still an enormous challenge to environmentalists to convince ordinary people in developing countries that there need be no conflict between conservation and development; that without conservation of the basic elements of the environment – the productivity of soils, the hydrological cycle, the purity of water and the fertility of pastures – there could be no development. Within such an approach, conservation of wild-life, of genotypes, of biological diversity, could find a place. The message that had to be got across was that, if development and conservation were on a collision course, then development would fail. Linked to this problem was the point, also made by **Lang**, that development required the transfer of the best technology, and that meant for developing countries, the best technology on affordable economic terms to make it usable by them. That led naturally to the matter of cooperation. We should not beat about the bush: the future of the world environment – real cooperation – demanded more resource transfer from the currently affluent nations to those who would have to become affluent in the future if they were to grapple with poverty and provide a decent life for their people, and so create, eventually, a harmonious world. The notion that an award of something like 0.7% of GNP for development aid, mostly with strings attached so that it was as much to the advantage of the investor as to the recipient, was fatuous and pitifully inadequate. If real progress was to be made, then we should accept that a level of development aid of something like 5% of GNP, annually, for 30 years, without bearing immediate loan interest, would be necessary. Only then would we ensure the much greater, long-term, loan interest of future harmony and stability for humanity, or put another way, only in that way would we prevent the alternative 'opportunity cost' of social disruption that, otherwise, would ensue!

Westing [in a written communication submitted after the Conference, Eds.] summarised the general reaction. He observed that, sadly enough, it was all too readily demonstrable that our world had become gravely threatened – in fact, that the human environment was deteriorating substantially in a number of disparate ways. **Holdgate** had substantiated this with overwhelming clarity – a dismayingly dismal picture, indeed. To counter these diverse threats to the integrity of our environment with any real hope for success would require the fulfillment of at least the following five broad tasks:

(1) that the threats themselves be identified and their magnitudes determined (questions now largely behind us);

(2) that the causes of these threats be uncovered, both the immediate (proximate) causes and the underlying (ultimate) causes (the proximate causes now being largely uncovered, the ultimate ones to some considerable extent still eluding us);

(3) that means of ameliorating these threats be developed by addressing the causes at all levels, both the technical means and the institutional means (the technical means to a considerable extent being within our grasp, the institutional means not really as yet);

(4) that the resources necessary for coping with these threats – material, intellectual, and financial – be established (still requiring a major effort); and

(5) that public understanding and support on the one hand, and governmental support on the other, be created to bring about the required shifts in societal resources, both by educational means (including sound means of communication) and via the expansion of participatory democracy (this fifth set of tasks constituting the real challenge of our time).

We owed our gratitude to **Martin Holdgate** for having provided an eloquent overview of 'Our Threatened World' that set the stage for our attempt during these several days to seek ways of 'Surviving With The Biosphere'. He had sorted out the several tasks just outlined and put them into their proper perspective. Indeed, without having received such clarification and guidance it would have been impossible to deal intelligently with the goal we had established for ourselves.

Perhaps of greatest value in establishing our priorities was **Holdgate**'s insight that our environmental problems had been growing faster than the societal mechanisms needed to cope with them; and, thus, that we had become substantially dependent upon social transformations in order to realize the needed technical transformations.

Amongst the most necessary of these social transformations in our search for surviving with The Biosphere were the following:

(1) the development of a pervasive recognition that most global problems did not respect national frontiers, many being confined to a particular eco-geographical region involving a number of nations and some spreading even more widely. This basic fact, in turn, implied the necessity for obligate inter-state cooperation, both globally and, more often, regionally, i.e. the necessity for some relaxation of our deeply ingrained notions of national sovereignty;

(2) the development of a pervasive recognition that our environment was now being so heavily utilized for vital human purposes that such frivolous forms of environmental damage as that caused by warfare could no longer be tolerated, the more so because of the ever-growing potential for massive direct and indirect wartime disruption of the environment. This basic fact, in turn, implied the necessity for unconditionally accepting non-violent forms of dispute resolution, i.e. the necessity for some change in our deeply ingrained notions about using deadly force for achieving political goals; and

(3) the development of a pervasive recognition that our numbers and our aspirations had (at least at our present level of development) outstripped the ability of The Biosphere to provide for the sustained utilization of the renewable resources of the environment, and – equally important – to provide for the sustained discard into the environment of our solid, liquid, and gaseous wastes. This basic fact, in turn, implied the necessity of curbing both our numbers and our aspirations so that they matched the biospheric carrying capacity both globally and regionally, i.e.. the necessity for some change in our deeply ingrained notions about our presumed rights to multiply and to gild our lilies; and, collaterally, the necessities to achieve human equity (both in spatial and inter-generational terms) and to develop greater respect for the needs of the other living things with which we shared this planet.

Thus in conclusion (although one could, of course, quibble with a number of minor points of commission and omission – as one could with any presentation as sweeping and all-inclusive as this one), he was pleased to be able to reiterate that **Holdgate** had set us off in the right direction to seek means over the next few days for Humankind to develop new cultural norms that would provide the basis for achieving environmental security – an environmental security that, in combination with social security, would lead to true human security.

Annexe 1: The Environmental Situation in Bangladesh*

H. E. HARUN UR RASHID
Ambassador of the Permanent Mission of Bangladesh
to the United Nations, Geneva, Switzerland

Bangladesh, a country of 110 million people squeezed into 144,000 sq km of territory, faces in an extreme form the problems both of environment and of global climate change.

Bangladesh suffers not only from floods and cyclones but also from droughts. The floods in 1987 and 1988 did not follow the normal pattern. Each was sudden, quick and untimely. Moreover, besides the surface water, ground-water levels swelled and joined the surface water. Ordinarily, the three mighty rivers, the Ganges, Bramahputra and Meghana, do not rise together during the monsoons. But in 1987 and in 1988, the water-levels of these rivers rose at the same time to danger levels causing unprecedented flooding. This phenomenon is an aberration of Nature.

Many of the causes behind the terrible floods of Bangladesh in 1987 and 1988 are beyond Bangladesh's control. The country has been one of the principal victims of a change in the global climate which has lately manifested itself in extreme weather conditions from the US to Bangladesh.

The impact on Bangladesh of monsoons is increased by the high degree of soil erosion in the region. Much of the blame lies in the deforestation of large areas in the north beyond Bangladesh territory i.e. areas in Nepal and eastern India. With less vegetation to absorb the rains, erosion is naturally increased and the rivers in Bangladesh become highly silted. It is suggested through an empirical study that a strong link exists between deforestation and flooding.

For Bangladesh, the situation will be further exacerbated by the 'greenhouse effect'. Climate experts have now agreed that the global climate is warming and that this implies a raising of the sea-level. Some estimates put this increase at over 1 metre by the year 2030. For Bangladesh the implications are potentially disastrous. A third of the land is only 1 metre above the sea-level and higher tidal waves would salinate ground water-supplies, further damaging the country's water resources.

* [This contribution was submitted by the author and was available to participants at the Conference. Although it sets out the problems clearly for one developing country it is typical of many of them. Eds.]

Silt is, of course, Bangladesh's *raison d'etre*. Only 10,000 years ago, there was no land at all. Since then the entire Bangladesh delta has been built from silt deposited by the mighty rivers which converge in Bangladesh in a trident junction.

The problems facing Bangladesh are so complex that no single country would be able to deal with them effectively. They need a global reach and outlook supplemented by regional cooperative efforts. The Government and the people of Bangladesh are making a maximum effort in seeking to meet the situation. A comprehensive national programme has been undertaken concentrating specifically on three main aspects: immediate relief and rehabilitation measures, medium-term measures for reconstruction of the economy and infrastructure and longer-term measures with special emphasis on disaster preparedness and preventive schemes.

Like all other countries, the environment of Bangladesh is deteriorating due to gradual pollution by natural phenomena to some extent and to a greater extent by human activities.

Let me outline the different constituents of environmental pollution in Bangladesh. The first item is the soil. The soil is deteriorating through many factors, *e.g.* salinity, and the threat of desertification in the north. Second, is water. The waters are becoming polluted, not only because of population pressure but also on account of the dumping of industrial wastes in the upper reaches of the rivers, beyond the control of Bangladesh. The air is also getting polluted by the ever-increasing number of vehicles and industries. The forest resources are also in bad shape. Domestic and wild animals are decreasing in numbers partly due to slaughtering for meat. Lastly, the growth of population and poverty is leading to pollution of the environment. The present population density is about 1,900 per square mile (*c.* 1,187 per km^2) in Bangladesh which appears to be at a danger stage in respect of environmental pollution.

2. The Atmospheric Greenhouse: Ordeal by Scepticism

F. KENNETH HARE

Chancellor of Trent University, Peterborough, Ontario, Canada, K9J 7B8; Chairman of the Advisory Board, Institute for International Programs, University of Toronto, Ontario M5S 1A1, Canada; Emeritus Professor of Geography & Meteorology

ORDEAL BY SCEPTICISM?

The years 1988 and 1989 were, perhaps, those of the 'greenhouse'.* They were the time when environmental consciousness broke onto the world's political stage. They witnessed the arrival of Mrs Thatcher among the converted, together with President Mitterand. The 1988 World Conference on the Changing Atmosphere, held in Toronto, was attended by Prime Ministers Gro Harlem Brundtland and Brian Mulroney, as well as by some foreign ministers. A common factor in these conversions appears to have been the 'greenhouse' effect – or, rather, the recognition that world energy strategy may hold the key to the future of this planet's habitability. I would rank the 'greenhouse' with Mrs Brundtland's famous offspring, 'sustainable development'. These two ideas have galvanized international politics – as, I should add, have the competing Russian elements of *glasnost* and *perestroika*. To be frank, I am not sure what will emerge when we put these all together in a bag, and shake them out.

Why should I start by talking about an ordeal by scepticism? I need only point to our chairman, Reid Bryson, who has consistently declined to believe the usual 'greenhouse' story, and is sceptical of the model results that point to induced climatic change (even though one of those models, identified as OSU6, actually predicted the existence of the 14.7-months polar tide, often called the 'Chandler wobble', of the Earth about its axis; Reid had been one of the prime exponents of the impact of this wobble on world climate). He is not alone in this scepticism. Other well-known scientists have also questioned the validity of the modelling results, notably the Marshall Institute trio of William A. Nierenberg, Robert Jastrow & Frederick Seitz (1989), and likewise Richard S. Lindzen of MIT (1990).

This doubt as to the reality of the global warming has been prompted in some by the feared economic consequences. When *Forbes Magazine* recently ran an article on what it called *The Global Warming Panic*, it carried the

*[Using this term in the modern, colloquial sense and commonly retaining these 'quotes'. Eds.]

headline 'Apocalypse sells well in the media and even better on Capitol Hill. And that is why fears of the "greenhouse" effect threaten to push the US into a costly environmental mistake.' The author, Warren T. Brookes, asserted that the media make much of apocalyptic pronouncements, but he did not point out that journals which are read extensively by the business community do well by rejecting those same views (Brookes, 1989). He comes up with a judgement that echoes – in fact quotes with approval – the 1984 words of one Bernard Cohen: 'The coming debacle is not due to the problems the environmentalists describe, but to the policies they advocate'. Global warming, asserts Brookes, may well prove Cohen right.

The controversy boiled over into the popular media in 1988, when James A. Hansen, of the Goddard Institute for Space Studies in New York, testified before a US Congressional committee that it was time to admit that the 'greenhouse' warming was upon us. In the wake of this I found myself singled out in an editorial in *Nature,* entitled 'Jumping the Greenhouse Gun'. Persons who ought to know better, the editorial asserted, naming Hansen and myself, had prematurely announced that the 'greenhouse' warming was in progress. What I had said, at the Toronto Conference in 1988, was this: 'that we are indeed witnessing the beginnings of the process, and that the delegates did not come to this Conference to chase a will-o'-the-wisp' – adding, however, that this was only a personal opinion. It is still my opinion, though I agree that we are uncertain, and divided as to the outcome.*

As time has passed, the debate has increasingly focused on a penetrating question: can we reasonably hope to predict the future over periods of half-a-century when numerical weather prediction models, the progenitors of the climate models on which long-term predictions are based, lose most of their success in predicting actual events in a week or less? Arising from this reasonable question are others. Is dynamical modelling of atmosphere and ocean the only valid approach? Can we not learn from the geological and biological evidence? If climatic change does come, or is already in progress, can we predict the consequences for natural ecosystems? For crops? For resource-based economic activity? For human settlements?

If we lack predictive power, then governments, industries, and individuals, will indeed be loath to invest heavily in change. This is not, of course, a new experience for them. *All* public policy is put in place in an atmosphere of uncertainty, and in a clamour of conflicting advice. Some of this sounds like the counsel of wise old owls, and some of it like the braying of donkeys. Environmentalists are apt to assume that this Babel is peculiar to their field. It is not. It is all-pervasive. And economists find it just as difficult to predict the future as do climatologists, and surely often more so. Yet no one argues that we should abandon attempts at economic planning. Even free-market

*[Nevertheless at the Second World Climate Conference, which took place in Geneva in late October and early November 1990, there seemed to be general acceptance of an overall global warming of 0.5 to 0.6°C in recent years. Eds.]

enthusiasts, and those who live on market speculation, try to foresee the future. And in the field of politics how many commentators, authorities, bureaucrats, or even bookmakers, were able to foresee the astonishing political changes of the past two years in eastern and central Europe, and in South Africa – let alone the Gulf?

It is my job to assess the level of uncertainty that pertains in the arena of climate, which is properly seen as crucial to several other environmental domains. I shall not attempt to summarize either the history or the characteristics of recent climatic change, *i.e.* the events of the past century. This has been done in a series of international reviews, of which the Villach Assessment of 1987, the ensuing Bellagio Conference of 1987, and the Toronto Conference of 1988, are the most recent. I shall make frequent reference to a recent report of the Commonwealth Secretariat, commissioned by the Heads of Government, dated September 1989, and prepared by a panel chaired by one of this morning's speakers, Martin W. Holdgate (1989).

The Operative Processes of Change

Most of the observed atmospheric variations of temperature, precipitation and other elements, are due to internal processes of the climatic system. The atmosphere and ocean together form a non-linear dynamical system that is capable of endless variation, some of it periodic (such as the weather differences between day and night, or summer and winter) and some of it aperiodic, because of disturbances. The latter systems characteristically behave in a chaotic fashion, in which processes that are in principle deterministic nevertheless permit a never-repeating succession of the weather changes that are so familiar in all parts of the globe. Climate is the average state of this system, and the statistics indicate the modes of variation. Climatic change is said to occur if the average state and modes of variation change to a new condition that is statistically distinct from its predecessor. The change may be imposed by external factors (such as the incidence of solar radiation or the composition of the atmosphere) or by internal factors (such as interaction between ocean and atmosphere).

In the past two decades we have learned to add atmospheric chemistry to the properties connoted by the term climate. The composition of the atmosphere and the chemical interactions within it, are central to the functioning of climate, and hence of ecosystems. Two processes of major political concern, causing the ozone problem and the 'greenhouse' effect, involve the chemistry as much as the physics of the atmosphere. The disturbance in stratospheric ozone amount and distribution – and hence of radiative absorption of solar and terrestrial radiation – is thought to be due to the release of chemical pollutants. The 'greenhouse' effect arises from increased carbon dioxide and other infrared-absorbing gas concentrations that collectively tend to warm the Earth's surface. The 'greenhouse' effect has dominated recent discussions, but it was the ozone problem that led to the first international efforts to control global atmospheric changes.

The 'greenhouse' effect arises from the presence in the atmosphere of certain trace-gases that are largely transparent to incoming solar radiation, which hence can reach the Earth's surface (in the absence of cloud) in a little-attenuated state. But these gases also have the property that they are partially opaque to, and hence retard, the return flow of infrared radiation from the Earth and atmosphere to space.

The presence of the 'greenhouse' gases thus traps heat near the Earth's surface and in the surface waters of the oceans, whereas these same gases resist the net upward flux of radiative energy, and hence lead to a warming of the Earth's surface (while tending to cool the stratosphere). The gases that are active in this respect include water vapour (H_2O), carbon dioxide (CO_2), methane (CH_4), nitrous oxide (N_2O), and ozone (O_3), as well as certain synthetics – such as the chlorofluorocarbons (CFCs) – which are involved in ozone instability. All but ozone are increasing in the atmosphere (Rasmussen & Khalil, 1986) – methane currently by 1% per annum, and carbon dioxide by about 4% per decade – presumably because of human interference. Recent calculations (*e.g.* Hansen *et al.*, 1988) suggest that the total 'greenhouse' effect is now due slightly more to the other gases. collectively, than to carbon dioxide alone. The total effect is, however, most often expressed in terms of the equivalent carbon dioxide concentration.

The naturally-occurring 'greenhouse' gases are involved in vital ecosystem processes. All pass to and fro between the involved biota, the atmosphere, and the ocean. Together they raise the surface temperature of the Earth by about 33 Kelvins (deg. C) above what it would be in their absence (about 255 K, or 18°C). Climatically, they render the Earth habitable by warm-blooded animals and chemically they make possible the existence of life outside the oceans. So they can scarcely constitute an environmental threat and it is only when their concentration is disturbed that a problem arises.

It is usual to distinguish between the other 'greenhouse' gases and water vapour, because the latter exchanges mass and latent heat with the ocean and moist continental surfaces (precipitation and evapotranspiration). A recent study by Raval & Ramanathan (1989), and an attendant commentary by Cess (1989), define as a measure of the 'greenhouse' effect (G) the difference G = E – F, where E is the infrared surface emission, and F is the emission to space at the top of the atmosphere. These writers distinguish between

(1) *direct forcing, i.e.* the heating effect of increased 'greenhouse' gas absorption, without change of surface temperature;
(2) *a temperature feedback*, arising from increased surface emission; and
(3) *a water vapour feedback*, due to the consequent added water vapour content from net evaporation.

Table 2.1, reproduced from Cess (1989), shows how these processes combine to produce a net heating of 1.73°C for an increased direct forcing of 4 W m^{-2}. If the atmosphere becomes more opaque, due to the addition of 'greenhouse' gases, surface temperatures and atmospheric humidity rise to

Table 2.1. Responses of the hypothetical planet's system to a change of direct forcing of 4 W m^{-2} due, say, to increased CO_2 (Kelvins, or degrees Celsius; Watts per square metre) (*Source*: Cess, 1989)

Process	Change in *Temperature*	Change in *'Greenhouse' Effect*	*Emission to Space*
1	0	4.0	–4.0
2	1.09	1.9	4.0
3	0.64	3.6	0
Total response	1.73	9.5	0

Key: Process 1 – Change due to direct increase in forcing due to increase in CO_2 (or other equivalent effects)

Process 2 – Change due to increased global warming effect

Process 3 – Change due to warming from water-vapour feedback.

the point where the Earth is able to return the added flux of energy to space. The water vapour feedback adds about 60% to the surface warming that would otherwise occur.

The Commonwealth group report (Comm-89, *cf.* Holdgate, 1989), using Wigley (1989) as a source, generalizes the various model forecasts to arrive at a predicted warming of between 0.5 K and 2.5 K by the year 2030 AD and from 0.7 K to 3.6 K by 2050 AD. The best estimate for the year 2030 is between 1.1 and 1.9 K, and for 2050 AD it is between 1.5 and 2.8 K (WMO *et al.*, 1986). These are more conservative than the figures of the 1987 Villach-Bellagio Assessment, which foresaw a range of between 1.5 K and 4.5 K (for a doubled concentration of 'greenhouse' gases), as possible by 2030 AD, and probable after the year 2050 (WMO & UNEP, 1988). But they still represent a radical change, being global averages. Their impact is bound to be uneven – bigger in polar regions, and somewhat smaller in the tropics, also with significant changes in precipitation.

How Good Are the Models?

All the critics lay major emphasis on the weaknesses of the models employed by the climatic forecasters. How valid are these criticisms?

In principle, the models merely depend on laws of classical physics and chemistry. Nobody questions that the climatic system obeys the laws of Nature. It is possible to write down mathematical equations that express the laws as they must apply to the atmosphere, the ocean, and the continents. We do so for the atmosphere's motion, and for the forces that maintain it against friction (dynamics); and we repeat the process for the ocean. We encapsulate the thermodynamics of the system – in effect, the use which it makes of solar energy. We stipulate that matter be neither created nor destroyed (the continuity equation). We try to write down the equations for the very large number of chemical reactions that affect atmospheric composition, notably in the presence of sunlight. The result is that, to the lay reader, our journals look like hieroglyphics – and the snob appeal of fancy jargon makes matters worse.

The mathematical representation of reality that emerges is not much in question. The physics and chemistry are not obscure. But the complete set of equations, *plus* the scale of the problem (Earth is big), dictates the use of supercomputers, and is extremely demanding both intellectually and technically. One of Theodore von Kármán's students is said to have said to his mentor: 'I find two subjects difficult, atmospheric turbulence [with which Kármán struggled all his life] and quantum mechanics.' Kármán replied: 'When you die and go to heaven, God will explain quantum mechanics.'*

We apply the methods to the problem with varying degrees of sophistication. We may solve the equations, for example, simply for the mean state of the entire global system. We may focus on the energy balance, or on the water balance. We may proceed separately for ocean and for atmosphere, or may link them together, as Nature does. We may choose to use zero, one, two, or three, spatial dimensions. And we may ignore time, as we do when we use a single average over time; or we may use time as our fundamental dimension, and solve the equations with respect to future instants or epochs. Obviously, the last case is the vital one today.

Early attempts to estimate the 'greenhouse' effect (*e.g.* Tyndall, 1861; Callendar, 1938; Plass, 1956) were zero- or one-dimensional, *i.e.* they were calculated for the entire planet as a single object, or for a hypothetical case in which properties might vary along the vertical axis, but not horizontally. Much was learned, and can still be learned, from such models. They agreed that the 'greenhouse' gases warmed the Earth's surface, and other arguments predicted a decrease of temperature with height until another influence – the absorption of solar radiation by stratospheric ozone – reversed the decline above the tropopause at an altitude of 10 to 17 km.

General Circulation Models

But the models now in question are mostly three-dimensional, *i.e.* they represent the full spatial extent of the lower atmosphere, and the ocean to a chosen depth. The most refined models now couple together the ocean and atmosphere, and allow ocean currents to bring about essential transports of heat and salt. The models contain simplified, but increasingly realistic, coastlines and mountain systems (orography). The representation of clouds remains problematic, but is becoming interactive, *i.e.* predicted by the model rather than specified by the modeller. By suitable choice of time-step and spatial resolution, the modeller can also eliminate what are thought to be irrelevant scales of motion, and emphasize the modes that seem to matter most. The modellers then proceed as follows:

*I heard this anecdote from Verner Suomi, the pioneer of meteorological satellite technology. A similar story is attributed to Joseph Lamb. F.K.H.

(i) *They use the explicit mathematical form for entities that they fully understand* (or think they do), such as Newton's laws, gravitation, and thermodynamic principles. The equations of motion express the Newtonian laws and gravitation. The thermodynamic energy equation does the same for thermodynamic constraints (such as the conservation of energy and the increase of entropy). The continuity equation makes sure that the other equations respect the conservation of mass. The chemical equations are less satisfying, as they contain empirical kinetic data that cannot yet be deduced; but they are reasonably firm, of simple mathematical form, and very numerous.

(ii) *They parameterize relationships and processes that they know exist, but cannot measure or fully understand.* Parameterization is just jargon for finding formulae that work even though they cannot be easily deduced, or replaced with accurate measurement. Examples are the effects of atmospheric turbulence, disturbances of the ocean surface, and the formation of cumulus convection in the lower troposphere. Most engineering and economic relationships involve parameterizations.

(iii) *They rely on observations of the Earth's, the atmosphere's, and the ocean's properties to put actual numbers into the equations,* as regards the initial and background states of the system (for example, the mean sea-level pressure, the magnitude of gravity and the rate of the Earth's rotation and orbital variation – including irregularities). They choose boundary conditions, to use the mathematical jargon, that seem realistic in the light of empirical knowledge, and vary these to fit hypothetical future situations (such as doubled 'greenhouse' gas concentrations).

When these models are applied, it is at once apparent that the best of the numerical weather-prediction models, starting from a specific weather situation, do well for a few days, but quickly depart from reality after that. This is because they are non-linear in character, and of the sort that is in part chaotic in behaviour; small initial errors, or small events that are undetected by the observational system, may amplify until they dominate the map. The equivalent real outcome is increasingly different as time goes by.

Attempted Long-term Simulations

However, when similar models are applied to long-term simulations, they may generate mean climatic maps that look quite like the modern reality. The statistics of disturbances may do likewise. There are now over a dozen such models, and others have been tried out, or are in the development stage. Models of this sort have been applied to the doubled and quadrupled 'greenhouse' states, and also to the remote past (for example, the remarkable COHMAP exercise, which has simulated past climates from the late Wisconsin (–18,000 years ago) until modern times, using the Community Climate Model of the National Center for Atmospheric Research (*see* Kutzbach, 1985 for background; and COHMAP, Members, 1988, also Prell & Kutzbach, 1987, for an even longer perspective).

These models can be so adapted that they do indeed simulate the broad outline of the modern climate – not only the surface distributions of temperature and precipitation, but also of levels well up into the upper troposphere. The most-often-used models reproduce the present 'greenhouse' effect quite well. Thus the Raval & Ramanathan (1989) study of the water vapour feedback, as developed by Cess (1989), shows that the clear-sky flux of energy to space varies with surface temperature by 2.38 ± 0.16 W $m^{-2}K^{-1}$ as averaged across 4 general-circulation model predictions. The Earth Radiation Budget Experiment (ERBE), available since 1985 (*cf.* Raval & Ramanathan, 1989), measures the same variation as 2.31 W $m^{-2}K^{-1}$, which indicates quite a close fit between model calculations and direct observations (*see* Table 2.1, p. 35).

However, and it is a major objection, intercomparison of the most-used models reveals deficiencies in their performance. These deficiencies are of three main kinds:

(i) The model predictions of future states do not correlate well with one another on sub-continental scales, *i.e.* in regional detail;
(ii) They similarly tend to differ significantly when applied to present-day distributions, *i.e.* from the real detail in world climate, which they reproduce only poorly; and
(iii) They differ notably in their predictions of the behaviour of the hydrologic cycle, *i.e.* as regards precipitation, evaporation, and streamflow, which are of crucial economic importance.

Intercomparison of predicted fields from some of the models has been attempted by Grotch (1988), whose results illustrate the above limitations. A study of fourteen atmospheric general circulation models by Cess *et al.* (1989) shows that there are large variations in global sensitivity among the models, and that these are largely attributable to what is conventionally (and confusingly) called 'cloud feedback'. Much of the criticism of the models hangs on the role of clouds in modulating the global 'greenhouse' effect.

Changes in Solar Irradiation Small

Other critics have speculated that variations in solar behaviour may invalidate 'greenhouse'-based predictions of warming over the next half-century. The climatic system is indeed sensitive to variations in the solar constant, or irradiance. Nierenberg *et al.* (1989) have suggested that there may be a decrease in solar irradiance in the next century, and that this would lead to cooling which would easily offset the alleged 'greenhouse' warming. They are also unimpressed with the evidence that a warming is already in progress.

On the other hand, there is little evidence of significant changes in the solar constant in the past century. Foukal & Lean (1990) have used satellite observations of incoming solar radiation since 1978 to calibrate a model relating the solar constant to changes in the photospheric activity of the sun (chiefly sunspot-cycle-related). This model, when applied to the period 1874–1988, shows variations in 12-months' running mean values of the solar constant of about 1 W m^{-2} (from 1367 to 1368 W m^{-2}), with a slow increase in mean values since 1945. These differences are enough to account for only 0.02 K in global mean annual temperature – an order of magnitude less than that observed since 1945, and even less than that observed since 1874. To match the predicted 'greenhouse' warming in the next half-century, the solar constant would have to increase by an order of 1%, or ten times the changes detected in the past record by Foukal & Lean (1990). The latter do not rule out other sources of altered solar output, not related to the sunspot cycles. But most students of the Sun's behaviour are obviously sceptical, and agree with Foukal & Lean that 'only a determined effort to measure the behavior of [the solar constant] over many decades can increase our understanding of this potentially important climatic parameter.' Until there is concrete evidence of variations of this magnitude in the solar constant, one must assume that the effect of such variations will be small in comparison with the 'greenhouse' effect.

Remember the Oceans

Yet another theme of the models' critics is that they do not effectively represent the role of the oceans. The latter are very cold, except for a shallow surface layer that interacts readily with the atmosphere. From the beginnings of 'greenhouse' modelling it has been expected that the oceans would serve as a heat-sink, retarding the anticipated heating, the so-called 'transient effect'. It has also been clear that transports of heat from tropical to polar latitudes by ocean currents – crucial to the temperature distribution in both ocean and atmosphere, are either ignored or only crudely parameterized in most models. Changes in these transports, and in the thermohaline circulation in the oceans, ought to be explicitly predicted if the models are to be effective in estimating future climatic change.

Early models of the 'greenhouse' process ignored these questions, or addressed the 'transient effect' only in crude, bulk terms. These were followed by models that took account of the obvious differences between land and sea (the so-called 'swamp' models) but still ignored the oceans' vast heat-absorption capacity, and their heat-transport by currents. But more sophisticated models are now appearing (*e.g.* Hansen *et al.*, 1988; Schlesinger & Zhao, 1989; Stouffer *et al.*, 1989; Washington & Meehl, 1989) in which some of these deficiencies are being rectified. In three of these cases, moreover, the analysis allows for the slow build-up of 'greenhouse' gas concentration (at 1% per annum, close to observed values), instead of solving for doubled-concentration steady-state. The Stouffer *et al.* analysis indicates (i) a marked asymmetry between the polar hemispheres in response to 'greenhouse' forcing, with the southern hemisphere responding much the more slowly; and (ii) a significant change in Atlantic thermohaline circulation, and in the formation of Antarctic bottom water (which together play a key role in maintaining the great cold [-2°C] of the deep, world ocean). But there are important differences between these four recent 'experiments', and it cannot yet be said that understanding of the role of ocean–atmosphere coupling has been properly reached.

Cloud Feedback

The same must still be said of cloud feedback (the extent to which 'greenhouse' warming may alter cloud type, height, amount, and distribution, and the effect that such change might have on any warming). But some progress is now being made, in both the modelling field (*see*, for example, Hansen *et al.*, 1984; Wetherald & Manabe, 1986, 1988) and in direct observation (Ramanathan *et al.*, 1989).

The study cited last is based on the analysis of ERBE data for April, 1985 (though the implications apply to much longer periods). Ramanathan *et al.* emphasize that one must properly speak of *cloud-radiative forcing*, rather than cloud feedback. The satellite observations show a cooling of the globe of about 45 W m^{-2} due to cloud *reflection* of solar radiation, and a warming of 31 W m^{-2} due to the 'greenhouse' effect of the clouds. The clouds thus produced a net cooling of 14 W m^{-2}. Such cooling was pronounced (up to 100 W m^{-2}) in cloudy mid-latitude and high-latitude areas. In tropical areas a near-balance existed between the solar reflection and 'greenhouse' warming effects of clouds.

But the truly important result of the Ramanathan *et al.* (1989) study is the demonstration that the net cloud radiative forcing (a cooling) is four times as large as the 'greenhouse' warming expected for a doubling of the concentration of the responsible gases. Hence a small change in the extent and location of cloud cover could, in principle, offset the 'greenhouse' warming. But it now appears that the role of cloud cover is more complex than is widely supposed. The level, water content, and form, of the clouds are all important, as is their microphysical state. Hence it is still impossible to deduce whether a 'greenhouse' warming will significantly alter the present role of cloud, which is a small global cooling. Continued satellite observation should move us closer to an answer. This illustrates a good general principle: in complex situations, make good use of monitoring and direct observation.

Although the major emphasis in the modelling experiments has been upon hemispheric or global implications of the 'greenhouse' effect, techniques exist for applying limited-area dynamical models to the question of regional impacts. Among the areas that have been examined are Sub-Saharan Africa, the Asian monsoon area, and Amazonia. The last area will be of special interest to us all, because of concern for the fate of the rainforests, and of the idea – probably erroneous – that changes in the regional climate of Amazonia must produce measurable effects on the global scale.

Amazonia Equivocal

Three atmospheric general-circulation model studies of the Amazonian question have been published recently, giving rather different results.

Dickinson & Henderson-Sellers (1988) used the Community Climate Model of the US National Center for Atmospheric Research to examine the consequences of the replacement of the rain-forest by poor grassland. They found reduced evaporation and higher surface-temperatures, but little change in precipitation. Lean & Warrilow (1989) similarly replaced the rain-forest by pasture, and applied a UK-designed Atmosphere General Circulation Model (AGCM) to Amazonia. Their results resembled those of Dickinson & Henderson-Sellers as regards albedo change, evapotranspiration, and temperature, but also projected significant precipitation and run-off decreases. A third study, by Shukla *et al.* (1990), also shows large decreases in precipitation, evaporation, and run-off, with a longer dry season than currently.

Here again, modelling yields equivocal answers. I nevertheless expect this class of modelling to become vital to the solution of policy questions, when and if the political community becomes persuaded that countermeasures to any 'greenhouse' warming are unavoidable. Obviously, regional modelling requires skills that differ in several ways from the global case. One of these is the ability to nest the limited-area models successfully within global models – a process in which Canadian and European modellers have become adept, most especially those of the United Kingdom. Another is the need to specify, by some system of spatial coordinates, the physical and biological characteristics of the land surface (where geographers such as Anne Henderson-Sellers have given a notable lead).

Conclusions about Modelling

I have indicated enough to justify the following generalizations about the modelling results:

(i) Modelling is a necessary, quantitative presentation of the ideas underlying the 'greenhouse' effects, and all other proposed mechanisms of change to the climatic system;

(ii) The procedure now preferred is to use coupled atmosphere–ocean models in which carbon dioxide and other 'greenhouse' gases are allowed to increase at realistic rates (*c.* 1% per annum in collective effect) – a procedure that is replacing the older concept of running the experiment until equilibrium is reached with a doubled CO_2 content;

(iii) Such models continue to predict a wide range of potential global warmings (of the order of 2 to 3 K by 2050 AD), but differ as to regional detail;

(iv) Incorporation of ocean models in a realistic way hints at the possibility of different responses between polar hemispheres as regards atmospheric temperature, thermohaline oceanic circulation, and ultimately the effects on deep and intermediate ocean temperatures and salinities;

(v) The cloud-radiative forcing effect has been confirmed, largely through the ERBE satellite results, as a major influence on future response to 'greenhouse' gas build-up; clear-sky response of the models is well-confirmed by the results;

(vi) Serious differences persist between models as to the future development of the hydrologic cycle; the latter characteristically displays details below the resolving power of the models, which do a poor job of predicting even present-day regional distributions;

(vii) There is convincing evidence that variations in the solar constant produce temperature changes only one-tenth as large as those observed, and as are predicted by the 'greenhouse' theory; but changes in the constant of the order of 1% (which would be required to match 'greenhouse' changes) cannot be ruled out from existing knowledge of the non-photospheric behaviour of the Sun; and

(viii) Regional-scale modelling of the 'greenhouse' changes will become vital if countermeasures are decided on by the world's nations. Such models are already being applied to areas of special concern, but not as yet with fully convergent results.

Evidence from the Climatological Record

In addition to the evidence derived from the modern instrumental record, chiefly of temperature, pressure, and precipitation, witness of past climatic fluctuations or persistent trends is available far back into geological history from a variety of proxy sources. Summaries are given by R. S. Bradley (1985) and Elsaesser *et al.* (1986). These proxy sources have been greatly enriched in the past four decades by the use of geochemical methods, notably when applied to stable isotopes of oxygen, hydrogen, and carbon. Systematic exploration of ocean-bottom sediments, of glacial ice, and of cave speleothems, has enormously enlarged the available evidence. In particular the multi-millennial ice-cores taken from Greenland and Antarctica have revealed that the parallel variation of carbon dioxide and methane concentrations with air temperature is detectable back for at least 150,000 years. Between them the geologists, geochemists, and palaeoecologists, have transformed the science of palaeoclimatology.

For our present purpose, however, I shall concern myself only with the question: is the signal of the 'greenhouse' effect yet visible in the modern temperature record? Several groups have reconstructed hemispheric and global time-series of mean annual surface-air temperatures, with attempts to remove the disturbing effects of inequality of record between land and sea, of urban growth, and of the poverty of records from the polar regions.

Analysis of this question by Wigley (1989), as presented in Comm-89, suggests that the recent observed warming lies near the lower boundary of the probable range of temperatures associated with the 'greenhouse' warming, as was predicted by the family of models discussed above. Wigley used the curve of mean annual global air temperature compiled by the UK Meteorological Office and the University of East Anglia (*see* Jones *et al.*, 1986; Jones *et al.*, 1988; Kerr, 1990). In its most recent guise, this curve (Fig. 2.1) shows the following features:

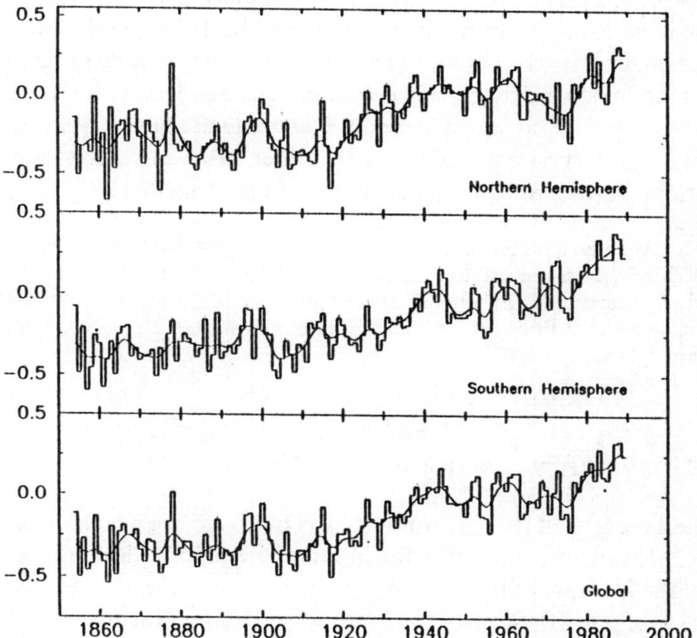

Figure 2.1. Curves of mean annual surface air temperatures (°C) for the two polar hemispheres and the global average since 1855. The zero in each curve is the 1951 to 1980 mean value. The global value for 1990, not plotted, was 0.05°C higher than that for 1988 and is therefore the highest on record. The values shown by the columns are the raw annual averages. The trend is indicated by the smooth curve. For all three curves values are based on:

1. weighted averages for land areas, as calculated by the Climate Research Unit of the University of East Anglia;
2. air temperatures over marine areas supplied by the United Kingdom Meteorological Office in connection with the Climate Research Unit;
3. a weighting process to allow for inequalities for data availability over the globe and to eliminate records vitiated by urban warming effects.

Curves kindly supplied by T. M. L. Wigley and P. D. Jones, University of East Anglia, private communication.

(i) Widely fluctuating global temperatures between 1856 and 1900 of about 0.2 K below 1951–80 'normals';
(ii) A cold first decade of the twentieth century, about 0.1 K colder than 1856–1900;
(iii) A rise of about 0.4 K from this value to a peak in the early 1940s;
(iv) A cooler phase from the late 1940s until the mid-1970s at values close to the 1951–80 'normals'; and
(v) A pronounced warming from 1975 until the present day. The 1980s produced six of the ten warmest years in the record, going back nearly a century-and-a-half.

In effect the evidence shows that overall warming of the global surface has been a feature of this century. It has not been smooth and linear. In particular the pause in warming from the late 1940s into the 1970s, which was world-wide (though most marked in the northern hemisphere), is difficult to reconcile with any simple dependence on 'greenhouse' gas build-up (which appears to have proceeded, with no equivalent pause, at a rate of about 0.4% per decade for CO_2, with no estimate available for the other gases). Lower-stratospheric temperatures appear to have fallen significantly (Angell, 1988).

Table 2.2. Average Annual Increase in 'other' (i.e. apart from CO_2) Greenhouse Gases, 1975-85 (parts per million millions by volume, except parts per thousand millions for N_2O and CH_4). *Ninety* per cent confidence limits given below, based on observations at South Pole, and in the Pacific North-west of the US (*Source*: Rasmussern & Khalil, 1986)

CCl_4	CH_3CCl_3	CCl_3F (F–11)	CCl_2F_2 (F–12)	N_2O	CH_4
2.4±0.3	8.2±0.6	10.6±0.8	18.5±0.9	1.04±0.14	17.5±1.3

Detailed records of the 'greenhouse' gas build-up are available only back into the 1970s. Table 2.2 (after Rasmussen & Khalil, 1986) shows what is known of the increase of the main performers (apart from CO_2) for the eleven years 1975–85, at the South Pole and in the Pacific North-west of the US. The most striking increase is that of methane, CH_4, where the concentration is mounting at about 1% per annum (Blake & Rowland, 1988). Allowing for the fact that some of these gases (including methane) are much better infrared absorbers than carbon dioxide, the annual increase in 'greenhouse' gas concentrations in the atmosphere (and hence effects on The Biosphere) is equivalent to an almost 1% increase in carbon dioxide, and is likely to attain that figure shortly, even if fossil-fuel use is curtailed.

In summary, the climatological record (in the widest sense) shows a continuing buildup of 'greenhouse' gases, at a rate equivalent to nearly 1% per annum; little prospect that such increases will be quickly reversed; a rise of global annual mean surface-air temperatures since 1905 of about 0.5 K, with a reversal to slightly lower values in the 1950s, '60s, and early '70s; very high interannual variability; and a very recent – in the 1980s – apparent acceleration of the warming, such that we have had recent experience of years approaching those of the postglacial thermal maximum some 9,000 years before present (BP).

The increase of temperature cannot, with full assurance, be referred to the 'greenhouse' effect because it has been spasmodic, with three decades of cooler conditions following the peak in the 1940s. In addition, the warming appears to have been concentrated in regions different from those predicted by the available models.

On the other hand, the observed overall increase during this century is consistent with the more conservative model predictions. More warming might have been expected, but what has been observed is more than could have been expected from other known sources of heating (such as a variable solar 'constant').

Have Physical Impacts of the Warming Been Observed?

The past decade has seen a mounting flood of evidence, much of it equivocal, that the small and halting warming so far observed has had measurable effects on other natural features and on the human economy. This flood has been matched by similar allegations that acid deposition has affected forests, buildings, human health, crop production, and above all freshwater ecosystems. A mood has been created that anticipates such impacts, and actively looks for them.

Sea-level Rise

The first impact is clearly on sea-level, whose expected rise has immense implications for countries with extensive low-lying territories. The measurement of mean sea-level is difficult. Values may be affected by tectonic rise or fall of the land, which is especially marked in recently deglaciated areas and along certain continental margins, as well as by changes of marine origin – changes in ocean volume, due to additions or losses of water, or to altered temperature and salinity – and to altered wind-stresses on the ocean surface.

Comm-89, quoting Wigley (1989)*, gives a 'best estimate' for future sea-level rise of between 17 and 26 cm by the year 2030, with an absolute range of estimate between 5 and 44 cm. Here again there is clearly major uncertainty as to the future outcome. Comm-89 (Holdgate, 1989) gives approximately equal weight to additions of water volume due to thermal expansion of the ocean column and the melting of alpine glaciers (many of which have been in rapid decline for decades). A much smaller contribution is predicted from the Greenland and more limited ice-caps, whereas Antarctica may contribute a net loss of water, *i.e.* a growth in the volume of ice on land.

Elsaesser *et al.* (1986) tabulated fourteen different estimates of the current rate of global sea-level rise, after elimination of differential land movement. The estimates were made at various dates from 1941 until 1985, and mostly gave rates of sea-level rise between 10 and 15 cm per century. Comm-89's 'best estimate' range is thus higher than recent values by 3 to 6.5 times; obviously they expect a rapid acceleration of sea-level rise if the accelerated warming of the 1980s continues.

A recent detailed re-examination of this issue by Peltier & Tushingham (1989), noteworthy for its careful assessment of the effect of tectonic movements, concludes, however, that the present rate of sea-level rise may already have reached 24 cm per century – almost double the typical estimate

*[In answer to our query about this, and adding some micro-history, Hare replied (*in litt.* 28 January 1991): 'As to Wigley (1989), this was a personal briefing he gave to Margaret Thatcher at very short notice. He didn't publish a paper, but used the illustrations as input to the Comm-89 (Holdgate, 1989) reference. I asked him about this in Norwich last July. Wigley was the primary author of the meteorological section of Holdgate, 1989 – Martin [Holdgate] was compiler, editor, and part-author.' Eds.]

of a few years ago. Similar values were published by Barnett (1983*a*, 1983*b*). Another Canadian authority, Stewart (1989), is much less optimistic that the effects of glacial melting and water-column expansion can be separated from tectonic effects. He expects the 'greenhouse' signal to be small. We do not yet have firm estimates of secular temperature changes of the ocean water-column, though spectacularly large rises of surface-water temperature have been alleged for the 1980s from some satellite data.

It has been assumed by most authorities that the rise of sea-surface temperature will increase vapour pressures over the oceans, and that this will necessarily increase mean precipitation. Recent analyses of precipitation over the continents (J. R. Bradley *et al.*, 1987) do indeed show such increases in recent decades over parts of the continents, though not over others. It has been further assumed that the major glaciers – alpine and continental – would be likely to experience increased snowfall on their upper levels, partially or completely offset by increased melting in marginal areas. But there has been, and remains, considerable uncertainty as to the relative size of these competing effects.

Zwally *et al.* (1989) have recently reported, from analysis of Seasat and Geosat satellite data, that the absolute surface level of the Greenland ice-sheet south of 72°N was *increasing* at a rate of 20 + 6 cm per annum in 1978 (Seasat), and at 28 + 2 cm in 1985–86. These figures depend on radar altimetry. Zwally *et al.* (1989) conclude that ...'Greenland ice-sheet growth is consistent with the generally warmer temperatures experienced in this century. If climate continues to warm, enhanced precipitation in polar regions may offset increases in melting.' They add, however, that the much lower levels of mass input (*i.e.* snowfall) will make Antarctica's behaviour much harder to detect, and recommend laser altimetry to assess its contribution. Their analysis has been challenged by Douglas *et al.* (1990), who argue that (i) there were variations in satellite altitude which were not allowed for by Zwally *et al.*, and (ii) observations of the Earth's axial wobble (and hence position of the North Pole), and of satellite-deduced sea-levels, invalidate their procedure.

Yet another uncertainty concerns the permafrost of the Northern Hemisphere. Recent data on the concentration of methane (CH_4) from 80°N to 50°S (Blake & Rowland, 1988), show high values in high-northern latitudes in September 1987. These and other observations have prompted the hypothesis that the melting of permafrost is releasing methane from hydrated forms of it occurring in and below the frozen layers. A major Global Change project, involving US and Canadian programmes, is investigating the source of methane excesses in Arctic and sub-Arctic settings.

Temperatures have apparently risen in permafrost sites in Alaska and northern Canada (*see* Elsaesser *et al.*, 1986 Table 5, for a summary) by amounts that, in most cases, exceed the observed change in air temperature. A. Lachenbruch and his associates in Alaska have shown, through deep borings in permafrost, that the upper part of temperature profiles are consistently warmer than theory predicts. Borings at Prudhoe Bay revealed the anomalously warm layer to be 160 m deep, with a surface-soil temperature rise of 1.8 K in the past 67–117 years. A. Lachenbruch's private assessment communicated to Elsaesser *et al.* (1986) was that, in the Arctic Coastal Plain of Alaska, most boreholes indicate a warming of the permafrost of 2 K or more 'in the last several decades to a century or so'. Canadian sources (*e.g.* Paterson, 1968; Cermak, 1971) also indicate large soil warmings near and just south of the permafrost border.

There is thus evidence that both the ice-masses and the permafrost of northern latitudes have already responded to climatic warming. But the mechanics of this warming, and the relation between atmospheric, oceanic, and cryospheric, changes is not yet firmly established. It is to be hoped that the IGBP and associated programmes will clarify matters.

Conclusions

Where the scepticism to which I refer in my title comes from should now be obvious. There is, firstly, major uncertainty as to the confidence in the model results that predict an intensified warming. There is, secondly, the fact that the observed change of temperature during the past century has been irregular and rather small, in spite of a uniform and roughly exponential increase in 'greenhouse' warming. And, thirdly, the physical correlations of the warming, such as sea-level rise, permafrost melting, and glacial-mass balance, are also very uncertain.

In my judgement, the governments of the world, and the rest of the scientific community, ought nevertheless to *assume that the warming will continue, and probably accelerate*, because the augmented 'greenhouse' effect seems the likeliest cause. The fact that the response of the climatic system to the observed buildup is irregular, does not alter my judgement; I should have been amazed if the complex system had followed the buildup in a simple, parallel fashion. But equally it is my personal view that we ought to stress the uncertainties, and at all times to admit our fallibility. I am aware that politicians dislike the phrase 'on the other hand' in the mouths of advisers. But we have no moral choice: the uncertainty is large and real.

Yet, the uncertainty involved in this issue is much lower than one finds in the socio-economic domain. Social and economic indicators are even less reliable than ours, and the supporting disciplines at least as insecure. Politicians can and do base sweeping action in such fields as the budgetary process and energy policy on the basis of evidence that is flimsy in comparison with that reviewed in this paper. Politics involves making inspired guesses on the basis of very partial evidence.

I have been encouraged by the obvious signs that many politicians and civil servants now agree with me, and are moving towards action. Business leaders similarly have the matter high on their agenda, notwithstanding the apprehensions they have as to the cost of the necessary remedial measures (with which I have not dealt in this paper). We seem at last to be moving towards coming to grips with a problem that I have lived with during most of my life. In closing, let me urge three things as essential for further progress *inter alia* in tackling climate-related questions:

(1) *The modelling exercises* to which I have referred, in spite of their obvious limitations, need to be extended and intensified. On a world scale, a mere handful of research workers are at work, and a mere fistful of pounds, dollars, francs, roubles, and so on, are at their disposal. More people need to be trained, more centres put into motion, and more resources provided to each. This must be an international collaborative exercise specifically under the auspices of the World Climate Programme. Modelling is just another word for the application of sound physics, chemistry, and mathematics, to analysis and prediction of real environmental processes. It is not a luxury, but a core method of geophysical and geochemical science. Modelling will never lead to certainty; but it can diminish uncertainty.

(2) *Environmental monitoring* is an essential component of the world effort. Monitoring is an absolute necessity in the study of complex systems, which usually defy simple analysis. It means the patient, long-term measurement of key variables in the environment, again in accordance with internationally coordinated norms. Ocean, land areas, and the whole depth of the atmosphere, need such scrutiny, as do the affected parts of biotic communities and human economic activity (Wood, 1990).

(3) *Much closer rapprochement between the physical and biological sciences is required.* I have not had time to review the question of ecosystem response, but this does not mean I undervalue its study. The climatic system includes crucial biological control processes – for example, transpiration over land; photosynthesis and respiration; and the precipitation processes for carbon species in ocean waters. Tradition and upbringing have tended to keep geophysical scientists apart from their biological cousins. They must come together world-wide, above all within the framework of the International Geosphere/Biosphere Programme, and within the moral authority of the objectives of the World Council For The Biosphere.

References

Angell, J. K. (1988). Variations and trends in tropospheric and stratospheric global temperatures, 1958–1987. *Journal of Climate,* 1, pp. 1296–13, illustr.

Barnett, T. P. (1983*a*). Recent changes in sea level and their possible causes. *Climatic Change*, 5, pp. 15–38, illustr.

Barnett, T. P. (1983*b*). Some problems associated with the estimation of 'global' sea level change: Report to the National Climate Program (cited by Elsaesser *et al.*, 1986, *q.v.*). [Original not examined.]

Blake J. R. & Rowland, F. S. (1988). Continuing worldwide increases in tropospheric methane, 1978 to 1987. *Science*, 239, pp. 1129–31, illustr.

Bradley, J. R., Diaz, H. F., Eischeid, J. K., Jones, P. D., Kelly, P. M. & Goodess, C. M. (1987). Precipitation fluctuations over Northern Hemisphere land areas since the mid-nineteenth century. *Science*, 237, pp. 17–25, illustr.

Bradley, R. S. (1985). *Quaternary Paleoclimatology*, Allen & Unwin, Boston, Massachusetts, USA: xvii + 472 pp., illustr.

Brookes, W. T. (1989). The global warming panic. *Forbes Magazine*, issue of December 25th, pp. 96–101.

Callendar, G. C., (1938). The artificial production of carbon dioxide and its influence on temperature. *Quarterly Journal of the Royal Meteorological Society*, 64, pp. 223–40.

Cermak, V. (1971). Underground temperature and inferred climatic temperature of the past millennium. *Palaeogeography, Palaeoclimatology, Palaeoecology*, 10, pp. 1–19.

Cess, R. D. (1989). Gauging water-vapour feedback. *Nature* (London) 342, pp. 736–7.

Cess, R. D., Potter, G. L., Blanchet, J. P., Boer, G. J., Ghan, S. J., Kiehl, J. T., Treut, H. Le, Li, Z-X., Liang, X-Z., Mitchell, J. F. B., Morcrette, J.-J., Randall, D. A., Riches, M. R., Roeckner, E., Schlese, V., Slingo, A., Taylor, K. E., Washington, W. M., Wetherald, R. T. & Yagai, I (1989). Interpretation of cloud-climate feedback as produced by 14 atmospheric general circulation models. *Science*, 245, pp. 513–6.

COHMAP members (1988). Climatic changes of the last 18,000 years: observations and model simulations. *Science*, 241, pp. 1043–52.

COMM-89 (1989). *See* Holdgate, M. W. (1989).

Dickinson, R. E. & Henderson-Sellers, A. (1988). Modelling tropical deforestation: a study of GCM land-surface parameterizations. *Quarterly Journal of the Royal Meteorological Society*, 114, pp. 439–62.

Douglas, B. C., Cheney, R. E., Miller, L., Agreen, R. W., Carter, W. E. & Robertson, D. S. (1990). Greenland Ice Sheet: is it growing or shrinking? *Science*, 248, p. 288.

Elsaesser, H. W., MacCracken, M. C., Walton, J. J. & Crotch, S. L. (1986). Global climatic trends as revealed by the recorded data. *Reviews of Geophysics*, 24, pp. 745–92.

Foukal, P. & Lean, J. (1990). An empirical model of total solar irradiance variation between 1874 and 1988. *Science*, 247, pp. 556–8.

Grotch, S. L. (1988). *Regional Intercomparisons of General Circulation Model Predictions and Historical Climate Data.* US Department of Energy, Washington, DC, USA: 291 pp., illustr., tables.

Hansen, J. E., Lacis, D., Russell, G., Stone, P., Fung, I., Ruedy, R. & Lerner, J. (1984). Climate sensitivity: analysis of feedback mechanisms. Pp. 130–63 in *Climate Processes and Climate Sensitivity* (Eds J. E. Hansen & T. Takahashi). (Geophysical Monograph Series, 29.) American Geophysical Union, Washington, DC, USA: 336 pp.

Hansen, J. E., Fung, I., Lacis, A., Rind, D., Lebedeff, S., Ruedy, R. & Russell, G. (1988). Global climate changes as forecast by Goddard Institute for Space Studies three-dimensional model. *Journal of Geophysical Research*, 93(D8), pp. 9341–64.

Holdgate, M. W. (1989) (Chairman of Comm-89.) *Climate Change: Meeting the Challenge.* (Report of a Commonwealth Group of Experts) Commonwealth Secretariat, London, England, UK: 131 pp., illustr., tables.

Jones, P. D., Wigley, T. M. L. & Wright, P. B. (1986). Global temperature variations between 1861 and 1984. *Nature* (London), 322, pp. 430–4.

Jones, P. D., Wigley, T. M. L., Folland, C. K., Parker, D. E., Angell, J. K., Lebedeff, S. & Hansen, J. E. (1988). Evidence for global warming in the past decade. *Nature* (London), 332, p. 790.

Kerr, R. A. (1990). Quotation from second annual report on global temperature, by P. D. Jones & D. E. Parker, *Science*, 247, p. 521.

Kutzbach, J. E. (1985). Modelling of paleoclimates. *Advances in Geophysics*, 28A, pp. 159–96, illustr.

Lean, J. & Warrilow, D. A. (1989). Simulation of the regional climatic impact of Amazon deforestation. *Nature* (London), 342, pp. 411–3, illustr.

Lindzen, R. S. (1990). Some coolness concerning global warming. *Bulletin of the American Meteorological Society*, 71, pp. 288–99, illustr.

Nierenberg, W. A., Jastrow, R. & Seitz, F. (1989). *Scientific Perspectives on the Greenhouse Problem.* George C. Marshall Institute, Washington, DC, USA: 35 pp., illustr.

Paterson, W. S. B. (1968). A temperature profile through the Meighen ice-cap,

Arctic Canada. *IASH-AISH Publications*, 79, pp. 440–9, illustr.

Peltier, W. R. & Tushingham, A. M. (1989). Global sea-level rise and the greenhouse effect: might they be connected? *Science*, 244, pp. 806–10, illustr.

Plass, G. N. (1956). The carbon dioxide theory of climatic change. *Tellus*, 8, pp. 140–54, illustr.

Prell, W. L. & Kutzbach, J. E. (1987). Monsoon variability over the past 150,000 years. *Journal of Geophysical Research*, 92(D7), pp. 8,411–25, illustr.

Ramanathan, V., Cess, R. D., Harrison, E. F., Minnis, P., Barkstrom, B. R., Ahman, E. & Hartmann, D. (1989). Cloud-radiative forcing and climate: results from the Earth Radiation Budget Experiment. *Science*, 243, pp. 57–63, illustr.

Rasmussen, R. A. & Khalil, M. A. K. (1986). Atmospheric trace-gases: trends and distributions. *Science*, 232, pp. 1,623–4, illustr.

Raval, A. & Ramanathan, V. (1989). Observational determination of the greenhouse effect. *Nature* (London), 342, pp. 758–61, illustr.

Schlesinger, M. E. & Zhao, Z.-C. (1989). Seasonal climatic changes induced by doubled CO_2 as simulated by the OSU atmospheric GCM/mixed-layer ocean model. *Journal of Climate*, 2, pp. 459–94.

Shukla, J., Nobre, C. & Sellers, P. (1990). Amazon deforestation and climate change. *Science*, 247, pp. 132–5, illustr.

Stewart, R. W. (1989). Sea-level rise or coastal subsidence? *Atmosphere-Ocean*, 27, pp. 461–77, illustr.

Stouffer, R. J., Manabe, S. & Bryan, K. (1989). Interhemispheric asymmetry in climate response to a gradual increase of atmospheric CO_2. *Nature* (London), 342, pp. 660–2, illustr.

Tyndall, J. (1861). As cited by R. D. Cess (1989). [Original not seen.]

Washington, W. M. & Meehl, G. A. (1989). Climate sensitivity due to increased CO_2: experiments with a coupled atmosphere and ocean general circulation model. *Climate Dynamics*, 4, pp. 1–38, illustr.

Wetherald, R. T. & Manabe, S. (1986). An investigation of cloud change in response to thermal forcing. *Climatic Change*, 8, pp. 5–23, illustr.

Wetherald, R. T. & Manabe, S. (1988). Cloud feedback processes in a general circulation model. *Journal of the Atmospheric Sciences*, 45, pp. 1,397–415, illustr.

Wigley, T. M. L. (1989). As cited in Holdgate(1989).

Wood, F. B. (1990). Monitoring global climate change: the case of greenhouse warming. *Bulletin of the American Meteorological Society*, 71, pp. 42–5, illustr.

World Meteorological Organization, United Nations Environment Programme & International Council of Scientific Unions [cited as WMO *et al.*] (1986). *Report of the International Conference on the Assessment of the Rôle of Carbon Dioxide and of Other Greenhouse Gases in Climate Variations and Associated Impacts.* Villach, Austria, 9–15 October 1985. Report No. 661, WMO, Geneva, Switzerland: iii + 78 pp., illustr.

World Meteorological Organization & United Nations Environment Programme [cited as WMO & UNEP] (1988). *Developing Policies for Responding to Climatic Change*, summary of discussions and recommendations of workshops held in Villach (28 September–2 October 1987) and Bellagio (9–13 November 1987) under the auspices of the Beijer Institute, Stockholm. Report WCIP-1, World Meteorological Organization, Geneva, Switzerland 1988: v + 53 pp.

Zwally, H. J., Brenner, A. C., Major, J. A., Bindschladler, R. A. & March, J. G. (1989). Growth of Greenland Ice Sheet: Measurement. *Science*, 246, pp. 1587–91.

Commentary on Chapter 2

CHAIRMAN: Professor Reid A. Bryson
PANELLISTS AND OTHER CONTRIBUTORS:

Ferguson, Bryson, Fearnside, Bazzaz, Fosberg, Harun ur Rashid, Hare, McCloskey, Bryson, Holdgate

Ferguson commented on three aspects of the climate change question: uncertainty in scientific understanding, evidence for climate instability and change, and the risk management approach to dealing with the problem.

The reports of the Intergovernmental Panel on Climate Change (IPCC) indicated that there was a consensus amongst several hundred leading scientists that climate warming due to an enhanced 'greenhouse' effect, caused by Man's activities, would occur. The IPCC estimated that an average annual global temperature increase of between 1.5°C and 3.6°C for a doubling of the atmospheric concentration of CO_2, or the radiatively equivalent increase in all greenhouse gases. This fell within the same general range as previous predictions, published over the last 10 years, of between 1.5°C and 4.5°C. There was also a consensus that the average temperature increases would be larger in mid-latitude and polar regions than in the tropical belt. From the specific work of the General Circulation Model (GCM) of the Canadian Weather Service together with comparisons of the results of various other circulation models – of which there were about 15 – **Hare** had illustrated that the average rise expected for North America was of the order of 4 or 5 degrees Centigrade, that in the Tropics, about 2°C, and at the Poles, about 10 degrees. It was worth remembering that almost a century ago the Swedish physicist, Arrhenius, without the benefit of a GCM and all the other data now available, calculated on the basis of basic physics and meteorology, that a doubling of atmospheric CO_2 would lead to an average temperature increase of 5°C. It had to be said that he was not far wrong! Indeed, one could argue that scientific predictions had been reasonably consistent over a long time-period. Allegations of scientific uncertainty, he noted, were sometimes exaggerated! On the other hand, no one would dispute that there was some uncertainty in our understanding of, and ability to predict, climate change. Efforts to reduce this uncertainty should be increased, but it would be quite unrealistic to expect 100% confidence in long-term climatic forecasting for the foreseeable future.

The problem was, therefore, he suggested, one of probabilities and risk management. Political leaders had to take decisions based on analyses and predictions which were much less certain. He commented, wryly, that long-term consensus predictions by expert economists and energy planners might well be less certain than those of climate experts, yet they seemed to be accepted more readily by decision-makers as a basis for action! Because of this kind of bias in the minds of politicians and others, he thought that there might be some merit in considering the problem of detecting climate change from the perspective of legal logic as well as by scientific logic.

In an ideal world, everyone would accept a result if science could provide absolute proof – 100% certainty – that climate change was occurring. That was the equivalent of the 'smoking gun' proof in a murder trial. But such absolute scientific proof was unlikely in the near future, as he had already implied. The only practical alternative to the 'smoking gun' in the world of law, therefore, was to base a case on circumstantial evidence. Such a case would be credible if there were sufficiently strong corroborating elements. One way to get at this for climate change was to go back ten years, when expert climatologists had become concerned about possible climate warming, and to ask what kinds of evidence, or indicators of change they would have considered and examined then. He suggested that the possibility that six of the warmest years in the century would occur in the 1980s would have been considered a remote possibility unless climate warming were indeed occurring. Such an indicator was not statistically conclusive but it was strongly suggestive. Much the same argument applied to other indicators such as the 1% annual increase in atmospheric methane, which was poorly explained unless there was a possible feedback to the melting of the permafrost, or the increasing desertification in certain regions and of increased droughts in others. The recent increase in the apparent frequency of hurricanes, or the fact that in early 1990, in the space of a little over one month, approximately 5 major storms had affected Northern Europe, two or three of which had wind-speeds associated with them of 100-year return periods, suggested that something was happening to climate. Could, he asked, all these events, taken together, be considered more than normal climate variability? He believed that they were indicative of, and supported the case for, climate change.

He then asked whether or what measures might be taken, given the scientific consensus on climate warming and allowing for the uncertainties, to deal with the problem? There was a consensus that certain strategic actions, which made economic and environmental sense and minimized the risks, could be taken now. Scenario studies could provide indications of relative costs of action or inaction. For example, a recent study by the US Electric Power Research Institute had examined the economic implications of climate warming for a power company in the southern USA. The scenario adopted predicted a 1°C average annual temperature increase by 2015 AD. Four situations had been examined and gave different results, viz.:

1. If the company had taken action on the prediction and it had *come true*, there would have been no loss to the company.

2. If the company had taken action on the prediction but it had *not come true*, the additional cost to the company would have been $US 10 millions by 2015 AD.

3. If the company had taken *no* action and the prediction had *not come true*, there would have been no loss to the company.

4. If the company had taken *no* action on the prediction but it *had come true*, then the additional cost to the company by 2015 AD would have been $US 55 million.

Thus, as this study had illustrated, apart from environmental considerations, there could be significant economic risks in disregarding climatic predictions. Action on some things should be taken as soon as possible where there was a high probability of change and high risk, others could, of course, be delayed for some years, and yet others would need to be considered very carefully because they might be dependent upon fine-scale predictions which were clearly still unreliable. For instance, he thought that the probability from modelling that the sea-level would rise over the next few decades was high, and so both coastal zone investment and human beings were at risk; planning for action should commence now. Techniques would have to be developed to estimate these risks and procedures developed to prioritize them.

The consideration of environmental costs and risks, in addition to purely economic ones, would add further complexity to the strategic studies that would be required for the protection of the environment and possibly for regulatory change: that could not be avoided. Putting a monetary value on environmental and resource factors was admittedly difficult but the need for 'environmental accounting' was becoming more urgent and more common (*see* also Chapter 16, Eds). He thought that, with the current rate of global depletion, the value of the environment and natural resources would increase just as surely as economic discounting decreased the future value of currencies.

He noted, as many had pointed out already, that certain regulatory actions could only be justified because they made sense both by minimizing the potentially harmful effects of climate change, and also for other environmental and economic reasons. He considered that risk management studies should take account of these multiple payoffs and, in order to improve the usefulness of such studies, he thought it would be helpful if, in future, scientists would express climatic predictions in terms of probabilities over ranges of possible change. That would be helpful to future actuarial practice.

Bryson (as Chairman) said that he wished to clarify some of the statements made for the benefit of those who were not aware of the conventions of meteorologists. When the results of a model study showed that the Earth would be *x* degrees warmer in the next 50 years, meteorologists implied that they had compared the results of a model run contemporaneously with the same model projected by 50 years. It was important, therefore, to know how accurately the model replicated contemporary conditions so that one could have confidence, or lack of confidence, in its predictive capability 50, 20, 10, or even 5 years ahead. Secondly, he wished to comment upon errors with GCMS. The statement that all the models agreed with each other was almost meaningless because they tended to be clones of each other; so if they disagreed the differences were apparently very small. He considered an error such as a rainfall rate error for the summer, i.e. the predicted precipitation. The error might be 3, 5, or 15 mm per day in different situations. Such differences might not appear impressive but cumulatively, for a year, they needed to be multiplied by 365, giving errors of 1m at best up to 5m at the worst. In other words, models could be + or – 100% correct in predicting rainfall. Thus prediction of a drought in the USA or a flood elsewhere is + or –100% reliable, or + or –1m where the rainfall was around 750–1,000 mm (30–40 inches), and + or –5 m in monsoon areas, where the real errors occurred. Temperature errors were similar. The error in a simulation of Antarctica today was 20°C, in both summer and winter: those for the interiors of continents were about 5°C and, interestingly even although actual observed temperatures were used as input, they were *not* zero over the oceans! Thus meteorologists' statements must be examined in their proper perspective, as comparisons of model simulations rather than of the real world.

Fearnside wished to complement **Hare**'s contribution by extending information about the climatic effects of deforestation in the Brazilian Amazon. Two major effects were expected from the continued deforestation of the Amazon region. One was a major effect on the water-cycle, resulting in a decreased rainfall not only in the Amazon but in the rest of Brazil also, including the major agricultural areas in the south-central part of the country; that he would not elaborate further. The second was the contribution of deforestation to the 'greenhouse' effect: this was what he proposed to comment on.

Estimates of the rate of deforestation were crucial to determining the magnitude of the effect. Many estimates existed but he considered that a loss of approximately 20,000 sq.km per annum of the tropical forest proper, or 38,000 sq.km per annum, if the *cerrado* – the scrub savanna in south-central Brazil – was included. These estimates were lower than many, e.g. Norman Myers used 50,000 sq.km per annum, and was greater than others, but **Fearnside** believed that good reasons could be provided to support his figures.

To estimate the contribution of this deforestation to the 'greenhouse' effect, data were needed both for the amount of deforestation and the forest biomass in different regions, and for how much of the biomass was converted to charcoal, how much decayed, and how much was burnt totally at each site. In the states around the lower edge of the nine states which constituted Brazil's legal Amazon, there had been rapidly increasing trends of deforestation. These states were on the border between the *cerrado*, the savanna, and the forest region and had a lower biomass than the average for the whole area. The areas with the highest biomass, not surprisingly, were the forests in the centre of the region but these were least affected by deforestation. He and his colleagues had made several studies to measure the biomass directly: by cutting and weighing, by estimating volume over large areas, by harvesting plots and weighing the biomass before and after burning, by improving the plot design, and so forth. The conclusion they had come to was the obvious one if one examined the forest after it had been burnt, namely, that most of the biomass was still there! Quite a lot of trunks were left standing , although their leaves and and branches were burnt, and some were on the ground, as well as charcoal. This last was very important when calculations of the carbon balance were made, because some of this carbon remained in the long-term pool, through being transferred to the soil, rather than being released immediately into the atmosphere. About 70% of the biomass was left, 30% was burnt off in the initial burn and its carbon lost to the atmosphere. Of course, the 70% was eventually lost in the same way in subsequent burns when the ground was burned repeatedly to provide cattle pasture. This was the fate of most deforested areas. They might be used first for annual crops but, once they were planted for pasture, they were burnt every 2 to 3 years and so, of course, were the logs left on the ground. Indeed, logs could smoulder for many days until they had become converted entirely into ash, releasing 'greenhouse' gases. These often had a greater effect than those released in the initial burn because smouldering released a higher proportion of methane to carbon dioxide, as well as carbon monoxide. As the instantaneous effect of methane was 20 times more potent per tonne of carbon than its equivalent of CO_2, but its atmospheric life-time was shorter, the net effect over its full life-time was about 3.7 times greater than that of CO_2. Nevertheless, this raised the relative impact of deforestation, compared with fossil fuels, on the 'greenhouse' effect. The carbon monoxide removed OH radicals from the atmosphere, but since they had a major role in 'cleaning' methane, CFCS and other 'greenhouse' gases in the atmosphere, the CO released also enhanced, indirectly, the 'greenhouse' effect. Between successive burns the timber also decayed and this too released CO_2. The relative contributions of different decay agents to the process were not known but that due to termites resulted in release of methane as well as CO_2: he had made a range of estimates of this contribution. In summary, after an initial burn, there was a continuous input of a variety of 'greenhouse' gases to the atmosphere spread over many years.

In terms of gross carbon, uncorrected for equivalent contributions of methane and CO, the total deforestation of the legal Amazon would release 51 gigatonnes (i.e. 51 billion metric tonnes) of carbon at an annual rate of 270 million metric tonnes. This annual release was almost three times the annual Brazilian release from burning fossil fuels, but the difference in terms of cost and benefits was tremendous. The latter powered all the industry and transportation in the country, the former simply destroyed the forests. Moreover, the carbon released was approximately 5% of the total annual release in the world to the atmosphere. The conclusion which had been drawn by the recently retired Minister of the Interior, that since 95% was released by the rest of the world, Brazil need not do anything, was false. In reality, that 5% was highly significant. It should be compared with the recent calculation by the Environmental Protection Agency of the USA that, by doubling the efficiency of all the world's vehicles, the 'greenhouse' effect would only be reduced by 3.7%! Brazil was an obvious place where a beginning could be made in reducing 'greenhouse-effect' emissions; indeed, benefits would accrue if even the rate of deforestation were to be reduced. That, he thought, could be achieved although it would require firm action by the Brazilian government to remove the present motives for deforestation.

Bazzaz suggested that there were two important biological issues in connection with global climatic change. One was potential climatic change due to the increase in the quantity of 'greenhouse' gases in the atmosphere leading to temperature rise. That condition was the one most frequently discussed but its magnitude was still controversial. The other was quite certain and perhaps even more important, namely, the actual increase in concentration of CO_2 in the atmosphere. Even without any concomitant temperature rise that could have a significant effect on ecosystems. The importance of this ought to be underlined. Enhanced CO_2 increased the photosynthetic rate of C3 plants, which were in the majority, decreased stomatal conductance, thereby increasing the efficiency of water use, and led both to increased growth and to its reallocation. These changes were subject to interaction with environmental factors, e.g. an increased effect in the presence of nutrients. Because not all species in a community responded similarly to enhanced CO_2 there was likely to be a change in competitive relationships. In particular there was a change in tissue quality, the Nitrogen:Carbon ratio being reduced. Two important implications followed from this. Firstly, there would be an increased consumption by herbivores because, when they were fed, or found high C: low N tissue, they selected it, and ate more of it to obtain sufficient nitrogen. But increased consumption could result in a decline in herbivore populations. Secondly, the low N: high C ratio in the tissues would be reflected in the litter and, in most ecosystems, decomposition rates would decline because the rate of decomposition was largely dependent on the nitrogen concentration. If that were to happen, then litter could accumulate on the forest floor, for example, and there would be a slowing down of nutrient circulation. In other words, the direct biological effects of enhanced atmospheric CO_2 on ecosystems are significant, measurable and capable of documentation, and so should be considered very seriously in any discussion of the effects of climatic change, whether or not accompanied by a rise in temperature.

Fosberg reminded the Conference that they should be cautious about assessing odds. He recalled that 15 years ago, at the 1st ICEF, odds of at least 30 to 1 had been suggested against a depletion in atmospheric ozone, and that had not been considered worth bothering about by some when considering the fate of the human race!

Harun ur Rashid remarked that, however sceptical or not one might be about predictions, the fact was that in 1987 and 1988, in Bangladesh, two unusual events had coincided. Firstly, the three rivers from the Himalayas which converged on Bangladesh did not usually rise at the same time. Nevertheless, despite meandering across Nepal and India for 1,700 miles, or India and Nepal for 1,600 miles, their water-levels rose simultaneously leading to unprecedented flooding. Secondly, in Bangladesh, ground-water levels which never normally rose or reached surface water, did so in 1988, enhancing the flooding. Whatever the accuracy of predictions that scientists might make, it was not good enough to describe them as aberrations of Nature and be defeated by them!

Hare intervened to respond to some of the comments. He agreed that whether or not the 'greenhouse' effect would develop according to the predictions of the modellers, the changes in atmospheric constituents was a fundamental one. He agreed that the increase in CO_2 was fundamental to the performance of ecosystems, but more than that, the entire chemistry of the trace species in the atmosphere had been disturbed with revolutionary consequences for atmospheric chemistry. Even if the 'greenhouse' effect had never been mentioned, the changes in the concentration of the really active chemical species in the atmosphere was something that no biologist could afford to neglect. It was essential that the biological impact of these changes should be studied quite independently of the possible warming effect. For example, what was happening to the free hydroxyl in the atmosphere, and what effects would a change in that component have.

On the matter of sceptics and scepticism, he was not surprised that there was confusion. He recognized three varieties. The venal sceptics, who were very common in the White House and in his own capital (Ottawa), just hoped that the effect would go away because their scepticism arose from fear of the consequences rather than from an appraisal of the evidence. The second kind were good scientists, such as **Bryson**, or R. S. Lindzen at MIT, who were fully entitled to their scepticism and performed an invaluable and essential role of informed criticism. That was as fundamental to science as to good literature. The last were the multitude of naïve sceptics who neither knew about the subject, nor understood it. Unfortunately they occupied a good deal of the time of, and on, the media.

He had absolutely no doubts in his own mind that there had been a real disturbance of the world's climate and that governments ought to be desperately concerned about it. They should not take the forecasts of the modellers literally, for the reasons given by **Bryson**, but they *should* adopt a defensive posture, because the empirical signs of change were sufficiently alarming to make that sensible and desirable. Certainly the Government of Bangladesh, or the Canadian government, both of whom had important land potentially at risk, indeed, any government whose economic or human security was near a climatic margin should adopt a defensive strategy. He wished that he could have devoted a complete contribution to this theme.

McCloskey, continuing the theme asked **Bryson**, in view of his convincing scepticism, what he would advise President Bush to do?

Bryson responded by observing that there were perfectly good reasons for conserving carbon, whether locked up in rain-forests, or as coal and oil – which keep very well in the ground! Even if global warming were not to occur, some other kind of climatic change would come about. He spoke with the authority of one who had spent 40 years studying climatic change, which, indeed, changed all the time. Whether or not enhanced CO_2 was causing global warming there were other good reasons for conserving stocks of hydrocarbons, coal, or rain-forests: it would be stupid not so to do.

Incidentally, as a sceptic he had noted that **Hare** had shown a graph in which both CO_2 and temperature rose in parallel and implied that the correlation between the two had a 'cause and effect' relationship. If that was assumed to be true, and there was no logical reason why it should be so, then he wondered why the temperature had dropped between 1550 and 1650 AD when the CO_2 had not? He remarked that until one could explain the past one could not understand the future!

Holdgate concluded the discussion by observing that in the general field of environmental pollution control we had learned the value of the precautionary approach. If there were sensible and reasonable grounds for believing that something would be detrimental to The Biosphere then it was wise not to do it and both to plan and to take precautionary measures. The Commonwealth Expert group, referred to by **Hare**, had taken that view about the atmospheric input of 'greenhouse' gases.

He also wished to support the views of **Bazzaz** concerning biological effects by remarking that the perturbations of atmospheric chemistry were not impact-free in themselves and that more needed to be known about them. Even if some of the more simplistic projections were considered, the consequences were alarming. A 1°C rise in mean temperature, theoretically, could displace zones of tolerance of plants of the order of 120 km polewards, or 150 m up a mountain, but the rate of re-adjustment of species in Nature could well lag behind the rates of some of the projected changes. This could stretch out ecotones and have knock-on effects that had not yet been thought through fully. The agricultural impacts were equally alarming. The zones within which particular crops could be grown, at least without a considerable degree of adaptive breeding, would be changed and the impact on traditional cultivation systems and dietary habits would impose yet further stresses. It would be sensible to try to work out what these changes might be and then adopt a reasonably precautionary approach, despite the uncertainty. One of the lessons of today's world was that, in any event, one had to plan for action in the face of uncertainty, and for scientists there was nothing unusual in that.

3. Optimal Land-use: Forestry

M. E. Duncan Poore

Senior Consultant, International Institute for Environment and Development, 3 Endsleigh Street, London WC1H 0DD, England, UK

Introduction

What is meant by the best use of land? In human terms it must mean that men and women should get the greatest benefit (in the broadest sense of the word) from the land that is consistent with maintaining the potential of The Biosphere to provide future benefits. But this begs a whole series of questions, such as: what will future generations want, and what can they reasonably expect? How far can The Biosphere be simplified or overloaded but still function?

To answer these questions would require a book, and I have only thirty minutes. All I can hope to do is to set before you some of the perplexing issues which I have experienced in trying to decide how humanity might be guided, coaxed, or bullied, into planning wisely for the future – and to make some suggestions.

There is a story told of a traveller going, shall we say, to Rome, who asked from someone that he met by the roadside what was the best route. And the answer was: 'Well, I can tell you the way, but if I were you I wouldn't start from here.' But we have to start from here, however unfavourable this may be, and also start with the realization that some other parts of the framework for our course towards the middle of the next century can hardly be altered. Thus first, by the middle of the next century the human population of the world is projected to reach 10 thousand millions, with by far the largest proportion in tropical and subtropical countries. Second, the maximum area of available land is fixed; it may indeed decrease because of rising sea-levels. And third, present land-use in many parts of the world is far from satisfactory. Over large areas, land resources are already degraded; in other areas they are deteriorating rapidly; and it is only in some resource-rich areas that full scope still remains for the wise development of these resources.

Moreover, we are not looking for a static Utopia. If sustainable development means anything, it means that the process of development itself should also be sustained – that we should aim to accommodate a continual progression of aspirations which should be met; for what generation of Mankind has not wished to better its lot!

To illustrate the points I wish to make, I shall concentrate on tropical rain-forest countries and, in particular, on the future of their uncultivated or marginal lands. From this study I hope to be able to draw some conclusions which are applicable to other ecobiomes and other situations.

Three Main Conditions

If we review the present state of land over the Earth's surface, we can identify several significantly different conditions. There are three in particular, which one might call under-used, fully used, and over-used. The fortunate nations which still contain under-used land, notably some of the forest-rich tropical countries such as Brazil and Gabon, can still, if they wish, choose how they will use their land. The countries with fully-used land, mainly in temperate climates, have been able to develop their land without too much degradation; an increase of overall yield or value can come from further intensification, or the substitution of one land-use for another. The third group, consisting of those countries with widespread over-used and degraded lands, have little choice; they must either undergo further degradation or adopt the long, hard path of land rehabilitation.

If the world is to approach the optimum, countries in the first group need to exercise their choice wisely and those in the third group must be helped to rebuild the potential of their land. It should be possible to use any excess production of food in the second group to ease the transition of the others. But the necessary intensification raises questions of possible ceilings imposed by use of energy or by the capacity of the environment to absorb pollutants. And there are also serious constraints imposed by poor distribution and by arguments against any form of dependence.

The present situation gives no grounds for complacency. Uncultivated land in the tropics is under great pressure – for conversion or because of over-use already. The findings of the study, covering 117 'developing' countries, made by FAO, the United Nations Fund for Population Activities (UNFPA), and the International Institute for Applied Systems Analysis (IIASA), in 1983 (FAO, 1984) reveal the magnitude of the problem:

- In 1975 there were 54 countries that could not support their existing population from their own lands, using low inputs.
- In 1975 about 2.5 thousand million hectares (38% of the total land area) were carrying more people than they could sustain on a long-term basis at low inputs.
- If no long-term conservation measures are taken, food production may be depressed by about one-fifth because of land degradation.
- The situation is expected to deteriorate in most countries by the year 2000, with population growth outstripping the increase of potential production.
- Only 1.6 times the expected population in the year 2000 could be

supported by the entire potentially cultivable land in 117 countries, even if this land were used only for food-crops or for grassland supporting livestock.

- If one-third were deducted for non-food-crops etc. (a reasonable deduction), the land could only support 7% more than the number of people expected in the year 2000; and the population is expected to grow by another 43% by 2025 AD!

It is, therefore, urgent that we move in the right direction, and move fast. But what should that direction be?

Tropical Deforestation Dilemma

In the last few-years, tropical forests have become a centre of world attention. News items and television programmes abound and, as about all matters that become world issues, opinions – often strongly and even intemperately held – tend to take precedence over facts, so that it becomes very difficult to distinguish truth from fiction. But, of course, these opinions are often based on facts, and we can learn much from them especially about the values that people hold to be important. Perhaps naturally, the focus of this international attention is tropical deforestation, rather than optimal land-use in the tropics. Raging fires and the felling of trees provide more dramatic pictures than the painstaking business of building up a profitable livelihood with good prospects for our descendants.

Let us now look at the arguments on both sides – the critics of deforestation and the defenders of development based on exploitation of the forest. The purpose of such an examination is to identify, on one hand, the values of tropical forest that are considered important by those who criticize deforestation and, on the other, the reasons why certain people feel that there are compelling reasons to use or transform the forest. In this way we may perhaps see how far it is possible to reconcile these conflicting views. This is not a simple issue of opinion in the less-developed countries *versus* that in developed ones, nor of one political persuasion against another; both sides of the question are argued in both groups of countries, and are a faithful reflection of a genuine concern that tropical forests should be used in the best long-term interests of the countries in which they lie.

The arguments used by the critics of tropical deforestation are now all-too-familiar. Essentially they are these. The rain-forests are the richest ecobiomes on Earth. Their destruction will lead to the extinction of millions of species (the vast majority with qualities as yet unknown) and loss of biological diversity, with consequent serious effects on the development of future potential in agriculture, forestry and medicine. In addition, the loss of forest will lead to serious erosion, exaggerated floods and droughts, great distress to local peoples living in or near the forest, and the possible loss of their traditional knowledge, also disruption of local, regional and, perhaps, global, climates, and a significant increase of carbon dioxide in the atmosphere.

The protagonists of forest-based development deploy different arguments. Forests and forest lands are their main resource. It is necessary to clear forest to provide land to feed their increasing populations. The timber which can be obtained from the forest is their main source of foreign exchange and of internal revenue. Without forest clearance they cannot develop; nor can they provide for the basic needs of their people, some of whom should be helped to leave the forest and join the mainstream of modern life. Forest clearance and the harvesting of timber can be carried out carefully and skilfully. What they are doing is no different from the forest clearance in the now-developed countries which financed and gave impetus to the earlier stages of their own development.

There are those on either side who can hear no good of their critics: those for example in the tropical countries who suggest that the environmental movement in the developed 'First World' is all part of a conspiracy to hinder their Third World development; or the environmentalists who blame excessive consumption and commercial adventurism in the rich countries, greed and corruption in the developing countries, and self-interest in the international financing agencies. There is of course some truth in these charges, but there are also many on both sides who are working with great integrity and sincerity for a sustainable solution. Between these, constructive compromise is possible. And I believe that it is by combining their views that one can begin to find a recipe for a land-use that is environmentally and economically sound. If it can also carry with it the support of both bodies of moderate opinion, it stands at least some chance of being politically acceptable.

What, then, needs to be done is to reconcile the preservation of environmental values with enhancing the production from the resources of the forest, against the background of a pressing need in tropical countries to produce more food and to increase their purchasing power.

Let us once again enumerate the points on both sides. The significant environmental values are these: biological diversity, the volume and quality of soil and water, the moderation of climate, carbon storage, the quality of life of forest peoples, and the knowledge to be derived from local peoples and from the study of natural ecosystems. The economic values reside in crop production on land cleared from forest and the yield of marketable products from the forest itself.

Two Bases for Compromise

Where, then, is the room for compromise? It lies in two facts: first, that it is not strictly necessary to maintain the forest intact to preserve all its values; and, secondly, that it is technically possible to harvest from the forest while maintaining many of its values. This leads to the general conclusion that it is possible to have a pattern of uses, some exclusively for each primary purpose, but others enabling the coexistence of different uses which are mutually compatible.

In striving to reach a constructive compromise along these lines, one needs to look at the various values of natural, uncultivated lands and examine to what extent these values are amenable to substitution (in other words, to what extent would their disappearance lead to irreversible environmental damage or deterioration). For in an ideal world we should, of course, give priority to preserving those features for which substitutes cannot be found, or whose destruction is not irreversible.

Highest on the list comes the conservation of biological diversity – part of the insurance for the future of Mankind. It is no doubt technically possible to preserve this in gene-banks, seed-banks, zoos, and botanical gardens, but only with enormous investment. These artificial means have their places; but the most practical and economical way of preserving most biological diversity is by protecting areas of natural ecosystems in the field. If we follow the logic of this argument, the effective preservation of these areas should have very high priority indeed. But how much land does it require, and where? I do not accept the contention that diversity is destroyed *pari passu* with habitat – an argument favoured by those talking about the destruction of the tropical forest. There is good evidence that, with well-chosen reserves and protective measures, most biological diversity can be preserved in a small proportion of the total area; 10% might well suffice (Sayer & Whitmore, 1991).

Next comes the protection of soil and water. Here some substitution is possible. Although natural vegetation may be the best (and cheapest) protective cover, there are other possibilities. Grassland dominated by rhizomatous grasses is as good as, if not better than, forest. Tree and bush crops can all provide effective protection, though the effectiveness will depend upon the method of conversion and upon the quality of soil conservation measures practised. High investment of either labour or machine-time is needed, both in conversion and in maintenance. In extreme circumstances, the most effective cover for a water catchment may even be concrete. A problem with most artificial systems, however, is their fragility if they are abandoned; catastrophic erosion can be caused by the collapse of terrace systems, for example. And there is historical and current evidence that abandonment is frequent, as people migrate from the countryside – either forced by their inability to maintain a livelihood, or tempted by prospects of economic development elsewhere.

Different Needs for Climate Regulation and Carbon Storage

Different criteria apply to climate regulation and to the values of alternative land-uses as carbon banks. For climate regulation the characteristics of vegetation that are important are concerned with energy and water balance – albedo, surface roughness, and rooting depth, for example. The significant measure for the withdrawal of carbon from atmospheric circulation is the total reached by adding together the long-term standing biomass, the amount stored in the soil as humus or peat, and any long-term storage of harvested wood or wood products. Both natural forest and plantations can play a role; and temperate forests are comparable in importance with tropical forests. Storage is the critical attribute.

If one applies this scale of precedence, the use of those lands whose values are unsubstitutable should be strictly regulated, recalling that the uses which are environmentally acceptable will be few. By contrast, in those lands whose values are substitutable (generally on richer soils), many possible uses are environmentally acceptable. One can view these categories of land as resembling the sectors of a fan, with restricted scope for the parts near the base and much wider possibilities for those near the edge. It is in these last, the better lands, that the main possibilities for increasing production must lie. In these better lands, market forces can be allowed much greater influence in determining land-use in ways that are productive and sustainable. This will apply to countries at all stages of development.

Growing Recognition of Benefits from Protection

At the protection end, there is fortunately a growing recognition in most countries, especially those that can afford it, of the extra-market benefits to be derived from the protection of catchments, from the preservation of biological diversity, and from climatic regulation. So, one may reasonably expect the market to take care of the production end of the spectrum, and interventionist policies of one kind or another to provide for the protection end. Also, it is clear that many of the environmental functions are compatible with a certain degree of economic use. But intervention and control cost money and effort; it is therefore encouraging to note signs of an increased readiness of the richer half of the world to assist the poorer in environmental protection.

I believe that the most recalcitrant problems lie between the two extremes of lands of high value for protection and of high value for production; they lie in the huge areas which are used for the harvesting of forest produce, including timber, and for shifting cultivation. This is the type of land whose management is described so deceptively simply in the World Conservation Strategy (1980) as the 'sustainable utilisation of species and ecosystems'. It is these ecosystems which are so often the target of overexploitation, of misuse, or of unwise conversion, and which contain species that are over-hunted and over-collected. Similar problems occur in most of the habitable parts of the world that are not under permanent agriculture – and cropland is at present estimated to cover only 11 per cent of the 13.1 thousand million hectares of land area.

Problems of Sustainable Management

The sustainable management of such uncultivated land faces a number of problems. Where it is carried out by government, it requires a stability of policies and their application, and a standard of control, which are rarely attainable. Many such areas suffer from being a 'common resource'; often ownership and land rights are obscure, and management is frequently technically complicated and difficult to apply consistently. All of these make such areas subject to over-use and degradation.

In the case of tropical forests, there are a number of possibilities for management, the rival merits of which are constantly being hotly debated. Among them are the following: use by hunting and gathering communities of native peoples; shifting cultivation supplemented by hunting and gathering; the harvesting of timber from the forest for domestic use or for export; and harvesting of the many other non-timber products of the forest – such as rattan, nuts, gums, game, fish, fruits, etc. All of these kinds of harvesting can be carried out sustainably. Should an ideal pattern of land-use provide for all of them?

I believe that it should; but I have one strong reservation. It is, in principle, possible to manage these forest ecosystems in perpetuity for a sustained yield of specified forest products. But is this the real issue? Many things are possible but do not take place because nobody particularly wants them to.

I can accept that there are native peoples who have been accustomed to harvest forest produce in a genuinely sustainable way, or to practise a truly sustainable system of shifting cultivation, although I do wonder if they are as common as some authorities suggest. There may also be governments who have the sincere intention to manage their forests for timber for a sustainable yield in perpetuity. But I suggest that the forest peoples in question are ones who have been largely insulated from outside contacts, as have those governments from the full impact of the global economy.

Changes to Modern Situation

The situation now is very different from what it used to be. Hunting–gathering communities obtain new tools and weapons and greater incentives to market their produce. Roads or outboard motors bring shifting cultivators into closer contact with markets and their vagaries: chain-saws make it easier to clear land. Radio and television make these people aware of how others live. Products collected from the wild may have a stable market for a time; but, if they are sufficiently valuable, they are taken into cultivation and eventually even synthesized.

Silvicultural operations are carried out today on a crop which will mature in 25–50 years' time. But much timber is now a commodity in which there is a global market and which is subject to many possibilities for substitution. When the harvest is due, the market may be very different and the kinds of management which are economic today may no longer be economic. Moreover, if science were to be applied as effectively to wood production as it has been to other crop production, it would be possible to grow all the wood the world is likely to need on about 300 million hectares, or less than one-tenth of the present forest area.

Perhaps even more significant are the changing priorities and perceptions of the people concerned. The kinds of activity, and the economic activities, which are preferred today, are most unlikely to be those desired by the next generation but one.

In making these points I am not arguing against land being devoted to the various kinds of potentially sustainable cropping that are mentioned above. I am arguing that these may only be appropriate for particular places at particular times; none of them is necessarily more right than the others. I am also making a plea that planners should be aware that change is likely, even if they cannot predict the nature of the change with any precision; and that they should look forward and attempt to take account of the environmental consequences of such changes. In Sarawak, for example, lands that were subject to shifting cultivation about 100 years ago have been found to be practically indistinguishable from virgin forest; and the same could be true even of badly-logged forest. But, as mentioned above, the erosion on steep terraced farmland, if abandoned, can be catastrophic.

Uncultivated Lands – Most Threatened

My conclusion is that uncultivated lands which are accessible and habitable are the most threatened by degradation and misuse. It is with reference to these that some fundamental thinking about optimal land-use is most crucial. Their management often poses complex technical problems. But, even more important, they – more than any other lands – are the potential victims of a fluctuating economic environment. And they constitute the areas in which the fruits of wise development could be most evident.

To conclude: I have become convinced that it is impossible to define with precision any general system of optimal land-use. This is because the 'best' at any place depends not only on the ecological and social conditions there, but also on the dynamics of development in that particular situation. And while the first two are in principle easy to describe, the dynamics of development are exceptionally difficult to predict.

Although I have illustrated this thesis by concentrating on the management of ecosystems for a sustainable yield of products, the same tendencies also affect land at the two extreme ends of the spectrum that I have identified. For example, at the Nature reserve end of the spectrum, the land usually owes its original protection to its inaccessibility and its low economic value; but as a country develops, the justifications for such protection change – for tourism, for scientific research, or for the preservation of biological diversity. And such an area will remain protected provided there is no discontinuity in the reasons that society can find to preserve it. For the fertile land, at the other end of the spectrum, there is generally no problem about finding economic justification, though in detail the use may frequently change to meet different social objectives.

I conclude that there is a fundamental paradox in the idea of 'optimal land-use' within the context of 'sustainable development'. The 'Brundtland Report'(WCED, 1987) described the terms thus: 'Sustainable development is development that meets the needs of the present without compromising the ability of future generations to meet their own needs', and 'the satisfaction of human needs and aspirations is the major objective of development.' Development therefore implies changing needs and aspirations which, given human nature, are always likely to be for expansion. The ideas of sustainable management and optimum land-use, on the other hand, both imply a state of equilibrium. Yet, given the scale of our present problems, it is unlikely that development will ever be sustainable unless we attempt to achieve optimum land-use at each stage, and organize the transitions from one use to another in ways that cause little if any environmental damage.

References

FAO (1984). *Land, Food and People.* (Based on the FAO/UNFPA/IIASA report 'Potential population-supporting capacities of lands in the developing world'.) FAO, Rome, Italy: 96 pp.

Sayer, J. A.. & Whitmore, T. C. (1991). Tropical moist forests: Destruction and species extinction. *Biological Conservation,* **55** (2), pp. 199–213, tables.

World Commission on Environment and Development (cited as WCED) (1987). *Our Common Future.* Oxford University Press, Oxford, England, UK: 383 pp.

World Conservation Strategy (1980). *World Conservation Strategy: Living Resources Conservation for Sustainable Development.* IUCN–UNEP–WWF, 1196 Gland, Switzerland: pack containing various parts, illustr. but pages unnumbered.

Note

[Commentary on this chapter is combined with that of the next (see p. 92). Eds.]

4. Optimal Land-use: Agriculture

JOHN BURNETT
c/o Department of Plant Sciences, University of Oxford,
South Parks Road, Oxford OX1 3RB, England, UK

INTRODUCTION

The optimal use of land is that which most effectively reconciles the competing uses to which it may be put without irreversibly reducing its potential productivity. It is convenient to recognize six broad categories of potentially competing land-use, namely:

a. Rural living-space;
b. Food production for subsistence, *i.e.* to prevent starvation;
c. Crop production for profit, whether of food or other commodities, excluding timber;
d. Forestry;
e. Land conservation of all kinds, including recreation; and
f. Urban and industrial development and its infrastructure.

This chapter will be concerned principally with the optimal use of land in connection with items *b* and *c*, comprising agriculture, but it is only sensible to consider agriculture and forestry as complementary activities, both of which are limited by the availability of water, of land, and by the fertility of the soil. A fuller picture of land-use can only be obtained, therefore, by considering this chapter with the preceding one (*see also* Commentary, p. 92).

Although optimistic, the estimates made about the future availability of land in *World Agriculture Toward 2000: An FAO Study* (Alexandratos, 1988) will be accepted as a first approximation: they will not be discussed further. However, the availability of water and soil fertility are interrelated and much more controversial, so they will be discussed later. In addition, there are two social problems which bear upon agricultural productivity and forestry, especially in the less-developed countries, namely, poverty and landlessness. Both are related to population pressure. Here, their impact on crops, stock, and fuel-wood, will be considered briefly; but the wider issues of population control, and social and political activities, will be addressed elsewhere (*see* Chapters 13, 15, & 17–9).

Table 4.1. Numbers Below the 1.4 Basic Metabolic Rate (BMR) Threshold in 1983-5 and Estimated for 2000 AD. (Data for 89 less-developed or developing countries; adapted from Alexandratos, 1988.)

Region	1983–5		2000	
	Millions	% Population	Millions	% Population
Sub-Saharan Africa	142	15.2	194	28.7
Nr East/N Africa	24	9.2	29	7.6
Asia	291	21.8	246	13.9
Latin America	55	14.2	62	11.6
Total	512	21.5	532	15.6

General Considerations

Six facts – and a seventh, speculative consideration – are basic to any discussion of the future of world agriculture. These are:

1. The world population of humans in the 21st century is expected to reach 7–10 thousand millions. The staple diet of about half the people will be rice, of about another one-third it will be wheat, and of the rest it will be maize and coarse grains, *i.e.* barley, millets, sorghum, etc. Root and tuber crops will play a lesser, more localized part.

2. In 1983–5 it was estimated that about 512 million people (21.5% of the population in the less-developed and developing countries) were undernourished, *i.e.* below the 1.4 kilocalorie, Basic Metabolic Rate (BMR) threshold. Their numbers and proportions have continued to rise. Undernourishment was unequally distributed, being greatest in Asia and sub-Saharan Africa and least in the Near East/North Africa region (Table 4.1: *cf.* Alexandratos, 1988). These inequalities have persisted since 1985 and are expected to continue.

3. Poverty has increased and is increasing throughout the world. In 1980, the World Bank estimated the numbers of absolute poor to be 780 millions, 90% of whom were rural poor (except in Latin America, where 40% were urban poor). About half the rural populations of Asia (excluding China) and of Latin America, and over 65% in sub-Saharan Africa, were considered to be living in poverty (FAO, 1987). An earlier study (FAO, 1986) suggested that some 30 million agricultural homes and 138 millions of the poor were landless, mostly in South Asia, but that landlessness is still on the increase in SE Asia, Africa, and Latin America.

4. Since about 1985, world grain production has remained more or less constant after 40 years of fairly steady increase. In 1988 the wheat crop in the USA was 190 million metric tons, which was 12 million tons less than that required for self-sufficiency. Canada and China, too, experienced serious

shortfalls. Consequently, after withdrawals to make good these deficiencies, the world's overall wheat reserve fell to its lowest – 54 days – since the end of World War II (IWC, 1989).

5. There are doubts throughout the world, to varying degrees and for different reasons, as to how far existing technology and practices can ensure sustainable agricultural production. In the developed world the principal concerns are the effects of industrial pollutants, excess N fertilizers, and pesticides, on soil fertility and yield. In the less-developed countries, the concerns are more with soil erosion, desertification, salinization, loss of soil fertility, and the availability of water. In fact, all these factors apply to some degree throughout developed, developing, and less developed, countries alike.

6. The land available for cultivation of *all* kinds is limited. Optimistic FAO estimates in 1985 suggested that the equivalent of about 115 million hectares of *additional* land would be needed in developing countries by 2000 AD, *i.e.* about 8 million hectares per annum. This needs to be compared with estimates of the annual losses, mostly in the same regions, of 6 to 8.5 million hectares of productive land per annum – 5 to 7 millions through desertification and 1 to 1.5 million through salinization. Moreover, the additional land required will encroach further into existing tropical forest, because of the agricultural need for high soil-fertility (Alexandratos, 1988). Such an expansion will further exacerbate the problems both of water conservation and erosion. A further competing claim on land for cropping is for fuel-wood, which currently supplies about two-thirds of the basic energy requirement in rural Africa and about one-third in Asia. Major scarcities are said to occur throughout the developing countries, especially in the most-heavily-populated areas. Other competing claims for land are for cash-crops and forestry.

7. Additionally, it is not yet clear what the consequences will be for world agriculture if the predicted rise in temperature due to the 'greenhouse effect' comes about. Some speculations suggest that a combination of increased temperature and reduced moisture in the Northern Hemisphere will seriously reduce the yields of wheat in their principal areas of production. The most recent assessment by Adams *et al.* (1990), however, suggests that wheat and other yields will be increased! Yet other speculations suggest that the timing, location, and amount, of rainfall in the monsoon, on which much of Asian agriculture depends, will be changed. All of these, including global warming itself, are unproven speculations built on inadequate scientific foundations (*cf.* Chapter 2); but it would seem prudent to anticipate a period of unusual extremes of weather over the next 50–100 years, at least, which will almost certainly be disruptive of present patterns of agriculture.

In summary, therefore, the technical problems of feeding the world in the future will have to be achieved against a background of increasing numbers of undernourished, poor, landless people living on increasingly limited land areas on which the control of erosion, salinization, and loss of fertility, will have to be far more efficient than at present. This will have to be achieved against a background of some scepticism about the efficacy of existing technologies and practices which, in any case, may be unsuitable if major climatic change does, indeed, occur.

These problems take no account of adverse social or political situations, which certainly impede food production at present (*vide* the present situation in the Horn of Africa; *Annexes 5.1–5.3* to Chapter 15). We must also remember that all countries have similar problems which differ more in degree than kind; but, in any case, as all countries are inextricably linked together through world trade and the prevailing economic and political situations, there will be, inevitably, adverse global effects.

The first and major question to be addressed now will be, therefore, whether appropriate technology exists (or can be developed) for sustainable food production. Secondly, how can technology best be transmitted, particularly to poor farmers; and lastly, what kinds of agricultural system will best ensure optimal sustainable production?

Agricultural Research

Farming patterns are well-defined in most developed countries. There they are supported both by substantial subsidies and by well-developed, complex networks of local and central research and advisory agencies which prosecute fairly clearly-defined objectives, namely:

(a) initially, high-yielding monocultures with good specific disease resistance but,

(b) latterly and increasingly, for improved quality, e.g. improved nutritional or vitamin content in cereals or texture of cooking in potatoes, etc.

The reverse of such organization is the case in most of the less-developed countries. To deal with this situation, the major International Agricultural Research Centres (IARCS), such as the Centro Internacional de Mejoramiento de Maiz y Trigo – CIMMYT (1966, originally the Rockefeller Foundation Mexican Programme [1943]), or the International Rice Research Institute (IRRI: 1960), which have jointly been responsible for breeding the cereals of the 'green revolution', were established. Since 1971, they have been coordinated by the Consultative Group on International Agricultural Research (CGIAR), whose stated aim is:

> 'through international agricultural research and related activities, to contribute to increasing sustainable food production in developing countries in such a way that the nutritional level and general economic well-being of low-income people are improved' (CGIAR, 1987).

This is a much wider aim than that of the original IARCS, and it is now pursued in a larger-than-formerly number of institutes, many of which CGIAR has founded, e.g. the International Crops Research Institute for the Semi-arid Tropics (ICRISAT: 1972), the International Livestock Centre for Africa (ILCA: 1974), and the International Centre for Agricultural Research in Dry Areas (ICARDA: 1976), to mention a few. This should enable the most advanced agricultural research techniques of developed countries to be brought to bear on the problems of other countries where it was intended they should be complemented by local agencies on the responsibility of the individual countries concerned. Not surprisingly, this has not always been the case, *e.g.* where the production objectives appropriate to the major developed countries have dominated the work of the IARCS.

The less-developed countries differ greatly in their effectiveness at local levels, and also in the extent to which objective research is supported by their governments. Moreover, almost without exception, less-developed countries obtain subsidies from aid agencies more readily for practices which mimic those of developed countries, and whether or not they are really appropriate to the local conditions.

Plant Crops

The maintenance of major food-crops is clearly the first priority, and there can be no question but that the modern rice, wheat, and maize, varieties developed by IRRI and CIMMYT have played a major role in avoiding world famine on a gigantic scale. Over 50% of those crops that are now grown in developing countries are derived from the work of the IARCS, and indeed never before in human history have such increases in yield been achieved in so short a time. However, they do have some drawbacks.

The basic philosophy, not surprisingly, has been that adopted in the developed countries, but with some modifications. The new varieties have tended, therefore, to be very uniform and short-strawed (in the cases of rice and wheat), with appreciable requirements for both fertilizer and water (often involving a need for irrigation), with pest and disease resistance based on so-called 'vertical' *(i.e.* major gene-) resistance, on a narrowed genetic base *plus* use of pesticides and herbicides, but with a good average yield.

This policy has proved remarkably effective in advanced farming systems, at least in relatively equable environments where the farms are relatively large and the farmers relatively much richer. It has also proved successful with the, comparatively rare, larger farmers of less-developed countries. Inevitably, practically all farmers in these and in developing countries have perforce had to adopt these new varieties, but the high cost to them, although subsidized in many countries, has left most of them little better off financially as a result of such use. The vast majority of farmers in such countries are small and poor, and more of the rural population is landless in them than elsewhere. Traditionally, they have relied upon labour-intensive, mixed cropping involving several crops in a year, preferably with a safe but lower yield, rather than on a higher average yield which might well be distributed erratically over several years. The former, well-tried policy was an insurance against environmental fluctuations: if one crop failed, another might succeed.

This traditional policy involved a continuity of demand for labour, a continuous supply of cheap food, and the avoidance of undue reliance on extended periods of credit. For these farmers, stability, sustainability, and solvency, with mixed cropping leading to a cheap and continuous supply of food, are the preferred objectives. But, despite their notable successes, the IARCs have not yet met these requirements adequately. Indeed, it has not hitherto been their policy so to do: increased production above all has been their goal. Now it is clear that there will have to be change.

Change is, in fact, coming slowly. For example, trials have often been, and still are, carried out under conditions very different from those on farms – *e.g.* with optimal water regimens or pure stands of cultivars – but ICRISAT now conducts trials with mixtures of varieties or crops to simulate mixed cropping systems, *e.g.* of millet (various species of *Panicum* and *Sorghum*) and chickpea (*Cicer arietinum*) (Monyo *et al.*, 1976).

The long-continued policy of involving nationals of different countries at the IARCs before they return to their various domiciles, has begun to pay dividends at the previously inadequate local level of research-support. For example, *H–4*, a fairly stiff and fertilizer-responsive, medium-yielding but reliable Rice (*Oryza sativa*), was developed locally on badly-drained soils in Sri Lanka, using IRRI techniques (H. Weeraratne, unpub., quoted in Lipton & Longhurst [1989]). In many regions where hybrid Maize (*Zea mays*) has been introduced in Africa, local observations have shown that local varieties of sorghum and millets can suffer a check before flowering, but with less effect on final yield than hybrid maize can have. This is because they can compensate for reduced yield per panicle through an increase in tillering and/or panicle number under medium water-stress. Attempts are now being made to improve such local crops in order to replace the hybrid maize.

This situation reflects a general criticism that has been made of the IARCs, namely that they have concentrated on too narrow a range of staple crops; but this, too, is now changing. In particular, the emphasis put on major cereal crops has been to the detriment of root and tuber crops which, together with millets, sorghums, and unimproved maizes, are grown as widespread mixed crops – especially in sub-Saharan Africa and elsewhere.

Root and tuberous crops such as yams (*Dioscorea* spp.), Sweet Potatoes (*Ipomoea batatas*), and cassava (*Manihot* spp.), are a particularly rewarding field for development, for three reasons. Firstly, there is an immense range of such crops from which to make selections. Secondly, they are amenable to modern techniques such as meristem culture and heat treatment, which can lead to improvement in yield through eliminating persistent, chronic virus infection. Lastly, they include valuable cash-crops which can provide poor farmers with a chance to increase their income. There has been a neglect of starch crops generally, some of which could become of considerable importance in Asia – *e.g.* Sago palm (*Metroxylon* spp.), or in S. America *e.g.* arrowroot (*Maranta* spp.). Sugar-cane (*Saccharum officinarum*) too, could be selected for stem starch rather than sugar, quite apart from its possibilities as a bio-fuel.

Although yields have been increased dramatically, there have been major areas of concern, notably the narrowing of the genetic base of the new varieties and the extensive use of 'vertical' resistance (*see* p. 70) to disease. Wheats such as *Sonalika* came to dominate the wheat areas in North India, Pakistan, and NW Bangladesh, largely because the 'vertical' resistance bred into it to rust diseases held up for almost twenty years. Susceptibility to new races of rust now causes great concern (Saari, 1985).

Similarly, in South India and many parts of Bangladesh, rice *IR–20* proved resistant to all major rice pests – which of course are legion – and diseases for more than 15 years until the advent of brown plant-hoppers which also transmit virus diseases. At first, plant-hoppers were controlled by chemical pesticides which also killed their natural predators; but the plant-hoppers developed resistance to pesticides more rapidly than did their natural predators and, in the absence of the latter, rendered *IR–20* useless. Nevertheless, provided the genetic material is available, replacement can be achieved. For example, successive vertical-resistance genes were introduced into rice in Indonesia in the sequence of varieties: *IR–26, IR–36*, and *BP–56*, to control, successively, different races of brown plant-hoppers.

In most cases, resistance in wheats and rice has lasted for far shorter periods than in *Sonalika* and *IR–20*, respectively, which are, unhappily, the exceptions rather than the rule. The effort put into searching for vertical-resistance genes, followed by rapid-breeding programmes to get them into varieties, has had the effect of deflecting the search for 'horizontal' or so-called 'durable resistance' and, indeed, successful vertical resistance, because of its masking effect, has made the detection of horizontal resistance in the same varieties of plants more difficult. There is now a much clearer policy of breeding for multiple vertical resistance, *e.g.* in wheat (CIMMYT, 1983) and millet (ICRISAT, 1984). In addition, both horizontal and 'durable' resistance are now being sought, and methods are being developed to utilize them most effectively in breeding programmes.

Such utilization in breeding programmes will, however, require an immense technical and administrative effort – firstly because it is not yet clear how horizontal resistance operates, or how it can be controlled genetically. Secondly, effort will be required because a very considerable search will be needed to find effective sources of horizontal resistance and then to conserve it. Gene-banks already impose considerable problems of maintenance, and these will be exacerbated by such new demands. It is, however, recognized that present breeding practices are resulting in a dangerous narrowing of the genetic base of staple crops. This has been made worse in some crops by other practices, such as the use of cytoplasmic male-sterility derived from a single plant of the wild rice strain *Wild Abortive* in breeding almost all the contemporary hybrid rices of China and the USA.

All these problems also apply to pests. At present, major reliance for their control is still placed on the use of pesticides. This has three undesirable consequences for developing countries. Firstly, it is very expensive and, therefore, hardly commensurate with the overall, widespread poverty. Secondly, to apply pesticides in an effective *and* environmentally acceptable manner requires a good deal of sophistication, such as is uncommon in developing countries. Thirdly, most pesticides that are made available to developing countries are biologically and environmentally undesirable (Holl *et al.*, 1990), even though the ephemeral, synthetic pyrethroids used nowadays are far less objectionable than many of the earlier, persistent pesticides. This third, undesirable feature has been exacerbated by the deplorably reprehensible practice of exporting environmentally less-desirable pesticides to those countries, even after they may have been restricted, or even banned, in the developed countries from which they have originated.

Weeds are another major problem *inter alia* through their rapid rate of growth in the tropics and, consequently, high demands on scarce plant nutrients and on labour for their clearance with, so far as proves practicable, the return of those nutrients to the soil. Too much reliance has been, and still is, placed upon the use of herbicides, and too little on breeding crops that are resistant to weeds. A special example is the considerable and successful effort put into breeding against the weedy root parasite, *Striga* spp., widespread in Africa and spreading to India. In general, however, breeding for weed resistance involves modifying the crop's growth-habit, or rate of development, and this, like horizontal resistance to disease, is likely to be difficult because of its polygenic basis. However, alternative techniques such as herbicide-resistant crops – thereby permitting lower dosages to be used (Sun, 1986) – or alley cropping, are now being developed (*see* pp. 88–9).

Damage to crops by birds, too, is important, and has been neglected, although resistance to such damage is now being successfully bred into millet in the Sudan, albeit in order of importance after drought resistance and resistance to insects and disease (ICRISAT, 1984).

Other major problem-areas continue to defy solution, of which many are related to problems of adverse moisture conditions. The basis of drought-resistance, both physiological and genetical, is still in need of more detailed analysis before it can be used rationally in crop improvement. Slow progress in breeding upland rice and semi-arid winter crops reflect this ignorance while, at the other extreme, there is a need to produce good, deep-water, or flood-tolerant, rice varieties, which are so necessary for low-lying Thailand and Bangladesh (IRRI, 1982). Once again the genetic basis of the desirable qualities is polygenic, with all its attendant practical problems.

An alternative, potentially simpler, solution which has been adopted with Pearl Millet (*Pennisetum typhoideum*), is to plant mixtures of cultivars differing in their drought-tolerance. There is then a chance that some plants will survive, even in fluctuating water regimes, which are not uncommon. This is essentially a more sophisticated version of classical peasant techniques, made more effective by the increasing range of cultivars selected from collections and gene-banks. It has been argued that the IARC's have given too little attention to the use of improved composites and varietal mixtures as a strategy both to meet fluctuating conditions and the poor farmers' philosophy of mixed cropping. Incidentally, there is growing evidence from developed countries that such systems are useful in protecting crops against fungal diseases, as would indeed be expected.

Progress in tackling some problems has been hindered because they are still technically difficult to combat. Crops that can match the high-nitrogen fertilizer demand of the earlier varieties, without making such a demand, can be and have been bred. This has usually been achieved, often inadvertently, by breeding for root systems that extract soil nutrients more effectively than others, rather than for more efficient nutrient partitioning within the plant. Hence such cultivars deplete low-nutrient soils more rapidly than other varieties.

Although the physiology and genetics of nutrient extraction, partitioning, and conversion, have been studied in rice (IRRI, 1984), there is not much evidence that the resulting information has actually been used in developing varieties. It is not yet clear, therefore, whether it is possible to develop varieties which utilize nutrients more economically within the plant than do others. The problem is further complicated by interactions between nutrient utilization and other factors – for example, water balance. The relationship between this factor and available nitrogen can have significant effects on yield (IRRI, 1984). For example, in rice even 20 kg per acre (0.408 ha) of nitrogen can increase yield, per day of moisture stress, by 5 kg per acre (Wickham *et al.*, 1978).

Inevitably, 'natural' nitrogen nutrition has received much attention, and a great effort has been put into the promotion of nitrogen fixation, especially by symbiotic systems. The new understanding of the process which bacterial genetics has brought has explained many previously seeming anomalies. But the pursuit of transferring N-fixation to crops other than legumes and the actinorhizic, *i.e. Frankia*-symbiotic, species is tending to deflect attention from other, potentially more immediately practicable, strategies. A far greater effort than hitherto is needed in the study of the microbiology and ecology of tropical and semi-tropical soils, with the object of utilizing both the symbiotic and free-living N-fixing Bacteria. For instance, it would seem to be possible to improve N-nutrition through the exploitation of the surface association between the roots of many tropical grasses and N-fixing Bacteria, *e.g. Azotobacter paspali* and *Paspalum notatum*.

Another area, already under investigation, is the supply of organic N by the growth in rice paddies of water-ferns (*Azolla* spp.) which contain symbiotic N-fixing Cyanobacteria. Not only do organic compounds leak from the fern in life but, after death, its remains provide a further source of nitrogen. On the other hand it needs to be recognized that its decay also contributes methane – a very effective 'greenhouse' gas – to the atmosphere!

Three other weaknesses in development have been evident. The first is a general failure to develop other, more locally important, food-crops. The second is the unevenness of individual national responses, and the last is the failure to integrate cash-crops into the farming patterns to accommodate the new food-crops. In a sense these reflect more on local efforts than on those of the IARCs who, it may be fairly claimed, have had to concentrate their efforts on the major objective of producing sufficient food *per se*. They are now, especially in some of the more recently-founded IARCs, turning their attention to widespread, but more regional, crops of lesser importance such as the coarse grains.

As mentioned earlier, the ties developed with national organizations at the IARCs are now beginning to pay dividends at the regional levels, *e.g.* the development of *cv. H–4* of Rice in Sri Lanka. Despite such successes, it is very evident that the policy of training different nationals at the IARCs can only be effective if the nation concerned provides a well-found local research organization for them to return to, and develops a positive, flexible and adaptive, research ethos. Much, politically and administratively, remains to be done in those respects.

Domesticated Animals

World starvation is most likely to be alleviated through plant crops rather than animal stock. Nevertheless, even though domestic animals are far less efficient energy-converters than plants, they play a significant nutritional role and, perhaps of even greater importance, a significant social role in most human societies. Physiological, genetical, and husbandry, information concerning European livestock or Marino sheep is immense, compared with that available for tropical or semi-tropical domestic breeds. However, compared with temperate-zone beasts, yields, whether of meat, milk, or hides, are frequently low. This is one reason why there has been a tendency to export and adapt European breeds to the tropics, or to cross them with tropical breeds. Together, these tendencies have amounted to a policy of real but limited success; however, it would be wiser to concentrate far more in future on indigenous breeds in developing countries.

There is no doubt that ruminant nutrition is limited in most developing countries by an imbalance in the Protein/Energy (P/E) ratio. Tropical forages are usually low in protein and even in digestible carbohydrates, comprising, as they so often do, crop residues, weeds, and track-side vegetation, and only rarely highly nutritious fodder-crops. This is especially true in sub-Saharan Africa and India. Imbalance of the P/E ratio leads to ineffective biomass growth of the rumen microflora and its ineffective utilization, leading to reduced food intake. Provision of bypass protein such as urea, or fishmeal, especially if associated with more digestible carbohydrates, improves yields rapidly (Table 4.2, p. 76: from Preston & Leng, 1987).

Fortunately, urea is relatively cheap, and fishmeal is both the cheapest source of animal protein and is, or can often be made, readily available. The importance of a good P/E ratio cannot be overestimated. Imbalance in the rumen leads to acetate accumulation and to excessive heat production –

Table 4.2. The Effect of Bypass (BP) Protein on Live-weight Gain and Supplemental Conversion Rate of Cattle. (Data from Preston & Leng, 1987.)

BP Source	Basal Diet	Growth-Rate[1]		Conversion[2]
		without BP	with BP	
Fishmeal	Molasses	370	1,000	0.7
Fishmeal	Rice Straw[3]	100	400	0.17
Cottonseed meal	Pasture Hay	-320	220	1.9

[1] kg per day.

[2] kg live-weight gain per kg additional live-weight gain compared with unsupplemented diet.

[3] Ammoniated by urea ensiling method.

a situation to be avoided when heat exhaustion may arise in any event. A nutrient-lick block based on molasses, urea, macro- and micro-minerals, and a binding agent, has been developed by the National Dairy Board of India. Tethered cattle or buffaloes increased their forage intake when the block was provided, while lactating buffaloes, accustomed to long-term block-licking, adjusted their intake to optimize milk-yield (G. Kunju, in Leng, 1990).

Another common condition in tropical cattle is delayed onset of puberty and lactational anoestrous (sexual quiescence). An author cited in Leng (1990) has shown that a group of 8–12 months' anoestrous heifers and cows kept under village conditions in India, but provided with a molasses/urea block, were induced to cycle within a three-months' period. Moreover, pregnancy can be promoted and inter-calving periods reduced by 45–70 days if bypass protein or molasses/urea blocks are provided for cattle (Hennessy, 1986; ILCA, 1987). The reduction in the age of puberty to 3 months, and of the inter-calving period to 18 months, effectively doubles the number of milking animals in a herd. Thus the importance of these nutritional supplements for growth, heat control, and reproductive control, is undoubted and can improve the profitability of cattle enormously.

A major problem with cattle in the less-developed countries is disease and parasites of very many kinds. Insects are still best controlled by pesticides, with the same drawbacks as have been described for plant crops. Nevertheless, the synthetic pyrethroids, which have a minimal environmental effect, have been targeted successfully at some insects, notably *Glossina*. 'Screw-worm' flies (*Cochliomya* spp.) have been eliminated from the Southern USA and Mexico by the release of genetically sterilizing males at the breeding period, and it is proposed to use the same technique in the recent Libyan outbreak. The technique is, however, costly and demands a high degree of technical and operational efficiency. It is difficult to believe that it could be used effectively to control such widespread insect scourges as the

Trypanosoma-carrying Tsetse-flies (*Glossina* spp.) or, elsewhere, hard ticks (Ixodidae), which not only inflict direct losses on cattle but also transmit a range of virus, protozoal, and bacterial, diseases to them.

Protozoal and bacterial infections are largely controlled by drugs or antibiotics of varying efficacy, but always at a high cost to poor farmers, who therefore either do not use them or use them only in inadequate doses for insufficiently long periods. Viral diseases can only be controlled by eradication unless, as with rinderpest (which invaded Africa through a consignment of Indian Zebu cattle), immunization by vaccination of calves lacking colostrum-transmitted antibodies is possible. Although this brought some control locally, a pan-African rinderpest eradication campaign had to be launched in 1986 and a similar campaign is planned for Asia.

Helminth infections can be partially controlled by the newer, broad-spectrum anti-helminthic drugs, and they are particularly useful in reducing infections. In addition, alternate or rotational grazing, control of water-holes, and other management techniques, need to be practised. Frequently, however, this is not done and, only too often, the less visible worm infections go unnoticed by herdsmen.

European cattle are particularly susceptible to tropical diseases, but there is a good deal of low-level resistance to the local race of pathogens in indigenous stock, and appreciably higher levels in the indigenous wild animals. Unfortunately, there has been little development or exploitation of indigenous genetic resistance to tropical diseases, although this could change with new techniques which are becoming available (*see* p. 79).

The fundamental problem of animal disease in the less-developed countries, with the exception of parts of Latin America such as Argentina, is the acute shortage of veterinarians, coupled with the poor organization of the veterinary services. This is a matter, in the first place, for local governments to alleviate; but it has to be said that, as the professional rewards in these countries are unlikely to match those in developed countries, there is little incentive to attract expensively trained personnel. Aid directed specifically to the establishment of more training schools for veterinarians in developing countries would undoubtedly improve matters to some extent. There is also the possibility of training the equivalent of 'barefoot doctors', or, rather, veterinarians – a project which has been attempted in central Africa with some success (CTA, 1984).

Pigs and poultry are widespread in developing countries but their yields are generally low compared with those in developed countries. Nevertheless, there is great scope for the development of local, or national, breeding schemes for improvement. The techniques are well known, readily available, and comparatively cheap. They need to be applied as in China and Indonesia, for example, where they have resulted in considerable success.

In general, therefore, animal production in developing countries lags far behind plant crop production both in its achievements and in research. The establishment of ILCA is welcome but its efforts are only a 'drop in the bucket', and do little directly for the rest of the developing world (*i.e.* outside Africa). There is, however, much to commend the further development and application of an *improved nutritional approach* throughout all developing and less-developed countries. It is simple, well-established, and could produce rapid gains in a comparatively short time. It has the advantage that stock are regarded by many farmers as adjuncts to their main agricultural effort, so that such improvement could provide both additional income for a fairly small outlay and, in many societies, could improve their wealth through social custom. It is even possible to envisage some reduction in the actual numbers of beasts if individual productivity was greatly increased – a necessary concomitant to reducing overgrazing in many countries (*see* Water and Soils, p. 80 *et seq*).

Fishes

The development of aquaculture (especially in Asia where it is a long-established activity with a yield of up to 8.4 million tonnes per annum), would do much to improve the cheap animal protein available. The principal drawback is that the product, despite curing or smoking, is highly perishable, especially in transit. There is still surprisingly little appropriate freshwater research into aquaculture outside the Middle East (where the Israeli effort is outstanding), although there is somewhat more on mariculture. Traditional methods still largely prevail elsewhere. Change is a matter for national initiatives, and everything possible should be done to promote it. However, the international community could help in the development of survey methods of fish-stocks – particularly marine fish-stocks.

The effects of inshore environmental pollution, resulting from uncontrolled industrialization and/or pesticide runoff, are apparent on a wide scale, as are the consequences of over-fishing, *e.g.* in the Gulf of Thailand. But there is remarkably little quantification of such damage, and few attempts are made to control it. There have also been apparently inexplicable reductions of fish-stocks, *e.g.* of anchovies (various Engraulidae) off Peru, and of *Sardinella* spp. off W. Africa. These may reflect over-fishing but they could equally well reflect little-understood changes in oceanic climates. Phenomena of this latter kind affect developing and developed countries – *e.g.* Sardines (*Sardinops sagax*) off California – alike. They can only be resolved through international, cooperative marine research (*cf.* Chapter 6).

Biotechnology

Much is often made of the potential benefits to tropical agriculture of the application of modern biotechnological methods. Undoubtedly there will be benefits, but there is also the danger that such methods may do no more than speed up the application of basically unsuitable practices! For example, modern methods of storing frozen eggs and embryos, and of *in vitro* fertilization on top of the long-standing and well-tried applications of artificial insemination, can greatly assist reproductive efficiency and animal production. All these techniques are becoming commonplace in developed countries. If, however, they are used, as has been the case already, to promote cheaper, more rapid, and more effective, transfer of European breeds to the tropics, then the disadvantages already set out (*see* pp. 75–6) will merely be exacerbated. Similarly, if gene-cloning and -transfer is used to transfer single, alien, resistance genes, they will only provide short-term protection for crops or stock and ultimately lead to an extension of the range of the pathogen concerned.

By contrast, there is a real possibility by means of gene-transfer, of transferring 'durable' resistance far more rapidly than by classical breeding methods as, in some cases, it appears to be associated with blocks of genetic material. Such techniques might also be useful in transferring resistance from indigenous wild animals to domestic stock.

Quite apart from using biotechnology to speed up breeding programmes, it could provide wholly novel possibilities. For instance, it is already clear that viruses exist which are lethal to many insects, and their use as specific, biologically harmless, control agents has not only been mooted but tried out. As they are 'natural', highly host-specific agents, the expectation, which has proved to be the case in some instances, is that they will be environmentally neutral; but there is uncertainty. The concentrations that are likely to be released, even under test conditions, are probably higher than are to be expected in Nature.

There is great ignorance of the ecological consequences of totally eliminating a pest (as assessed by human standards) from an ecosystem; and there is concern that the pest's specificity could change – so, for example, widening the host-range. In particular, it has been cautioned that spontaneous mutation could render such viruses pathogenic to Man, particularly as some of them belong to groups whose supposed relatives are already human pathogens. In practice, it should be possible to build in 'self-destruct' genetic material into any virus, likely to be released, which would reduce such potential hazards. Some use has already been made of such agents against insect pests of temperate-zone trees, with a quite remarkable degree of success; but the technique is still in its infancy and should, obviously, only be applied with the greatest degree of caution.

In summary, then, biotechnology has much to offer. Its techniques can be used to speed up breeding methods and to bring about gene transfers that have, hitherto, either been impossible, or immensely laborious and with low rates of success. Wholly novel applications are also possible which, by their nature, are unpredictable at present. That should result in emphasizing the importance of assessing as early as conceivable, and far more fully than is done at present, the possible advantages and disadvantages accruing from their application – in particular on natural ecosystems as well as domestic ones. This is all the more important because, at present, the scientific sophistication required, together with the cost of applying biotechnological methods, is far greater than developing countries either possess, or could afford, respectively. Unless the immediate advantages are of immense value to the country concerned, the case for biotechnology in developing countries is likely to be a matter for technological aid from developed countries for some considerable time to come.

Water and Soils

'Food security requires water security!'. Erosion control and irrigation systems are essential in much of the arid and semi-arid tropics, and of equal importance to crops, as is breeding. It will be recalled that approximately as much land is lost annually through desertification and salinization as will be required to meet the needs of the expanding world population (*see* pp. 68). A regrettable feature of the present situation is that, too often, water and erosion research is not sufficiently linked to other agricultural research. With the introduction of more intensive rather than traditional, former, farming systems, and under the influence of developed agricultures but on increasingly fragile soils, there is a likelihood of accelerated erosion.

It needs to be remembered that, in a developed agriculture, there was a dust-bowl in N. America in the '30s which is still a living memory – acutely revived by the drought of 1988. Even in as apparently stable an agricultural system as that of the United Kingdom, intensive farming has promoted the exposure and extensive drifting and loss of glacial sands in N. Lincolnshire, while there has been a drop of more than 4.5 m in the level of the surface of Cambridgeshire fens through peat shrinkage over three centuries due to environmentally undesirable irrigation practices which still continue! Such observations are vivid reminders that no agricultural operation, however sophisticated, is without dangers to the sustainability of the basic resources of soil and water.

Soil erosion is of particular importance where land is denuded of trees – either to provide more land for farming, for fuel-wood, through overgrazing, or, what is increasingly common in Africa, because of all three activities. In Tanzania, for example, an early study over a two-years' period on 50 m^2 plots of red sandy-loam soil with a gradient of 3.5°, demonstrated a dramatic difference between plots covered with ungrazed thicket, grass, cultivated millet, and bare fallow (Table 4.3; *cf.* Rapp *et al.*, 1972).

The problem is both particularly acute and has been well researched in Kenya. Rangelands cover some 87% of that country and subsistence farmers represent about 80% of the population, but 50% of these farmers are dependent upon off-farm income which is most frequently achieved by migration of males into urban employment areas. Consequently there is a further maintenance burden on female labour which is itself exacerbated by the women's widespread practice of undertaking piece-rate labour for the richer, larger farmers, to increase their own household income. Thus small-farm management is neglected by both males and females on their own farms. In particular, terracing, mulching, and manuring – all labour-intensive activities are neglected, and erosion is exacerbated by tree-clearance

Table 4.3. Ground-cover, Loss of Soil, and Water Runoff, in Tanzania. (Surface gradient of 3.5°.) (Data after Rapp *et al.*, 1972.)

Ground-cover	Soil Lost (Tonnes per ha)	Water Runoff (as % of rainfall)
Ungrazed thicket	0	0.4
Grass cover	0	1.9
Millet crop	78	26.0
Bare fallow	146.2	50.4

due to the high demand for fuel-wood, especially around expanding rural settlements. This is where the best agricultural land exists, and here utilization of fuel-wood exceeds supply by about 7 million metric tonnes per annum (Western & Ssemakula, 1980). In fact, production in Kenya as a whole exceeds utilization, but most of it is in the rangelands, from which it cannot readily and economically be extracted and transported to the points of consumption.

Nevertheless, remarkable progress has been made in erosion control under the inspired direction of G. Wenner (1979), who demonstrated that the cheapest and simplest way to terrace was either to leave an unploughed strip across a slope, or to make ridges of dead crop-residues in a line across it. These catch and check water as it flows down the slope and, because they are not cultivated, they become spongy with holes formed by insects, worms, and perennial roots, and so act as areas of infiltration. In high-rainfall areas the downward edges of these terraces can be planted with perennial crops, elephant grasses (*Pennisetum* spp.), or fruit trees.

'Biological terracing' is much preferred to engineered terracing as it is less expensive and just as effective. The total cost of this programme has been remarkably low, some US$ 15.2 m in all, as by 1985 some 490,000 farms had been terraced out of the total 1.1 million in Kenya needing treatment at an average cost of US$ 28 per farm. The Swedish International Development Authority (SIDA) has acted both as adviser and, in part, has financed this long-term programme with a loan. Not only has the Kenyan government given the project a high priority, but SIDA is now contemplating extending the system to Zambia, Tanzania, and Ethiopia. However, just as in Kenya, success will be crucially dependent upon government cooperation.

In the more humid tropics, weeds, themselves a problem, can be cut and used for terracing – or, where alley cropping has been developed, cuttings or rougher brashings from the woody components can be used in the same way. The alleys of trees themselves act, of course, as terracing agents (*see* pp. 88–9).

Further research is needed to promote soil stability. Detailed information on soil types is still deficient in most of the less-developed countries, particularly concerning those soils which have a low organic content and high sand content, and consequently are most liable to erosion. Survey work for this and, indeed, land-use capability as a whole, is an urgent requirement. Controlled grazing schemes need to be developed, such as the migration trails in Botswana, which were chosen to minimize erosion. Much of this work should properly fall upon national government agencies, but assistance from the developed world on techniques and in training is also very desirable.

A major hydrological problem which is almost universally disregarded in developed and less-developed countries alike, is the increasing scale of utilization of non-renewable, or only slowly replaceable, water resources. It is regrettable that, so long as water can be obtained from an adjacent region, little consideration is given by the users as to whether or not it can be replaced. This is as true of Missouri and Iowa as it is, on a more spectacular scale, of those countries that depend upon the Nile for their water requirements. In many developing countries, and in a surprisingly large number of developed countries, the true situation is unknown or unrecognized owing to a lack of appropriate surveys or measurements – and in contrast to the south-western United States, where the future of the Ogallala Aquifer is causing grave concern (Glantz & Ausubel, 1984).

Water control is another difficult problem. Evaporation, from 2.5–3 m below the soil surface, of mineralized ground-water leads inevitably to salinization. The rate of salinization is proportional to the volume of evaporated ground-water *plus* the concentration of salts in it. The principles of control, as with erosion control, are simple in theory but difficult in practice. They are to reduce evaporation as much as possible and to hold the ground-water level as low as possible. This is particularly difficult in the widespread areas where irrigation is essential to crop growth, e.g. eastern and western China, India, Asia Minor, N. Africa, and the Middle East, amongst less-developed or developing regions, or where hydromorphic evaporation is due to soil type, *e.g.* on the high plains of S. America, Central Asia, or the Caucasus.

Maintaining a cover of vegetation is important, together with practices of minimum tillage, so that the evaporative surface of the soil is minimally disturbed. Mulching and the retention of humus, which reduces the evaporative susceptibility of the soil, are also important. The lining of wells and irrigation ditches, if it can be afforded, is a further valuable aid. There is great scope for applying the basic principles of soil conservation to particular situations, but this requires local cooperation and, above all, active and localized training.

In some cases a simple return to what were once widespread practices will suffice. A fascinating example is the plateau regions of Bolivia and Peru. The utilization of circular field basins (*qochas*) of the high Andes, and of the raised fields in waterlogged areas at all altitudes, have, apparently, been forgotten. Erickson & Candler (1988) have excavated the ancient raised fields near Huata, on the Peruvian side of Lake Titicaca (3,800 m above sea-level) which ceased to be cultivated about 400 AD. Parallel canals were dug, and the earth between them was piled to form mounds 1 m high, 4–10 m wide, and 10–100 m long. These have been re-created and planted with potatoes, of which the sustainable yields are twice those of control plots using modern methods of cultivation!

Other work (Kolata & Ortloff, 1989; Ortloff & Kolata, 1989) has shown that a Man-made system of terraces on the hill slopes, aqueducts, and raised fields, formed an integrated system. Salinization, heat storage, and the water-table, were effectively controlled. Training of the present population in these ancient methods has now begun and is proving effective. There is clearly a lesson here for other areas in S. America – there are over 38,000 hectares of abandoned agricultural land around L. Titicaca alone, and some 50,000 hectares of abandoned raised fields in the Caribbean lowlands that seem suitable for renewal. Indeed, there are many other comparable areas.

Conclusions

In conclusion, there is every reason to suppose that present techniques of, and trends in, agricultural research are potentially capable of providing sufficient staple crops to feed the human population of the world at the present levels. There is a reasonable expectation that the application of modern molecular biology, thoughtfully applied, will materially assist in speeding up plant crop-breeding programmes, while much the same applies to modern developments in animal reproductive control, embryo transplant, and the like, in cattle and other domestic animals.

Increased speed of breeding will, it seems, become particularly important if the suggested major climatic changes actually come about. Improved control of water resources, and more widespread and better control of erosion and salinization, are areas which require a great deal more, and more intensive, investigation and application than they are currently given. Their importance cannot be stressed too highly to all governments – especially, but not exclusively, to those of the less-developed countries. In particular, all countries need to practise the utmost economy in the use of their non-renewable water resources (*cf.* Glantz & Ausubel, 1984), while finally, there is little point in improving crops and stock if there is insufficient land in good heart on which to raise them!

Four further, major changes in agricultural and allied practices are needed, however. These are:

1. Due recognition of the needs of small farmers and, especially, of the landless, whose survival will depend upon their being able to hire out their labour. Thus the widespread objective of major monocultures over large areas, protected by expensive pesticides and using costly, fossil-fuel-derived fertilizers, will need to be modified. This will also apply particularly to developed countries in order to render their agricultures more sustainable.

In developing countries, however, there will have to be a deliberate policy of promoting diverse, stable, improved, and sustainable-yielding, crops that can be grown in mixed cultivations, possessing the maximum innate, biological protection to pests and pathogens. Compared with the current, comparable situation, stock will need to be better fed through cheap N-supplementation, numbers of pigs and poultry will need to be increased, and aquaculture for food and fishmeal will need to be increased more effectively and widely. Every measure should be taken to ensure a plant covering of soil for as long as possible. Minimum tillage, effective mulching, and the promotion of water conservation, should be practised by all agriculturists and through well-planned forestry. All these things are technically possible with present knowledge and research in the future to keep up with changed circumstances; but political will and appropriate economic practice are crucial, respectively, to their adoption and effective application.

2. More attention than currently will need to be paid to local improvement of so-called minor cereals as well as to the improvement of root-crops, tubers, etc.

3. The effective transfer of the kinds of information discussed here to small, poor farmers at the local level is needed, as also is securing their widespread cooperation.

4. A far greater commitment than at present will be required to meeting the needs of the rural communities in less-developed countries, with agriculture, forestry, and soil and water conservation, as primary objectives.

Transmitting Technology to Poor Farmers

Extension services are the standard means of transferring technological advances to farmers. There are, however, a number of drawbacks to them as currently established in many developing countries. Reasons for this have, increasingly, come to be recognized as including:

a) The organization and administration of the service may not be 'user friendly'.

b) Extension workers are usually educated to college or university standard. They then tend to use the language of their training, are not always well trained in communication skills and, because of their specialist training, may well have lost touch with everyday farming and its problems. Sadly, some extension workers may even regard farmers as ignorant, and may fail to respect their undoubted interest and practical experience, which the extension worker may himself or herself well lack!

c) There is always a danger that the inherent ability of farmers may be underestimated even by the most committed and experienced extension workers. The consequences are that communication is ineffective, and that the farmers will fail to heed advice, which they do not fully understand, or will even deliberately disregard it, because of the lack of empathy and sympathetic understanding on the part of the extension worker even when the advice is understandable.

Extension services, therefore, need to be so organized that they can actually reach the farmers on their own territory – geographically, linguistically, technically, and as regards mutual understanding and sympathy. Extension workers need to possess three primary attributes, namely:

1. To be well-trained in communication, in simple and plain speaking in a language that is at least widely understood locally, and be capable of adapting their approach to individuals from different kinds of background.

2. To be good judges of character. It is also important to recognize with whom he or she can communicate most effectively. In many societies it will be the women who do the farming, although the advice may well have to be delivered *via* the men! Sensitivity to local custom is an essential element in picking those with whom to communicate.

3. To be willing to listen and learn. The first attribute will improve the quality of the message, will enable the extension worker to pick the farmers who are most likely to benefit from the message and apply it actively, while encouraging other farmers to do likewise and transmit it more widely. The second attribute will assist in promoting confidence and in adapting the message to the local situation. There are now many examples of successful transmissions of ideas where the first step was to understand thoroughly current practices and the reasons for them – before even broaching any suggestion for change.

There is one further attribute in a good extension worker, namely, that he or she should be capable of bringing back the lessons he or she may have learned, or the beneficial practices he or she may have discovered – both to his or her seniors and to the research workers. There is then a greatly improved chance of research being adapted to the everyday needs of farmers.

In this respect, the system which prevailed until quite recently in Scotland was a model for any country – developed, undeveloped, or developing. Advisers (extension workers) were attached to districts, or given particular technical specialities, for which they were made personally responsible. Each was attached to an Agricultural College which gave practical training to the farmers and farm workers. Thus advisers met farmers both at the Colleges and in the districts where they lived and worked, and so had ample opportunity to get to know them and their problems.

The Colleges are themselves adjacent to, or affiliated with, the University Departments of Agriculture; the former carry out applied research, and the latter the basic and strategic research. There was thus a continuous chain from the farm *via* the advisers to Colleges, and thence on to University Departments, i.e. from practical needs to the best contemporary research. Communication was thus *two-way*. This is, of course, but one way to organize an extension service, but it illustrates the essential principle of effective two-ways communication which is a fundamental necessity if the service is to be truly effective.

An important but too-often neglected technique of communication is the use of radio. The transistor radio is now a universal domestic chattel, and so information can be transmitted at extremely low cost compared with that of providing trained extension workers. However, great attention needs to be paid to the form in which information is broadcast. Reading from a technical manual – which has been employed – is likely to have no, or even a negative, effect. But a technique such as that used in Bengal, where farming information is put to music and song in an attractive chat-show, has proved highly successful. A comparable technique, namely the use of tapes to relay extension information in drama form, often including roles for local personalities, has proved popular and effective in Zambia. Both techniques need to be employed more widely.

It is becoming increasingly clear that a key change – other than macro-organizational change – is taking place in the relationship between the extension workers and groups of farmers. Ideally, both should be practising new techniques together, and discussing freely the changes proposed, with both parties teaching and learning by example. Taken together, all these things demand a new approach, in many countries, to the selection and training of extension workers. The employment of individuals with more limited training as intermediaries, and the importance of knowing the social milieu of the farmers, as well as the science, are matters calling for new experiments and new thinking. Here, developed countries may perhaps act as catalysts or sounding boards; but they must also be prepared to learn, so that, if they are to give practical aid in this general area, they will do so sensibly and wisely.

Optimal Systems

It will have become apparent from the earlier discussion above that, in undeveloped and developing countries, the primary problem is to ensure the sustenance of, and improve the living conditions of, those most numerous, agriculturally-dependent individuals, namely the poor small farmers and landless labourers, who utilize mixed cropping as their basic technique. Advances in the yield or other attributes of crops or stock need to be integrated into new and effective farming systems that meet those constraints.

There are two other conditions that must be met. The first is the promotion of additional income, so as actually to raise the standard of living above mere subsistence. Three possible means exist of ensuring this, namely (1) a common solution, which is the migration of the male (usually) to urban employment for at least part of the year; (2) investment in good years, or from additional earnings in urban employment, in stock, which is a common practice in southern Africa; and (3) the most desirable, to obtain profit from cash-crops grown in addition to the basic staples. This last means should surely be originated at the national level rather than being an international initiative, if only because cash-crops on a small scale can only be for local markets, at least in the first instance, though cooperatives and national, or even international, trade may follow. This kind of initiative should help to alleviate the burden of debt which otherwise only too few poor farmers can avoid, and also should provide more work for the unemployed.

The problem of adding cash-crops to promote income above the subsistence level is still perhaps one of the greatest challenges of the many outlined here. Although by such means food shortages may have been contained, the new varieties have resulted in little cash-gain to such individuals and, indeed, by making greater cash demands on them, may have increased their dependence and poverty. Their standards of living have, at best, remained much the same.

The development of major cash-crops, *e.g.* cotton and cocoa, has followed similar lines to that of the major food-crops, and has often involved increased mechanization, particularly at harvest. This, however, has reduced the need for labour and such cash-crops are, in any case, unsuitable for small or pastoralist farmers. Appropriate crops have not been developed to provide them with the additional income which they need, either to meet the increased costs of the new food-crops, e.g. the cost of annual purchase of hybrid seed and/or fertilizers, or to raise their standard of living.

Filling such needs is sometimes attempted through the keeping of stock, the sale of fuel-wood or charcoal for cooking, or by fishing or aquaculture. The first two of these tend to promote erosion, directly or indirectly, while the last two are not always possible because of the absence of suitable bodies of water or overcrowding of those that exist. Solutions to these problems are diverse, but they are sometimes soluble at the local level, although that does not necessarily prevent but rather may encourage their widespread adoption in similar situations elsewhere if they prove successful. Two examples, rope production at Nada in N. India, and a dry-season gardening project in Niger, illustrate this.

Nada was a dispersed, poverty-stricken hill village. The low-caste Harijan Nada, many of whom were landless, were persuaded to give up the uneconomic grazing of the denuded hillsides in order to plant Bhabbar Grass (*Eulaliopsis binata*) for rope-making by the women. This, with some aid, proved immensely successful and later, improved irrigation allowed wheat and vegetables to be developed also. Sustainable development and an improved standard of living now seem assured (P. R. Mishra & Madhu Sarin, in Conroy & Litvinoff, 1988).

The Niger example demonstrates the value of improved water-utilization during a prolonged dry period. Areas of low-lying moist sand, enriched by the downward erosion and transport of fertile topsoil, were used to grow Okra (*Hibiscus esculentus*), Manioc (*Manihot aipi*), and Tomatoes (*Solanum lycopersicum*), to subsistence level. Shallow wells of a novel, cheap, long-lasting design were developed. Chick Peas (*Cicer arietinum*) were introduced to promote soil nitrogen, and some shade-trees were planted which also helped to conserve water. Dry-weather vegetable gardening then became possible, and this was strengthened by the introduction of a wider range of vegetable seeds. Thus all-year-round production became possible, with enhanced economic independence. Aid had been provided in the design of the wells and the provision of the new seeds. This development is now being copied in comparable areas in Mali, Senegal, and W. Sudan (Cottingham, 1988).

These two examples have three features in common. They depended upon the extension of the farming practice to new cash-crops, they were enhanced by improved water regimes, and they resulted from intelligent aid applied at the local level. These, indeed, are likely to be practices which could be applied successfully elsewhere.

The other condition (besides the promotion of additional income) that must be met, more in rural Africa than in rural areas in other developing regions, but also in all the burgeoning urban developments of Africa, Asia, and Latin America, is the provision of fuel for cooking and, to a lesser extent, for heating. This problem requires local solutions. Fuel-wood provides 20% of the total energy requirements of Latin America, 10% in Asia, but 60% in Africa. There are even wider local variations, however; for example, 95% of Tanzania's total energy requirement is met by fuel-wood.

It has been claimed that the demand for fuel-wood is a serious cause of deforestation, but this is generally not the case; clear-felling to provide more agricultural land is the main culprit. Urban demands for charcoal, used in inefficient stoves, results in even more destruction of small trees than do loppings, which generally meet rural cooking requirements. On the other hand, deforestation preceding agricultural expansion, coupled with the seemingly inevitable subsequent population growth, does mean that increased time and effort are spent foraging from the larger settlements for tree-loppings, because of the tree shortage in the immediate vicinity. Hence, time for farming activities is further curtailed.

Rural requirements can be met by planting small, quick-growing trees as part of the farming regime. When pollarded, these can supply fuel-wood requirements. Such trees can be incorporated in biological terracing, or in alley farming. A remarkable example is the system of alley cropping developed in Ghana, where the major agricultural problem is weed growth. Initially, land is cleared of bushes or weeds two or three times by hand early in the growing-season – a labour-intensive activity. The weed residues are retained as a mulch over seeds, sown in broad rows, of the principal staple crop *e.g.* Maize.

With the coming of the rains, growth of both the crop and the weeds erupts, and the latter are briefly controlled by herbicides – their remains adding to the surface mulch. Leguminous trees, especially improved varieties of *Leucaena leucocephala*, are planted in rows on either side of the cropping strip. Their foliage, growing over the crops, provides shade and shelter, and symbiotic N-fixation in root nodules results in enriched soil-nitrogen. Later the branches, which have helped to shade out the weeds, are cut back, to allow unimpeded crop growth and harvesting. The cuttings, depending on their size, are either added to the mulch, or used for fuel-wood. The leaves are used for fodder. In addition, yams can be trained on the pruned stumps as a catch-crop, while re-growth of the trees after harvesting of the crops maintains weed control without further hand-weeding, and the cycle can be repeated (Fedden, 1988). Thus soil conservation, weed control, the provision of a staple food-crop and a catch-crop (which provides additional income), as well as a supply of fuel-wood, can all be provided in a potentially sustainable manner.

Fuel-wood for urban needs is best met in the vicinity of towns by establishing dedicated plantations of trees that are suitable for charcoal production. This is essentially a forestry operation and so will not be dealt with here. Even so, the problem is not purely a technical one, for any such development must lead to a potential loss of income which many of the urban poor rely on through foraging for, and selling, fuel-wood. No completely clear solution exists, except in a few local areas.

Incidentally, a major effort to design more efficient cooking stoves for urban populations is desirable, as existing designs are inefficient. This would result in a modest but by no means negligible economy in the use of fuel-wood.

General Conclusions

What are the chances in the future of using our planet's land, so far as agriculture is concerned, both optimally and sustainably to support the health, welfare, and happiness, of Mankind? The evidence available suggests that the necessary technology either exists, or, through developments such as those outlined in this chapter, could be achieved in the foreseeable future. It is now, essentially, a matter of will and of commitment, primarily by governments; but it will have to be driven by totally committed people. Human selfishness and ignorance will have to be overcome.

The developed countries will have to adopt a new morality, a very different economic stance from currently, and apply their intellectual efforts and concern to developing aid in a way which is far more effective than at present. The peoples of less-developed countries will, at the same time, have to curb their fears of exploitation, greatly increasing their national efforts, while all peoples will have to work together for the truly common good. Humanity, not technology, is likely to be the weakest link, so that I can only echo the remark of the late Sir Frank Fraser Darling when asked about the likely outcome of the world in the future: 'I am', he said, 'reasonably pessimistic!'.

References

Adams, R. M., Rosenzweig, C., Peart, R. M., Ritchie, J. T., McCarl, B. A., Glyer, J. D., Cuny, R. B., Jones, J. W., Boote, K. J. & Allen, L. H., Jr. (1990). Global climatic change and US Agriculture. *Nature* (London), 345, pp. 219–24.

Alexandratos, N. (1988). *World Agriculture Toward 2000; an FAO Study.* Bellhaven Press, London, England, UK: xvi + 339 pp., illustr.

Browder, J. O. (Ed.) (1988). *Fragile Lands of Latin America: strategies for sustainable development.* Westview Press, Boulder, Colorado, USA: 301 pp.

CGIAR (Consultative Group for International Agricultural Research) (1987). *Annual Report 1986/7.* Washington, DC, USA.

CIMMYT (Centro Internacional de Mejoramiento de Maiz y Trigo) (1983). *Report on Wheat Improvement 1980.* El Baton, Mexico.

CTA (The Technical Centre for Agricultural and Rural Co-operation) (1984). *Le role des auxiliares d'elevage en Afrique.* Rapport de seminaire. GTZ-IEMVT, Paris.

Conroy, C. & Litvinoff, M. (Eds.) (1988). *The Greening of Aid: sustainable livelihoods in practice.* Earthscan Publications in assn. with IIED, London, UK: xiv + 302 pp., illustr.

Cottingham, R. (1988). Dry-season Gardening projects, Niger. Pp. 69–73 in Conroy, C. & Litvinoff, M. (Eds.) *The Greening of Aid: sustainable livelihoods in practice.* Earthscan Publications in assn. with IIED, London, England, UK: xiv + 302 pp., illustr.

Erickson, C. L. & Candler, K. (1988). Pp. 230–49 in Browder, J. O. (Ed.) *Fragile Lands of Latin America: strategies for sustainable development.* Westview Press, Boulder, Colorado, USA: 301 pp.

Fedden, M. (1988). *Forest Farm Husbandry.* Technology Consultant Centre, Kumasi, Ghana: iii + 69 pp., illustr.

FAO (Food and Agricultural Organization) (1986). *The Dynamics of Rural Poverty.* FAO, Rome, Italy: 40 pp.

FAO (1987). *The Fifth World Food Survey, 1985.* FAO, Rome, Italy: 73 pp.

Glantz, M. H. & Ausubel, J. H. (1984). The Ogallala Aquifer and Carbon Dioxide: comparison and convergence. *Environmental Conservation,* 11, pp. 123–31, map.

Hennessy, D. W. (1986). Supplementation to reduce lactational anoestrus in first calf heifers grazing native pastures in the subtropics. *Proc. Australian Soc. Animal Production,* 16, pp. 227–30.

Holl, K., Daily, G. & Ehrlich, P. R. (1990). Integrated Pest Management in Latin America. *Environmental Conservation,* 17, pp. 341–50.

ICRISAT (International Crops Research Institute for the Semi-Arid Tropics) (1984). *Annual Report.* Hyderabad, India.

ILCA (International Livestock Centre for Africa) (1987). *Annual Report.* ILCA, Addis Ababa, Ethiopia.

IRRI (International Rice Research Institute) (1982). *Symposium on Drought Resistance in Crops with Emphasis on Rice.* IRRI, Los Baños, Philippines: 414 pp., illustr.

IRRI (1984). *Annual Report, 1983.* IRRI, Los Baños, Philippines: xv + 495 pp., illustr.

IWC (International Wheat Council) (1989). *Market Report.* IWC, London, England, UK.

Kolata, A. L. & Ortloff, C. J. (1989). Thermal analysis of Tiwanaku raised field systems in the Lake Titicaca Basin of Bolivia. *Journal of Archaeological Science,* 16, pp. 223–63.

Leng, R. A. (1990). Ruminant nutrition in the tropics. Pp. 221–38 in Speedy, A. (Ed.), *Developing World Agriculture.* Grosvenor Press International, London, England, UK: 286 pp., illustr.

Lipton, M. & Longhurst, R. (1989). *New Seeds for Poor People.* Unwin & Hyman, London, England, UK: xiv + 473 pp.

Mishra, P. R. & Sarin, M. (1988). Social security through social fencing, Sukhomajri and Nada, North India. Pp. 22–8 in Conroy, C. & Litvinoff, M. (Eds.), *The Greening of Aid: sustainable livelihoods in practice.* Earthscan Publications in assn. with IIED, London, England, UK: xiv + 302 pp., illustr.

Monyo, J. H., Ker, A. D. R. & Campbell, M. (1976). *Intercropping in Semi-arid Areas.* IDRC–o76e, Ottawa, Ontario, Canada.

Ortloff, C. J. & Kolata, A. L. (1989). Hydraulic analysis of Tiwanaku aqueduct structures at Lukurmata and Pajchivi, Bolivia. *Journal of Archaeological Science,* 16, pp. 513–35.

Preston, R. D. & Leng, R. A. (1987). *Matching Ruminant Production Systems with Available Resources in the Tropics and Sub-tropics.* Penambul Books, Armidale, Australia: 245 pp.

Provost, A. (1982). Scientific and technical bases for the eradication of rinderpest in intertropical Africa. *Rev. Sci. Techn. OIE*, 1, pp. 687–704.

Rapp, A., Beny, L. & Temple, P. (1972). Erosion and sedimentation in Tanzania. *Gografiska Annaler*, 54A, pp. 105–379.

Saari, E. E. (1985). South and South-east Asian Region. In *CIMMYT Report on Wheat Improvement, 1983.* CIMMYT, El Batan, Mexico.

Sun, M. (1986). Engineering crops that resist weed-killers. *Science*, 231, pp. 1360–1

Western, D. & Ssemakula, J. (1980). *The present and future patterns of production and consumption of wood energy in Kenya.* Beijer Institute, Stockholm, Sweden.

Wenner, G. (1979). *An outline of soil conservation in Kenya.* Ministry of Agriculture, Nairobi, Kenya.

Wickham, T. H., Barker, R. & Rosegrant, M. V. (1978). Complementarities among irrigation, fertilizer and modern rice varieties. Pp. 211–40 in *IRRI Economic Consequences of the New Rice Technology. IRRI, Los Baños,* Philippines: v + 402 pp., illustr.

Commentary on Chapters 3 and 4

Joint Chairmen: Professor Valentin I. Kefeli (ch. 3) & Reid A. Bryson (ch. 4)

Panellists and Other Contributors:

Bailey, Cloudsley-Thompson, Oza, Trompf, Poore, Fearnside, Poore, Wasawo, Batisse, Poore, Ramakrishnan, Furedy, Fosberg, Wasawo, Westing, Burnett, Poore, Simon, Kefeli

Bailey observed that land-use management deals with productivity systems that are ecosystems from which a product, such as wood or water, was extracted. Optimal, or sustainable land-use must be consistent with a capacity for renewal of the system.

The assessment of optimal land-use required estimates of response to utilization and, in order to do this, relationships or, more accurately, rules, had to be developed to describe the inter-relation between the necessary response information and the ecosystem classes. These rules took many forms ranging from simple, experience-based judgements to multivariate regression models and complex mathematical simulations. Their application was based on transfer by analogy, i.e. the rules were extrapolated from experimental sites to analogous areas defined by classification.

In order to extend information to other sites it was necessary for them to be within ecologically similar land-types, and this required the land to be subdivided into ecosystems. That approach was based on the questionable hypothesis that all replications would show a similar response, but correlations between types of ecosystem and behavioural response were generally low. This was because the criteria for classification had been applied over large areas without any consideration of possible compensating factors. These might, for example, have produced the same ecosystem type but for different reasons. Thus climatic effects could be modified by soil factors, e.g. moisture-demanding species were well known to extend often into less humid regions on moist sites.

It was, therefore, necessary, if a reliable ecosystem-behaviour relationship was to be established, to divide the land into 'relatively homogeneous' ecological regions where similar ecosystems had developed on sites with similar problems. For example, similar sites occurred in several climatic regions and, within a region, would support the same vegetation community but, in other regions, different communities could develop. Beach regions in the tundra climatic region supported low-growing shrubs and forbs, whereas beaches in the sub-arctic regions usually supported dense growths of Black Spruce (*Picea mariana*), or Jack Pine (*Pinus banksiana*). Soils displayed similar trends.

The theory was that climatic regions created regional unity. They delimited patterns of site-level systems with similar relationships between climate, landform and biota, which were repeated throughout the region. So, by monitoring the behaviour of representative sites it should be possible to predict the behaviour elsewhere, i.e. to extend data, spatially, from a limited sample.

There was now an increasing necessity in ecosystem monitoring of international scope, for the development of global ecosystem regionalization to serve as a basis for a reliable cross-national exchange of information. The value of such a system had been endorsed by the International Union of Forestry Research Organizations who had recommended international support and implementation for such a scheme. He and his colleagues, with the encouragement of several international organizations had prepared a new world map to show the distribution of such terrestrial eco-regions. It had been adapted from existing information and regional boundaries still needed to be verified and refined. The mapping of the eco-regions of the oceans was envisaged as the next logical step.

[The map has been published in *Environmental Conservation* (1989) 16 pp. 307-10 and accompanying folded, coloured map. Eds.]

Cloudsley-Thompson reminded the Conference that some 60 years ago, R. Pearl, G. F. Gause and T. Park, using experimental populations of animals such as *Paramecium* and *Tribolium*, demonstrated that they invariably increased in a sigmoid curve whose asymptote was limited by food shortage, conditioning or density. Extrapolated to human populations, these factors were represented by famine, pollution, disease and warfare. Famine had already extracted a toll in many Third World countries, whilst AIDS, as already mentioned (Commentary 1, p. 25), might become an increasing threat to human survival everywhere: war would be discussed in another session.

Sustainable land-use was crucial for future survival. Nevertheless, a sustainable form of land-use, nomadism, was, even today, being actively discouraged in many Sahel countries, and more land was lost annually, world-wide, through water-logging and salinization than was gained through the construction of new dams. Overhead irrigation too was especially harmful when applied to stabilized dunes. Even poverty, the greatest of evils, was relative. He recalled that a British official in the Government of the then Anglo-Egyptian Sudan, wrote of the north of the country in the 1930s, that he knew of no other country in which peoples' desires were so nearly achieved: the diversity of desert habitats was often overlooked by planning authorities! As M. Kassas had emphasized elsewhere, one should not think in terms of 1,000 feddens (acres) but rather of 1,000 farms, each of 1 fedden.

On forestry, it should be remembered also that trees, unlike farm crops, produced not only firewood and timber but a diversity of foods and medicines as well.

He could also support the importance of the transistor radio as an agent for disseminating useful ecological information simply; for even in the remote Nuba Mountains in the Sudan in 1967, the local dancers at a *kambala* were singing about recent events in Egypt, their information having been acquired in that way!

Oza offered two comments. Firstly, on optimal land-use, he remarked that Urban Man had invaded the forest, driven out the forest dwellers and virtually shattered the forest ecosystem. In SE Asia the tropical forest had been replaced by rice cultivation and it was there that the population explosion had been greatest. He did not wish to suggest that it was the rice which had led to this explosion, but his second point was that, if the situation was to be rendered sustainable, then the remaining forests would need to be conserved and only agroforestry could take care of the environmental degradation.

Trompf wished to reveal the fact, then only two weeks old, that the Minister for the Environment in Papua-New Guinea had declared that his government was happy for 80% or more of the tropical rain-forest region of his country to remain intact in perpetuity as a world heritage, *provided that* the international community paid for the implicit consequences of the disparities between the First and Third Worlds! He wondered whether **Poore** would recommend that that kind of action should be taken by other governments. Such conservation was associated with the whole business of the sale of genetic materials from tropical areas to the industrial countries. He had heard recently that there might be a grass in New Britain which would provide a cure for AIDS, since two people had emerged from the jungle apparently cured, but that was only an example of the kind of economically valuable products that such a region could provide. Were there any new ideas, he asked, as to how poor countries could actually exploit such natural capital in non-agricultural or forestry ways?

Poore responded to **Oza**'s points by first remarking that there were many different situations in SE Asia and certainly in parts of India and Thailand. In those areas where rice had replaced forests, with accompanying land degradation, he believed that international aid would be needed to begin to restore resources by forestry plantations, agroforestry and so on. But there were still forest-rich countries inhabited by forest dwellers in that region where the question was, 'What was to happen to the forests?' It might be possible to introduce food-crops into the forest so that its structure was not destroyed but rather altered to provide a much richer food base. The nearest he had seen to this was the fruit orchards of the former Dyack shifting cultivators in Sarawak where they planted durians or other fruit trees when they abandoned shifting cultivation. These eventually developed into a fruit orchard, for the areas were not cultivated further and the people, who owned the trees individually, returned to them year after year. He had seen one such area a few weeks ago which was 100 years old and where he was told how grandparents had planted them and they had been looked after ever since by the families.

The matter of the proposal concerning Papua-New Guinea opened up the whole question as to what extent the developed countries should pay for the conservation, or wise use, of the environment in the developing countries. There was a lot of thinking going on about this. For instance there were, of course, the Debt for Nature exchanges, and other ways in which debt conversion was manipulated to produce investment in good land-use. Then there was talk of a carbon tax and he wondered whether there was the possibility of a carbon incentive to be offered internationally to those countries which kept, or established, a substantial, natural 'carbon bank'. He rather doubted whether the suggestion that organizations in developed countries should buy areas in developing countries for conservation was acceptable: it had implications for sovereignty. On the whole he was doubtful if things were going to be done solely for the benefit of the developing countries themselves. Although advice on wise management would, on the whole, be to their advantage, he thought that money was likely to go, very selectively, only to those things where the costs were likely to fall. Moreover, it seemed to him likely that developing countries would incur costs through not doing things which the developing countries wanted to be done.

Fearnside said that he detected some unease in responding to the problem posed by Papua-New Guinea's action, and he thought a frank response was called for. The question of how far blackmail came into any of these negotiations had to be faced squarely. It worked both ways: either a developing country with tropical forests asked for so much money not to cut down their forests, or international agencies suggested that no loans would be forthcoming if the forests were not preserved. Both propositions were unacceptable ways of negotiating, but it remained important that significant sums should still be transferred from the developed countries without the implication of blackmail by either side.

Poore accepted the strictures about blackmailing, or even conditionality. He believed that codes of conduct were evolving and that, gradually by example, case-histories would demonstrate how such matters were best conducted so that there would be a significant flow of funds when a good case was made for them. Whether, for example, it was in the best interests of Papua-New Guinea to conserve 80%, 60%, or 70% of its remaining tropical forests, he could not say since he did not know whether the areas included those most important to conserve. Such questions were still open and the international community would have to feel its way carefully through such situations; there was no simple formula or solution.

Wasawo stressed the importance of effective leadership in order to ensure optimal land-use. He gave two contrasting examples of the problems in Africa. In Uganda Idi Amin, having declared what he described as an 'economic war', instructed the people to settle the forest reserves, which they did. In one area the rich replaced the trees by banana plantations bringing in cheap labour from other areas, made their profits, and moved on after 5 years. Soil fertility was lost but the labourers, although possessing land elsewhere, had remained, eking out a bare subsistence. The present government now wanted them to return, although their presence in the forest was not of their choosing, but they were reluctant. By contrast, in western Uganda, the people from the heavily populated Tigezi region, because of the population pressure, spread into and settled in the forest. Now, despite the law against forest settlement, they were not being forced to leave! This inconsistency demonstrated clearly the expediency of government policy, and reduced confidence in the leadership. His second example was from Ghana. He had been told by the Director of Forest Research that it had been found that their most important crop – cocoa – did poorly if the forests were removed. Much forest had already gone; its removal was the norm. How were people now to be persuaded not to clear their forest because cocoa productivity would suffer? Conservation of the forests in these two countries posed different, practical problems of leadership; there were no easy solutions in practice.

Batisse noted that **Poore** had presented two contrasting attitudes – to cut the forest down, or to conserve it intact – but had also suggested that it was not necessary to conserve the whole forest to ensure that biological diversity was maintained. It seemed to him that the main question was whether there was a real rationale to determine what should be conserved and what could properly be utilized, preferably in a sustainable way. Was the apparent lack of any such rationale used in the Amazon, or central Africa, for example, to avoid protecting the forests? Was the kind of exercise that had been carried out recently in Manaus by Conservation International and the World Bank, in which GIS was used to try to define a series of layers, some more important for conservation, others which could be neglected, a good one? Finally, he commented that by working with the local people and determining different zones for conservation etc. one was coming close to the idea of a Biosphere reserve.

Poore said he would attempt to answer the first question by talking about the way nature conservation had developed in Britain. Initially, a group of scientists had studied the distribution of natural communities, flora and fauna, and drawn up a list of proposed nature reserves. Their judgement had stood the test of time even though their data were incomplete, although, as expected, with the acquisition of more knowledge the reserves had been increased, or extended. He thought this gave some guide. A GIS applied to Amazonia, using the best available information, should enable reserves to be indicated immediately, even though the expectation would be that they would prove insufficient as more data became available. Nevertheless, he thought the initial choices would be at least 90% correct. Queensland had adopted this technique successfully.

He entirely accepted **Batisse**'s suggestion that much of what he had suggested was, in effect, similar to the creation of Biosphere reserves; it was a model of what might be done world-wide. This reminded him that he had omitted to point out that although the commonest objective throughout the world of much of government policies, support and research, and of finance, effort and incentives for agriculture and forestry, had been directed solely towards increasing production and productivity, that was not enough. The objective should be broader, namely, to support a complete pattern of land-use, farming systems and social systems. Those two different objectives needed to be sorted out everywhere, not least in the European Community!

Ramakrishnan thought that one of the problems of land management in the tropical rain-forest areas, particularly where there was shifting cultivation, had been the total disinterest shown by agricultural scientists in understanding these traditional systems and how they operated, coupled with the attempt to impose on local communities an unacceptable, and at times, unsuitable, technology that was often inappropriate to the prevailing ecological conditions. He therefore supported fully the notion that there should be a far closer interaction between the natural sciences and the social sciences in the area of land management.

Furedy pointed out that one of the commonest problems encountered in rural areas, especially in Asia, was the indebtedness of poor farmers. That indebtedness might be perpetuated from generation to generation with no hope of it being discharged by the family. So, small farmers had been drawn into cash-cropping in the hope of release from debt but that often merely increased it because of the further demands that cash-cropping imposed.

Fosberg asked for more information about how, what he preferred to call 'tree gardening' rather than the less-appropriately named agroforestry, fitted into the proposals made both for forestry and agriculture.

Wasawo added that in Malawi the ordinary people had always recognized certain trees as valuable for their crops. When they cultivated land they left trees and also looked after their seedlings. Similar practices were followed in the Kilimanjaro region and he thought it important that modern agroforestry should learn from such traditional practices rather than create new conditions which would negate centuries of experience.

Westing raised a further question about land-use for one renewable resource that had not been mentioned, namely, fibre production. He had the impression that about half the fibre produced in the world ended up, in the form of wood, as fuel. Most of the references to fire had been to undesirable fires for deforestation, but he wondered about the significance of the use of wood for cooking in relation to sustainable development

Burnett acknowledged **Ramakrishnan**'s support. There were, indeed, far too many attempts to impose, with the best possible will in the world, inappropriate technologies on poor farmers by governments acting on the advice of agricultural scientists. That did not mean that the scientists' contribution was not of importance but, too often, research was carried out without knowledge of actual farming conditions – in developed as well as in developing countries!

Indirectly, unsuitable crop systems affected the situation described by **Furedy**. However, he thought that another important aspect of the switch to cash-cropping, which she had not mentioned, was that it was often associated with a complete change-over from staples to cash-crops. What ought to be promoted was a mixed farming system with cash-crops *plus* staples, if indebtedness was to be overcome. This did not necessarily mean that more land would be needed and the solution could lie in the 'tree gardens' of **Fosberg**, as he had already described for Ghana. There the so-called alley cropping, developed from traditional mixed farming, was ideally suited to the humid tropics: it promoted trees, provided a mixed farming system extending returns over more than one season and also dealt with the principal problem, weed control. Other systems could be developed elsewhere, as in the 'fruit orchards' of Sarawak described by **Poore**, or the traditional practices described by **Wasawo** for Malawi and Kenya. He agreed entirely with the notion that agroforesters could learn much from such practices but he also observed that the Ghanaian method showed that it was still possible to improve on older systems.

Lastly, he commented that he thought the contemporary, fuel-wood crisis had been somewhat exaggerated, although, of course, it was a real problem. The actual needs for cooking were relatively small in forested areas and small-scale plantations for this purpose usually solved the problem. A more difficult problem was the needs of the urban poor who often saw provision of firewood as a means of livelihood, despite the distances which often had to be travelled for its collection. That practice also led to total deforestation spreading out and around cities, which was not desirable. The fuel problem was exacerbated because the charcoal and other wood-burning stoves in use were extremely inefficient. A considerable saving could be effected by improving their design. However, in the short term the provision of controlled 'fuel plantations' in the semi-urban penumbra would probably meet existing needs, but he did not think that solution would meet the future needs implicit in the projected growth of urban connurbations in Asia, Africa and Latin America. There were perhaps two decades left in which to try to solve the problem.

Poore agreed with **Burnett**'s comments on the fuel-wood situation, particularly in the urban situation. City fuel-lots had, in some areas, increased urban poverty because the plantings had been made on what had, formerly, been common land, thereby reducing the possibility of its use for grazing or other purposes. In the rural situation, he added, considerable use was made of agricultural residues as well as wood for fuel and, conversely, trees planted for fuel-wood could also be used for other things such as shade and fodder.

Simon added the comment that one of the major problems with peri-urban wood lots was that they were frequently by-passed in favour of the collection of 'free' wood from the forest. That raised the problem of what were 'free' resources, a problem he hoped would be addressed by the economists later in the Conference.

He then pointed out that the discussion had centred almost exclusively on the problems of rural areas but, in fact, the fastest growing use of land, in the Third World certainly, was a whole set of urban uses. It was not only a question of absolute scale but of the relative rate of change in the sequence, closed forest to open forest, to urban uses and loss of forest, and the spread of urban areas into previously highly productive agricultural areas. This was one of the fundamental sets of issues and problems which would have to be tackled seriously by every country in the Third World and, indeed, beyond. Europe and North America had somewhat different problems but, in principle, the issues were pretty well universal.

Kefeli, finally, reminded the Conference of the basic integration of plant and soil which provided the basis for all biological productivity and hence the use of land for crops, stock, or timber. In Pushchino there was a small soils museum; some samples were of almost completely mineralized soil, others from adjacent sites had been associated with plants for a long time and had developed a high humus content. Humus reflected the biological product of interaction between the atmosphere, through the plant via photosynthesis, and the soil, which provided nutrients – a fantastic range of autotrophic and chemoautotrophic reactions.

A 1m profile of one of these highly humified black soils – a chernozem – had been sent from Russia to Paris at the turn of the century but had been destroyed during the student unrest in 1968. It had not proved possible for the Russian Academy to replace it because the chernozems of the Ukraine had suffered enormous degradation. That typified the problem of soil regeneration. Much of that loss reflected the demands made by government to produce higher and higher crop-yields resulting in the loss of soil fertility. His Director, Professor **Kovda**, who had been unable to attend the Conference because of illness, had studied this problem. Two approaches were possible, either the study of the polymeric biosyntheses of humus and soil, both *in vitro* and *in vivo*, or a reduction of intensive production to allow the soils to reconstruct themselves. Some compensation could be provided by developing industrial photo-biochemical systems with autotrophs such as mass production of algae. In the USSR they were also concerned with the problems of soil pollution by heavy metals and pesticides which depressed soil microbiological activity and hence humus development. Heavy-metals and acid rain – industrial pollution – also affected photosystem II of photosynthesis adversely. Thus pollution affected the whole of the atmosphere-plant-soil interactive processes which resulted in productivity, and so, ultimately, problems of land-use.

Annexe 2: Environmental Problems of Soils and Land-use in Hungary*

GYÖRGY VÁRALLYAY

Professor of Soil Science, Director, Research Institute for Soil Science and Agricultural Chemistry, Hungarian Academy of Sciences, H-1022 Budapest, Herman Ottó ut 15, Hungary

Soil – as reactor, transformer and integrator of the combined influences of other natural resources – is the most important 'life-medium' for microbial activities, for natural vegetation and cultivated crops. Soils are the most important means of production in agriculture and the most significant – conditionally renewable – natural resource of Hungary. Consequently, their rational utilization has particular significance for the Hungarian national economy and their conservation is an important task for biospheral conservation, environmental protection, and the prevention and control of various environmental hazards (Várallyay, 1989*b*).

LIMITING FACTORS OF SOIL FERTILITY IN HUNGARY

The *main factors limiting soil fertility* in Hungary are the extremely coarse texture, acidity, salinity-alkalinity, extremely heavy texture, waterlogging, water- and wind-erosion, and the shallow depth of soils. A simplified distribution map of these factors is shown in Figure 4*a*.1, and their territorial limits are summarized in Table 4*a*.1. (Szabolcs & Várallyay, 1978).

SOIL DEGRADATION PROCESSES

Soil degradation is usually a complex process in which several features of soil deterioration can be recognized to be contributing to the loss of land or its 'productive capacity', to the limitations of normal soil functions, and/or to the decrease of soil fertility due to unfavourable changes in soil processes and, consequently, in soil properties (FAO, 1983; Várallyay, 1989*c*).

In Hungary the large extension of various undesirable soil degradation processes represent serious environmental problems, both in the directly affected area and in its surroundings (Hinrichsen & Enyedi, 1990).

The *main soil degradation processes in Hungary* are as follows (Várallyay, 1989*c*):

*[Further discussion of soil factors was not possible but this Annexe sets out in some detail the kinds of environmental problem which will need to be faced by countries of eastern Europe to improve land-use. Eds.]

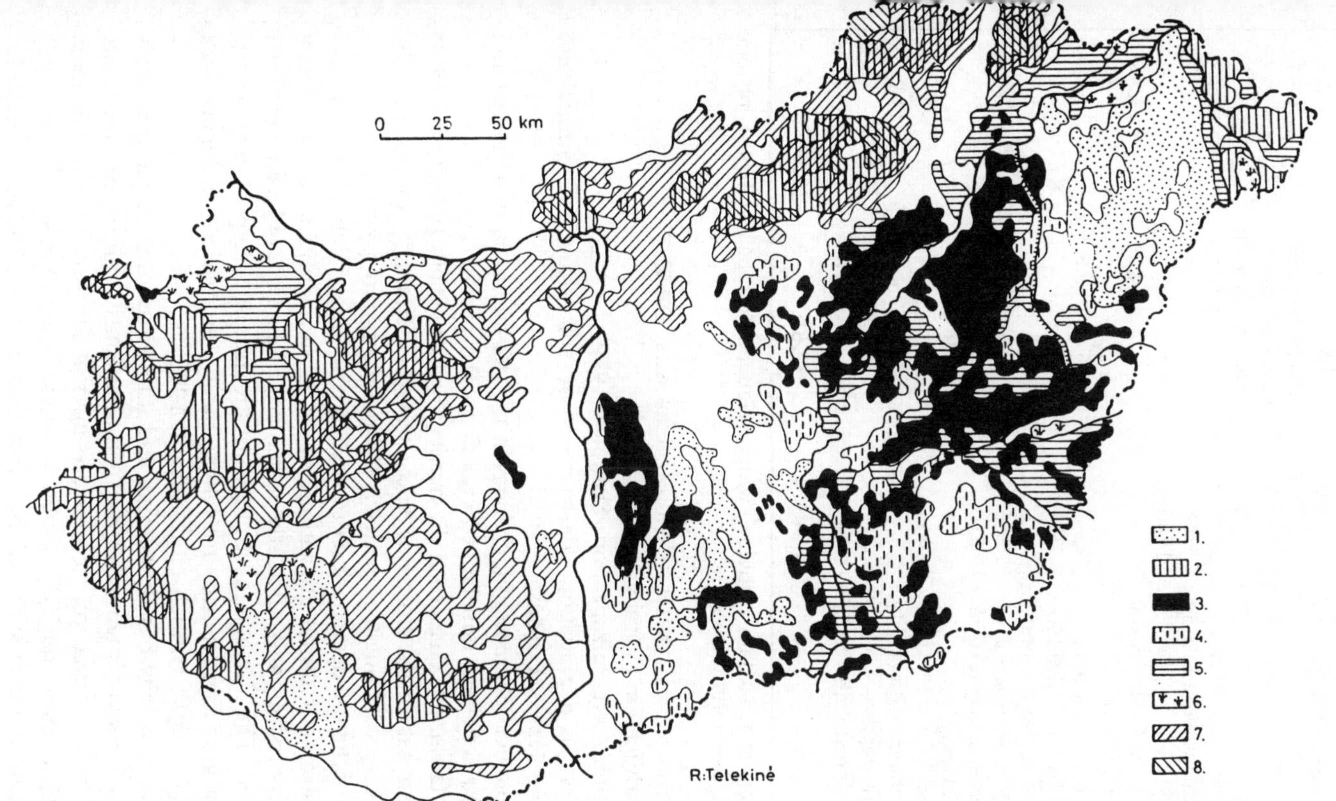

Figure 4a.1. Map of the limiting factors of soil fertility in Hungary.
Original scale: 1: 500,000. Main limiting factors of soil fertility. 1. Extremely coarse texture; 2. Acidity; 3. Salinity and/or alkalinity; 4. Salinity and/or alkalinity in the deeper layers; 5. Extremely heavy texture; 6. Water-logging; 7. Erosion; 8. Shallow depth.

Table 4a 1. Limiting Factors of Soil Fertility in Hungary.
(Based on the 1: 500 000 scale map)

Limiting factors of soil fertility	*Area 1000 ha*		*Area – as percentage of the total agriculture and forestry area*		*Area – as percentage of the total area*	
1. Extremely coarse texture	746		8.9		8.0	
2. Acidity	1200		14.3		12.8	
a) combined with erosion		348		4.2		3.7
b) combined with solid rock near to the surface		67		0.8		0.7
3. Salinity and/or alkalinity	757		9.0		8.1	
4. Salinity and/or alkalinity in the deeper layers	245		2.9		2.6	
5. Extremely heavy texture	630		7.5		6.8	
6. Waterlogging	161		1.9		1.7	
7. Erosion	1455		17.4		15	
c) combined with acidity		348		4.2		3.7
8. Shallow depth	217		2.6		2.3	
c) combined with acidity		67		0.8		0.7
Total	**4996***		**59.6***		**53.5***	

*In the case of soil acidity combined with erosion, or with shallow depth, only one factor was taken into account.

1. Soil erosion by water or wind (Stefanovits, 1964);
2. Acidification (Várallyay *et al.*, 1989);
3. Salinization – alkalization (Szabolcs, 1979);
4. Physical degradation: destruction of soil structure, compaction (Várallyay & Leszták, 1990);
5. Biological degradation: changes in the soil microbial populations and their activities, unbalanced organic matter regime.
6. Unfavourable changes in the nutrient regime of soils (Várallyay, 1985);
7. Decrease of the buffering and 'stress-moderating' capacity of soils, leading to drought-sensitivity, soil pollution and toxicity.

Most of the soil degradation processes are not unavoidable consequences of intensive agricultural utilization and social development, but obviously can be successfully and efficiently prevented, eliminated, reduced, or at least moderated. The Hungarian 'strategy' for the necessary step-by-step actions for soil degradation control are schematically summarized in Figure 4*a*.2 (Várallyay, 1989*c*; Hinrichsen & Enyedi, 1990).

The scientifically-based planning and implementation of rational land-use and soil management, ensuring, simultaneously, normal soil functions and the maintenance of, or increase in, soil fertility, require adequate information about the soil: exact, accurate, quantitative data on well-defined soil characteristics and their variability in space and time (Arnold *et al.*, 1990). A large amount of such information is available for Hungary as

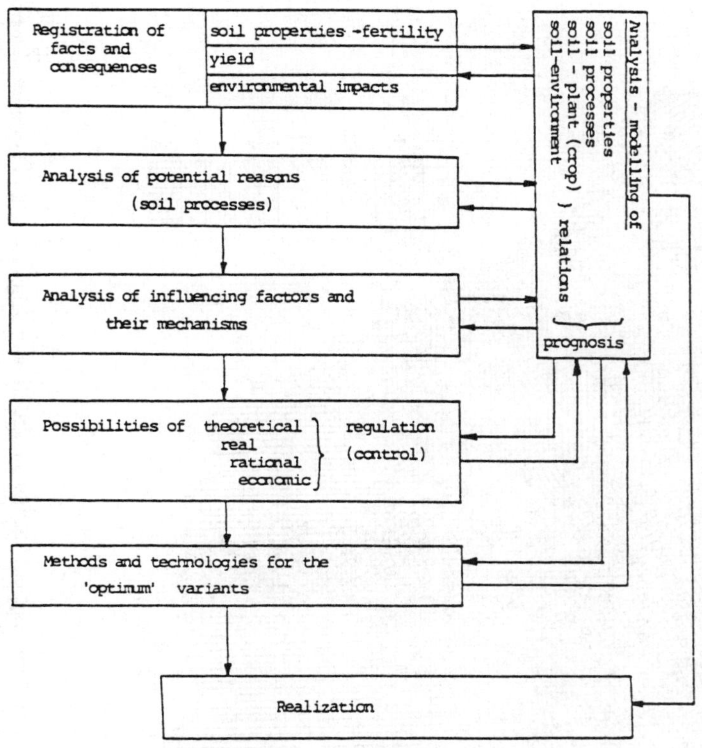

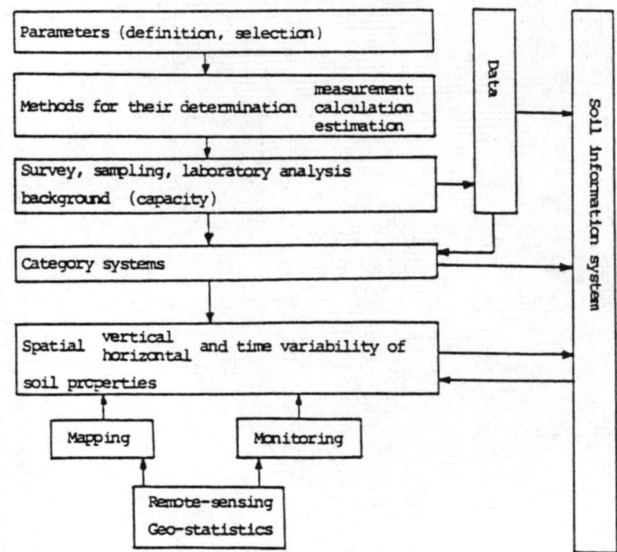

Figure 4a.2. Conceptual model for the control of soil degradation processes.

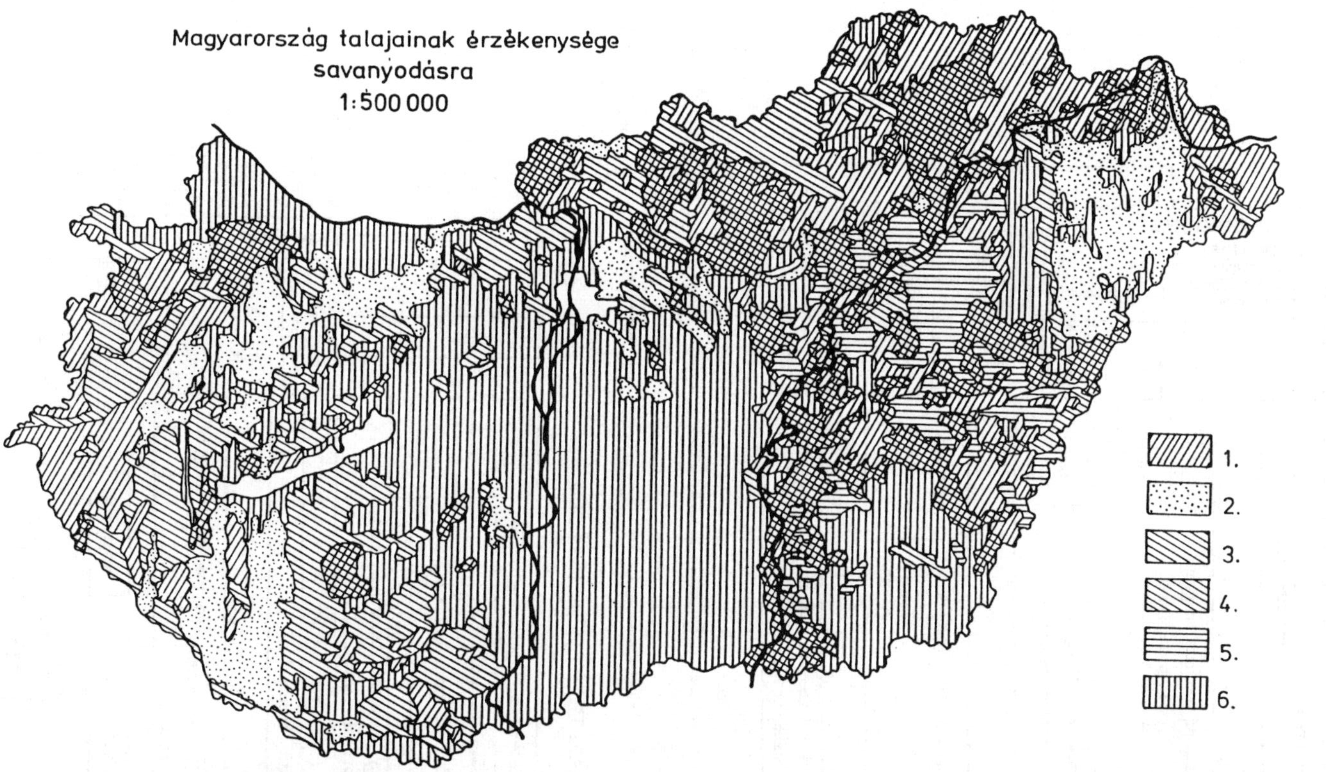

Figure 4a.3. Map of the susceptibility of Hungarian soils to acidification.
1. strongly acidic soils (13.0%); 2. highly susceptible soils due to their low buffering capacity/slightly acidic soils with light texture and low organic matter content (14%); 3. susceptible soils due to their medium buffering capacity/slightly acidic soils with medium texture and organic matter content (5.0%); 4. moderately susceptible soils with heavy texture and/or high organic matter content (23.0%); 5. slightly susceptible soils/salt affected soils non-calcareous from the surface (4.0%); 6. non-susceptible soils/calcareous from the surface (41.0%).

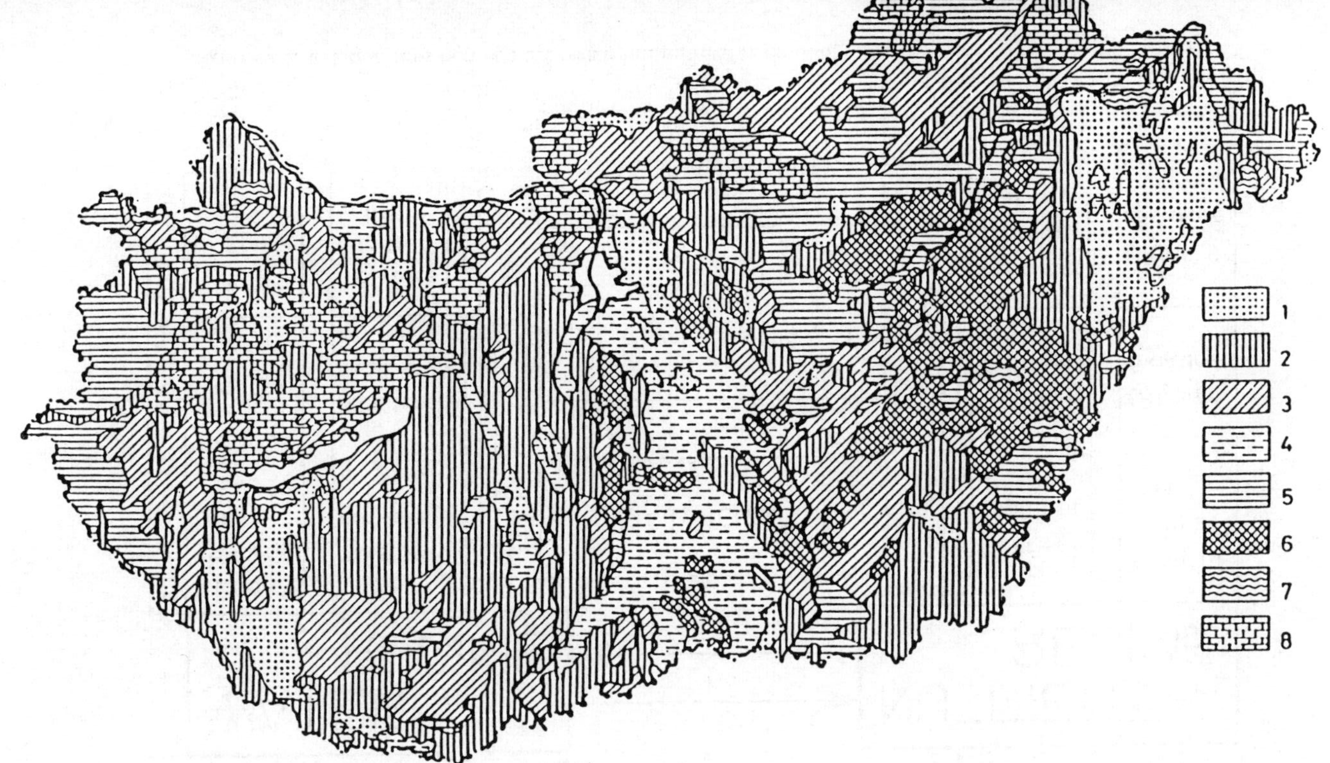

Figure 4a.4. Map of the susceptibility of Hungarian soils to physical degradation.
1. non-susceptible soils/sandy soils without structure and with a low content of cementing compounds as carbonates or sesquioxides (10.5%); 2. slightly susceptible soils/medium-textured soils with well-developed structure and high aggregate stability (23.0%); 3. moderately susceptible soils/medium-textured soils with moderately developed structures and low aggregate stability (17.8%); 4. soils susceptible to compaction and surface crusting but not to structural damage/sandy soils without structure but high amount of cementing compounds, mainly carbonates (11.0%); 5. soils susceptible to structural damage and compaction/heavy-textured soils of swelling-shrinkage character and low structural stability (12.9%); 6. soils susceptible to both structural damage and compaction due to salinity-alkalinity (9.6%); 7. organic soils/peats (5.7%); 8. shallow soils/solid rock or cemented layer near the surface (9.5%).

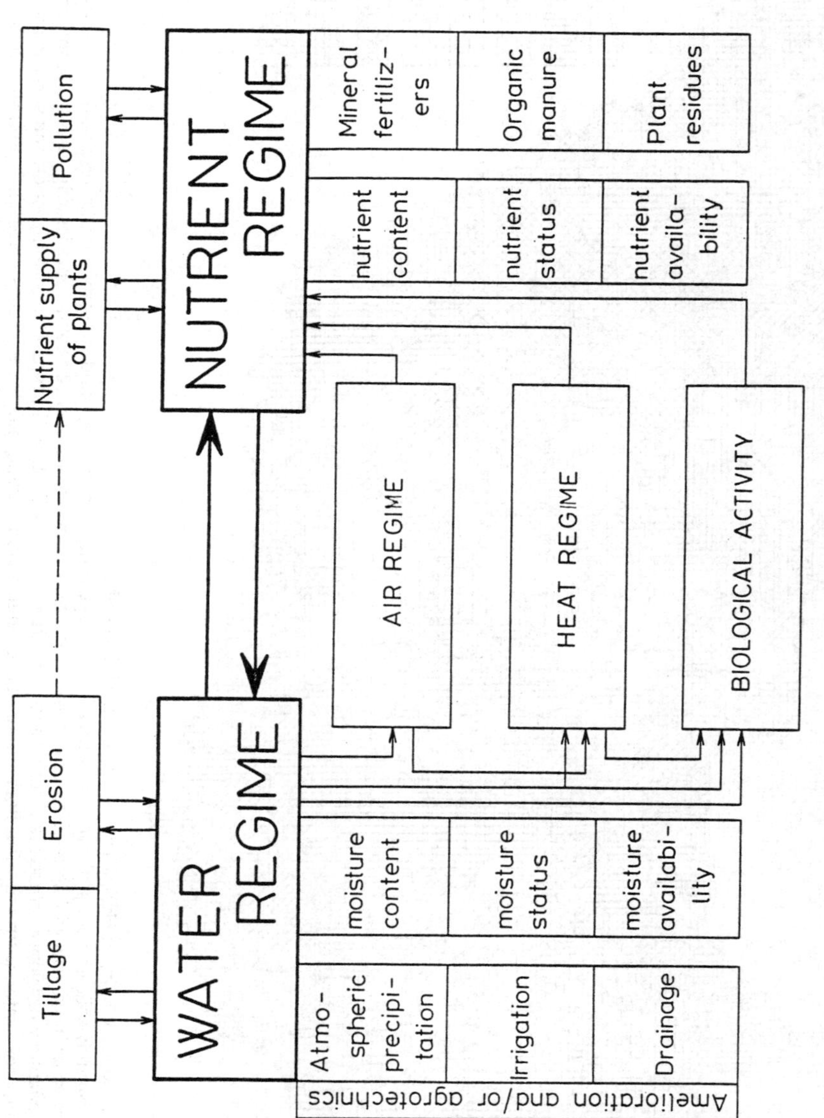

Figure 4a.5. Relationships between the water and nutrient regimes of soils.

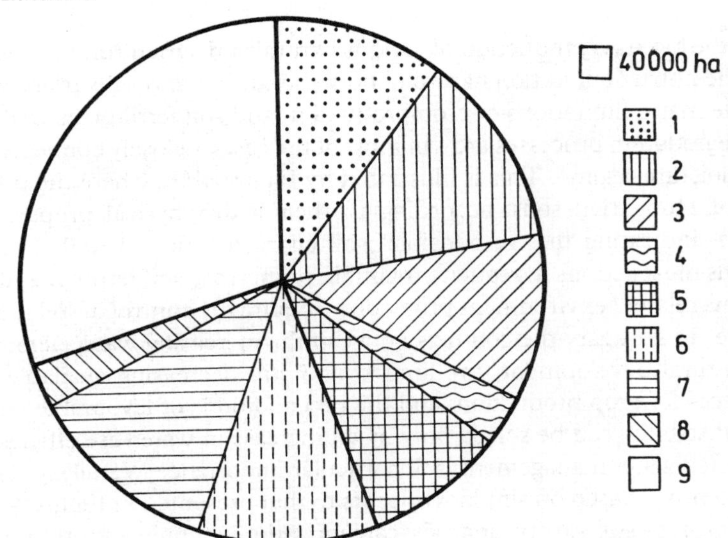

Figure 4a.6. Distribution of soils in Hungary with good, moderate, and unfavourable hydrophysical properties.
1-5. Soils with unfavourable hydrophysical properties: 1. due to very coarse texture; 2. due to very heavy texture; 3. due to strong salinity-alkalinity; 4. due to peat formation; 5. due to shallow depth. 6-8. Soils with moderately unfavourable hydrophysical properties: 6. due to coarse texture; 7. due to heavy texture; 8. due to moderate salinity-alkalinity in the deeper layers. 9. Soils with good hydrophysical properties.

a result of various soil surveys, analyses and mapping on various scales carried out during the last 50 years (Várallyay, 1989*b*). In recent years all the existing data were organized into a computerized geographical information system HunSIS: *Hungarian Soil Information System* (Csillag *et al.*, 1988).

In addition to the database (descriptions, data, maps) on the present status of soil properties and soil processes, category systems were established and maps were prepared on the *susceptibility of soils* to water and wind erosion (Stefanovits, 1964), to salinization/alkalization (Szabolcs, 1979; Szabolcs *et al.*, 1969), to acidification (Várallyay *et al.*, 1979) and to structural destruction and compaction (Várallyay & Lesztâk, 1990). The simplified version of the last two maps (prepared on a scale of 1: 100,000) are presented, as examples, in Figures 4*a*.3 and 4*a*.4, giving the territorial extension of the categories distinguished (as a percentage of the total area of Hungary) as well.

Soil moisture regime as the factor of soil fertility

Soil moisture regime has particular significance in both soil formation and soil degradation processes. It determines the dynamic ratios between solid, liquid and gaseous phases. It influences the air and heat regimes, biological activity, and plant nutrient status of the soil. The quantity, state and movement of soil moisture determine not only the water supply of plants, but regulate their nutrient uptake and metabolism as well (Figure 4*a*. 5). In most of the cases, soil moisture regime strongly influences, sometimes determines, the ecological potential and agricultural productivity of a given area, the biomass production of various natural and agricultural ecosystems, and the nutrient pollution hazard of both the surface and subsurface waters.

The major limitations of crop production and soil fertility, as well as the soil degradation processes are – in most of the cases – closely connected with soil moisture regime. This is illustrated by Figure 4*a*.6, where the distribution of Hungarian soils, according to their hydrophysical properties are shown, indicating the 'responsible' soil characteristics as well. For these reasons most actions to regulate/maintain or increase soil fertility, and those for 'soil-related' environment protection, require the control of soil moisture regime. In Hungary the conflicts between the increasing water demand for agricultural development, the limited and still decreasing available water resources for crop production, and the preconditions of favourable environmental stability can be solved only by an increase in water-use efficiency, in which soil water management is of particular importance (Várallyay, 1989a).

Recently – based on similar concepts to that presented in Figure 4*a*.2 – a *comprehensive* soil survey–analysis–categorization–mapping–monitoring *system* was developed for the exact *characterization of hydrophysical properties*, modelling *and* forecasting of *water and solute regimes of soils*. (Várallyay, 1985; 1989*a*, Várallyay *et al.*, 1980). The system represents an exact scientific basis for soil moisture control and it has been used efficiently for practical soil water management in Hungary. Its potential elements are, at the same time, effective measures for environmental control, as are summarized in Table 4*a*.2.

Environmental aspects of land use and soil management

Changes in Hungary's *land use* and cropping patterns between 1950 and 1988 are shown in Table 4*a*.3. As can be seen, the following changes are evident:

a. The amount of uncultivated land (e.g. settlements, roads, railways, open-pit-mines, industrial enterprises, etc.) increased from 7.8% to 11.4%. It means the loss of about 335,000 hectares of agricultural land, an area which is equal to the size of one small Hungarian county (administrative region).

b. The loss of arable land amounted to nearly 790,000 hectares, dropping from 59.3% to 50.8%. Vineyards also decreased from 2.5% to 1.5% (by about 90,000 hectares).

c. Grasslands diminished from 15.9% to 13.0% amounting to 240,000 hectares. Two factors contributed to this change: more land put to the plough, and an increase in the area of grasslands converted to fish-ponds, reed-beds, goose and duck farms, etc.

d. There has been a considerable increase in the amount of land used for gardens (1.0→3.6%), orchards (0.6→1.0%) and forests (12.5→18.0%!) Afforestation was carried out mainly for recreation and for environment and landscape protection. It was the only, but considerable, positive change in the land-use pattern.

e. The *cropping pattern* indicates stability (Table 4*a*.3) with the exception of a radical reduction in potato production (4.3→0.9%) and a slight decrease in barley (8.6→5.5%) on the one hand, and a sharp increase in

Table 4*a*. 2 Elements and Methods of Soil Moisture Control with their Environmental Impacts

Elements		*Methods*	*Environmental impacts* Favourable	Unfavourable
Surface runoff	Moderation of these elements	Increase of the duration of infiltration (moderation of slopes: terracing, contour ploughing; establishment of permanent and dense vegetation cover; tillage); improvement of infiltration; soil conservation farming system	1, 1a, 5a, 8	–
Evaporation		Help infiltration (tillage, deep loosening); prevention of run-off and seepage, water accumulation	2, 4	–
Feeding of ground-water by filtration losses		Increase of the water storage capacity of soil; moderation of cracking (soil reclamation); surface and subsurface water regulation	5b, 7	–
Rise of water		Minimization of filtration losses (↑); groundwater regulation	2, 3, 5b, 5c	–
Infiltration	Increase of	Minimization of surface runoff (tillage practices, deep loosening) (↑)	1, 4, 5a, 7	–
Storage in the soil in available form		Increase of the water retention of soil; adequate cropping pattern (crop selection)	4, 5b, 6, 7	–
Irrigation		Irrigation; groundwater table regulation	4, 5c, 7	9, 10
Surface / Subsurface } drainage		Surface / Subsurface } drainage	1, 2, 3, 5c, 6, 7	11

Environmental Impacts

Favourable

Prevention, elimination, limitation or moderation of the following phenomena:

1 Water erosion
1a Sedimentation
2 Secondary salinization, alkalization
3 Peat formation, waterlogging, overmoistening
4 Drought sensitivity, cracking
5a Surface runoff (→surface waters → eutrophication) of plant nutrients
5b Leaching (→subsurface waters) of plant nutr.
5c Immobilization of plant nutrients
6 Formation of phytotoxic compounds
7 'Biological degradation'
8 Flood hazard

Unfavourable

9 Overmoistening; waterlogging; peat and swamp formation; secondary salinization/alkalization
10 Leaching of plant nutrients
11 Drought sensitivity

Table 4a.3 Land-use and cropping patterns in Hungary

Land-use	1950	1988	Main Crops	1951–60	1988
Arable land	59.3	50.8	Wheat and rye	32.2	30.4
Gardens	1.0	3.6	Barley	8.6	5.5
Orchards	0.6	1.0	Maize	23.4	24.2
Vineyards	2.5	1.5	Pulses	1.4	2.2
Grasslands	15.9	13.0	Sugar beet	2.1	2.2
Forests	12.5	18.0	Sunflower	2.8	8.4
Reeds	0.3	0.4	Potatoes	4.3	0.9
Fish-ponds	0.1	0.3	Silage maize	1.7	6.2
Uncultivated areas	7.8	11.4	Alfalfa	4.6	6.6
			Vegetables	2.1	2.2
			Other crops	16.8	11.2
	100	100		100	100

Expressed as percentage of the total 93,032 km² surface of Hungary — Expressed as percentage of the total arable land

sunflowers (2.8→8.4%) and silage maize production (1.7→6.2%) on the other.

The greater proportion of the arable land (14.5 + 81.0%), forests (68.9 + 30.6%), grasslands (17.8 + 77.7%) and even the orchards (24.4 + 58.9%) and vineyards (14.7 + 54.5%) were owned by state farms and used by cooperative farms. These were characterized by their large size, 5,000 and 3,500 hectares, on average, respectively; their *large-scale, high-input agricultural production*; their irrationally large, consequently heterogeneous agricultural fields with a size of 50-70, sometimes 100-120, hectares; the irrational, lack of, or slightly flexible, land-use and cropping pattern; by monocultures, mechanization with heavy machinery, a high rate of application of chemicals as mineral fertilizers, pesticides, growth-regulators etc.; and misguided soil management – improper tillage operations, fertilization and irrigation practices, etc. – which resulted in many cases of environmental side-effects, such as various soil degradation processes, soil toxicity, pollution of surface and subsurface waters, and so on.

In the smaller fields of the private sector, subsidiary farms or household farm-plots of the cooperatives, greater opportunity was provided for a more rational, microscale utilization of the land with higher flexibility. But here the usually uncontrolled use of pesticides and fertilizers often led to overdosage, causing soil and water pollution problems (Hinrichsen & Enyedi, 1990).

Realizing these facts and the high variability in spatial and temporal yields of the main crops, a National Programme was initiated by the Hungarian Academy of Sciences in 1979 for '*The Assessment of the Agro Ecological Potential of Hungary*' (Láng *et al.*, 1983). On the basis of a comprehensive and precise multidisciplinary analysis of natural factors, their territorial distribution, temporal variabilities, and yield potentials, the final conclusion was drawn that, total crop production in Hungary could be increased by 60-70% without any additional inputs, simply by rational land-use, i.e. the territorial co-ordination of agro-ecological conditions, land-site characteristics, and the

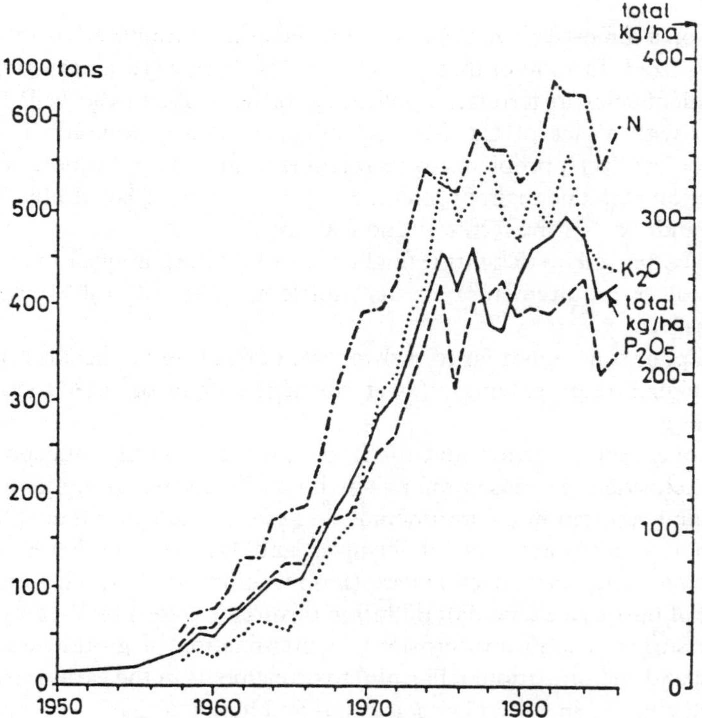

Figure 4a.7. Use of mineral fertilizers in Hungary.

ecological requirements of the cultivated crops. In spite of the fact that the results, conclusions and recommendations were presented to the policy-makers and the tasks formulated in numerous official documents, little initiative was taken to implement them, and the lack of flexibility in the state-controlled economy regulators (prices, subsidies, credits, taxes, etc.) did not stimulate rational land-use practices. On the contrary, it sometimes prevented it, e.g. the economic 'pressure' for maize production in the cooler and more humid NE hilly lands instead of soil-protective grassland farming gave rise to low yields with high costs and a high risk of water erosion.

Environmental aspects of fertilizer application

Before World War II, the plant nutrient status of Hungarian soils was rather poor, due to the negative nutrient balance: more nutrients were taken up by the cultivated crops and taken away from a given territory as yield, or biomass, than were put back in the form of organic and green manures, or fertilizers.

From 1955, as can be seen in Figure 4*a*.7, there was a rapid increase in fertilizer application. This tendency was one of the reasons for the substantial yield increase during the same period. Another consequence of it was that, due to the positive nutrient balance, the nutrient status of Hungarian soils was significantly improved. In the early 1970s well-equipped agrochemical laboratories were established in each county, a regular soil-test system, on a three-year cycle, was introduced and a national advisory service was organized. In spite of these developments there were serious problems and inadequacies in fertilizer application technology, in the N-P-K ratio together with a lack of Ca, Mg and micronutrient supply and a limited variety of fertilizers; problems in their storage, time of application, mode of distribution etc. The main problem, however, was an unfavourable 'polarization-tendency' in the fertilizer application:

(a) In better soils→ rich farm→higher rate of fertilizer application (in spite of the lower requirements)→better nutrient status of soils→net effect: overdosage;

(b) in poor soils→poor farms→lower rate of fertilizer application (in spite of the higher requirements)→lower nutrient supply of soils→net effect: underdosage.

The over-generalization and the imperative 'maximum-concept' led to false conclusions, decreased the effectivity and efficiency of mineral fertilization, and resulted in environmental side-effects such as soil acidification (due to non-adequate type of fertilizer and lack of simultaneous lime application) and its consequences (mobilization of toxic elements and fixation of nutritive elements); pollution of surface waters by P-compounds (due to surface runoff and erosion); contamination of ground-waters by nitrates; and accumulation of harmful toxic elements in the various stages of the 'food chain' – in soils, plants, animals and human organs – according to their solubility, mobility, and availability (Várallyay, 1990).

These side-effects, however, are not inevitable consequences of fertilizer application. They can be prevented by the *rational* use of fertilizers through taking into consideration the nutrient requirements of cultivated crops, the nutrient status and other properties of soils, as well as the climatic and hydrologic conditions of the site (Sarkadi & Várallyay, 1989).

Alternative agriculture and future trends

In recent years there has been a considerable change in the concept of Hungarian agricultural development. Instead of the global quantity aspect, quality (exportability), efficiency and economy (based on a real and exact cost-benefit evaluation), as well as environmental consequences became more and more important. The rational re-privatization of land and market-oriented production have made potentially possible the establishment of a flexible, environment-friendly, alternative agriculture. It requires the following elements:

1. Territorial coordination of the agro-ecological conditions and the ecological requirements of cultivated crops, taking into consideration both the production and the environmental aspects on short-, mid- and long-term time-scales (rational land-use).

2. Homogenization of agricultural fields through rationalization of the field size, or by local amelioration and/or differential agro-technics.

3. Precise and scientifically-based crop production technology with 5 fundamental elements:

a. adequate cropping pattern and crop rotation;
b. minimalization of 'production' wastes (recycling);
c. improvement of water use efficiency;
d. rational use of fertilizers;
e. rational plant protection systems with minimum use of chemicals.

The rate, direction and technologies of crop production are economy driven. By contrast, the maintenance of soil fertility, the quality of surface, and sub-surface, water resources, and the protection of the natural environment, The Biosphere, are not economy-dependent, but imperative tasks. The effective realization of these tasks must be jointly guaranteed by the State, as well as by farming units and landowners. These are the primary pre-conditions for successful agricultural development and environment protection in Hungary.

References

Arnold, R. W., Szabolcs, I., & Targuljan, V. (Eds.) (1990). *Global Soil Change.* International Institute for Applied System Analysis, Laxenburg, Austria: 110 pp.

Csillag, F., Szabó, J., Várallyay, G., Zilahy, P. & Vargha, M., (1988). Hungarian Soil Information System (TIR): a thematic geographical information system for soil analysis and mapping. *Bull. of the Hung. Nat. Comm. for CODATA,* No. 5, 13 pp.

FAO (1983): *Guidelines for the Control of Soil Degradation.* UNEP-FAO. Rome.

Hinrichsen, Don & Enyedi, Gy. (Eds.). (1990). *State of the Hungarian Environment.* Statistical Publ. House, Budapest, Hungary: 143 pp.

Láng, I., Csete, L. & Harnos, Zs. (1983). [*The Agroecological Potential of Hungarian Agriculture at 2000.*] Mezõgazdasági Kiadó, Budapest, Hungary: 265 pp.

Sarkadi, J. & Várallyay, Gy. (1989). Advisory system for mineral fertilization based on large-scale land-site maps. *Agrokémia és Talajtan* 38, pp. 775–89.

Stefanovits, P. (1964). [*Soil Erosion in Hungary.*] (Hung.) OMMI Genetikus Talajtérképek Kiadványai. Ser. 1. No. 7., Budapest.

Szabolcs, I. (1979). *Review of Research on Salt Affected Soils.* Natural Resources Res., XV. UNESCO, Paris, France: 137 pp.

Szabolcs, I. & Várallyay, Gy. (1978). [Limiting factors of soils fertility in Hungary.] (Hung.). *Agrokémia és Talajtan* 27, pp.181–202.

Szabolcs, I., Darab, K. & Várallyay, Gy., (1969). Methods for the prognosis of salinization and alkalinization due to irrigation in the Hungarian Plain. *Agrokémia és Talajtan,* 18 (Suppl.), pp. 351–76.

Várallyay, Gy. (1985). [Moisture and substance regimes of Hungarian Soils.] (Hung.). *Agrokémis és Talajtan,* 34.

Várallyay, Gy. (1989*a*). Soil water problems in Hungary. *Agrokémis és Talajtan,* 38. pp. 577–95.

Várallyay, Gy. (1989*b*). Soil mapping in Hungary. *Agrokémia és Talajtan,* 38, pp. 696–714.

Várallyay, Gy. (1989*c*). Soil degradation processes and their control in Hungary. *Land Degradation and Rehabilitation,* 1, pp. 171–188.

Várallyay, Gy. & Leszták, M. (1990) Susceptibility of soils to physical degradation in Hungary. *Soil Technology,* 3.

Várallyay, Gy., Rédly, M. & Murányi, A. (1989) Map of the susceptibility of soils to acidification in Hungary. Pp. 79–94 in *Ecological impact of acidification.* Proc. Symp. '*Environmental Threats to Forest and Other Natural Ecosystems*', *Oulu, Finland, No. 1–4., 1988.* Budapest, Hungary.

Várallyay, Gy., Szücs, L., Rajkai, K., Zilahy, P. & Murányi, A. (1980). [Soil water management categories of Hungarian soils and the map of soil water properties 1: 100 000.] (Hung.). *Agrokémia és Talajtan,* 29, pp. 77–112.

5. Uses and Abuses of Ocean Space

EDWARD D. GOLDBERG
Professor of Chemistry, Scripps Institution of Oceanography, University of California at San Diego, La Jolla, California 92903, USA

INTRODUCTION

The renewable resources of the oceans can help immeasurably towards providing an increasing world population with enriched lives. However, renewable resources can become non-renewable if abused. In the following paper I examine the potential uses of ocean space that appear most attractive for improving our human living-standards: these uses are, particularly, mariculture, waste disposal, recreation, and transportation.

Coastal space is actively sought for all of the above and some other activities, and its relative value is comparable rather with that of land in downtown Tokyo. The open ocean space, although often perceived as having a worth more similar to that of the Sahara Desert, still can be employed in one way or another for all of the above pursuits – and also others, such as mining and warfare, which we do not propose to consider here.

MARICULTURE

The ranching and farming of fish, shellfish, and Algae, are ventures that are expanding in both the developing and developed world. Mariculture has been so effective that farmed shrimp and prawns now account for 26% of the world market (WSF, 1990). As a consequence, the price has dropped substantially. In many countries, Salmon (especially *Salmo salar*) grown in pens now exceed in weight those caught in the wild (Isaksson, 1988). As a result of such successful mariculture, both shrimpers and salmon fishermen may become 'endangered species'.

In the face of all the success of maricultural activities, however, many scientists are apprehensive and engender concern about the loss of marine coastal quality. Thus finfish and shellfish farms are producers of huge amounts of organic wastes arising from uneaten food and metabolic waste-products. For example, a 0.816 ha (two acres) salmon farm containing from 50,000 to 100,000 fish produces 100 tonnes of faeces and uneaten food annually (Barinaga, 1990). As a consequence, the environs can become anoxic. For example, Hong Kong has degraded environments under mariculture rafts that occupy waters in what were formerly some of the cleanest and most beautiful areas (Morton, 1989). This situation has disturbed the Hong Kong Yachting Association, which represents conflicting user communities who would much prefer to see the space dedicated to surfing, canoeing, wind-surfing, motor-boating, swimming, sailing, snorkelling, and SCUBA diving. Mariculture accounts for some 50% of the total live fish consumption in Hong Kong and has an annual value of US $25 millions.

To alleviate the problem which maricultural wastes pose on environmental quality, the siting of cages for farming marine organisms in areas of strong tidal flows has been suggested (Frid & Mercer, 1989). Generally, sea cages have been placed in sheltered zones having only low tidal forces. But where the strategy of placing them in areas of strong tidal forces has been adopted, healthy communities of benthic organisms have prospered downstream from the activity.

Another problem involves the extensive use of antibiotics. Marine mariculture is a form of monoculture and, as such, is subject to crop failures due to the invasion of predators. The 1988 Black Tiger-shrimp (*Penaeus monodon*) devastation in Taiwan may have been a consequence of such a phenomenon. The inhibition of such disasters is sought with therapeutants that include the powerful antibiotics tetracycline and streptomycin. There are few investigations concerned with their effects on non-target organisms, but the fear that tragedies are likely haunts many naturalists.

Then there is the problem of competition for food with natural populations, especially among ranched organisms. An example involves the predation by salmon after their introduction to the ocean system from rivers or estuaries. Fish-ranchers clearly wish to use fully the carrying capacity of the oceans for the benefit of their products, *e.g.* by taking advantage of the availability of natural foods. It is difficult to ascertain whether that capacity is being exceeded or not. Increasing salmon releases by ranchers to the Columbia River are associated with declining adult returns (Isaksson, 1988). Moreover, the Atlantic Ocean may be supporting a smaller Salmon (*Salmo salar*) population today than it did in the 1700s.

Also, there is the worry about habitat loss. The depletion of mangrove forests, a valuable but declining marine ecocomplex (concerning which *see* the note in *Environmental Conservation,* 17(3), p. 274, 1990), is associated with 'shrimp' farming in the southeastern Pacific (Escobar, 1988). In Panama, there is an apparent areal loss of 1% of mangroves annually. In Ecuador, 60,000 of the 177,000 hectares of mangroves have already been turned over to salt-water 'shrimp' farming.

Finally, there is the unusual consequence of maricultural activities through the introduction of alien species, especially, of Algae, by the importation of bivalves from outside of the maricultural area (Rueness, 1989). For example, the continued spread in European waters of the introduced brown seaweed *Sargassum muticum,* came about from the importation to France of the Japanese Oyster (*Crassostrea gigas*). The ecological impacts of such transplantations are difficult to predict. Some alien species, although widespread, have little or no impact upon the indigenous organisms. Alternatively, other alien species can displace natives, with potential impacts upon higher levels of the food-web.

Waste Disposal

The period extending from the mid-1950s to the mid-1970s saw marine scientists and engineers pursuing studies of how best to protect the oceans from pollution which could affect public health and the integrity of communities of marine organisms (Goldberg, 1976). The initial entry of concerned scientists was [*e.g.* Goldberg *et al.*, 1978 – Eds.] made when recognition that promiscuous releases of artificial radionuclides from nuclear-energy facilities could result in unacceptable human exposures. Then the extensive uses of DDT and other pesticides, which were destroying non-target organisms, became of great concern. These pursuits were taken up by the lay public, who constituted themselves into environmental groups and pushed forward their own perceptions as to how to protect the oceans' systems.

One of the consequences of the lay public's activities has been a nearly complete ban on all ocean disposal. It was initiated, *inter alia*, by national and international regulatory agencies. Wastes were not to be put into the oceans – only disposed of on land and to a lesser extent released into the atmosphere by combustion and incineration. As a consequence, terrestrial resources were threatened and in some cases degraded There is a continuous loss of potable underground waters (which constitute some 50% of the US supply) through contamination by entry of hazardous wastes of one type or another. But the critical question is whether or not the cessation of ocean disposal is a rational course to protect the entire environment. I submit that the answer is *No*.

Scholars who carry out multi-media assessments with respect to waste disposal conclude that, on the basis of science, engineering, sociology, and economics, there can be made a rational decision as to which of the options, land *versus* sea, is more reasonable for a given waste in a given area (NAS, 1984). Even though the last of the US ocean dump-sites (No. 106, 106 miles [*c.* 170 km] off the coast of New Jersey) successfully accommodated both industrial and domestic wastes for decades without unacceptable degradation of the environment, it is to be closed down by an act of the US Congress in the early 1990s. The limiting factor is the effect on the biota of surface waters. The disposal of industrial wastes created no significant changes in the populations of the indigenous organisms, especially of the abundant zooplankton. Sub-lethal effects were noted in the waste plumes at the site, but appear to involve only a very small percentage of the organisms in the water-mass. At the rates of dumping in the 1980s, long-term consequences were judged minimal (Capuzzo & Lancaster, 1985).

Going one step further, however, the disposal of toxic wastes in the oceans is prohibited by the London Dumping Convention (Bruce, 1986), and the thought of such discharges clearly horrifies world citizenry. Yet disposal of 'subsealed' high-level radioactive waste has been seriously proposed by eminent scientists (Hollister *et al.*, 1981). The fine-grained sediment deposits of the deep-sea floor have been suggested as sites to accommodate those highly-toxic substances. The areas of such deposits cover about 20% of the Earth's surface. The radioactive waste would be placed in a chemically-stable, solid form, encased in canisters, and introduced to the sediments either through gravitational or by explosive boosting systems that would be activated by a sonic sensor when the wastes were just above the sediment/water interface. The great difficulties in identifying land sites may direct world citizenry to consider the oceans as a repository for some of the ever-increasing amounts of high-level radioactive wastes.

Even hazardous wastes have been successfully disposed of in ocean waters. The classic case involved the ocean disposition in 1970 of around 67 tonnes of nerve-agents loaded aboard an obsolete World War II Liberty ship (Linnenbom, 1971). The vessel was towed to a site 400 km east of Cape Canaveral, Florida, and sunk in 5,000 m of water. The ship did not break up upon hitting the bottom and there was no evidence that marine biota were subsequently affected by any leaking of toxic agents.

Still, some waste-disposal practices may be jeopardizing ocean resources in unacceptable ways, for example by the continued entries of domestic and agricultural wastes with nutrients such as phosphate and nitrate which are necessary for plant growth. Perhaps one of the most serious problems of the coastal ocean and estuaries is the trend towards more and more eutrophication as a consequence of the introduction of increasing amounts of the biostimulants by industrial, social (detergents), and agricultural, practices. It appears that, on a global basis over the past decades, there has been a 15-fold increase of nitrate and a 5-fold increase of phosphate entering the oceans.

As a consequence, there has been an increase in phytoplankton production and an alteration of their communities' species composition. The diatoms, which are at the base of the food-chain for filter-feeding fishes, and zooplankton which largely support some commercial fisheries, are being displaced by the dinoflagellates. These latter organisms are a poor food-substitute for the diatoms. Furthermore, alterations in the community structure can occur. The biostimulants can also enhance the production of macrophytes (*e.g.* sea-grasses and attached Algae).

Furthermore, with increased discharges of industrial and domestic wastes to the coastal waters, and with deforestation and other land-use changes, there has been a significant entry of anthropogenic organic carbon. The increased flux of organic matter to the sediments in some areas is increasing the areas of anoxia. All of these concerns have been observed world-wide in coastal waters and estuaries. But increasing fluxes of materials that have been produced or mobilized by human activities, particularly emphasize the need to control them.

Recreation

Residents and tourists are placing more and more demands upon the resources of coastal areas for recreation as a consequence of increasing amounts of leisure time. Tourism now accounts for around 10% of the world's gross national product. There is a multitude of activities taking place in the coastal zone, some of which, such as water skiing and sailing, require large areas. Furthermore, there is often a desire for fresh fish and seafoods, free from any polluting or preserving substances.

However, the suitability of many coastal areas, especially in the less-developed parts of the world, can be compromised by the entry of enteric microorganisms from domestic waste disposal. Entry to humans can take place either by consumption of seafoods or through exposure on beaches or in waters.

The United Nations Environment Programme (UNEP) has brought together a substantial database that reveals widespread microbial contamination in coastal waters. The association of mortalities and morbidities, through contact in one way or another with these organisms, indicates that this may be the most important marine pollution problem facing world society. It is curious that this concern is only weakly addressed at international marine pollution meetings and by the authors of books on the parent subject.

Residents and tourists look to coastal areas as a source of recreation and rest as well as a place to enjoy seafoods. They would be directed away from such regions if there were any hint that their health might be endangered by frequenting them.

Conflicts in use can also arise from the recreational activities themselves. Perhaps the 'textbook example' of this from recent experience involves the impact of tributyltin (TBT) compounds, used in antifouling paints (applied to the bottoms of recreational craft in 'marinas'), upon maricultured oysters. These anthropogenic compounds are among the most toxic substances that are apt to be introduced into the marine environment. They cause physiological damage to such organisms as various dog-whelks (popular name of nassariid gastropod Mollusca – Eds.) and other marine snails at levels of parts per million millions in coastal waters.

The first TBT episode took place in Arcachon Bay, France, where 10% of the country's oysters are produced. In the late 1970s, the oysters (Pacific Oyster, *Crassostrea gigas*) were producing deformed shells, to the extent of being unmarketable. Healthy oysters, transplanted to the area, suffered a 50% mortality within 130 days. The diffusion of the highly toxic TBT from the marinas to the oyster farms was identified as the cause of the trouble. Subsequently, the use of the antifouling agent was strictly regulated. Within several years, oyster production returned to normal.

Transportation

The rising living-standards anticipated by increasing numbers of world citizens may be achieved by extending intercontinental marine transport. With the varying abilities of countries to produce needed goods – be they agricultural, industrial, medical, etc. – the interdependence of national economies emphasizes the need for ready trade. For those countries producing or requiring high-volume, low-cost commodities, such as coal, oil, timber, and wheat, large-bulk ocean-traversing carriers (150,000 deadweight tonnes or more) appear to be economically reasonable.

At the present time such vessels are primarily involved in the transport of petroleum and petroleum products. They require deep-water ports for berthing (water depths of 55 feet [= 16.7 m] or more), facilities that are often not available. The only Asian ports meeting these specifications are in Japan. Yet it is those deep-water-port-deficient countries that may be in particular need of large quantities of low-bulk imports in the future. The expense of constructing megaports will limit them to very small numbers. The criteria for upgrading existing ports depend upon a number of factors. First of all, they must interface with the land transportation system on which they impinge. Are there adequate rail facilities? Can the associated highways accommodate the increased traffic and accompanying noise? Are there appropriate terminal areas with associated storage facilities?

A sense of the future may be seen in the trade in grains. Present assessments suggest that North America (Canada and the United States) will be major exporters of grains in the future, primarily to Asia. Asia has become a major importing region as a consequence of its small and shrinking cropland area per person, whereas agricultural technologies (including integrated pest-control and improved plant species for crop production) promise to maintain the North American countries as producers. Large-bulk carrying vessels with associated megaports may provide the wherewithal for further trade, and may well become crucial if major famines are to be avoided in a foreseeable future world.

There can be unacceptable environmental impacts as a consequence of dredging operations and the disposal of resultant spoil materials in order to construct and maintain the megaports, with dredging operations appearing to be the more serious. The resultant changes in the geometry of the harbour-to-be can affect the local hydraulic regime, while circulation patterns can be altered with disturbances to the prevailing composition of sea-water. Even the biological productivity of the harbour can be significantly altered.

There are few examples in the literature that warn of the seriousness of toxic materials in dredged spoils. Yet the concerns can be addressed before the sediments are removed, as was done at the Crystal Mountain Workshop (NOAA, 1979) which considered, among others, the problem of the toxic metal cadmium in dredged materials and its consequential transfer to humans through the consumption of shellfish, such as various clams and oysters – known concentrators of the metal. For the worst possible case scenario, oysters living in the spoil region could accumulate cadmium to levels higher than those which were deemed acceptable by the US Food and Drug Administration for human consumption. On the other hand, in some cases, the dredged materials can be considered a resource – for use as construction aggregate, sanitary landfill, beach replenishment, and the creation or enhancement of wetlands.

Thus there is concern that the quality of the coastal environment could be compromised by the creation of deep-water ports. Furthermore, these ports may compete, or otherwise interfere, with other activities such as recreation and mariculture. Perhaps the greatest conflict will involve obtaining appropriate funding for their construction in competition with other supplicants at the public trough.

Overview

More than twenty years ago Garrett Hardin (1968) introduced the paradigm 'the tragedy of the commons' to illustrate the exploitation of the environment. He argued that an increasing population, resulting from the freedom to breed, leads to the loss of renewable resources for human society. He recognized the over-use of the resources of the 'wet commons' – the fresh- and salt-waters – of the world through over-fishing and contamination by chemical and radioactive wastes and heat. His concern for this example of pattern, now widely recognized as classic, can be further extended by instances cited in this presentation. But what general strategies can be put forth to protect the hydrosphere?

Perhaps the allocation of property at sea, comparable with that on land, might be a first step. Bowden (1981) characterizes three types of property: private, public, and common. The first encompasses rights that are given by sovereign authorities (governmental) to individuals or groups. Public property is similar, except that it is held by a public agency. With common property the rights involved are in a natural resource that is held by a class of users whose rights are co-equal.

For example, plots of the coastal zone might be introduced to the private sector for aquaculture. The liabilities and rewards from such property would be similar to those enjoyed by landholders. In salmon ranching, governments or commerical firms release large numbers of smolts (young salmon migrating to the sea for the first time) for subsequent capture in the course of commercial or recreational fishing. The sea is then a common property in which the fish remain until caught. Then they may become private property. Conflicts do arise, however, and we have the situation in which large corporations release smolts and then capture all the adult fish they can. Clearly such enterprises can jeopardize the livelihoods of fishermen, who object to the use of public property for private gain.

There is also concern relating to the use of the oceans with the 200 nautical miles zone (*c.* 371 km) as a waste receptacle. I believe such use will increase with time. Again a common-property resource is being used. Here the resolution of use conflicts may be necessary, especially where waste discharges interfere with aquaculture or recreation.

The tragedy of the land commons can be avoided in the wet commons, where appropriate actions can mitigate over-use and over-abuse. Recognition and avoidance of potential conflicts in mariculture, waste disposal, recreation, and transportation, are crucial to effective resource management – as they will be with other uses of the coasts, seas, and oceans, that will doubtless become important in the future.

Acknowledgements

I am grateful for the travel support furnished by the Soros Foundation, and to readers of earlier drafts of this paper for helpful comments.

References

Barinaga, M. (1990). Fish, money and science in Puget Sound. *Science,* 247, p. 631.

Bowden, G. (1981). *Coastal Aquaculture: Law and Policy.* Westview Press, Boulder, Colorado, USA: xiv + 241pp.

Bruce, M. (1986). The London Dumping Convention, 1972: First decade and future. Pp. 298–316 in *Ocean Yearbook 6,* (E. M. Borgese & N. Ginsberg, Eds.). The University of Chicago Press, Chicago, Illinois, USA.

Capuzzo, J. M. & Lancaster, B. A. (1985). Zooplankton population responses to industrial wastes discharged at Deepwater Dumpsite 106. Pp. 209–29 in *Wastes in the Oceans,* Vol. 5. (D. Kester, W. V. Burt, J. M. Capuzzo, P. K. Park, B. H. Ketchum & I. V. Duedall, Eds.) John Wiley & Sons, New York, NY, USA: 346 pp., illustr.

Escabar, J. (1988). The southeast Pacific. *Siren,* 36, pp. 28–9.

Frid, C. L. J. & Mercer, T. S. (1989). Environmental monitoring of caged fish farming. *Mar. Poll. Bull.*, 20, pp. 379–83, illustr.

Goldberg, E. D. (1976). *The Health of the Oceans.* UNESCO Press, Paris, France: 172 pp., illustr.

Goldberg, E. D., Bowen, V. T., Farrington, J. W., Harvey, G., Martin, J. H., Parker, P. L., Risebrough, R. W., Schneider, E. & Gamble, E. (1978). The Mussel Watch. *Environmental Conservation,* 5(2), pp. 101–25, 10 figs and 12 tables.

Hardin, G. (1968). The tragedy of the commons. *Science,* 162, pp. 1243–8.

Hollister, C. D., Anderson, D. R. & Heath, G. R. (1981). Subseabed disposal of nuclear wastes. *Science,* 213, pp. 1321–6, illustr.

Isaksson, A. (1988). Salmon ranching: a world review. *Aquaculture,* 75, pp. 1–33, illustr.

Linnenbom, V. J. (1971). *U.S. Naval Research Laboratory Memorandum Report 2273.* US Naval Research Laboratory, Washington, DC, USA: 40 pp. illustr.

Morton, D. (1989). Hong Kong's pigs in the sea. *Mar. Poll. Bull.*, 20, pp. 199–200.

NAS [National Academy of Science](1984). *Disposal of Industrial and Domestic Wastes: Land and Sea alternatives.* National Academy Press, Washington, DC, USA: 207 pp., illustr.

NOAA [National Oceanic and Atmospheric Administration] (1979). *Assimilative Capacity of U.S. Coastal Waters.* US Department of Commerce, Environmental Research Laboratories, Boulder, Colorado, USA: 284 pp., illustr.

Rueness, J. (1989). *Saragassum muticum* and other introduced macroalgae: biological pollution of European coasts. *Mar. Poll. Bull.*, 20, pp. 173–6, illustr.

WSF (1990). *World Shrimp Farming 1989.* Published by Aquaculture Digest, 9434 Kearny Mesa Road, San Diego, California, USA.

6. Uses and Abuses of the Sea: Fishing

JOHN A. GULLAND*

Renewable Resources Assessment Group, Imperial College of Science & Technology, 8 Princes Gardens, London SW7 1NA, London, England, UK

INTRODUCTION

Fishing is Man's oldest use of the sea, and despite the rise of other uses – including navigation, oil and mineral extraction, and waste disposal – remains one of the most important. The current annual harvest of fish (taken here to mean all living creatures, including marine mammals, invertebrates, and seaweeds) is around 80 million tonnes, worth (by the time they reach consumers' tables) some US$100–200 thousand millions. This figure, however, understates the importance of sea-fish and fishing in many parts of the world. In most of the developed world, fishing accounts for only a very small fraction of the GNP (Gross National Product), and with some exceptions, e.g.. Japan, fish is not an important part of the diet. In much of the developing world, however – especially in isolated coastal areas – fishing may be the largest source of employment, and the principal source of protein in human diet.

Over the past 40 years, the world harvest of sea-fish has increased fourfold. During most of this period the growth has been fairly steady, at around 7% per year, though with a period of slower growth in the 1970s. The growth has been most marked away from the traditional centres of large-scale fishing in the North Atlantic and North Pacific, having been mostly in the tropical areas where the need for increased food-supply tends to be greatest.

In a general way it may be said that the use of the oceans as the main source of fish has been largely successful; against that there are, however, the collapses of what in their day were some of the greatest fisheries in the world – Antarctic whaling, the Peruvian anchovy, etc. These would suggest that marine resources have often been abused, rather than used wisely. It is the aim of this paper to seek the truth as between these opposing views. This will

*[Died 24 June 1990, to our great regret and the marine world's loss. At the Author's request this paper was presented at the Conference and has otherwise been handled by Dr N. V. C. Polunin, Centre for Tropical Coastal Management Studies, Department of Biology, The University, Newcastle upon Tyne NE1 7RU, England, UK. Eds.]

be done in two stages: first by looking at how fishing affects marine ecosystems (some impacts of other human activities are discussed by Goldberg, Chapter 5), and then by looking at how fishing has been controlled to prevent abuse, and the successes (or failures) of those controls.

THE IMPACT OF FISHING ON THE MARINE ECOCOMPLEX

Fishermen are fortunate in that it is very much harder to cause a major disturbance to the ecocomplex as a whole in the ocean than on land; it is also generally harder to have a serious and lasting impact on fish stocks than on land animals. On land a few days' work with a chain-saw or a plough can destroy most of the plants in a rain-forest or a savanna on which the rest of the system depends. In the ocean the system starts with microscopic plants (the phytoplankton); and it is not until several stages along the food-chain that the animals are reached – the herring, cod, tuna, etc. – which are the prime targets of fishermen. Even if these animals are affected – and most of them are very resilient under fishing pressure – the effect on the system as a whole may be very small. It has been said – light-heartedly, but with a fair degree of truth – that if all marine mammals were removed from the oceans, it would be difficult for the average marine ecologist to see any difference in the objects of his study. After a century of heavy fishing in the North Sea, and several centuries of lighter fishing before that, the fish stocks there, and the rest of the ecocomplex, are very similar to those that prevailed before fishing started.

The reason for the general resilience of fish stocks under exploitation lies in the great fecundity of most fishes. The average female will produce from a few thousand eggs up to more than a million, depending primarily on the species. Nearly all of these will die before reaching maturity – on the average in a stable population only two will survive. It only takes a reduction in this mortality from, say, 99.998% to 99.996%, to make up for halving of the adult stock due to fishing, and to arrive at the same number of young fish (the recruits) entering the fishery. It is the general experience that the abundance of adults can be greatly reduced by fishing without any change in the average numbers of young reaching a fishable size, though any such relation is often obscured by the high natural year-to-year variation in the numbers of young produced in many stocks.

This happy condition cannot hold in extreme cases. If the adult stock is reduced enough, in due course the point must be reached where the production of recruits to the fishery cannot be maintained. This will happen soonest if the fecundity is low, and for the marine animals with the lowest fecundity (the mammals, which mostly produce only one young at a time – annually in the case of seals, but at intervals of two or more years for most whales) – it happens very quickly. In these, even a small reduction in adult stock means a loss in the source of recruitment, and the stock is unable to sustain more than a very light harvest-rate – no more than a few per cent per year, compared with rates of 50% or more that many fish-stocks can sustain.

It is consequently not surprising that many stocks of marine mammals have collapsed when they became the target of commercial harvesting. The collapse of the stocks of large baleen whales in the Antarctic in the middle of this century is the best-known example, but more dramatic was the fate of the 'fur seals' in the same area a century earlier. Bonner (1982) describes how the South Shetland Islands (and their rich stock of 'fur seals') were discovered in 1819, followed by an uncontrolled slaughter which left no seals to be found by 1834.

Examples among fishes are much fewer, but elasmobranchs (sharks and rays) also have low fecundities, many producing only a few tens of potential young per year, or in some cases even fewer. This makes them vulnerable to overexploitation, which can be particularly serious when the fish are caught incidentally, in fisheries directed at other, and more resilient, species. Thus the Common Skate (*Raia batis*), which used to be reasonably common in the Irish Sea, has a very low fecundity and is caught by most of the types of gear (trawl, line, etc.) that are used for bottom-living fish. Over the last 50 years it has declined steadily in abundance, and is now virtually extinct locally (Brander, 1981). However, as the Skate has little popular appeal, its disappearance has attracted little comment.

Fortunately it seems that most of these threatened fish-stocks can recover if left alone, though the process can take time. The last significant harvest of 'fur seals' at South Georgia was around 1870, and for nearly half-a-century virtually no 'fur seals' were seen there by the occasional visitors. From the 1920s onward they were sighted in small numbers, and by 1956 Bonner (1982) found small but thriving colonies. Since then the numbers have been carefully monitored, and the recovery has been dramatic. Numbers are doubling every five years, and by the early 1980s had reached about a million – comparable with the original population. The Right Whales (*Eubalaena glacialis*) in the southern hemisphere seem to be doing the same thing. Hunting of this species had died out for lack of Whales towards the end of the last century, and for half-a-century Right Whales remained very scarce, with no obvious sign of recovery. In the last 20 years, however, good counts of Right Whales off South Africa have been made from the air, and these have shown that the numbers are increasing steadily at some 7% per year; moreover, Right Whales are being seen more and more regularly around Australia and New Zealand.

Fish-stock Collapses often Unexplained

The collapses of marine mammal stocks are easily explained, even though they were not prevented. The same is not true of the collapses that have occurred in many fish-stocks, notably a number of the stocks of small pelagic fish (various anchovies, sardines, and herrings) that have supported some of the biggest fisheries in the world. Collapses have followed periods of heavy fishing often enough (the Peruvian Anchovy [*Engraulis ringens*] in 1971–2, the Californian Sardine [*Sardinops caerulea*] in the 1930s, etc.) to make us fairly sure that fishing has had a share in these events. However, when any case is examined in detail, other explanations emerge as possible, if not probable. For example, the Peruvian Anchovy collapse coincided with a strong El Niño event, which event has often been put forward as the prime cause of the collapse. The case of the Californian Sardine collapse was long a matter of controversy (Murphy, 1966) – a controversy that is still not wholly resolved.

Wider examination of the history of these stocks of small pelagic fishes (e.g. Csirke, 1988) shows that large changes in abundance, with periods of several decades of high abundance alternating with similar periods of low abundance, are common, and many changes occur in the absence of fishing. Thus Soutar & Isaacs (1974) showed, from the examination of fish scales in bottom deposits, that the populations of both the sardines and anchovies off California have undergone big changes over periods of centuries.

Based on these stocks of varying abundance, it is natural that commercial fisheries will develop, other things being equal, during the periods of high abundance. Thus it is likely that the first big change in the level of natural abundance observed during the history of commercial fishing will be a 'collapse', occasioned by a shift from a period of high to one of low abundance. However, as longer periods of observation accumulate, it is possible to observe expansions or 'explosions' of fish-stocks that are as striking as the 'collapses'. Among the most striking has been that of the Japanese Sardine (*Sardinops melanostictus*). This provided catches of around 2 million tonnes in the 1930s, collapsed to the point where only 17,000 tonnes were taken in 1970 (though Japanese fishermen caught it whenever they could), but exploded in the late 1970s. Catches in the 1980s have been of around 4 million tonnes. The collapses of these pelagic stocks therefore provide some dramatic examples of how much fish-stocks can change; but they also provide much-less-clear evidence of the degree to which Man has abused natural resources by overexploitation.

Bottom-living Fishes Relatively Stable

At the other extreme lie some of the stocks of bottom-living fishes, such as the Plaice (*Pleuronectes platessa*) in the North Sea, where natural variation is small, and the impact of heavy fishing can be more easily demonstrated, even though this impact is much less dramatic than in some of the above cases. For these demersal species the main impact of fishing is that the additional mortality greatly reduces the abundance of older and larger fish. The recruitment is not noticeably affected by these changes in adult stock, at least over the ranges observed in practice (down to perhaps 20% of the average unexploited level). The result is that, at high levels of fishing, the catch is made up of large numbers of very small fish, so that the total weight caught (and also the total value of the catch) is less than at some intermediate level of fishing when fewer, but on the average larger, fish are caught.

Most stocks are intermediate between the above two extremes. Many, while lacking the large long-term changes of some pelagic stocks, show great

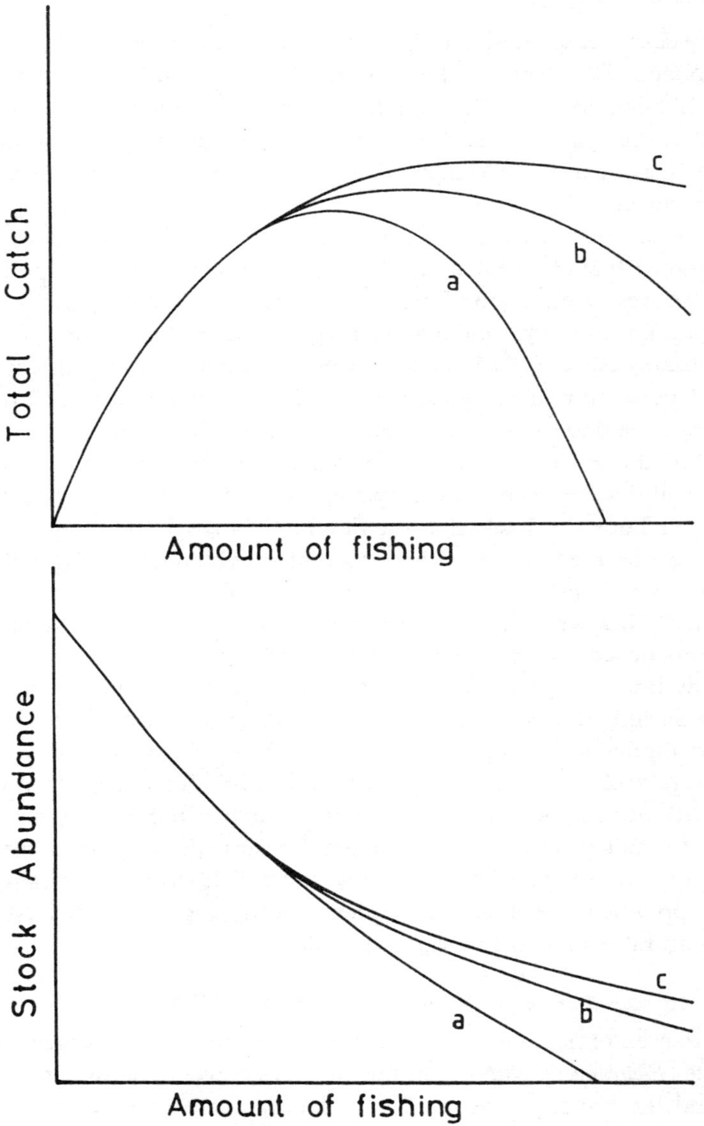

Figure 6.1. The relation between the amount of fishing and the total catch (above) and stock abundance (below). Curve (a) corresponds roughly to marine mammals, curve (c) to flatfish, and curve (b) to intermediate species.

year-to-year variation in year-class strength. A noticeable example is the North-sea Haddock (*Melanogrammus aeglifinus*), for which consecutive year-classes typically differ by a factor of three, and occasionally – as in the 1962 and 1963 year-classes – by nearly a thousand-fold. This variation in recruitment, clearly unconnected with any differences in adult stock, often makes it very difficult to see whether changes in adult stock brought about by fishing have any effect on the average recruitment. Most stocks are also intermediate in respect of their response to reduced adult stock as between whales (clear and immediate response) and Plaice (clearly no appreciable response over quite a wide range of adult stocks). Thus for most individual stocks, the evidence of mean recruitment being affected by reductions in adult stock is poor, but the evidence for increased survival between egg production and recruitment is equally poor.

In brief, the impact of fishing on marine resources can be represented by two families of curves relating the sustained yield and the abundance of the fish-stock to the amount of fishing (strictly the fishing mortality) (Figure 6.1). As fishing increases, so too at first does the total catch, but then this reaches a maximum. Thereafter, if recruitment is seriously affected (as for whales), increased fishing may cause a dramatic fall in catch. If recruitment is not seriously affected (as for Plaice), the decline may be moderate, and even intense rates of fishing give moderately high sustained catches. The abundance always declines with the amount of fishing, but there are differences at the higher rates. In the first case (whales), high rates will lead to a steep decline, and beyond a certain level sustained heavy fishing will lead to extinction (curve a). In the second case, which includes most fin-fish, even very high rates are sustainable, though the stock abundance may be very low. In most of these cases poor catches, and poor economic returns, will prevent the amount of fishing from increasing to the point at which the existence of the stock is threatened. Depending on the stock, there will be a greater or smaller variation about these curves of average conditions, over periods ranging from a few years (*e.g.* for Haddock) to decades or longer (for same pelagic stocks).

The curves in Figure 6.1 are largely based on analysis of the dynamics of a single species, treated more or less in isolation. Of course every species in fact interacts closely with many other species as predator, prey, or competitor for space or food. In the case of fish, the complexity of these interactions is increased by the fact that an individual fish changes its position in the food-web greatly as it grows from a newly-hatched larva to an adult. Thus in its first few weeks of life a baby Cod (*Gadus morhua*) is potential food for a juvenile or adult Herring (*Clupea harengus*), which in turn becomes food for an adult Cod.

When the interactions between species are taken into account, the relations between the total amount caught, the abundance of the stock, and the amount of fishing, will differ from the single-species situation. In practice the difference may not be much. If the fishery is not selective and, in the manner of many trawl fisheries in tropical waters, takes most bottom-living species in the area, the relations between the amount of fishing and the total catch or overall abundance will look like one of the flatter curves in Figure 6.1 (*see*, for example, Figure 13.2 in Pauly, 1988).

If the fishery is selective, the effects may be more interesting than otherwise. Thus a fishery of a predator such as Cod may allow more small of other fish (such as Haddock) to survive, consequently increasing the yield from them. In such cases, where there is also a fishery on the prey species, very heavy fishing on the predators may give a smaller yield from them but a greater yield from all species together. Simplistic calculations based on food-chain theories and trophic levels, would suggest that the increase in gross yield achievable by changing from a fishery targeted on predators to one on prey, could be very large (perhaps times ten), but in practice it seems to be much smaller.

Competition Could Also Be Important

Among pelagic species, competition may be important. The collapse of one species, e.g. of anchovy, often occurs at about the same time as the rise of another, e.g. of sardine, in the same area. It is tempting to see this as the replacement of one species (perhaps depleted by heavy fishing) by another with similar ecological requirements. However, detailed examinations of the magnitude and timing of these changes give only moderate support to this hypothesis (Dann, 1980). Nevertheless, to the extent that the decline of one species is matched by the rise of another, and provided the fishermen can switch from one to the other, the effective relation between the amount of fishing and the yield or total available biomass-take will be like one of the flatter curves of Figure 6.1, rather than the sharply-peaked curve which may be applicable to a single species.

In summary, considerations of interactions – whether the purely biological interactions between species, or the interactions between fisheries targeting on different species – complicate the picture, and the yield or biomass involved will no longer be a single-valued function of the amount of fishing. Rather will it depend on how the overall amount of fishing is targeted on different species. However, the general pictures provided by the curves of Figure 6.1 still apply. In particular it is true that even a small amount of fishing will cause some reduction in standing stock biomass, and that the greatest catch will be taken with a moderate amount of fishing.

Fishery Management

Why Management is Needed

Figure 6.1 shows that levels of fishing above that giving the maximum in the yield curve (the 'Maximum Sustainable Yield' – MSY) are pointless. Less will be caught, and costs will be greater. In fact the optimum level of fishing is somewhere on the left-hand shoulder of the curve. The gross yield will be less, but there will be other advantages – lower costs, greater stability in the yield, etc. However, in the absence of explicit controls, the amount of fishing will often exceed the optimum – sometimes grossly so. This is because most fisheries are good examples of 'the tragedy of the commons' (Hardin, 1968). If each fish-stock had a single owner, he or she would presumably fish somewhere near the optimum. But nearly all fish-stocks are common property, and so usually, in the absence of special measures, anyone is free to fish – any national in the case of waters under national jurisdiction, and anyone at all in the cases of stocks on the high seas.

If fishing is too intense, no one has any incentive to take unilateral action to reduce the amount of fishing, even if the harmful effects of over-fishing are obvious. If anyone reduces his or her activities, he or she will catch less, while the fellow-fisherfolk will reap the benefits and catch more. Controls have therefore to be applied to all participants in a fishery. This implies either an authority capable of imposing controls, or an agreement among participants that they will implement whatever measures are agreed upon as needed. As, at least until the changes in the accepted Law of The Sea in the late 1970s, most of the major marine fisheries were carried on wholly or partially beyond the bounds of national jurisdiction (*e.g.* beyond the 12 nautical miles [*c.* 22km] which was the common extent of territorial seas), international agreements were an essential part of management of most marine stocks. In discussing the management of pertinent fisheries, it is convenient to look first at the institutions involved, then at the different strategies that can be used, and finally at the mechanics of deciding on the precise measures to implement those strategies.

Management Institutions

The earliest international agreement to manage and conserve a living marine resource was that between Japan, Russia, the USA, and the UK (also on behalf of Canada), established in 1911 to conserve the Fur Seals in the North Pacific (apparently *Callorhinus ursinus*) which has continued, in different forms, to this day. Since 1911, a large number of other agreements have been reached, many of them establishing International Commissions or similar bodies. Some of these have responsibility for particular groups of species, either globally (as in the case of the International Whaling Commission (IWC), established in 1946) or in particular areas (e.g. the International Commission for the Conservation of Atlantic Tuna – tuna in this case including 'billfish' [many species involved] and other related types). Others deal with fisheries in general, but in more restricted areas (e.g. the north-west Atlantic).

One criticism of these bodies has been that they take too restrictive a view of the ecological problems, focussing too narrowly on the commercial species. To meet this criticism, the convention setting up the Commission for the Conservation of Antarctic Marine Living Resources (CCAMLR) deliberately took a wider 'ecosystem' approach. However, the experiences of CCAMLR and of its Scientific Committee to date indicate that the wide approach has to be based on a sound knowledge of the dynamics of the main target species (in the Antarctic, mainly Krill, *Euphausia superba*).

Another criticism of these international bodies is that they lack power. Thus, in many cases, decisions have to be taken by consensus, and where a majority can take decisions, there are usually procedures whereby dissenting parties can object to the decision and thus avoid having it applied to them.

This is a valid criticism, and because of this weakness the decisions taken by international bodies have usually been weak and late. However, this is not the fault of those drawing up the terms under which the Commissions were established. Membership cannot be made compulsory, and the choice is between a body with relatively weak powers, but with comprehensive membership (and thus effective in exerting those powers), and a body with stronger powers, but ineffective because it is unlikely to have comprehensive membership. Any country not wishing to accept the rules can leave – or even not join at all.

Various Strategies

The importance of international arrangements has decreased since the 1970s when, as a result of the discussions at the UN Conference on the Law of the Sea, most countries extended their jurisdiction over fisheries to 200 nautical miles, usually in the form of an Exclusive Economic Zone (EEZ). These 200-miles' zones cover areas in which well over 95% of the world catch was taken, even though some of the areas, *e.g.* off Namibia (South-west Africa), have not so far come under coastal state control. However, in view of the large catches that are currently being made there by foreign fleets, it is a reasonable supposition that an early act of a newly independent Namibia will be to establish an EEZ. Nevertheless there are still important stocks – tuna, whales, Krill, etc. – that are taken wholly or partly beyond 200 miles, and for which international arrangements are still needed. Also, many stocks move between the EEZs of two or more countries, who need to agree on the strategies to be favoured and the measures to be taken. The problems of these shared stocks are less severe than those of the earlier high-seas fisheries, as only a limited number of countries are involved, and it should be easier for Canada and the United States, for example, to agree on what should be done about the stocks moving between their zones off eastern America than when a dozen or more European countries were also involved.

In practice the extension of limits by reducing the number of countries involved to one or only a few, goes only a little way towards better management. In theory the Ministry of Fisheries, or corresponding department of government, usually has the powers to impose whatever measures may appear 'necessary'. In practice the ability of governments tends to be limited by political pressures from the fishing industry. The general establishment of EEZs eased one problem, but many difficulties still remain.

Choice of Measures

Measures to manage fisheries fall into two main categories, namely those that control the amount of fishing, and those that control the kinds of fish caught. In the latter case this means especially those measures that give protection to small, immature fish. Minimum-size restrictions on the fish that can be landed, or controls on the mesh-sizes of nets, do not, however, with some exceptions (such as closure of nursery ground), interfere with the ability of a fisherman to fish when and where he wants. They are therefore usually more acceptable than controls of the amount of fishing, and most of the early regulations – especially those introduced by the international Commissions in the North Atlantic – were of this type.

Unfortunately, though often beneficial, controls of the kinds of fish caught do not tackle the problem of too much fishing, and cannot, by themselves, ensure a healthy fishery and fish-stock. For this, some direct control on the amount of fishing (strictly, the fishing mortality-rate) is needed. The fishing mortality cannot be controlled directly, but has to be controlled indirectly, either by limiting the inputs (the number and power of vessels, etc.) or the outputs (the catch, usually in the form of quotas). Both methods imply that at least some fishermen cannot fish whenever they wish to, and there is an important distinction between open-access systems, under which anyone can fish for a time, until the target level of catch or amount of fishing is reached, when everyone stops (*e.g.* for very short open seasons). Modern theory therefore points to some kind of limited access system as being desirable, either limiting the number of vessels, or allocating shares in the total quota to individual fishermen or fishing enterprises (effectively the system used by the IWC in terms of national quotas). A particular form of allocated quota that has strong economic attractions is the Individual Transferable Quota (ITQ), which allows each holder of an ITQ to arrange his operations to take his allocation in the most profitable manner, and is now in use in New Zealand.

Whatever system is used, it will be without effect unless the number of vessels, or the total quota, is set at the right level. This is a two-stages' process – first a political decision as to the level of fishing mortality to be attained, and then the scientific process of calculating what fishing effort (*e.g.* number of vessels) or total catch will correspond to this level. Unless the yield curve for the stock in question, i.e. the relevant curve of those in Figure 6.1 is very sharply peaked, the political decision is not simple. For many fisheries the curve is flat over quite a considerable range near the maximum yield. Any point on the curve is sustainable, and there are few biological grounds for preferring any one particular level near the maximum to another. In the long run, lower levels of fishing should be more attractive; but in an intensive fishery, achieving such levels may require that many of the present participants leave the fishery. This is seldom politically attractive, and it is not surprising that a relatively high level of fishing is chosen, or that an explicit choice is avoided in the hope that the decision (and the blame) will be taken by the scientists.

If it were easy to know just how many fish there are in the sea, and how many young fish would enter the fishery next year, or by how much the fishermen would increase their efficiency, the scientific stage would be easy. With such knowledge it would be possible to adjust the total allowable catch, or the number of vessels for the forthcoming year, to correct for changes in stock abundance or in vessel efficiency, and thus maintain the fishing mortality at the desired level. In practice both types of knowledge are incomplete, and this incompleteness is particularly important for stocks that show marked year-to-year natural variation, and are subject to control by catch limits (e.g. Haddock in the North Sea). The result is that, despite considerable research programmes (*e.g.* surveys of young fish), the catch limits for these stocks are subject to considerable uncertainty, and hence last-minute adjustment. This does nothing to increase the confidence of the fishing industry in the scientists! In general it has to be accepted that all pertinent scientific advice has to be subject to greater or lesser uncertainty, and management policies have to take this into account.

Aquaculture

There is a school of thought that basing an industry on harvesting fish from wild stocks is outdated, and that fishery management of the kind described in the previous section is merely tinkering with the problem. Just as on land Man has changed from being a hunter–gatherer to being a farmer, so on water must he change from being a fisherman to being a cultivator of fish. This may be true for many freshwater and brackishwater areas, but for 99% or more of the oceans it is nonsense. However, this ignores the differences between the useful species on land and in the sea. On land the useful species are the plants or the animals that feed on them. By clearing the forests or ploughing the plains and planting crops, the farmer can benefit from a high proportion of the primary production, and this proportion is reduced by only one step in the food-chain if he prefers to raise cattle or sheep. In the sea, with the exception of seaweeds in a very narrow coastal strip and some pelagic drifts such as those of the Sargasso Sea, the plants are microscopic, and so, too, are nearly all the animals that feed on them. Though there are some fish, *e.g.* the Anchoveta (*Centengraulis mystecetus*) of Peru, that feed at least partly on phytoplankton, in most cases it is only after several steps along the food-chain that the fish are big enough to interest commercial fishermen. The proportion of the original primary production that is available for harvest is therefore very small, and the prospects of 'farming' the North Sea for Cod are not dissimilar to those of 'farming' the Serengeti plains for a harvest of lions and hyenas.

This does not mean that aquaculture has not got a good future. Indeed some of the recent great success-stories in the world's fish business have been of aquaculture – Salmon (*Salmo salar*) in Norway and Scotland, and various shrimps and prawns in Thailand and other tropical countries. Statistics recently published by FAO (1989) show that the current aquaculture production probably exceeds 10 million tonnes, namely well over 10% of total world fish production – though accurate figures are difficult to obtain, and in fresh waters the distinction between culture and the harvest of wild fish is not always easy.

The FAO figures also show the great extent to which culture is restricted to certain products. Half the production takes place in fresh water (with carps accounting for over 3 million tonnes) or brackish water (650 thousand tonnes in 1986 – mainly involving the traditional Milkfish (*Chanos chanos*) culture in south-east Asia, and the rapidly-growing shrimp industry). Of the 5.5 million tonnes taken from marine waters, some 5 million tonnes are of seaweeds (which are included in the FAO statistics) and molluscs. The latter (oysters, mussels, and clams) are exceptional in that they are filter-feeders, and thus feed either directly on phytoplankton or on the next step along the food-chain. The culture of molluscs does, therefore, to that extent resemble the raising of domestic animals. In those areas where this type of culture is carried on, it is possible to talk about farming the sea. However, these areas are limited to sheltered waters such as the *rias* of northern Spain, and the total area for farming of this type is only a very small proportion of the area of the continental shelf, let alone of the whole ocean. The methods used are also very labour-intensive, and production in many of the traditional centres in Europe and north America is at best increasing slowly, being maintained largely by the increase in the real price – a century ago oysters (*Ostrea* spp.) were a cheap food in London.

The kinds of aquaculture that have recently been most successful have been the intensive rearing of high-valued species in ponds or cages. Recent figures indicate that shrimp and prawn culture, particularly in southeastern Asia, has been expanding at some 25% per year and by 1988 had reached some 500 thousand tonnes, worth, at the farm-gate, some $2.5 thousand millions. By any account, this is a major industry, and is contributing significantly to the exports of several of the countries involved. Farming of Salmon in Norway and Scotland has been almost equally successful. These industries, that depend largely or wholly on the supply of artificial feed, are much more closely related to the rearing of battery chickens than to traditional agriculture. Salmon in particular require feed with a very high protein content, and usually with a high proportion of fishmeal. Thus salmon farming can be considered as just a way of converting low-value fish, such as Capelin (*Mallotus villosus*), into a product that the consumer will pay a high price for. This type of culture, which accounts for almost all of the 300 thousand tonnes of finfish cultured in the sea, leads to a more efficient use of the sea only to the extent that it encourages the more intense harvest for fishmeal of those low-valued species which would otherwise be only lightly exploited.

There remains one form of positive intervention that is not included in the formal statistics of aquaculture. This is stocking or ranching, namely the release into the wild of young fish to replace the losses due to fishing or otherwise. Because of their high fecundity, it is possible to hatch vast numbers of fish, and if most survived, hatchery releases would maintain fish populations at a high level. However, survival of any small fish is very low, and it seems that fish released from hatcheries usually have even less chance of survival – a young fish which has learnt that a disturbance marks the arrival of food is an easy prey for predators. If the young fish are reared past the stage when mortality is high, the costs of food etc. make the activity uneconomic except for the most highly-priced fish, and then, for both salmon and shrimp, it is worth keeping them until they reach market size. Although releases of small fish or other forms of stocking can be useful in small bodies of water, they do not offer a solution to the problems of depleted stocks in the open sea.

There is one important exception. Salmon undergo a major change in their life when, as 'smolts', they migrate from fresh water to the ocean, where they achieve most of their growth. Young Salmon, reared past the stage of highest mortality and released as 'smolts', are adapted to change in their habitat, and experience shows that the survival in the wild of 'smolts' reared in hatcheries before release can be similar to that of 'natural' smolts. Hatchery production of Salmon smolts has proved a very economically rewarding activity. The value of the returning adult Salmon can greatly exceed the costs of the hatchery. As a result, hatchery operations have expanded very greatly on the west coast of north America – especially for the more valuable species of Pacific Salmon (Chinook and Coho, respectively *Oncorhynchus tshawytscha* and *O. kisutch*), as well as in the Baltic for Atlantic Salmon (*Salmo salar*) – to the point at which the majority of catches in some areas are of hatchery fish.

These operations do provide the only significant example of positive action to increase the effective use of the primary production of the open sea; but they have their problems. As the fishes return to their home-waters, hatchery and natural fish become mixed, and inevitably suffer similar fishing rates. However, hatchery fish can sustain a higher fishing rate, so that when they are fished at this rate the natural stock declines, and in many areas the rise of hatchery operations has seen the decline of the natural stocks in absolute as well as relative terms.

How Well Has Management Performed?

Marine Mammals

When one looks for examples of mismanagement of marine resources, many of the obvious examples are of marine mammals – especially whales and seals. Certainly the collapses and near-extinctions of many stocks have been dramatic, and the reason has been pointed out earlier – their fecundity is low and they can sustain only a very low harvest-rate. The disasters that have happened to marine mammal stocks are not due to unusual weaknesses of those institutions responsible, but to the biology of the animals. In fact bodies such as the IWC have probably performed rather better than the corresponding bodies dealing with finfish. Because the effects of mismanagement are much more drastic, and the links between the rate of exploitation are much clearer (those dealing with marine mammals seldom face the uncertainties of why a stock declined that are at all similar to those surrounding the decline of, say, the Peruvian Anchovy), the history of marine mammal management gives much clearer lessons, and examples of both failures and successes.

Seals

Overall, the failures have been much more obvious than the successes. Though the extermination of Stellar's Sea-cow (*Hydrodamalis gigas*) in the 18th century is the only certain example of absolute species extinction, many stocks of seals and whales have been reduced at various times to levels at which the risk of extinction must have been appreciable. Populations of 'fur seals' in the southern hemisphere probably came closest to extinction, but it is interesting to compare the fate of populations where there was an authority in potential control, and those which were exposed to the worst aspects of an open-access resource. Thus Bonner (1982) points out that the seals in the River Plate have been harvested more or less continuously since the early 16th century. Management may not have been sophisticated, but the ability of the local authorities to cut harvests when it seemed the stock was going down, and allow expansion when prospects improved, has ensured the long-continuation of the stock and consequently of the industry. In contrast, when the sub-Antarctic islands and their rich 'fur seal' stocks were discovered, there was no authority, and the slaughter was uncontrolled. Within less than 10 years of the discovery of the South Shetlands in 1819, the 'fur seals' there had virtually disappeared (*idem*).

The subsequent recovery of some of these stocks, often after a period during which the numbers stayed at a very low level, has been among the more striking examples of the resilience of natural populations – even those with low fecundity. The 'fur seals' at South Georgia have increased at around 15% or more annually (i.e. doubling every 5 years) – certainly during the 1960s and 1970s, and probably for most of this century (Payne, 1977; Bonner, 1982), and from virtual disappearance at the end of last century to number now a million or more. This rapid growth demonstrates the size of the sustained harvest that might be taken if these stocks were properly managed.

A close approach to such management has been attained for the North Pacific Fur Seal (*Callorhinus ursinus*). At the time of the international agreement in 1911, the stocks had been grossly depleted by uncontrolled high-sea kills. By restricting the harvest to juvenile males on the breeding islands, the stocks were steadily built up from about 300,000 in 1911 to well over a million breeding females. The harvest was also increased to a peak of over 60,000 animals annually between 1940 and 1955. Changes in harvesting practice may have accounted for moderate declines in the harvest since 1955. Up to that time the Fur Seal could be pointed to as an example of highly successful management. Since then an attempt to increase catches by including a proportion of females has shown that the dynamics of the stock, and the effects of density-dependent factors, are not quite as simple as had been hoped. Recently the stock has declined for reasons that are not clear, although the balance of evidence points to the entanglement, especially of juveniles, in lost or discarded fishing-nets. This is a problem which has become more serious with the use in the nets of synthetic materials. Because of the unexplained decline, Fur Seal management cannot be claimed as a complete success, though the causes of the failure seem to lie outside the direct harvest.

Whales

The case of whales is more complex, and illustrates more of the factors that contribute to success or failure. Though public attention has focused on the IWC only in the last ten to fifteen years, the more interesting period was earlier. When the IWC was established in 1946, and set limits (16,000 Blue Whale Units) on the international harvest of baleen whales in the Antarctic, it represented for its time a very positive move towards more rational use of marine resources. However, it had several flaws. In particular the figure of 16,000 BWUs had little scientific foundation (though based on the best knowledge available), there was no automatic mechanism for revising this figure, and there was not the quantitative research on which revisions could be authoritatively based.

In fact, 16,000 BWUs exceeded the joint sustainable yields of the stocks, so that their abundance declined at an accelerating rate as the gap between the harvest and the declining sustainable yields widened. By 1960 it was clear, from even a crude examination of the catch-rates of the commercial operations, that the Blue Whales (*Megoptera musculus*) had declined to a very low level, and that the Fin Whales were also declining fairly rapidly. At this point the IWC called in independent experts (the Committee of Three, later Four, Scientists), who for the first time gave the Commission the sort of quantitative advice on which decisions could be taken. They showed that the sustainable yield curve for whales was most like the sharply-peaked curve in Figure 6.1, that the stock of Fin Whales (*Balaenoptera physalus*) had fallen below the MSY level, with catches grossly exceeding the current sustainable yield. Blue and Humpback (*Megaptera novaeangliae*) whales were even more depleted than the others. Catches of Fin Whales would have to be cut by at least two-thirds to stop further decline, and by even more to restore the stocks to the MSY level.

If this kind of advice had been available ten years earlier, the Fin Whales would have been nearer MSY, and the necessary cuts in catches would have been smaller. It is also possible that the Commission (and the whaling industry) would have acted quickly enough to stabilize the catches and stock somewhere near the MSY level. As it was, the amount of reduction was so great that it took several years of hard negotiation to get agreement to the necessary reductions, by which time the stocks had declined still further to well below the MSY level. Shortly afterwards the Commission adopted a policy of protecting all stocks that had become depleted below MSY, and killing of all baleen whales, except the small Minke whale, was ended.

Clearly, so far as the main Antarctic stocks – by far the biggest potential resource of whales in the world – are concerned, management has failed. However, the key factors in this failure occurred a long time ago – largely in the 1950s, when the needs to keep the BWU quota under careful review, and to undertake the quantitative studies which were needed to provide the scientific backing to that review – were ignored. It is tempting to hope that we can now do better; indeed the history of Minke whaling suggests that this hope has some foundation. The Minke Whale (*Balaenoptera acutirostrata*) is much smaller than the other baleen whales, and while the main value of a whale carcass was in its blubber, it had no economic significance. Now that the main value is in the meat, and other whales are protected, it has become commercially attractive.

When the question of harvesting Minke Whales arose, the IWC set precautionary quotas based on conservative estimates of the stock abundance and on the sustainable rate of harvest. It also initiated a series of surveys to provide improved estimates of abundance. These surveys have shown that the Minke Whale is very abundant, with a population in the Antarctic of the order of a million animals. The initial quotas were clearly conservative, and could well be increased; and if other things were equal, one could now point to a significant sustained Minke Whale harvest (albeit very much smaller than that which might have been attained from Blue and Fin Whales) as an example of successful management. Other things have not been equal, and decisions in the IWC are based on many more factors than merely the best estimate of sustainable yield.

Shifts of Emphasis

This is not the place to describe the history of the IWC since 1965, and the influence of different pressure-groups. The balance of power within the IWC has shifted from active whaling countries to those that are basically opposed to any whaling. One effect has been a change in the burden of proof from 'don't apply restrictions unless they are clearly needed' to 'don't catch any whales unless it is proved that it is safe to do so'. Another effect has been the introduction in 1982 of a moratorium on all commercial whaling, to be reviewed not later than 1990. Though some whaling on a commercial scale has continued under various pretexts, catches are now small, and it is unlikely that Japan (now the only substantial catcher of whales) would resume significant catches of Antarctic Minke Whales unless the IWC had agreed on quotas at the appropriate level. The MSY of Minke Whales may not be known, but it is quite clear that catches of several thousand of them annually – quite enough for a commercial operation – are sustainable. However, it is highly unlikely that a proposal in the IWC to change the present zero quotas (the mechanism for applying the moratorium) would receive the necessary majority. Though I believe (Gulland, 1988) that Man now has the knowledge and (on the part of the whaling countries) the will, to manage whale stocks sensibly, it is unlikely that this belief will be demonstrated in the near future by a commercial Antarctic Minke Whale hunt carried on at a sustainable level.

Much the same applies to sealing. The strong protectionist attitude of the US Marine Mammal Act of 1972 has brought harvesting of the Fur Seal on the Pribilof Islands almost to an end. The collapse of much of the market for fur has ended much other sealing, including that for Harp and Hooded Seals (respectively *Pagophilus groenlandicus* and *Cystophora cristata*) in eastern Canada, which became notorious because of the strong anti-sealing campaign carried out by environmental groups. Although the success of that campaign was largely due to the public appeal of a blue-eyed white seal-pup, a large part of the scientific argument put forward by the anti-sealing groups was that the catches were endangering the stocks. In 1970 this was true, as the situation resembled that of Antarctic whales in 1960 (*see* p.134). Little quantitative research had been done, however, though catches had exceeded the sustainable yield, and stocks had declined to the point where a high proportion of all the Harp Seal pups were being killed. This situation changed in 1971 when quotas were set. Initially these were too high, but they were cut by 1972 to a level (150,000) which was close to the sustainable yield. The subsequent history of the stock, and intensified research, have shown that the quota levels have, if anything, allowed a slow increase in the stock. By the mid-1980s the management of this stock could be described as successful, but then the collapse of the market ended most of the harvest (Anon., 1986). It is now possible to point to the Harp Seals as an example of ongoing successful management.

In brief, marine mammal resources were badly mismanaged – or not managed at all – for centuries until the last decade or two. During former times their fate was largely ignored by the public. But now there is the ability in terms of both the scientific knowledge available and the public willingness, to use that knowledge to manage marine mammals better and better. However, the public opposition, arising from past management and what is seen as the special position of marine mammals, makes harvesting almost impossible. For the foreseeable future these animals are unlikely to be treated as a resource, whether to be abused or used sensibly.

Finfish Management

It is more difficult to reach a clear judgement of how well finfish stocks have been managed. In only a few cases can we be sure that fishing has brought a stock near to extinction, and these mostly concern species with little economic value or public appeal. No one seems to have worried at the disappearance of the Common Skate from the Irish Sea (Brander, 1988). The causes of the collapses of many pelagic stocks are less clear. Although those in authority can be criticized for acting too slowly (Saetersdal, 1980), it is far from certain that earlier action would have prevented the collapses.

What is clear is that, very often, management has not maintained the economic health of the fishery even when the stocks themselves have remained in good condition. Experiences throughout the world have confirmed the predictions of Hardin (1968) and many others of what happens to common property resources. However successful a fishery may be in its early stages – and some, as in Peru, have been spectacularly successful – the amount of fishing will, it seems inevitably, expand until falling catch-rates make the fishery unprofitable. This equilibrium may often be well to the right of the peak in the yield-curve (*cf.* Figure 6.1), especially if the curve is relatively flat. In practice the situation is worse. Fish populations are variable and in good years fishing effort is 'sucked in', so that fishermen often find it difficult to leave in poor years. On average, fishermen tend to lose money, being often supported by government subsidies of one kind or another.

In contrast, if the amount of fishing can be controlled and kept at a moderate level, somewhere a little to the left of the peak in Figure 6.1(a), then catches will be high, costs low, and profits very good. In a few cases the necessary controls have been applied, giving considerable benefits which have been realized in different ways. In some places, e.g. Western Australia, the number of vessels is limited that are licensed to go fishing, and the benefits have appeared as greatly-increased capital values of vessels with a licence. Some coastal states limit the number and entry of foreign vessels, and also charge licence fees. Off the Falkland Islands these fees have been sufficiently large to become by far the major source of income for the Falklands Government, and so transform the local economy. Despite these high fees, the limits set on the total number of vessels keeps stocks and catch-rates high, and there is strong demand for licences.

These successes are exceptions. More typical of intense fisheries are those of the North Sea. Here the stocks, especially the demersal stocks, have been heavily fished for a century, and for the last 20 years the fisheries involved have been subject to an increasing number of controls – initially to mesh- and minimum-size regulations, and later to limits on the Total Allowable Catch of most of the important species. With the possible exception of the Herring, these controls have allowed the amount of fishing to go on expanding to a level where the productivity of the stocks is seriously endangered, with the amount of fishing on most stocks from two to four times the optimum. Currently the amount of fish that is being harvested from the North Sea is not very different from what it might be under good management, but the cost of catching it is twice or more than twice as much as it needs to be. In purely biological terms, the resources may not be badly abused, but in economic (and also social) terms they are. So are many other fishery resources around the world.

Incidental Kills

From the above account it appears that directed harvesting is not at present a major threat to marine resources – great damage has been inflicted on marine mammals, but deleterious levels of harvest have been stopped, and fish stocks have enough biological resilience to withstand current levels of harvest even when these are far too high if judged by sensible economic standards. The situation of some of the animals taken in fishing operations directed at other species is less good. Much public attention has been given to the fate of mammals and birds. Three forms of impact can be distinguished, namely (1) the tuna-porpoise problem, where purse-seiners deliberately set their nets on schools of porpoises (dolphins) in order to catch the tuna swimming below them; (2) the taking of birds and small mammals in operating gill-nets and other gear, and (3) the entanglement (principally of mammals) in discarded fish-nets and other gear.

In the first years after the purse-seining technique for tuna was developed in the late 1950s, very large numbers of dolphins – probably a few hundred thousands – were taken by the (almost exclusively US) fishery in the eastern Pacific. Up to about 1970 this incidental kill was virtually unknown outside the fishery, but when it became more generally known, great pressure was put on the industry to reduce kills, and their consequent efforts were commonly successful. Due to modifications of the gear and of the methods of operation, incidental kills dropped sharply to a few tens of thousands annually by the late 1960s. Although these and other measures, largely introduced by the US in respect of their fishing fleet, have failed to achieve the long-range objective, spelled out in the 1972 Marine Mammal Protection Act, of incidental kills approaching zero, the level of kills presented little threat to the dolphin stocks.

The more recent picture is less satisfactory. For various reasons, of which the US dolphin regulations are one, the proportion of purse-seiners in the total fleet flying the US flag has become small. Most of the other countries involved (*e.g.* Mexico) are poor and have higher priorities than protecting marine mammals. Regulations to give this protection are weaker, and are less rigorously enforced, for example in Mexico, than in the US. Now there is some evidence that the incidental kills are increasing, though it is not clear whether they are approaching the level at which some stocks of dolphins might again be endangered.

Accidental killing of birds and mammals by fishing gear mainly concerns gill-nets, though there is a possibility that the decline of some stocks of albatrosses may be caused by the birds being hooked on long-lines that are set for catching southern Blue-fin Tuna (*Thunnus thynnus*). Purely from the point of view of conservation of fish-stocks, the use of gill-nets is usually beneficial, being highly selective – normally of the larger fish. However, modern monofilament synthetic nets are almost invisible in the water, and are very efficient killers of mammals and birds as well.

As in the tuna/porpoise problem, for some time the fishermen cared not at all and little was known by the general public. This has recently changed, and there is great concern over some of the fisheries, particularly in the South Pacific. Despite this concern, the information on the numbers of animals killed is in most cases poor, and even less is known about the possible long-term impact on the stocks. The main fisheries that may be having an effect are those in the North Pacific for 'salmon' (where it is known that quantities of Dall's Porpoise [*Phocaenoiles dalli*] are killed) and 'squid', for Albacore Tuna (*Thunnus alalunga*) in the south Pacific, and for Salmon around Greenland (where various species of auks are the most affected).

The management and conservation of these resources is not likely to be achieved by the normal means of a management body taking decisions based on advice from a scientific group. The information base is poor, and with many of the fisheries occurring on the high seas beyond the 200 nautical miles limit, there is no authority with either the responsibility or the capability of resolving the conflict of interests between fishermen and those interested in mammals and birds. This conflict is particularly sharp because much of the public, after the disasters to marine mammals were publicized in earlier years, believe that no killing of marine mammals should be allowed.

The matter is most likely to be resolved by the strength of public pressure on each side, either within countries or between countries, without too much attention to the details of the scientific evidence on the strength of the threat to marine mammals or birds. In general, especially within the countries of the western world, this gives the advantage to the conservation side. Any fishery in western Europe or North America that can be shown to be killing significant numbers of mammals or birds has a cloudy future. In other countries, notably in Korea and Taiwan, which are particularly active in some of the gill-net fisheries, the conservation groups are less powerful, and within-country action alone is unlikely to end the incidental kills.

In the countries stressing gill-net fisheries, international action is important. In the south Pacific, the fact that the wide scatter of islands brings most of the fishing grounds within 200 nautical miles of one state or another, gives the advantage to the coastal states, who have naturally adopted a conservational attitude. However, in the north Pacific it is less easy for those interested in the mammals and birds, even in such powerful countries as the US, to bring pressure on the smaller fishing countries. Incidental killing is likely to continue in these fisheries for some time, though now that there is much more awareness than formerly of the potential damage and increased pressure to collect information, there is less likelihood that the killing will continue at a volume and for a period sufficient to pose a very serious threat to the stocks.

Entanglement in discarded fish-nets and other gear or materials engenders less direct conflict than the above. It is now widely agreed that dumping of wastes at sea is not acceptable, and there are consequently international agreements prohibiting such dumping, with increasingly effective national implementation of these agreements. Although still far from complete, this implementation ought to ensure that the threats to marine mammals and birds from dumping of wastes should steadily decline.

References

Anon. (1986). *Seals and Sealing in Canada.* Report of the Royal Commission, Ministry of Supply and Services, Ottawa, Ontario, Canada: 3 vols.

Bonner, W. N. (1982). *Seals and Man: a Study of Interactions.* University of Washington Press, Seattle, Washington, USA: 170 pp.

Brander, K. (1981). Disappearance of Common Skate (*Raja batis*) from the Irish Sea. *Nature* (London), 290, pp. 48–9.

Brander, K. (1988). The multi-species fisheries of the Irish Sea. Pp. 303–28, illustr., in *Fish Population Dynamics*, 2nd ed. (Ed. J. A. Gulland). John Wiley & Sons, Chichester, Sussex, England, UK: xi + 422 pp., illustr.

Csirke, J. (1988). Small shoaling pelagic fish-stocks. Pp. 270–302 in *Fish Population Dynamics* (2nd edn, Ed. J. A. Gulland). John Wiley & Sons, Chichester, Sussex, England, UK: xi + 422 pp., illustr.

Dann, N. (1980). A review of the replacement of depleted stocks by other species, and the mechanics underlying such replacement. *Rapp. Procès Verb. Reun. Cons. Int. Explor. Mer,* 177, pp. 405–21.

FAO (1989). Aquaculture production (1984–86). *FAO Fisheries Circular 815.*

Gulland, J. A. (1988). The end of whaling? *New Scientist*, 636, pp. 42–7.

Hardin, G. (1968). The tragedy of the commons. *Science*, 162, pp. 1243–8.

Murphy, G. I. (1966). Population biology of the Pacific Sardine (*Sardinops caerulea*). *Proc. Calif. Acad. Sci.*, 34, pp. 1–84.

Pauly, D. (1988). Fisheries research and the demersal fisheries of Southeast Asia. Pp. 329–48 in *Fish Population Dynamics,* 2nd edn. (Ed. J. A. Gulland). John Wiley & Sons, Chichester, Sussex, England, UK: xi + 422 pp., illustr.

Payne, M. R. (1977). Growth of a fur seal population. *Phil. Trans. Roy. Soc., London*, B279, pp. 67–9, illustr.

Saetersdal, G. (1980). Review of past management of some pelagic stocks and its effectiveness. *Rapp. Procès-verb. Reun. Cons. Int. Explor. Mer*, 177, pp. 505–12, illustr.

Soutar, A. & Isaacs, J. D. (1974). Abundance of pelagic fish during the 19th and 20th centuries as recorded in anaerobic sediments off the Californias. *Fish Bull. US*, 72, pp. 275–94, illustr.

Commentary on Chapters 5 and 6

CHAIRMAN: Professor Norman B. Marshall
PANELLISTS AND OTHER CONTRIBUTORS:

McCloskey, N. V. C. Polunin, Clark, Oza, Harun ur Rashid, Fosberg, Vallentyne, Batisse, Tisdell

McCloskey wished to comment upon the contributions of both **Gulland** and **Goldberg** (Chapters 5 & 6) and also to introduce the notion of marine wildernesses.

Gulland had commented that the commercial taking of whales resulted in driving populations of most species of the great whales to near-extinction. For example, the International Whaling Commission (IWC) had estimated in 1989 that about 500 Antarctic Blue Whales (*Balaenoptera musculus*) were left (Anon. 1, 1989)*, from an estimated original population of some 225,000. Since 1948 exploitation has been conducted under rules of the IWC.

There were major flaws in the operation of the Commission, as the Blue Whale example illustrated so forcefully. First, not all whaling nations had been members of the Commission during the years of heaviest killing. Second, both the database and the scientific methods used by the Scientific Committee were hopelessly inadequate to the task of assessing stock abundance, determining annual net recruitment, acquiring natural history information on whales and their roles in the marine habitat, and devising management schemes that would insure sustainable yields. Early efforts at management had been derived from finfish models, which could not be applied to marine mammals. Third, not until the 1970s did the IWC get enough new member nations that considered whales and whaling from views other than economic interest, to bring about a reorientation of the Commission. In 1982 the Commission adopted an indefinite moratorium on commercial whaling. It was designed to take effect in 1986, allowing industries and workers in commercial whaling countries an opportunity to phase out. To change the moratorium required a 75 percent vote of those nations voting in the IWC.

In 1972, after defeating the first attempt to adopt a moratorium on commercial whaling, the IWC began to adopt more responsible measures, such as abolishing the infamous Blue Whale Unit. In succeeding years the science consistently improved, which revealed the extent of uncertainties. Uncertainties still existed for every stock of every species.

It should be noted that the Commission moved to protect depleted species only when there were too few of them left to make it pay to send the fleet out!

The point was that schemes which sought to regulate taking marine resources had little hope of providing reasonable restraint if those who made the rules were the same as those who benefited economically from exploitation of the resource. It was more profitable to exploit the resource rapidly, rather than to stretch out the process by taking only a sustainable yield (if that number could be determined with certainty). Even today scientists reported that they still did not know enough to be able to determine what that yield might be. The prevailing view within the IWC today was that because of the uncertainties, no commercial taking should be authorized.

The IWC today was quite a different body from when it held its first meeting in

*[*See* p. 146. Eds.]

1948. In fact, most groups in the conservation community recognized it as being essential in dealing with the world's whales. But the economic incentive was always there. In spite of the moratorium, three countries (Japan, Norway, and Iceland) continued to do commercial whaling, although they called it research whaling. Their research programs had yet to be accepted by the IWC Scientific Committee. Those three countries were expected to ask for certain stocks of Minke Whales (*Balaenoptera acutorostrata*) to be reclassified in order to allow commercial whaling on them to be resumed. When the Commission adopted commercial quotas, it did so by assigning allowable take numbers to stocks of species in designated geographical area. The Commission was prohibited by the 1946 International Convention for the Regulation of Whaling from assigning quotas to nations.

There was no international agency that regulated the taking of small cetaceans (dolphins and porpoises). The IWC so far had chosen not to do it, although there was no language in the Convention of 1946 that prevented the commission from exercising that authority.

Small cetaceans suffered from incredible destruction at the hands of fishermen using modern technology. Dolphins and porpoises were set on deliberately with massive purse seines in order to catch yellowfin tuna. Six million of these cetaceans had died from this cause in the last thirty years in the Eastern Tropical Pacific. In the past the dolphin 'bycatch' was done primarily by US fishermen. Now, the US was responsible for about one-fifth of the annual take of up to 100,000 as more countries took up purse-seining on dolphins in the Eastern Tropical Pacific.

Marine mammals also suffered hundreds of thousands of mortalities from entrapment in monofilament drift gillnets on the high seas and in waters of national jurisdiction. World opinion has been aroused over the bycatch of dolphins, as well as of non-target fish species, seabirds and sea turtles.

The vehemence of public outrage had forced a major food company (Heinz, that sells StarKist brand canned tuna) in the United States to announce that it would cease to buy tuna from national and international suppliers in order to make any product if the tuna were caught by methods that harm dolphins. The company had also pledged to label its tuna products as 'dolphin-safe' so that the consumer could make a choice in the market-place. The two other major sellers of tuna in the US market had also joined this conservation effort, and were in the process of developing dolphin-safe standards.

In the meantime, US conservationists were pressing for passage of federal legislation (H.R. 2926) that would require labels on products containing tuna to inform consumers whether the product was 'dolphin-safe', or had been caught using methods that harm dolphins. Good intentions must be backed by laws and regulations. There must be strict standards with stiff penalties. A number of factors made that necessary. One, the standards announced by companies could differ, and unless the *minimum* standards were the same, the consumer would not know with any certainty what was being offered for sale. Second, the high seas nature of the fishery made it very difficult for anyone but official observers to verify that the tuna had been taken in a dolphin-safe manner. Third, the tuna boats were so huge that the catch from each voyage was worth many millions of dollars. The temptation to cheat would be very high. Dolphin-safe tuna sold to canners for more than half as much again as dolphin-unsafe tuna, or a ratio of 11 to 7. By June 1990, H.R. 2926 had been amended to ban the sale in the US of all products derived from fish caught with purse seines in the Eastern Tropical Pacific, or with driftnets anywhere.

The UN General Assembly (1989) had adopted resolution 44/225 which recommended that fishing nations should prohibit the use of large-scale pelagic driftnets (*see* p. 143) on the high seas by 30 June 1992 and to cease such activities by no later than 1 July 1991 in the South Pacific. The recommended moratoria, if implemented by fishing nations, would remain in effect until effective conservation and management measures had been taken based upon a number of considerations including statistically sound analysis. The burden of proof would be on the driftnet fishing nation to convince the international community that the conservation standards had been met before pelagic driftnet fishing could be resumed. The UN was to be commended for taking this step that alerted nations to the destruction of living marine resources.

Some coastal states conducted direct fisheries on small whales as they migrated past their shores. Japan, for example, had recently increased its direct take of Dall's Porpoises (*Phocaenoides dalli*) to more than 40,000 from a population estimated to be about 105,000. There was no effective management programme. Even Japanese scientists had warned that the stock would be destroyed if the take was not stopped now.

The loss of great and small whales was too drastic a price for the natural world to pay for mankind's rapacious appetite for marine resources. Besides, there were excellent arguments to support the idea that cetaceans should have the right to live out their full lives without harassment and decimation by people. They were the highest form of life in the oceans, as we liked to think people are on land. **Gulland** was right when he said that harvesting of marine mammals was unacceptable to the public.

Gulland had given an optimistic interpretation of the impact of commercial fishing on fish stocks. Except for those species of low fecundity, he had stated that fish stocks were generally resilient under exploitation, even when taken at rates of 50%, or more, annually because of the great fecundity of most fishes. He then commented that if the adult stock was reduced enough, in due course a point would be reached where the production of recruits to the fishery could not be maintained. From her reading of various reports of present-day commercial fishing activity, it appeared that the point of reduced production of recruits had already been reached in significant areas of the world's oceans.

Her concern stemmed, in particular, from the use of vast driftnets on the high seas. The invisible monofilament nets hung down like curtains of death suspended from a series of floats and were held in a vertical position by weights. They drifted on the surface of, or in, the water, and caught anything that swam or dived into them. Large-scale pelagic driftnet fishing had been developed by the UN Food and Agriculture Organization. FAO had promoted their use all over the world and it was still so doing. There were now an estimated 1,500 of the massive driftnet vessels worldwide. The South and North Pacific had been especially hard hit.

Driftnets were from 35 to 60 kilometres long and 10 to 15 metres deep (Anon. 1, 1989). These figures should be multiplied by the numbers of vessels operating in the South Pacific alone! Japan sent 60 in 1988, and Taiwan sent 130! The nets were set every night, and the incredible harvest of target and non-target fish taken up the next day certainly added up to dangerous overexploitation. Yet, the official stance of the Japanese at a South Pacific Forum Fisheries Agency meeting in 1989 was that, until there was scientific proof that drift gill-net fishing was threatening tuna stocks, there was no need to stop (Anon. 1, 1989). This attitude was the reverse of that of the present-day IWC in managing the great whales. That response also ignored the wasteful casualties of non-target fish species, and of sea turtles. It had been estimated (Anon. 2, 1989) that one million seabirds and 200,000 marine mammals, in particular, dolphins and porpoises, were killed each year in driftnets.

Besides direct killing, driftnets also presented another terrible problem in the marine environment. When monofilament nets were lost or abandoned, they became ghost executioners of fish, marine mammals, sea turtles, and seabirds. The nets continued to drift around in currents, finally coming to rest on the ocean floor laden with the remains of their catch. The nets were not degradable.

A third method of controlling fishing – by regulating fishing techniques – should be added to the two methods discussed by **Gulland**, which were limiting the inputs – number and power of vessels, or outputs – namely quotas.

Some efforts to prohibit or control driftnets were under way. UN Resolution 44/225 recommended that pelagic driftnets should be prohibited as mentioned earlier. Nations of the South Pacific agreed on November 29, 1989 to a moratorium on the use of long driftnets in the South Pacific, but neither Japan nor Taiwan had agreed to abide by it.

Legislation was moving through the US Congress (S. 1025 and H.R. 2061 to amend the Magnuson Fishery Conservation and Management Act – 16 USC 1826) that would place a permanent ban on the use of indiscriminate and wasteful fishing practices, such as driftnets, beyond the exclusive economic zone of any nation. If passed, the US. Government was directed to seek international agreements to implement the act. It provided for sanctions against those nations which did not abide by the Act.

The UN resolution and most conservation groups were of one accord in urging all nations to pass and enforce laws to prohibit the use of the following harmful fishing techniques:

1. Stop large-scale pelagic driftnets that have incredible impact on all life that ventured into them;
2. Stop purse-seiners from setting on dolphins in order to catch yellowfin tuna;
3. Stop the use of explosives or poisons to catch fish.

Some further actions that nations could take were:

1. Coastal states could prevent the use of their ports by vessels using the fishing techniques listed above. New Zealand had done this recently.
2. All states could ban imports of fish products that were caught by those harmful fishing methods.
3. All states could require labels on packages of fish products that would inform the consumer if the fish was caught using methods that harmed marine living resources.
4. States could establish sanctions that would ban the import of all fish products from countries whose practices undermined international fishery agreements.

Goldberg had advocated using the deep sea-beds for disposal of toxic and hazardous wastes. He had further suggested that scholars could make rational decisions on whether to dispose of those wastes on land, or at sea, on a case-by-case basis. There were two points that merited further discussion.

The first was that these decisions were too important to be left solely to the experts. To do that assumed that politicians and industry would remain neutral while the experts went about their work. That rarely, if ever, happened. It also assumed that politicians would establish, without disproportionate input from industry, a perfect scheme for assigning to scholars the task of deciding where to dispose of certain wastes. There was a major role for the public in decisions that directly impacted on the wet commons of national jurisdiction and beyond, to the open seas. The public should be involved in meaningful ways in decision-making processes which affected the wet commons.

It was fitting that the London Dumping Convention prohibited dumping toxic waste in the ocean in spite of the fact that eminent scientists had seriously proposed that the sea-bed should receive high-level radioactive waste.

The second point to consider was whether all those wastes needed to be produced in the first place. More emphasis should be placed on reducing or eliminating waste; next, on re-using or recycling the waste. More capital investment and effort was needed to improve technologies in order to reduce the generation of waste, and to seek new uses for it. Only after intense commitment to reducing the production of waste had run its course should disposal at sea be considered. In the meantime, people, industries and countries that produced waste must dispose of it (if disposal was the only alternative) in their own backyards! It was unacceptable to transfer it to impoverished 'developing countries' that only saw the advantage of receiving hard currency, or to dispose of it in the wet commons.

It was essential that national and international programmes for the development of systems of marine protected areas were started, or enlarged where they already existed. These systems should provide authority and mechanisms to identify representative biological communities, critical and sensitive areas, outstanding geographic and scenic features, and areas suitable for benign recreation. As of 1985 some 69 nations had designated 430 marine protected areas. A recent example was the declaration by Ecuador that the entire Exclusive Economic Zone (EEZ) around the Galápagos Islands was a whale sanctuary. Earlier Ecuador had said that tuna fishermen could no longer catch dolphins in its waters.

McCloskey also suggested that wilderness, as we understood and applied it on land, should be applied to the marine environment. Wilderness was usually a very large tract. Some had suggested a minimum of at least 25,000 hectares in size. It was unaffected by modern man, where people were visitors only for benign recreation or for research by permit. Wilderness values could apply to the watery, fluid ocean deeps: on the floor, in the water column of the high seas, and on the continental shelves. Designated wilderness would aim to prevent harmful human intrusions and uses of the identified areas. They would also serve as controls for measuring human impacts on non-wilderness (McCloskey, in press).

While the marine environment occupied some 70% of the Earth, it too could be compromised and destroyed by thoughtless, unplanned and unregulated human actions. The GESAMP (Joint Group of Experts on the Scientific Aspects of Marine Pollution) Report on the State of the Oceans (1990) had stated that 'The open sea is still relatively clean'. The Report had also noted that 'oil slicks and litter are common along sea lanes.'

GESAMP also said that the margins of the sea, however, were severely impacted by human use and misuse. **Goldberg** had referred to the general trend towards eutrophication of the coastal ocean and estuaries and cited a 15-fold increase in nitrate and a 5-fold increase of phosphate entering the oceans, on a global basis. The coastal marine surface and floor were littered; waters were polluted; natural features were destroyed by platforms, megadocks, anchors, trawls and explosives used for fishing; while marine areas essential for breeding and as nurseries for valuable species were converted to human uses. Many of those uses were not critical to human welfare. Coastal nations should closely regulate human use of their EEZs, and establish protected areas in them.

Public opinion must also be rallied to exert pressure on governments and on international agencies, such as the United Nations, to undertake development of enforceable systems of marine protected areas, including wilderness. These vital steps should be taken while there was still time for people to act as stewards of the marine environment. If current trends in uses of the high seas continued, they could no longer be described as being relatively clean, as the GESAMP report stated. Further, Mankind's ingenuity in developing the technology for exploiting the living and mineral resources of the high seas (for example, the pelagic long-range driftnets) would undoubtedly produce the means to exploit ocean resources that seem unexploitable now.

For these reasons, she suggested that now was the time for people of the world, acting through their governments, to address the problems and potential of the marine environment. It would be necessary to build on existing international agencies, such as UNEP, or to build new international institutions; existing scientific bodies could be called upon. The public, in the form of nongovernmental organizations, should have an important role.

If a coordinated international programme was not developed soon, the opportunity to prevent degradation of the high seas would have been lost. After-the-fact regulation and clean-up efforts were costly and of limited effectiveness. It was better to avoid or closely regulate the abuses. [*cited*: Anon. 1, 1989. 100 years to save whales. *The Pilot*, Newsletter of the Marine Mammal Action Plan of UNEP; Anon. 2, 1989. Commonwealth Countries Urge End to Drift Net Fishing. *The Siren*. Newsletter of the Oceans and Coastal Areas Programme of UNEP; GESAMP, 1990. *The State of the Marine Environment*. UNEP Regional Seas Reports and Studies No. 115; McCloskey, M. E. A wilderness concept for the oceans. (in press); UN General Assembly, 1989. Large-scale driftnet fishing and its impact on the living marine resources of the world's oceans and seas. Resolution 44/225.]

N. V. C. Polunin said that it was very important to realize that the ecology of the seas was different from that of the land in many significant respects. The average marine population was recruited via long-distance larval dispersal, the nature of which was such that there was a particularly poor relationship between adult-stock size and recruitment. Orderly growth in numbers did not occur in local populations in the sea. The environment offered physical support, a 'bathing' solution in which nutrients were dissolved and, thirdly, relative constancy of variables such as temperature, relative to the land. These features had enormous consequences for the size, mobility, chemico-physical make-up, and functioning of the average marine organism, relative to its terrestrial counterpart. The 'typical' food-chain was basically different: thus in the sea grazing usually took up a large part of the primary productivity directly, which was not at all the case on land. Ecosystems as a whole thus tended to be different in nature between land and sea. These fundamental ecological differences were very important because in the sea, being hunter-gatherers (as against cultivators on land), humans depended almost entirely on that aspect. The areas most productive to us in the sea were in a narrow inshore zone, which was also that most susceptible to pollution. This meant that waste-disposal, or fishery management, was not governed so much by a firm concept of maximum sustainable capacity or yield, but rather by perceptions of what an associate, John Pope, had referred to as the 'minimum sustainable whinge'. Dissenting views would always exist, but the object was to keep these to as low a level as possible!

Why was it important to recognize differences between the sea and the land? He gave an example, which was also a form of confession. For years people, including himself, had been setting up marine reserves, but the more he thought about these protected areas in the sea, the more he believed we needed to go back to first principles to figure out whether they were appropriate and whether they actually worked.

To begin with, what were the objectives? Past documents such as the World Conservation Strategy had told us that our aims in conservation were primarily to deal with issues such as habitat protection, species extinction and the maintenance of genetic resources. It was easy to argue that these were not matters of a high priority in the sea: there was really nothing as yet to compare with rain-forest burning and clear-felling in the sea, only a handful of marine higher vertebrates faced conceivable extinction in this age, and there was little cultivation.

Did marine protected areas 'work'? Remarkably, and ironically, some of the only evidence which we had on this score arose from the draconian prevention of fishing in the North Sea in both world wars. After both closures over a huge area, catches per unit effort of fishing increased dramatically over pre-war levels. But who would accept such drastic controls now? Would small areas, covering only a fraction of a stock's distribution, 'work', and how would that be? As he had mentioned, local sub-populations were open, and not self-recruiting, so that replenishment of numbers could not be expected to work. There were special cases where small-area closure would have a beneficial effect on recruitment, particularly where the fishing process had not only depleted numbers but also, as with heavy trawling gear, damaged recruitment habitat. In addition, where exploitation disturbed animals, these might aggregate in closed areas, but this increase in numbers would be due to migration and not to recruitment. Very often, and perhaps typically, area closure would not help problems of 'recruitment over-fishing'. 'Growth over-fishing', however, could be expected to respond to area closure, and increases in average size of fish were being recorded in response to area regulations, particularly in a few Mediterranean sites where controls had been effective over some years.

He would like to have recommended, in the face of this ignorance, combined with the potential importance of such management to the rehabilitation of exploited stocks, that we should use existing and planned reserves and any other closed areas in the sea, to find out more about the dynamics of these measures as a management technique. At present we knew too little, and that, he suspected, was because there had been an erroneous assumption that marine ecology was the same as terrestrial ecology.

Clark noted that **Goldberg** had introduced the question of disposing of wastes in the sea. Some of his remarks were controversial, but this was a subject that caused great concern and such was the heat generated in discussions about it, that it was often difficult to get a hearing for a dispassionate examination of the problems caused by waste discharges into the sea. In those circumstances there was little possibility that wise solutions would emerge.

It was true that there were many 'hot spots', such as industrial estuaries, where the marine environment had been degraded by the input of wastes. This need not have been so, but such damaged areas were essentially local problems which could be solved, or not, as the local population decided. They had no significance outside the immediately affected area.

In a few cases, much greater areas were degraded by waste inputs and had a regional, or even international, impact. Notable examples were the eutrophication of the upper Adriatic and the eastern North Sea. In both situations, the damage was caused principally by very large river inputs into shallow seas with limited water exchange. Bringing these under control required regional, or international, action, and presented a formidable problem.

However, while it was easy to produce examples of damage to the marine environment caused by waste discharges, the sea could degrade organic material without harmful results, and it would be wasteful of other resources not to make use of this facility. The secret, of course, was not to overload the system. Sanitary engineers responsible for disposing of sewage into rivers had been aware of this for many years and were capable of managing their effluent discharges in a very acceptable way. There was no reason why this should not be possible for marine discharges.

Goldberg had also made what must be a much more controversial suggestion, that the deep sea might be the safest place to dispose of some dangerous, long-lived materials.

It was refreshing to hear such an open-minded view of how the sea might wisely be used. Most conferences dealing with the sea took the view that damaging wastes should not be discharged into that environment – often, indeed, that nothing of any sort should be discharged into the sea. Of course, that did not solve the problem of waste disposal, but merely transferred it to a different environment for someone else to take care of.

It had to be accepted that disposal of wastes inevitably had an environmental cost as well as a financial cost; if the environmental cost was not in the sea, then it would be in fresh waters, the atmosphere, or on land. In the past, the objective was to make the financial cost of waste disposal as low as possible; it was now generally accepted that we should minimize the environmental cost.

It was in response to this view that the UK Royal Commission on Environmental Pollution had developed the concept of the 'Best Practicable Environmental Option', namely the policy that the preferred route for disposing of a waste was that which caused least environmental damage (it being impossible to avoid environmental costs altogether). This implied that sea disposal, even if it involved some environmental damage in the sea, might sometimes be preferable to other options if they would cause greater damage in another environment.

This was not a view that had been generally accepted. The sea was somehow regarded as sacrosanct and should be removed from the available waste disposal options.

Holdgate, in his magisterial contribution (Chapter 1), had pointed out that action was achieved through the pressure exerted by groups of people. That was very true and nowhere more so than over environmental issues. But in countries where the population had become environmentally aware, people had become so frightened of pollution that almost all ways of disposing of a waste were rapidly becoming unacceptable.

He gave one example. The North-west Water Authority in the United Kingdom had responsibility for disposing of the sewage generated in Manchester, Liverpool and the surrounding conurbations. As much as possible of the sewage sludge was used for agricultural purposes, or put in land-tips, but appropriate land for disposal was in short supply and over 1 million wet tonnes of sewage sludge per year were deposited in Liverpool Bay where it was acknowledged to be damaging. The estuary of the R. Mersey, on which Liverpool stands, was the most polluted estuary in Britain. Work was in progress to improve that situation. It would generate even more sewage sludge for disposal. Several years ago, the North-west Water Authority realized that sea disposal of sewage sludge would soon become unacceptable and looked for alternatives. Land disposal could not be increased, and the only remaining option was incineration. However, when it attempted to construct its first incinerator, it was prevented from doing so by the strength of local public opinion!

Two conurbations in north-east England disposed of sewage sludge to the North Sea, a practice which was now to be phased out. Both recently proposed to build incinerators, but each proposal had aroused intense local public opposition and we were now in a position of stalemate.

This was clearly an unsatisfactory situation. Wastes, none more so than domestic sewage, would continue to be produced and would have to be disposed of somewhere, somehow.

If public concern about the dangers of waste disposal became so comprehensive, we would be in danger of choosing, not the environmentally safest option, but the one that caused least public outcry. He suspected that that had already happened over the disposal of low-level radioactive waste in Britain.

If, as **Holdgate** had suggested, pressure must come from below upwards to safeguard the environment, the general public would need to be much better informed and more sophisticated than it was at present.

Unfortunately, the general public relied on newspapers and television for most of its information. The media were notorious for being interested only in bad news and for having an extremely short attention-span. As a result, the general public had become frightened about pollution, but understandably it lacked the perspective to evaluate the various options.

There was no shortage of detailed analysis of political or financial matters in the serious press, but he saw little informed discussion of waste disposal and pollution problems. Yet it was only through the media that an informed public view would be generated and then **Holdgate**'s 'groups' would provide benign pressure.

He was not optimistic that this would be achieved quickly, even in developed countries with a literate, educated population. Developing countries might have different obstacles to achieving wise management of their environment and natural resources, but it was by no means certain that the developed world had yet found the right balance.

Oza said that we must have a campaign to save the seas; indeed, he had already published a manifesto for the Indian Ocean (Oza, 1983: *Environmental Conservation,* 10 pp.331–5). Whenever giant fertilizer, or petrochemical complexes were established they were either on the banks of rivers, or on coasts. The products of soil erosion and leakages from the complexes were transported to the river mouths and polluted the coastal zone, while dumping of all kinds in the sea appeared to be based on the notion that it was a free zone. The consequences were well known, pollutants accumulated on the gills of fish, they became de-oxygenated and died in their thousands. Oceanic islands were some of the worst danger spots, for they often sheltered relic species. Selected islands or areas should be declared as reserves and illegal and indiscriminate visits prohibited. What was needed were immense reserved areas of coastlines, oceans and their islands. This was especially true for the Indian Ocean, which, if declared a 'Zone of Peace for Nature', would not only provide a sanctuary for the great whales and smaller cetaceans such as dolphins and porpoises, but would also help in the conservation of marine turtles and dugongs and, of course, fish.

Harun ur Rashid drew attention to the particular problems of developing countries such as Bangladesh. In the Bay of Bengal, unauthorized dumping by stealth had grown over recent years. The government was well aware that there were ships with highly toxic cargoes in coastal waters and areas economically important to Bangladesh; some companies had even asked permission to dispose of these cargoes at inland sites. Even when permission had been refused, illegal dumping had then taken place in the coastal waters. Unfortunately, the government had neither the capacity nor competence to catch those vessels. A second problem, already discussed, was oil-spills in the Bay by supertankers. Thirdly, because the Bay was so rich in fish, there had been overexploitation by foreign trawlers which they were also unable to catch.

Fosberg suggested that simply because there was a past, sedimentary record it should not be assumed that it represented a 'natural' event or phenomenon. It could have been the record of an earlier pollution event. Equally, it should not be assumed that contemporary events were not pollution-assisted or determined, even where there appeared to be no obvious connection. He cited the current plague due to the 'Crown of Thorns' starfish on coral as an example.

Secondly, on the question of ocean wildernesses, he suggested that one of the more appropriate areas would be one including coral atolls, such as some of the smaller archipelagos in the Pacific, or Indian, Oceans. They included not only deep water, even close inshore, but also a complete range of habitats from the deep bottom to the top of the reefs, with a range of animal and plant populations which he thought was almost unequalled.

Vallentyne thought that more should have been said about fisheries management, although, in fact, the fish manage themselves satisfactorily and Man's job was rather to manage the uses and abuses of the fish!

So far as regarding the oceans as an infinite waste-dump that, tragically, had to deal with problems of terrestrial garbage. In part, at least, he saw that basically as a problem of how to relate Man's behaviour to the planet he inhabited. He had to say that **Goldberg**'s proposals to give a shot of this or that to the ocean was the same philosophy as that which had got Mankind into such a mess with the Great Lakes in N. America; first it was 'only into the river', then 'only into the bay' – 'the whole lake' – and finally, 'the whole lake system'! He wondered if **Goldberg** had really meant what he had said?

Batisse remarked that there were always neat and distinct discussions on the one hand about forests, erosion, population and so on, and on the other about marine matters, waste dumping in the ocean, whales and the like, but what **Goldberg** had said made it clear that the most serious problems, which affected the largest number of people, were in the coastal areas. Unhappily, those areas fell between two stools; they were neither considered seriously by marine scientists, nor by the terrestrial scientists. This had serious consequences both at the scientific level and at the administrative level so that, at the decision-making level, there was a split, and what was done on land was not coordinated with what was done for the sea. This had been very evident in the 'Blue Plan' for the Mediterranean, where the real problems were conflicts in the coastal fringes of the kind described by **Goldberg**. The problem for the future was to resolve the conflict between urbanization, agriculture with its runoff of phosphates etc., industry, the building of huge harbours, and fishing and leisure activities in waters polluted from the land. It was one thing to talk about the protection of the marine environment in general: quite another to protect coastal areas – which in England implied something to do with the sea, but in France, because they were a nation of peasants, was to do with the land! Those two meanings of 'coastal' had to be reconciled and something would have to be done for coastal regions globally, for they were amongst the most critical ecocomplexes in the world.

Tisdell recalled that, in the Brundtland Report, mariculture had been seen as a means of sustainable development, but he thought that this might have to be revised in the light of some of the issues raised by **Goldberg**. The conflicts in the coastal regions raised real doubts as to whether mariculture was a good means of sustainable development.

He saw **McCloskey**'s comments, ostensibly concerned with non-economic aspects of the seas, as balancing those of **Gulland** and **Polunin**, which had tended to be concerned with commercial fisheries. However, **McCloskey** was really concerned with marine animals of economic value but which were not marketed! It was a mistake to suppose that economics was only concerned with commercial matters; economists were concerned, for example, with the *existence* value of whales. Economics was concerned with the administrative resources provided to achieve best what the people wanted, and were prepared to pay for, within the limits of what was available but, if people wanted more conservation, and that was not being delivered by the resource allocation mechanism, then that would have to be altered if economic problems were to be addressed effectively. A second question which the whale issue raised was, how much weight should be given to existence value versus value in use, or, more specifically, how far should the views of those wishing to conserve whales prevail over those whose interest was in the value of whales in use? Could a compromise position be sought and obtained? For example, should some whales, *e.g.* Minke Whales, be caught subject to safeguards? He felt that at present we had no proper mechanisms for reconciling conflicts between conservationists and non-conservationists. It was a matter, put crudely, of the simple ascendancy of one over the other. It reflected the fact that mechanisms for resolving social conflicts were really rather poor and, of course, there was no easy way, as value judgements were involved.

7. Aspects of Overcoming Freshwater Constraints: A Compendium*

E. Barton Worthington, L. Somlyódy, J. W. M. la Rivière, B. Juel-Jensen, G. E. Petts & J. R. Vallentyne

Introduction

Fresh water is, perhaps, the most important natural resource on which terrestrial life depends. Without it, very little life would exist outside the oceans. In addition to providing the greater part of the mass of practically all living organisms, water is essential for the transpiration stream in plants and for disposing of waste products of animals. For Mankind in particular, water is essential for washing and cleaning, for agriculture and forestry in their many forms – especially for irrigation, for fisheries, and for practically all other supplies of food and fibre. It provides for the cheapest form of transport, for a significant proportion of power, and energy for recreation in many forms.

To make water available where it is needed in space and time, has involved widespread interference with the natural hydrological cycle: evaporation, mostly from the oceans, is followed by condensation of atmospheric water to be precipitated as rain, snow, or hail – not forgetting occult precipitation as dew, which can be particularly important in desert areas. On land surfaces, water then follows a three-ways passage into runoff back to the oceans, percolation into underground aquifers, and evaporation, of which the major part is usually through transpiration of vegetation.

In this overall scenario the chief constraints fall under five main headings:

The *supply* of water and how it came to be.

The *quality* of water, its purity and impurity.

The *storage* of water in reservoirs by dams.

The *conservation* of great lakes.

The *health* problems of water-related diseases. [E.B.W.]

Water supply: the example of freshwater resources in Hungary

More than 90% of the country's surface-water resources originate from abroad (Figure 7.1, p.152) thus giving to Hungarian surface waters a special downstream character. Subsurface water resources are dominated by bank-

*[Based on a conference Round Table Dicussion moderated by E. Barton Worthington. Eds.]

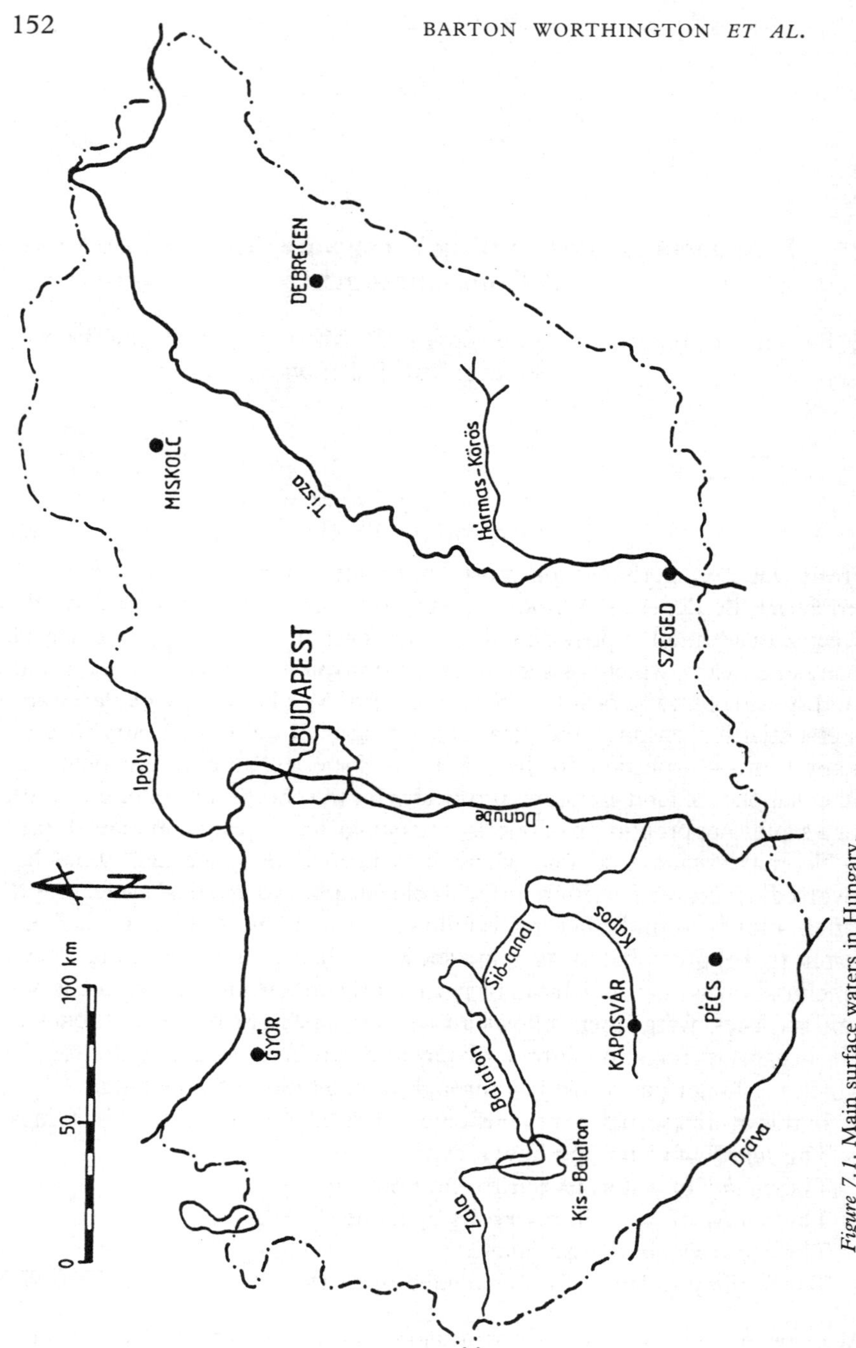

Figure 7.1. Main surface waters in Hungary.

filtered resources, three-quarters of which are found in the alluvial-terraces of the River Danube.

Between 1970 and 1985, water demand doubled, reaching an annual value of 6 thousand million cubic metres. Further increase, although at a reduced rate, is expected until the year 2000 AD. The most significant water user is industry (81%), including the electrical power-generation industry, the cooling-water demand of which amounts to 3.7 thousand million cubic metres annually. This is followed by agricultural and communal water-use. About half of the water demand of the latter is being met from bank-filtered water resources.

By comparing water uses with free-water resources, a favourable picture is obtained, and demand also seems to remain satisfiable to the turn of the millennium – at least when the various development plans are taken into consideration. However, a more cautious and less favourable description of the situation is justified by the fact that the spatial distribution of resources and demands is far from uniform. Thus while available free-water resources between the water systems of the R. Danube and R. Tisza (Figure 7.1) are 85% and 15%, respectively, the respective percentages for water demand are 59% and 41%. These proportions are even less favourable in periods of drought in the Tisza River Basin. Well-supported regional water-management policy is needed to enable water resource distribution difficulties to be resolved. Moreover, further decreases of water resources are to be expected, due to the downstream character of the country. This will be caused, potentially, by the reservoirs to be constructed by the upstream countries.

Water Quality

The quality of surface and subsurface waters is essentially determined by the magnitude of pollution, both in Hungary and abroad. Industry is the largest water-user and, at the same time, the largest wastewater discharger. Although the fraction of untreated wastewaters was decreased from 48% to 27% in the period 1965–85, the pollution-load still amounts to 84 million cubic metres annually. The discharge of organic and other pollutants – heavy-metals, oil, and oil products – is still substantial. The effect of agriculture on water quality mostly takes the form of diffuse pollution loads. The quantity of liquid manure and manure leachate (48 million cubic metres annually) is also significant; nearly 20% reaches the receiving surface and sub-surface water sources, meaning that safe disposal has not yet been secured.

The rate of development of public water-supply has been much faster, in the past decades, than that of sewerage and sewage treatment. Thus the former activity, called the public utility loop or ('scissors'), was 5% in 1945 and has risen to 38% today: 85 per cent of the total population receives a piped water-supply, while only 50% is connected to sewers. As a result of this, the communal sewage load has been steadily increasing, and about 1.0 million cubic metres of untreated sewage is being discharged onto the receiving surface and into sub-surface water sources annually. However, the actual situation is even worse, as two-thirds of the 'treated sewage' receives mechanical treatment only!

Following an individual survey made in the '50s, the regular nation-wide monitoring of water quality started in the '60s. The surface-water-quality monitoring network now consists of 250 sampling stations. Sampling frequency varies between 12 and 52 samples annually. The number of components used for qualification is, on average, about 25.

This regular monitoring enables water quality and its changes to be evaluated (for details *see* Hock & Somlyódy, 1990). For example, the six observed components of the oxygen and plant-nutrient 'household' provide information on the decomposition of organic matter and on tropical conditions as well as on the nitrification process. On the basis of measurements made between 1976 and 1985 on the 43 most important watercourses, it can be stated that, in about one-third of the stations, the water quality has improved; somewhat less than half of these stations fall into quality Class I, while in 15–20% of them the unfavourable quality (*i.e.* worse than Class I) is associated with excessively deteriorating conditions.

Comparing these results with those of the '50s, it can be stated that although – due to the control measures implemented – the rate of deterioration slowed down in several streams (and even improved in certain cases), the total number of those sections where 'worse than Class I quality' is associated with 'excessive rates of deterioration', is growing.

The Hungarian 'general picture' outlined may be supplemented by information on the state of the country's five most important river systems: the Kapos, Zala, Zagyva*, Danube, and Tisza, river systems (Figure 7.1). On the basis of data assembled over the past 15 years, it can be stated that the quality of water in the Kapos, Zala, and Zagyva*, rivers has been characterized, with the exception of a few components, by continuing deterioration.

At the Danube's entrance to the country, characterized as it is by water of quality Class I in respect of most of the traditional chemical water-quality components, a slight improvement can be demonstrated, with the exception of the nitrate ion. In the exit section even the nitrate ion exhibits a deteriorating trend. In the section where the River Tisza leaves the country, the changes in all water-quality components are continuously unfavourable (the situation is somewhat better at its entrance section). Thus, comparing the data of entrance and exit sections of the R. Danube and R. Tisza with each other, only irregular fluctuations occur in the case of the River Danube, whereas increasing rates of in-time quality-deterioration of all components characterizes the River Tisza along its longitudinal profile.

With regard to plant nutrients, the entrance section of the Danube can be best characterized by the data of Reinhard Liepolt, an Austrian biologist, through comparing the results of 1960 with those of 20–25 years later. The characteristic concentrations of the components of the nitrogen 'household'

*[Not shown on Figure 7.1. Eds.]

rose to 3–5 times higher values than previously, while the orthophosphate-ion concentration increased by nearly an order of magnitude. Simultaneously, the daily and seasonal fluctuations of dissolved oxygen concentrations increased, and the algal count rose to 5–10 times its former value. At present the nutrient concentrations are so high that the eutrophication process of the Danube (an observable phenomenon!) is limited neither by nitrogen nor by phosphorus, and light has become the factor that limits algal growth.

Budapest and Water Quality of the River Danube

On the basis of chemical components, the polluting effect on the capital, Budapest, can hardly be demonstrated, owing to the enormous dilution of contaminants. However, bacterial components, being much more sensitive, unambiguously indicate quality-deterioration. For example, coliform and *Streptococcus* counts are, respectively, five and ten times higher downstream of Budapest than are their respective upstream values, although even the latter exceed the limiting values set for water-associated recreation. It is worth while noting that the bacterial quality of Classes II–III of the early '70s dropped to that of Classes III–IV as of today.

With regard to heavy-metals, the wastewater discharges of Budapest increase the metal concentrations of the river water by about 20%. The maximum values of mercury and lead exceed the limit-values that are considered tolerable. Mercury, lead, chromium, and copper, contaminations can also be found in the bottom sediment of the river. Heavy-metal concentrations, measured along the Hungarian river-reach, exceed the natural background concentrations by 1–2 orders of magnitude, and the observed maximum values exceed, for all toxic compounds but copper, the concentrations that are allowed in soils as the tolerance limits of plants!

'Bank' of Well-filtered Drinking-water Resources

About 85% of the drinking-water supply of Budapest is provided by a 'bank' of well-filtered water resources. The wastewaters of Budapest (only 21% receive biological treatment) have an unfavourable effect on these resources as well. While the ratio of objectionable wells in the upstream, northern water resource is only 9%, the comparable figure in the downstream, southern water-base rises to 46%. Increased pollution-loads from the off-river background zone contribute to this ground-water contamination as well: in the Isle of Csepel, downstream of Budapest, the ground-water is equally rich in nitrate, organic carbon, iron and manganese. Nitrate contamination is especially dangerous when it exceeds 40 g/m^3, the 'tolerable' limit-value of the Hungarian drinking water standard. The limit is exceeded over nearly half of the Isle and, over more than 5 per cent of its area, it exceeds 200 g/m^3.

Owing to the increasing organic loads on to the channel sediment, near-anaerobic conditions prevail in some of the wells located along the Ráckevei arm of the Danube, which has a favourable oxygen supply. This situation can easily lead to an increase of iron, manganese, and ammonia, in the well-waters. The removal of these contaminants requires the introduction of expensive additional water-treatment technologies. The situation briefly outlined above could be worsened further in the future, by the seemingly insoluble sewage-treatment problems of the capital, at least in the short run, and also by the fact that the increase in the drinking-water supply must be based on abstraction, downstream of Budapest.

It should be noted that future changes in water quality of the Danube and associated bank-filtered water resources, depend greatly on the fate of the construction of the Gabeikovo-Magymaros river dam-complex upstream of Budapest – a joint investment of Czechoslovakia and Hungary. This large project was launched in 1977. During the early '80s, criticism within the community and new-born, alternative movements gradually became stronger. Concerns were raised about the impact on water quality, changes in ecosystems, genetic resources, nature conservation and aesthetic values. In 1989 Parliament suspended, temporarily, the construction of the dam-complex and this was followed later by the cancellation of the building of a downstream river barrage. The final fate of the hydro-project will be decided in the course of further interstate negotiations. For further details the reader is referred to Hock & Somlyódy (1990).

Other Sub-surface Waters

The quality of naturally protected sub-surface waters, other than bank-filtered resources, is stable, whereas that of the karstic waters and phreatic ground-waters (not having natural protection) is deteriorating. An especially severe problem is caused by nitrate pollution (the cause of which is partly the above-mentioned 'scissors of public utilities', and partly the non-point-source pollution of agriculture). As a consequence of this situation, about 700 smaller settlements have a drinking-water supply only from 'plastic bags'.

High salt, gas, iron, and arsenic, contents and the bacterial pollution of dug wells, create problems of differing magnitude in various parts of the country. The leaching of contaminants from inappropriate, or illegal, solid waste-disposal sites presents a serious threat to ground-water resources (several accidental ground-water pollution events associated with such disposal sites were discovered recently). The increasing occurrence of such cases in the future is highly probable.

The Water Quality of Lake Balaton

The long-term fate of the water quality of Lake Balaton (Figure 7.1) – one of the largest shallow-water bodies on the Globe – whose environs are the most important recreational area in Hungary, depends on the reduction of external phosphorus loads and on the (probably much delayed) decrease in the internal loads (i.e. releases from the sediment). After the Decree of 1983 of the Council of Ministers on regulatory measures for Lake Balaton, chemical phosphorus-precipitation technology was applied to 10 sewage-treatment plants. The construction of a regional sewage-treatment plant system is under way in the region, and two pre-filter reservoirs, with a 50% or more P-removal efficiency (those of Marcali and the Kis/Small/Balaton) have been put into operation. Thus the externally-originating phosphorus loads of the Lake have been cut by about half since 1982.

Nevertheless, no spectacular improvement in the Lake's water quality is expected, in spite of all these reductions of the externally-originating load. The reason for this is the high phosphorus release-rate of the bottom sediment (also verified by isotope-tracer studies) due to earlier accumulations of nutrients in the Lake's sediment. Another reason is that, owing to the excessive supply of nutrients, N-autotrophic heterocystic Cyanobacteria play a dominating role in the summer months in the determination of the water quality of the western parts of the Lake. The occurrence of these Cyanobacteria depends mostly on meteorological (temperature and light) conditions. Consequently a substantial and lasting improvement of water quality requires the renewal of the bottom sediment as well as suppression of the Cyanobacteria. These changes might take several years or even a decade to come about.

Greenhouse Effect and Water Quality

Scientific knowledge in this field only permits speculative statements. It seems to be fairly certain, however, that the estimated temperature rise of a few degrees in the Carpatian Basin will modify precipitation, run-off, snow-melt, the period of ice-cover of the surface waters, the stratification of deep reservoirs, the behaviour of contaminants that have been deposited in the bottom sediments, and also the composition of aquatic communities. Investigations of the interactions of all these effects has not been started yet – not even at the research level.

Expected Future Changes in Water Quality

Seemingly realistic plans for the development of urbanization, industry, agriculture, and the necessary infrastructure – all affecting the pollutant-load conditions of receiving waters – are available for the period up to 2000-2005 AD.

According to the plans available, 95% of the population will receive a piped water supply by the year 2005 AD, while the development of sewerage will not exceed 70%. This means that there will only be a beginning to closing the utility loop, or in other words that the pollution of our waters will continue. Industrial wastewater discharges, with their moderate estimated rise between 1990 and 2000 AD, will add to this problem.

Plans of the previous Ministry of Agriculture did not indicate substantial alterations in the use of fertilizers (and the associated non-point-source pollution). However, the structural changes already initiated by the new Government in 1990 may have a favourable influence in this respect.

All in all, the rate of water quality deterioration observed over the past few decades will continue further. To avoid this situation, the only possibility would be to increase the rate of construction of sewers and sewage-treatment facilities, in order to confine the utility loop to acceptable limits. But this sole action would require the investment of a couple of thousand millions of $US! Moreover, industrial wastewaters should be given appropriate treatments, and up-to-date agricultural production technologies should be applied, in order to reduce water pollution by agriculture. The precondition of achieving a reassuring long-term solution lies in intensive environmental management based on preventive actions. [L. S.]

Water Quality: Aspects of the Purity of Water

Practically speaking, there is no such thing as pure water. In the hydrological cycle, pure water may be generated by distillation; but this water will remain pure for a very short time only; as it rains down from the clouds, the water picks up gases and particles from the atmosphere. When it reaches the ground, it is further contaminated with natural and Man-made substances during run-off and as it percolates through the soils by leaching.

But why, when we only have impure water at our disposal, can we drink water safely from the tap in many parts of the world, as for instance in all cities of Hungary? The answer is, simply, because we can counterbalance contamination with water-quality management and, fortunately, only in a few cases, e.g. high-pressure boilers' water, has it to be as pure as distilled water. Each water-use (drinking-water, irrigation, fishing, etc.) has its own quality-standards, and we have physical and chemical technology, as well as biotechnology, to help us convert the raw water into a product that meets the quality criteria for a given use. By means of modern water-quality management methods, most pollutants can be eliminated, and, after all, we still have distillation as a last resort. It is essentially a matter of energy and money.

It is my task to analyse the constraints which we encounter in water-quality management. As a stand-in for Professor Stumm, I cannot do better than to take a formula which he developed with his colleagues in Dübendorf, that gives an excellent conceptual description of the counteracting processes of contamination and purification in a river basin. This formula is very useful for illustrating the constraints that occur under different conditions in various parts of the world. The formula reads, in a simplified form:

$$C = \frac{n \times GNP - E}{D}$$

where C = contamination load on the river ($\$/m^3$), i.e. money required for decontamination, n = number of inhabitants in the river basin, GNP = gross national product ($ per person per year), D = river discharge (m^3/year), and E = effectiveness of contamination ($\$/m^3$) i.e. money spent for decontamination.

In *industrialized countries,* where n is at a standstill, C can be kept reasonably low as there is a good capacity for water-quality treatment and E can be made as high as the voters want it to be. Beside the risks of spills of toxic chemicals, however, the production of new chemicals leads – in a river basin such as that of the Rhine – also to new waste chemicals entering the river, which forces the water engineers downstream to use more sophisticated treatment methods. Instead of continuing this battle of wits between chemical and water engineers, all parties are moving towards the idea that it is better to design the manufacturing processes in such a way that the waste discharges are minimized, and that it is more economical to invest in such clean technologies than in more complex treatment methods.

It is also important to note that, often, industry moves from a position of confrontation to a pro-active position in which it anticipates future environmental standards and sees a competitive advantage in designing its processes to meet those standards before the law so requires.

In the *Newly-industrialized Countries*, the population is often still growing, and first priority is given to increasing the GNP, i.e. industrialization and intensification of agriculture, without concomitant upgrading of water-quality management. In these countries serious water-quality problems are bound to occur, if only as a repetition of what has happened in Western Europe and the US. It would, therefore, be important to import to them not only clean technologies for CFCS, but also for other chemical industries.

In the *Less-developed Countries* the situation is much worse: n grows relatively fast and E is low. Although a low GNP – i.e. a low level of industrialization – means less pollution, the effects of the discharge of domestic waste on public health are disastrous, and meet insufficient response. The problem of the millions of people suffering from water-related diseases in tropical less-developed countries will be discussed in detail in the following contribution.

In winding-up I wish to emphasize three points of major concern:

1) *Acidification*, with its effects not only on soil and vegetation but also on water quality, especially in lakes: the extent of the problems caused in, e.g. Eastern Europe, and also in some tropical countries, is probably far from completely realized.

2) *Eutrophication* of lakes and coastal zones most fortunately is reversible, but only through time-consuming and costly procedures.

3) *Ground-water pollution by leachates from toxic-waste dumps:* as the containers, acting like time-bombs, are slowly rusting through, the chemicals which they contain will be gradually liberated, as shown in the well-known cases of Lekkerkerk and Love Canal. It seems certain that many instances of this kind of insidious pollution go still undetected, and even unsuspected, in many Third World countries, until it is too late.

In the tradition of Professor Stumm, I wish to conclude this contribution with a provocative statement of my own:

For counteracting the 'greenhouse effect', the planting of trees on a massive scale is often advocated. If we realize, however, how much faster photosynthesis – that is, CO_2 fixation – proceeds per gram of aquatic biomass, it appears worth while to examine how much carbon could be sequestred per square metre of *e.g.* a lake that has been made eutrophic and in which the fate of the bound carbon is directed towards rapid peat formation rather than mineralization back to CO_2.

[J.W.M. la R.]

Aspects of Health and Water-related Diseases

Water is essential for life. Man can survive for four weeks or longer without food, but if he has no water he would probably die before a week is out. Water does, however, carry a number of potential risks. Where I come from (Oxford), we drink recycled sewage. It has been heavily chlorinated, is safe, but if you think of its origin you are not tempted to drink it! Many buy bottled water. Some of you may be aware that the suggestion has been forwarded in some quarters that aluminium may be the cause of Alzheimer's disease. When recently the Oxford water had a high aluminium content, chlorination was temporarily suspended, and the result was that several people went down with infection with *Listeria monocytogenes* and others with *Cryptosporidium* infection.

Water carries a very large proportion of illnesses, particularly in underdeveloped countries. Only malaria probably accounts for more victims. It carries viruses: hepatitis A (infective hepatitis), and one form of non-A non-B hepatitis, hepatitis E. It makes it imperative that non-immune people from developed countries should be protected passively with gamma globulin before they venture into underdeveloped countries – unless they have already had the infection and so have antibodies which will last for life. Water carries enteroviruses, of which the most dangerous is polio. Paralytic polio is, alas, still very common in underdeveloped countries.

There is, in water, a rich 'botanical garden' of Bacteria. Vibrios abound, the best known being that which causes cholera. Bangkok has a great problem. The water that leaves the central pumping station is sterile. But the pipes are ancient, and the pressure is low. People put pumps on their water supply, and because the pipes are cracked, water from the canals, the *klongs*, is sucked into the main water supply. The water in the *klongs* is so polluted, that the larvae of the malaria-carrying mosquito cannot live in them. Therefore the incidence of malaria in Bangkok is near zero. On the other hand, over two thousand people are admitted to the infectious disease unit each year with cholera!

There are hundreds of different kinds of *Salmonella;* the majority just give a gastro-intestinal upset, but all can go into the bloodstream, and typhoid and paratyphoid practically always do so and cause severe illness. Water carries bacillary and amoebic dysentery.

One of the commonest of all organisms that is transmitted by water and food is *Giardia lamblia.* Giardiasis is one of the commonest diseases with which people who have been abroad present us in Britain. You can collect it in Leningrad or Moscow, and certainly in underdeveloped countries. We had a look at some Ethiopians and found that they all carried a fairly heavy load of *Giardia lamblia* as passengers. They had no symptoms, and undoubtedly had come to terms with their parasites. Occasionally the numbers rose to such a degree that they suffered from malabsorption, and became symptomatic. If a person from the West goes to that same country, the odds are that he or she will develop malabsorption. Lactose is not absorbed, and breakdown products such as butyric acid cause the symptoms of the condition.

In recent times it has become obvious that water, contaminated with *Legionella pneumophili*, can cause an unpleasant pneumonia, Legionnaire's disease, when aerosols (showers of fine water particles) are inhaled – particularly from shower-baths and from contaminated air-conditioning systems.

Cooling towers that spew out large amounts of aerosols have caused outbreaks of pneumonia. The water can be made safe by heating it to above 60°C.

We talk about safe water, but safe for whom? In many underdeveloped countries, vegetables are egged on with human dung. Most people have a little free acid in their stomach, and a small number of pathogens will be killed by the acid; but if the ingested microorganisms are too plentiful, some will get through and cause problems. Some people are at particular risk. The malnourished, very young children, and the very old, have not got the same immune defences that the average adult has.

In recent years an appalling problem has arisen with the spread of HIV (AIDS virus) infection throughout tropical countries. As many as 10% of many African nations are infected. They will react adversely even to a small number of pathogens. 'AIDS' was ill-defined, made to suit the circumstances that obtained in the United States. What you are attacked by depends entirely on which 'garden' you happen to be in. In Africa many immune-suppressed have a rich intestinal flora of potential pathogens (which in normal people are mere passengers) and they often get severe diarrhoea, and develop the so-called 'slim disease', which is practically unknown in the West. Tuberculosis is much more common in tropical countries than it is, say, in the United States, and therefore it is at least as common as, or commoner than *Pneumocystis carinii*, which attacks so many people with HIV infection in the West.

Elsewhere in this conference the problem of population explosion is being discussed; it has been accepted as priority number one by everyone. Contraception will help arrest the population growth, but it will incidentally also prevent the spread of HIV infection, which in tropical countries is spread heterosexually. In many countries in Africa 10% or more of the population are already infected. They will die; perhaps it may take ten or twenty years, but they will certainly die. The imperfect drugs we have at present to combat the condition at best prolong the duration of the illness. They are exceedingly expensive and Third World countries could not afford them.

It is quite unrealistic for anyone to invest money in underdeveloped countries, unless they devote a proportion of their budget to infrastructure, including health. In 1976 it was my privilege with a colleague from my College, a geographer and expert on Third World affairs, to undertake a survey, on behalf of the Saudi Foreign Ministry, of the needs of the Sudan for bilateral aid from Saudi Arabia. It was extremely revealing. We had been given a brief which said that the overriding considerations were l) economical, and 2) political. Humanitarian considerations were in very small print.

Let me give you two examples of how that can be disastrous. At Rahad there was a large irrigation scheme. There had been no provisions for adequate housing and, in particular, adequate water-supply and latrines. There had been no planning to prevent malaria and schistosomiasis. Most of the work-force were migrant workers, and some of them were already infested with *Schistosoma haematobium* and *S. mansoni.* Because they had no other way of relieving themselves, they defaecated and urinated into the irrigation channels and spread the cysts of *Entamoeba histolytica* (amoebic dysentery), and the eggs of *Schistosoma* spp. (causing Bilharzia). There were plenty of mosquitoes, no attempt having been made to prevent the larvae from living in the irrigation canals. The result at the end of the first year of operation was that the income fell short by 30% of the most pessimistic forecast. This was because of illness.

Another example: Kenana is the largest sugar plantation in the world. It has a huge refinery. Water from the White Nile is channelled along to an area fed by the Blue Nile, which has valuable silt from the Highlands of Ethiopia. Provisions for the workers were inadequate. Although the scheme was well engineered – the refinery with hiccups eventually worked – no one had thought about how the produce should be moved away. There were no roads, there was no waterway, there was no railway. The first year's crop had to be burnt.

I have mentioned these two examples to show that you do not have to be soft-hearted to invest in people. A hard-nosed financier will allow about 25% of his total investment to go towards infrastructure, and a large proportion of that will be for health. Only then will he get a return on his invested money. [B.J-J.]

The Storage of Water: Overcoming Damming Effects

The history of water-resource development is characterized by a progression of technological achievements, culminating in the mega-projects of this century. Since the mid-1970s, however, the mega-project has been attacked as often destructive and poorly conceived. Conflict has arisen because of a failure to recognize that water development is more than a technological problem; there are also important environmental and social issues (Cosgrove & Petts, 1990). This note examines the developments that are commonly necessary to achieve sustainable, and ecologically-sound, water development.

Dam-building Activity

The era of the mega-project for water resources development began in the second half of the nineteenth century. By the end of the 1930s, great multi-purpose dams, often grouped to form extensive multi-purpose projects, had come to symbolize social advancement and technological prowess (White, 1987). Less than forty years later, environmental groups were arguing that large-scale dam projects encapsulate practically every imaginable social, environmental, and economic, folly!

One view is that if there were accurate accounting of all environmental and social impacts, alternative ways would be found to provide the services which the large dams were intended to supply. Despite scientific studies during the 1970s and '80s (*e.g.* Petts, 1980) which demonstrated that dam-building could lead to non-sustainable resource development, large government-funded and internationally-financed water projects continue to be advocated by various bilateral and multilateral donors and leaders of several Third World countries.

It is still a popular belief that water represents a natural resource which is capable of development at a cost (economic and environmental) far below that of alternatives. This perception is based on four arguments:

a) Fresh water running to the oceans is wasteful;
b) Wetlands are wastelands and a hazard to human health;
c) Floods are an unacceptable threat to land and life; and
d) Hydroelectricity production is an environmentally sound option for energy development.

Although rapid and continuing urban development in drylands is placing a particular strain on water distribution systems (Beaumont, 1989), it is the last of these arguments that today is the main driving-force behind continued dam-building. World-wide, only *c.* 13.5% of the technically feasible hydroelectric power potential is exploited currently, and 40% of this is provided by only 150 large dams. Considerable opportunities therefore exist for carefully-planned hydroelectric power development. One scenario for developing countries requires 565,000 MW of installed hydropower capacity compared with an actual level of 154,000 MW in 1986; other scenarios envisage higher demand. Although critical capital costs of hydro-power development are high, and the lead-time can be 15–30 years, the profits eventually realized are usually favourable, owing to the long life-expectancy of installations and the simplicity of operating them. Average production costs are generally only a fraction of those of electricity from other sources. Nevertheless, economy demands the exploitation of a hydroelectric power source on the maximum feasible scale.

Opportunities exist to improve the efficiency of water-use, but in developing countries the large project remains the most attractive option to secure short-term economic growth and food security. Furthermore, water projects which have gestation periods of more than 15 years in countries with high population growth will have to proceed, because the risks of not harnessing these resources in a timely manner may be too high. The inescapable conclusion is that, for the next 50 years at least, large dams and reservoirs will remain the most attractive option to achieve energy and food security.

Towards the 21st Century

All large water-engineering projects change landscapes and the sustaining processes for those landscapes; they also cause social impacts. In looking to the future the problem is to develop the necessary tools, and approaches for a positive and cautious approach to river development.

Exploitation versus *Non-use*

Implicit in many approaches to water development is an acceptance to exploit, completely, landscapes that are thus degraded or denuded – as a trade-off to preserve a few 'natural' landscapes. For example, the World Bank's objective is to seek a balance between preserving the environmental value of the world's more important remaining wildlands, and converting some of them to more intensive, immediate human uses (Ledec & Goodland, 1988). Moreover, where developments are confined to less than 35% of a drainage basin, the effects on the fluvial hydrosystem may be negligible (Petts, 1980) or at least manageable. Clearly, the decision-making process requires methods to assess actual and potential values of the landscapes concerned, so that the location of proposed dams can be incorporated into an overall catchment management plan which gives due regard to conservation as well as other needs.

Secondary Regulation

A complementary approach is to develop water resources whilst simultaneously seeking to enhance degraded or denuded landscapes, and to preserve special ones (Petts, 1989). A range of management tools is being developed for impact mitigation and river restoration (Petts *et al.,* 1989; Petts, 1989, 1990). These include l) flow control, 2) structural design and maintenance, and 3) biological controls and regulations on human activity. In some cases, at least, natural restoration processes operating over time-scales of 10-100 years mitigate the effects of immediate impacts (Petts, 1987), and secondary regulation should seek to minimize these impacts and to accelerate the restoration process. Secondary regulation measures must become a primary component of all projects if sustainable, ecologically-sound, water development is to be achieved.

Project Administration

In many cases, impacts have not been caused by engineering works *per se* but by poor administration and coordination. The development and expansion of western modernization over the past three hundred years has been based not only on technological advances but also on improved coordination and effective administration of water and land management schemes (Cosgrove

& Petts, 1990). However, in many countries, and especially in the Third World, water developments have lacked the necessary coordination and administration, and there has been an excessive reliance on large-scale engineering. The establishment of effective management structures to administer and coordinate developments, especially within international river basins, is urgently required.

The post-modernist vision has concern for the quality of life. It focuses on investment in the environment, and is expressed by symbols of social advancement: landscape restoration and the protection of wilderness areas (Cosgrove & Petts, 1990). Sustainable development *requires* large water-engineering projects, but the goals of these projects, and their operation, must be radically revised. Any further large dams and reservoirs must be developed sparingly, and as part of a comprehensive catchment and national or regional plan, while secondary regulation measures must be a requirement for all schemes. The primary need is for the implementation of environmentally sound management goals – including the allocation of water to sustain environmental processes – and for the development of approaches to assess actual and potential environmental values. [G.E.P.]

A Comment by the Moderator

One of the most effective methods for conserving water is to reduce the amount flowing down streams and rivers – destined to be lost in the oceans – and to store it in reservoirs, most of which are impounded by dams. This process can improve the quality as well as the quantity of available water. It was therefore surprising that Dr Arthur Westing, in his paper to the session on Human Instability and again in a later discussion, picked on dams (and the water engineers who make them) as enemies of the environment.

It is fair comment that many dams, including some which impound large Man-made lakes, have raised plenty of environmental problems, many advantageous to humanity but some the reverse. It can be argued also that, in the USA, the Army's Corps of Engineers once had too free a hand in constructing very many small dams, as well as large ones, not all of which were successful in achieving their purposes. However, that was adjusted after the Stockholm environmental conference of 1972 and ICOLD's conference at Madrid in 1973, by arranging adequate ecological advice for the Corps which was then under the command of General Clarke.

There have, of course, been occasional disasters with dams, as in other branches of constructional engineering, which have caused serious loss of life and property, although these are counterbalanced by holding back water which, under the natural regime, could have caused major losses by floods. The disasters suffered in the Nile delta before construction of the original Aswan dam, and those of Bangladesh today, are examples. On balance, however, the millions of dams now operating around the world, down to the farm-pond size, and those which are yet to be made, are of benefit not only to humans but to many other species of The Biosphere as well.

Improvements in the methods of conserving and controlling water, as in any other enterprise, are always desirable. For example, until some 30 years ago, when additional water was required for a town or an industry, it was usual to build a dam near-by, and to convey water by pipe, allowing a small proportion as compensation water to flow down the original river bed. Fortunately it has now become widely recognized, as a result of representations by ecologists, that it is better, wherever possible, to store the water high up in the catchment, to use the river bed as the pipeline, with peaks and reduced troughs of flow, and to extract water from the lower reaches where it is needed. This system is much more 'friendly' to the environment. [E.B.W.]

The Conservation of Lakes: The Great Lakes

The Great Lakes Basin of North America covers an area of three-quarters of a million square kilometres and has a land:water surface ratio of 2:1. The Basin is inhabited by approximately 37 million people, 29 millions of them in the United States. The lakes contain 18% of all fresh water at the surface of the Earth and have theoretical water replenishment times ranging from 2–3 years for Lake Erie to 180–200 years for Lake Superior. They supply the drinking water for about 23 million people.

Before 1950 very little was known about the Great Lakes. Their depths had been charted for navigational purposes and historical records of catches of commercial fish were available; but, the sediments were uncharacterized, current systems little known, benthic life poorly understood, and it was said that a single sample of plankton from Lake Huron would almost certainly provide new records of plankton distribution, if not new species.

In 1964 the Governments of Canada and the United States requested the International Joint Commission to investigate pollution in Lake Erie, Lake Ontario and the connecting channels in the St. Clair, Niagara and St. Lawrence Rivers. This resulted in a comprehensive technical report in 1970 that paved the way for the first Great Lakes Water Quality Agreement in 1972. The Agreement committed the Governments to a multi-thousand million dollar clean-up programme that focused on improved waste-treatment facilities and a programme of phosphorus control to reverse rising trends of 'cultural' eutrophication.

In response to these measures conditions in the Great Lakes have improved dramatically since the 1960s. Raw sewage and industrial wastes are no longer permitted to be discharged directly or indirectly into the lakes. Unsightly refuse is no longer as common at the water's edge. Fish-kills are less frequent, and the once-rising trends of cultural eutrophication have been reversed. Lake Erie – which was never dead, as many claimed – is alive and well and with generally healthy fish populations. An ecosystem approach to environmental problem-solving has gained universal acceptance in the Basin. New and effective citizen groups have formed: for example,

Great Lakes United, a consortium of over 180 organizations with interests in the Great Lakes. Remedial action plans to clean up 42 'Areas of Concern' are in the making (Hartig & Vallentyne, 1989).

In short, the history of the Great Lakes over the past twenty years has been a success story – with some major exceptions: slowness in implementing the 'ecosystem approach'* (Vallentyne & Beeton, 1988), and contaminated sediments in 41 of the International Joint Commission's 42 Areas of Concern. A new problem has recently emerged: invasion of the Great Lakes by exotic species including the European Zebra Mussel (*Dreissena polymorpha*), probably from the ballast water of transoceanic ships (Hebert *et al.*, 1989). Each of these is briefly summarized below.

The 'Ecosystem Approach'

The Great Lakes Water Quality Agreement of 1972 was revised in 1978 to incorporate an 'ecosystem approach' to managing human uses and abuses of water in the Great Lakes Basin (Great Lakes Research Advisory Board, 1978). The foundation of the 'ecosystem approach' stems from recognition of The Biosphere as the context in which all human activities take place, and ecosystems (rather than environments) as subdivisions of The Biosphere. Accordingly, the Great Lakes Basin Ecosystem was defined in the 1978 Agreement as: 'the interacting components of air, land, water and living organisms, including humans, within the drainage basin of the St Lawrence River at, or upstream, from the point at which this river becomes the international boundary between Canada and the United States.' A 1987 protocol to the 1978 Agreement strengthened the requirement for an 'ecosystem approach' and added new sections dealing with sediments, remedial action plans in 42 Areas of Concern, ground-water and research.†

The characteristics of the 'ecosystem approach' as it has developed in the Great Lakes Basin are: knowledge of the Great Lakes Basin 'Ecosystem' and its relation to The Biosphere; a holistic viewpoint relating people to ecosystems that contain them rather than to external environments with which they interact; and actions that are ecological, anticipatory and ethical in respect to Nature (Christie *et al.*, 1986; Vallentyne & Hamilton, 1988). The best application of the ecosystem approach has been in the development of remedial action plans for the 42 Areas of Concern in the Basin (Hartig & Vallentyne, 1989). The requirement for public participation in development of these plans has enabled citizens, scientists, local politicians and members of environmental organizations to work cooperatively in examining problems and opportunities.

The adoption of an 'ecosystem approach' to water management necessi-

*[Strictly, the Great Lakes are a collection of ecocomplexes, each embodying several ecosystems – hence the quotation marks. Eds.]

† Information on Great Lakes Water Quality Agreements and activities of the International Joint Commission can be obtained from the Commission's Great Lakes Regional Office, 100 Ouellette Avenue, Windsor, Ontario N9A 6T3, Canada.

THE ECOSYSTEMS APPROACH
DRIVING SCHOOL

Figure 7.2. A cartoon by James Kempkes, Toronto, prepared for a workshop on the 'ecosystem approach' held in Hiram, Ohio, 1986.

tates integration of social, economic and environmental interests (Great Lakes Research Advisory Board, 1978; Christie *et al.*, 1986; Caldwell, 1988; Vallentyne, 1988). The shift from a water-based approach that excluded most social and economic concerns, to an 'ecosystem approach' that included social and economic concerns was revolutionary. It was a change of context from a system external to people, to a system where people are included; and from dealing with environment in a political context, to dealing with politics in an ecosystem context (Caldwell, 1988; Vallentyne, 1990).

In spite of the requirement for an 'ecosystem approach' in the 1978 Great Lakes Water Quality Agreement, implementation has been slow and uncertain. Governments have tended to operate in traditional ways - rather than to look beyond the water's edge into the human behavioural forces at work behind the scenes and into educational systems that could redirect those forces. On the other hand, citizens' groups have strongly endorsed the 'ecosystem approach' and called on the Governments to live up to the terms of the Agreement they endorsed. The question raised by cartoonist James Kempkes, in Figure 7.2, is whether the several steering wheels operated by various human interests are independent, interconnected, or disengaged!

In many ways, the 'ecosystem approach' is a bottom-up, grassroots equivalent of the top-down, global approach to sustainable development advocated by the Brundtland Commission (World Commission on Environment and Development 1987). It is of interest that the five characteristics listed by Holdgate (1992) for effective, sustainable development projects in developing countries pertain equally well to the 'ecosystem approach' as it has developed in the Great Lakes Basin; also that a seminar on ecosystem approaches to water management is being sponsored by the Economic Commission for Europe (Oslo, May 27–31, 1991).

One of the operational rules of the International Joint Commission is that members of its boards, committees and task-forces act for defined times and purposes in their personal and professional capacities rather than as agents of the organizations that employ them. This rule is equivalent to the principle of deferred judgment in brainstorming. It allows for more imaginative approaches to problem-solving than would otherwise be the case and deserves wider application, particularly in international circles.

Persistent Toxic Chemicals

The Great Lakes Water Quality Board (1985) identified 373 chemicals of concern in waters of the Great Lakes Basin – 11 on its 'primary track' and 362 on its 'secondary track'. The primary-track chemicals are: mercury, PCBS, DDT and its metabolites, mirex, dieldrin, toxaphene, 2,3,7,8-tetrachlorodibenzo-para-dioxin, 2,3,7,8-tetrachlorodibenzofuran, alkyl lead, benzo[*a*]pyrene and hexachlorobenzene. Often referred to as the 'elusive eleven', these remain in the system even though their uses have in many cases been curtailed or banned. It is significant that 8 of the 11 primary-track chemicals and more than half of the secondary-track chemicals are organohalogens.

During the past year (1989), 6 binational groups concluded that persistent toxic chemicals, mostly organochlorines, have harmed and continue to harm populations of 16 different species of fish, turtles, birds and mammals in the Great Lakes Basin. These organizations were: the Great Lakes Science Advisory Board (1989), the Council of Great Lakes Research Managers (unpublished), a workshop and conference on human health effects of toxic chemicals (unpublished), a workshop of the Health Committee of the Great Lakes Science Advisory Board (unpublished), a joint evaluation of the state of the Great Lakes undertaken by the US Conservation Foundation and the Canadian Institute for Research on Public Policy (Colborn *et al.*, 1990), and the International Joint Commission (1990). The effects included decline in populations, embryonic mortality, egg-shell thinning, tumours, birth defects, reproductive impairment, feminization of males, suppression of immune systems, and behavioural changes. All groups agreed that greater use should be made of the effects of toxic chemicals on populations of fish and wildlife as a basis for decisions relating toxic chemicals to human health.

A second conclusion reached by the above groups was more startling:

that the effects of exposure to toxic chemicals is greater for the progeny of exposed individuals than for the exposed individuals themselves. In short, human health specialists have been looking in the wrong place for effects; the chemicals primarily act as teratogens and have inter-generational effects. Because of their shorter generation times, wild populations of vertebrates may truly be what many have suggested: environmental equivalents of miners' canaries. If that proves to be the case, comparable effects on human populations from the high levels of toxic chemicals in the Basin in the late 1960s and early 1970s should reach a maximum in the 1990s. We may then discover that we have doubly jeopardized the lives of our children through the shameful legacies of pre-natally impaired health and toxic ecosystems (Great Lakes Science Advisory Board, 1989).

Comparable data on the effects of low levels of toxic chemicals on human populations are limited; however, one study in western Michigan showed that mothers who ate 500 g per month or more of fish contaminated with organochlorines from Lake Michigan produced children who, compared with non-fish-eating controls, weighed 160–190 g less at birth, had 6–7 mm reduced head circumference, had gestational periods shorter by an average of 4.9 days, and had discernible learning deficits that have persisted over five years (Fein *et al.*, 1984; Jacobsen *et al.*, 1990). These differences were significant at the 5% level.

Based on the above findings, the International Joint Commission (1990) concluded that persistent toxic chemicals are impairing the health of natural populations of vertebrates in the Basin and that there is a clear threat to human health. The Commission recommended that the Governments of the United States and Canada adhere more rigorously to the philosophy of zero discharge, and the goal of virtual elimination of toxic chemicals specified in the Great Lakes Water Quality Agreement of 1978.

The question arises: Do these conclusions about the health effects of persistent toxic chemicals pertain solely to the Great Lakes Basin, or is it only that the problem has been more thoroughly and systematically studied there? While no clear-cut answer can be given at the present time, the Great Lakes Science Advisory Board (1989) did conclude that it had no reason to believe that the Great Lakes Basin differed in any major way from other parts of North America in respect to the health effects of toxic chemicals. The problem is, in fact, global.

The Zebra Mussel

Hebert *et al.* (1989) first discovered the European Zebra Mussel (*Dreissena polymorpha*) in Lake St. Clair (located between Lake Huron and Lake Erie) in June 1988. By the summer of 1990 the mussel had spread to all the Great Lakes and to the St. Lawrence River. It is thought that the introduction occurred through the discharge of ballast water from one or more trans-Atlantic ships entering the St. Lawrence Seaway.

The Zebra Mussel is affecting profoundly the ecology of the Great Lakes and the economy of the region. During the spring of 1990 the transparency of water in the western end of Lake Erie was twice that of historic records – probably to be attributed to the shallowness of the water and the filtration efficiency of the Zebra Mussel. It has been estimated that the havoc created by the Zebra Mussel in water-intake lines of municipalities and industrial plants in the Great Lakes Region alone may cost US $4 thousand millions before the populations subside. In a few decades the mussels will probably have spread over the waters of North America as did European Starlings (*Sturnus vulgaris*) on land.

Exotic species are not new to the Great Lakes. Emery (1985) listed 34 exotic species of fish that were knowingly, or unknowingly, introduced into the Great Lakes from 1819 to 1974. More recently, the European cladoceran (*Bythotrephes cederstoemi*) was discovered first in Lakes Huron and Erie, and is now present in all the Great Lakes (Bur *et al.*, 1986; Garton & Berg, 1990; Makarwicz & Jones, 1990). A reproducing population of the European River Ruffe (*Gymnocephalus cernua*), has also recently been discovered in the western end of Lake Superior (Pratt, 1988). These findings could be indicative of further waves of introductions.

The International Joint Commission and the Great Lakes Fishery Commission are examining the scientific basis for regulations pertaining to the treatment and discharge of ballast water in oceanic ships before entering the St. Lawrence River. These range from mid-ocean discharge, to treatments with toxic chemicals or heat. At the time of writing, no firm regulations have been set in place. [J.R.V.]

References

Beaumont, P. (1989). *Environmental Management and Development in Drylands*. Routledge, London, England, UK: ix + 462 pp., illustr.

Bur, M. T., Klarer, D. M. & Krieger, D. A. (1986). First records of a European cladoceran, *Bythotrephes cederstoemi*, in Lakes Erie and Huron. *J. Great Lakes Res.*, 12, pp. 144–6.

Caldwell, L. K., (Ed.) (1988). *Perspectives in Ecosystem Management for the Great Lakes*. State University of New York Press, Albany, NY, USA: 320 pp.

Christie, W. J., Becker, M., Cowden, J. W. & Vallentyne, J. R. (1986). Managing the Great Lakes Basin as a home. *J. Great Lakes Res.*, 12, pp. 2–17.

Colborn, T. E., Davidson, A., Green, S. N., Hodge, R. A., Jackson, C. I. & Liroff, R. A. (1990). *Great Lakes Great Legacy*? The Conservation Foundation, Washinton, DC, USA, and the Institute for Research on Public Policy, Ottawa, Ontario, Canada: xliii + 301 pp., illustr.

Cosgrove, D. E. & Petts, G. E. (1990). *Water, Engineering and Landscape*. Bellhaven Press, London, England, UK : xiv + 214 pp., illustr.

Dixon, J. A., Talbot, L. M. & Le Moigne, G. J-M. (1989). Summary. In *Dams and the Environment: considerations in World Bank Projects*. World Bank, Washington, DC, USA.

Emery, L. (1985). Review of fish species introduced into the Great Lakes. *Great Lakes Fishery Commission Tech. ReP.*, 35, pp. 1–31.

Fein, G. G., Jacobsen, J. L., Jacobsen, S. W., Schwartz, P. M. & Dowler, J. K. (1984). Prenatal exposure to polychlorinated biphenyls effects on birth size and gestational age. *J. Pediatrics*, 105, pp. 315–20.

Garton, D. W. & Berg, D. J. (1990). Occurrence of *Bythotrephes cederstroemi* (Schoedler 1877) in Lake Superior, with evidence of demographic variation within the Great Lakes. *J. Great Lakes Res.*, 16(1), pp. 148–52.

Goodland, R. J. A. (1989). The World Bank's New Policy on the Environmental Aspects of Dam and Reservoir Projects. Unpublished paper given at the *Workshop on Environmental Implications of Large Dams.* World Bank, Washington, DC, USA.

Great Lakes Water Quality Board (1985). *Report to the International Joint Commission.* International Joint Commission, Great Lakes Regional Office, 100 Ouellette Ave., Windsor, Ontario N9A 6T3, Canada: 212 pp., illustr.

Great Lakes Research Advisory Board (1978). *The Ecosystem Approach: Scope and implications of an ecosystem approach to transboundary problems in the Great Lakes Basin.* International Joint Commission, Great Lakes Regional Office, 100 Ouellette Ave., Windsor, Ontario N9A 6T3, Canada: 47 pp., illustr.

Great Lakes Science Advisory Board (1989). *Report to the International Joint Commission.* International Joint Commission, Great Lakes Regional Office, 100 Ouellette Ave., Windsor, Ontario N9A 6T3, Canada: xiv + 92 pp.

Hartig, J. H. & Vallentyne, J. R. (1989). Use of an ecosystem approach to restoring degraded areas of the Great Lakes. *Ambio*, 18(8), pp. 423–8.

Hebert, P. D. N., Muncaster, B. W. & Mackie, G. L. (1989). Ecological and genetic studies on *Dreissena polymorpha* (Pallas): a new mollusc in the Great Lakes. *Can. J. Fish. Aq. Sci.*, 46, pp. 1587–91.

Hock, B. & Somlyódy, L. (1990) Freshwater resources and water quality. In *State of the Hungarian Environment* (Eds. D. Hinrichsen & G. Enyedi). Statistical Publishing House, Budapest, Hungary.

Holdgate, M. (1992). Overview: Our Threatened World. Pp. 1–21 in *Surviving With The Biosphere* (Eds. N. Polunin & J. Burnett). Edinburgh University Press, Edinburgh, Scotland: xxii + 561 pp., illustr.

International Joint Commission (1990). *Fifth Biennial Report under the Great Lakes Water Quality Agrement of 1978.* Part II. International Joint Commission, Ottawa, Canada, and Washington, DC, USA: 59 pp.

Jacobsen, J. L., Jacobsen, S. W. & Humphrey, H. E. B. (1990). Effects of *in utero* exposure to polychlorinated biphenyls and related contaminants on cognitive functioning in young children. *J. Pediatrics*, 116(1), pp. 38–44.

Ledec, G. & Goodland, R. J .A. (1988). *Wildlands: their Protection and Management in Economic Development.* World Bank, Washington, DC, USA.

Mandrak, N. E. (1989). Potential invasion of the Great Lakes by fish species associated with climatic warming. *J. Great Lakes Res.*, 15(2), pp. 306–16.

Makarawicz, J. C. & Jones, H. D. (1990). Occurrence of *Bythotrephes cederstroemi* in Lake Ontario offshore waters. *J. Great Lakes Res.*, 16(1), pp. 143–7.

Petts, G. E. (1980). Morphological changes of river channels consequent upon headwater impoundment. *Journal of the Institution of Water Engineers and Scientists*, 34, pp. 374–82.

Petts, G. E. (1987). Time-scales for ecological change in regulated rivers. Pp. 257–66 in *Regulated Streams: advances in ecology* (Eds. J. Craig & J. Kemper). Plenum, New York, NY, USA: vi + 257 pp., illustr.

Petts, G. E. (1989). Perspectives for ecological management of regulated rivers. Pp. 3–26 in *Alternatives in Regulated River Management* (Eds. J. A. Gore & G. E. Petts). CRC Press, Boca Raton, Florida, USA: x + 344 pp., illustr.

Petts, G. E. (1990). The role of ecotones in aquatic landscape management. Chapter 11 in *The Roles of Ecotones in Aquatic Landscapes* (Eds. R. Naiman & H. Decamps). Cambridge University Press, Cambridge, England, UK: (in press).

Petts, G. E., Imhof, J. G., Manny, B. A., Maher, J. F. B. & Weisberg, S. B. (1989). Management of fish populations in large rivers: a review of tools and approaches. *Canadian Special Publication of Fisheries and Aquatic Sciences*, 106, pp. 578–88.

Pratt, D. (1988). Distribution and population status of the ruffe, *Gymnocepalus cernua*, in the St. Louis estuary and Lake Superior. *Great Lakes Fishery Commission, Research Completion Report.* 11 pp.

Vallentyne, J. R. (1988). First direction, then velocity. *Ambio*, 17(5), p. 409.

Vallentyne, J. R. (1990). The person-planet connection in environmental education. *Environmental Awareness*, 13(1), pp. 3–10.
Vallentyne, J. R. & Beeton, A. M. (1988). The 'ecosystem' approach to managing human uses and abuses of natural resources in the Great Lakes Basin. *Environmental Conservation*, 15(1), pp. 58–62.
Vallentyne, J. R. & Hamilton, A. L.(1987). Managing human uses and abuses of aquatic resources in the Canadian Ecosystem. Chapter 19 in *Canadian Aquatic Resources* (Eds. M. C. Healey & R. R. Wallace). Canadian Bulletin of Fisheries and Aquatic Sciences 215, Ottawa, Canada: iv + 533 pp.
White, G. F. (1987). *Environmental Effects of Complex River Development*. Westview, Boulder, Colorado, USA: xi + 172 pp., illustr.
World Commission on Environment and Development (1987). *Our Common Future*. Oxford University Press, Oxford, England & New York, USA: xv + 383 pp.

8. Genetic Resources

Gábor Vida
Professor and Head, Department of Genetics, Eötvös Loránd Science University, Múzeum Körut 4/a, H-1088 Budapest, Hungary

Introduction

Many alarming changes are taking place in The Biosphere which have obvious direct effects on the well-being of humanity. Nobody would deny, for instance, the unfavourable consequences of pollution or the partial destruction already of the stratospheric ozone shield. Besides such drastic effects there are many less obvious, often sneaking, processes which affect us indirectly. These subtler effects, however, are not necessarily less important, only often more difficult to predict as regards their magnitude and consequences than such direct ones as pollution and overpopulation.

The loss of genetic diversity is an example of the sneaking category, with the additional burden of our high responsibility. For as Harvard's Edward O. Wilson claims in a recent *Scientific American* article (Wilson, 1989), 'it is the only process of environmental change that is wholly irreversible'. Conservation biologists have long recognized the impediments to open argumentation in our profit-oriented world. This is reflected in the '5th Iron Law of Conservation' formulated by Paul R. Ehrlich (*see* P. R. & A. H. Ehrlich, 1981) as follows:

> 'Arguments about the right to exist of nonhuman life-forms, or their aesthetic value and intrinsic interest, or appeal for compassion for what may be our only living companions in the universe, now mostly fall on deaf ears. Until ethical attitudes evolve further, conservation must be promoted as an issue of human well-being and, in the long run, survival.'

Dozens of splendid books and hundreds of sophisticated articles have been published in much the same spirit. Ideas, points of view, and further references, can be found, for instance, in Norman Myers's book *A Wealth of Wild Species, Storehouse for Human Welfare* (Myers, 1983), or E. O. Wilson's collection of papers under the simple title of *Biodiversity* (Wilson, 1988). Instead of compiling yet another list of potential benefits of biological diversity of immediate use to human welfare, let me emphasize only four aspects for consideration.

1. Recent development of molecular genetics, particularly the spectacular technical advancement in genetic manipulation, has made it possible to transfer genes and regulatory DNA regions (promoters) in any desired combination. Consequently any living creature can be regarded as a potential genetic source for, *inter alia*, agriculture, animal husbandry, or a microbial industry. Sophisticated procedures are available to isolate, propagate, and modify, pieces of DNA of any organism. Such DNA material can be incorporated into any other genome, thus creating transgenic plants, animals, Fungi, protists, or Bacteria. From simple practical aspects, therefore, the present Biosphere is an enormous storehouse of genes accommodating at least 10^{11} (one hundred thousand million) gene-forms.

2. Biotechnology, on the other hand, may result in an unprecedented decrease of genetic diversity of domesticated species. The world-wide application of modern, highly productive strains has already caused a drastic shrinkage of gene-pools in many species, in spite of the efforts to preserve the well-adapted but less productive local varieties. But even these modern strains have retained a fair amount of genetic variation. Breeding through genetic manipulation or 'engineering', however, means modification of an individual and not of a population – as all the previous practices did. The new manipulated, transgenic plant or animal strain is often the descendant of a single specimen, with all of its genetic consequences.

3. Although many impressive programmes and organizations have been launched (such as the International Board of Plant Genetic Resources), steps so far taken to maintain gene-banks as a source for future breeding cannot be regarded as completely satisfactory. Except for some major crops, where the situation seems to be more promising (*see* Duvick, 1984), they only preserve (rather than lastingly conserve) species and varieties as defined by Frankel & Soulé (1981). This means that, although the particular strain is kept alive, it has lost most of its evolutionary potential. Conservation in the full sense would involve retention of adapted genetic variation, which would require a large population-size with negligible sampling error in producing new generations. This condition usually cannot be met because of limited space. Although genetic drift can be avoided by storing seeds, spores, or germ-cells, this method creates other problems in a changing environment by having lost the ever-continuing race for adaptation (*see* the Red Queen model of Valen, 1973).

4. The genetic richness of The Biosphere, as contrasted with the difficulty of maintaining useful variation in artificial gene-banks, offers the easy solution of *in situ* conservation of natural or semi-natural communities. Here not merely a few but countless numbers of species can be naturally conserved simultaneously, with continuous adaptation to their physical and living environment, provided that the protected areas involved are large enough according to the basic rules of island biogeography (*see* Frankel & Soulé, 1981; Soulé, 1986).

Health of The Biosphere Imperative

So far, all my arguments have referred to human needs as if the only purpose of The Biosphere were to serve *Homo sapiens*. In fact, we are also dependent upon the healthy functioning of The Biosphere, as the title of this Conference already emphasizes. If I am to place my contribution into this frame, I have to discuss genetic resources from The Biosphere's point of view: is there any reason to worry from *its* side.

As a young and enthusiastic botanist (some 40 years ago) I took part in the vegetation mapping of Hungary, which was then being conducted by Professor Bálint Zólyomi. I was very much impressed by the regularity of species composition in the various plant communities and their strong dependence on some particular nutritional (*e.g.* soil or water) and climatic conditions. At that time, many of the communities in the mapped region were almost pristine; thus it was not too difficult to see the intricate regularities of natural community formation.

From the regularities of the vegetation it is possible to reconstruct a map of the original vegetation in the surrounding agricultural and other Man-made areas. Such reconstructed maps are available for Hungary, Europe, and of the whole Earth. With their help we can at least imagine how our undisturbed Biosphere used to look on and near the Earth's surface.

Taking Hungary as an example, we have drastically altered the vegetation (such as replacing the original forest by cropland) on more than three-quarters of the country. The remaining one-quarter has also been changed (degraded), although less markedly. Other countries have fared better or worse; but virgin vegetation is rare all over the world (*see*, for example, Dr Martin W. Holdgate's paper in this volume). Virtually all these changes have been made in one or two millennia, which is nothing but a tiny fraction (less than 0.0001%) of the history of The Biosphere as regards terrestrial ecosystems.

Long-term Tolerance by The Biosphere

One may say that The Biosphere has apparently tolerated this profound modification perpetrated by the presence in it of *far too many people*. We should also consider, however, the presumably long reaction-time of such an enormously large system. Besides, some alarming changes in the atmosphere are at least partly due to the changing Biosphere processes (atmospheric build-up of carbon dioxide and methane, for instance).

Actually, we are witnesses of a global-scale experiment, the outcome of which is at best uncertain. Concentrating now only on the replacement of the natural ecosystems by Man-made ones, a comparison will immediately uncover their contrasting characters and behaviours.

Natural communities usually harbour an enormous quantity of genetic variation. Their component populations are regulated, resulting in predictable behaviour. There is very little change in species composition on the scale of human life-span. In contrast, Man-made communities are poor in genetic variation and their behaviour is often unpredictable because of haphazard species composition and unbalanced interactions. They need extra energy (*e.g.* of Man- and/or animal-power, or power from fuel) to maintain their structure – a point which is currently gaining importance and, consequently, recognition.

It is not my task to analyse the reasons for this remarkable difference. Nevertheless, I should like to point to the fact that, at least in theory, it is possible to construct a completely different, nearly self-maintaining agro-ecosystem, with characteristics approaching those of a natural one. There is every reason to believe that an almost continuously harvested, complex plant community of suitable composition could produce much more biomass or energy than our conventional yearly- or seasonally-harvested crop monocultures. (It is a pity, however, that our need is not just for unspecified biomass but only for very special kinds of it.)

In the light of the rapid loss of pristine land, we only hope that we shall come to understand the intricate patterns and processes of natural ecosystems before they disappear completely. Otherwise, the huge amount of potentially valuable information that has accumulated in the long coevolution of natural communities will have been lost, and for ever. Are we sure that there is nothing useful to learn from such disappearing systems, and would it not be appropriate to concentrate more of our efforts on reading the lengthy genetic information that is available on the most abundant, pandominant species of The Biosphere, *Homo sapiens?* There is also virtually no end to our possibility of studying other species.

Natural Ecosystems are Inherently Heterogeneous

Fortunately, some nearly natural ecosystems still persist in our Biosphere – many of them merely or largely because they could not be easily used by Man. Examples are found among bogs, marshes, cliffs, deserts, etc. Original communities of the more typical habitats, such as scraps or strips of 'zonal vegetation', nowadays are largely restricted to the places that are least accessible to humans, constituting mere precious remnants of the once-dominating ecosystems. It is really worth considering one such example – a mixed forest of Beech (*Fagus sylvatica*), Fir (*Abies alba*), and Spruce (*Picea excelsa*), near Lunz-am-See in Austria, which is supposed never to have been exploited for timber (*see* Walter & Breckle, 1985).

The most striking feature of this and other virgin forests is its heterogeneous character – in species composition, age structure, and also in stands, which exhibit a mosaic arrangement of various stages of cyclic dynamics. Although local communities of this seemingly virgin forest cannot be regarded as stable in the strict sense, if one considers a larger region with this kind of dynamics, many of the ecological parameters, although changing locally (sometimes even chaotically), regionally give nearly constant equilibrium values of abundances of species and communities. Were such quasi-stable pieces perhaps responsible formerly for the whole Biosphere's behaviour?

Genetic Resources and The Biosphere

Let us return now to the previous question about genetic resources from the point of view of The Biosphere. If we were to create a much simpler Biosphere than the present one, dominated, shall we say, by Man and his 'useful' plants, animals, and microbes, would this also mean that some serious degree of the former stability would be lost? This point has been a subject of hot debate since Elton (1958) suggested, more than 30 years ago, that increase in the complexity of a community leads to reduced fluctuations of its population densities, and that they in turn become more resistant than formerly to invasions by other species. Robert M. May (1973), on the other hand, pointed out, in his book entitled *Stability and Complexity in Model Ecosystems*, that 'as a mathematical generality, increased complexity makes for diminished community stability'.

As for almost every disputed statement in biology, there is no shortage of pro- and contra-examples stimulating this controversy. Such confusion is partly due to the poorly-defined meanings of complexity and stability, as Stuart L. Pimm (1984) emphasized, and partly due to the inequality of the equally-treated units, such as species. (Botanical gardens, such as that of Kew, can be regarded as representing a biologically extremely complex system. If their stability is tested against perturbation, through removal of a single species, the outcome of the experiment will be entirely different according to whether this species is *Capsella bursa-pastoris* or *Homo sapiens.*) Of course, ecologists have long recognized that biotic communities cannot be treated as randomly-assembled collections of populations with fixed parameters of 'connectance' and of strength from their interaction. Even May himself (1975) emphasizes this point: 'Natural ecosystems ... are products of a long history of coevolution ... It is at least plausible that such intricate evolutionary processes have, in effect, sought out those relatively tiny and mathematically atypical regions of parameter space which endow the system with long-term stability'.

The structure of a food-web is a coevolutionary product. Several recent studies have shown that there is increased stability in natural ecosystems where 'connectance' and interaction strength exhibit a complex, blocking structure.* Moore & Hunt (1988), for instance, demonstrated that the below-ground food-webs of a North American shortgrass steppe are organized in compartments, thus resulting in improved stability at the bottom of the food-web (based on decomposers).

An even more interesting source of stability has been reported in a recent issue of *Nature*, where Pacala *et al.* (1990) report on the result of analysis of

*[In response to our request for clarification of this, Professor Vida replied (*in litt.*) 'The American authors obviously mean compartments of interacting populations as a semi-independent unit of a food-web.' Eds.]

various host–parasitoid systems, showing that variation in interaction can actually stabilize otherwise unstable systems. As Peter M. Kareiva (1990) comments in the relevant News and Views section in the same issue, 'rather than viewing environmental heterogeneity and variability as frustrating complications that threaten our ability to make sense of the world, we can now appreciate that this noise in our data may actually confer temporal order on the dynamics of species interactions.'. We should not, however, forget that, in the middle of the last century, Charles Darwin also made a similar point. Sir Roland Fisher's *Genetical Theory of Natural Selection* (1930) explicitly formulates that the success of adaptation (as measured by the increased average fitness of a population) depends on the additive genetic variance of that character.

Variation Imperative

In a changing world, the presence and amount of variation is of the utmost importance for the survival of The Biosphere. At the root of the intricate relationships within this large, complex system, genetic variation is widely responsible for the long-term viability.

Ever since J. B. S. Haldane's influential article on 'the cost of natural selection' (1957), or even before, population geneticists have been struggling with the difficult question of how the large amount of genetic variation is maintained in natural populations. There are many ways to explain the richness of a gene-pool, though clearly the most important limiting factor is the size of the population. (Unfortunately, by the term 'size', population genetic models usually mean 'effective' population-size which is often impossible to estimate in natural populations.) Consequently, many small Nature reserves are actually unable to keep enough genetic diversity, especially in populations of the less numerous 'top' predators, but also in functionally less-important other rare species. Relatively simple models of population genetics teach us how easily genetic variation can be diminished or lost. If the only source to regain genetic variation were spontaneous mutation, millions of generations would be needed to restore the original level (Crow & Kimura, 1970).

The theory of island biogeography, first developed by Preston (1962) and later extended by MacArthur & Wilson (1963, 1967), has been applied to such isolated island-like Nature reserves (*see* Frankel & Soulé, 1981). As the model predicts, immigration and extinction will approach a dynamic equilibrium, resulting in fixed numbers of species.

Here again, the kinds of species involved are important qualities for us to consider. We can imagine what an immigrant in a Nature reserve can be, while those which are likely to become extinct are usually those which are registered in the *Red Book*. We have to keep in mind, however, that extinction is not just a random process. Populations of equal sizes may have very different probabilities of extinction, depending in part on their genetic variation. The unfortunately very-well-documented vulnerability of island biotas can also be explained by depletion of their gene-pools. The small size of a population is a double handicap, being dangerous because of stochastic demographic events (as predicted by island biogeography) and dangerous because of the lack of adaptability (as predicted by population genetics). These facts have recently been taken into serious consideration at least in some cases of Nature Reserve Design.

Genetic Depletion for Biospheral Misfunctioning

It is well known that most of the natural communities of the temperate region show a log-normal distribution of species abundances (May, 1975). In other words, the majority of the species represented are rare. Isolated small remnants of the original ecobiomes are now on the way to genetic depletion, and hence large fractions of the species involved are condemned to extinction. (This process is often amplified by other effects. The Flora of the hills of Buda has lost, in this century, many of the attractive species through the attention of 'Nature lovers', and probably still more through 'plant lovers'!)

Destruction of the genetically very rich natural and semi-natural communities is progressing globally (Vida, 1978). Each year a country-sized piece of the most precious tropical forests is being lost. We should also look at our own country. I do not know of any country where this degradation from genetic loss has yet been reversed. Global depletion of genetic diversity is of great potential danger to the functioning of The Biosphere and hence to the well-being or very survival of our own species, which might be better referred to as *Homo 'pseudo-sapiens'*.

References

Crow, J. F. & Kimura, M. (1970). *An Introduction to Population Genetics Theory.* Harper & Row, New York, NY, USA: xiv + 891 pp., illustr.

Duvick, D. N. (1984). Genetic diversity in major farm crops on the farm and in reserve. *Economic Botany*, 38 (2), pp. 161–78, illustr.

Ehrlich, P. R. & Ehrlich, A. H. (1981). *Extinction: The Causes and Consequences of the Disappearance of Species.* Random House, New York, NY, USA: xiv + 305 pp.

Elton, C. S. (1958). *The Ecology of Invasion by Animals and Plants.* Chapman & Hall, London, England, UK: 181 pp., illustr.

Fisher, R. A. (1930). *The Genetical Theory of Natural Selection.* Clarendon Press, Oxford, England, UK: xiv + 272 pp., illustr.

Frankel, O. H. & Soulé, M. E. (1981). *Conservation and Evolution.* Cambridge University Press, Cambridge-London-New York-New Rochelle-Melbourne-Sydney: viii + 327 pp., illustr.

Haldane, J. B. S. (1957). The cost of natural selection. *Journal of Genetics,* 55, pp. 511–24.

Kareiva, P. M. (1990). Stability from variability. *Nature* (London), 344, pp. 111–2.

MacArthur, R. H. & Wilson, E. O. (1963). An equilibrium theory of insular zoogeography. *Evolution*, 17, pp. 373–87.

MacArthur, R. H. & Wilson, E. O. (1967). *The Theory of Island Biogeography.* Princeton University Press, Princeton, NJ, USA: 203 pp., illustr.

May, R. M. (1973). *Stability and Complexity in Model Ecosystems.* Princeton University Press, Princeton, NJ, USA: xi + 235 pp., illustr.

May, R. M. (1975). Patterns of species abundance and diversity. Pp. 81–120 in *Ecology and Evolution in Communities* (Eds M. L. Cody & J. M. Diamond). The Belknap Press of Harvard University Press, Cambridge, Massachusetts, USA: ix + 545 pp., illustr.

Moore, J. C. & Hunt, H. W. (1988). Resource compartmentation and the stability of real ecosystems. *Nature* (London), 333, pp. 261–3, illustr.

Myers, N. (1983). *A Wealth of Wild Species, Storehouse for Human Welfare.* Westview Press, Boulder, Colorado, USA: xiii + 274 pp., illustr.

Pacala, S. W., Hasell, M. P. & May, R. M. (1990). Host–parasitoid interactions in patchy environment. *Nature* (London), 344, pp. 150–3, illustr.

Pimm, S. L. (1984). The complexity and stability of ecosystems. *Nature* (London), 307, pp. 321–6, illustr.

Preston, F. W. (1962). The canonical distribution of commonness and rarity: Part I. *Ecology*, 43, pp. 185–215; Part II. *Ecology*, 43, pp. 410–32, illustr.

Soulé, M. E. (Ed.) (1986). *Conservation Biology, the Science of Scarcity and Diversity.* Sinauer, Sunderland, Massachusetts, USA: xiii + 584 pp.

Valen, L. van (1973). A new evolutionary law. *Evolutionary Theory*, 1, pp. 1–30.

Vida, G. (1978). Genetic diversity and environmental future. *Environmental Conservation*, 5 (2), pp. 127–32.

Walter, H. & Breckle, S. W. (1985). *Ecological Systems of the Geobiosphere, 1: Ecological Principles in Global Perspective.* Springer Verlag, Berlin-Heidelberg-New York-Tokyo: vii + 242 pp., illustr.

Wilson, E. O. (Ed.) (1988). *Biodiversity.* National Academy Press, Washington, DC, USA: xiii + 283 pp., illustr.

Wilson, E. O. (1989). Threats to biodiversity. *Scientific American*, 262 (3), pp. 60-6, illustr.

Commentary on Chapter 8

CHAIRMAN: Sir John Burnett
PANELLISTS AND OTHER CONTRIBUTORS:

McCusker, Szabó, Bazzaz, Kefeli, Stanton, Smith & Schultes, Oza, Tisdell, McCusker, Burnett, Poore, Burnett, Ramakrishnan, Szabó, Ramakrishnan

McCusker said that **Vida** had made a particular case for *in situ* conservation of genetic resources; she would like to examine the relative merits of *in situ* and *ex situ* conservation as complementary techniques.

The International Board for Plant Genetic Resources had concentrated up to now mainly on *ex situ* conservation but not entirely so. However, it had become increasingly interested and involved in the relationship and the interface between *ex situ* and *in situ* techniques. When IBPGR commenced its work, 15 or 16 years ago, *ex situ* conservation of crop germ-plasm was undertaken on a large scale because of the clear threats of genetic erosion in the field, resulting from land clearing and the widespread introduction of improved crop varieties. In situations such as this, where the land was required for other uses, *in situ* conservation was clearly not an option.

IBPGR became particularly concerned, therefore, with the *ex situ* conservation of crop germplasm, especially of ancient landraces being supplanted by modern varieties. Extensive collections were made to sample the range of genetic diversity present in the field. Such sampling often required collecting over a very wide geographic range and could bring together sources of diversity from areas well beyond the natural population range that could not feasibly be included in a single *in situ* reserve. One of the main advantages of *ex situ* conservation was that it offered the possibility to sample much wider areas than could realistically be conserved *in situ* and therefore it was a technique especially suitable to situations where diversity was very widespread. She distinguished here between genetic diversity within a single species – which is what IBPGR was attempting to conserve *ex situ* – and species richness within an ecosystem.

Ex situ conservation was not a suitable technique for conserving ecosystems because it was not generally practicable to reconstruct the vegetation element of an ecosystem from samples of the seeds of the species involved.

She stressed that the conservation of germplasm, *ex situ* in gene-banks designed for long-term storage, was not a substitute for *in situ* conservation but a supplement to it. In most contexts an integrated conservation strategy comprising both elements would be the ideal.

Whatever the strength of *in situ* conservation from a biological point of view, one of its great weaknesses was that it did not present germ-plasm in a packaged form ready for use. The most frequent of users of plant germ-plasm, the plant breeders, were looking for seed samples accompanied by data of the origin of the sample (i.e. 'passport data') together with data of the properties of the samples, the 'characterization and evaluation data'. They needed samples that had been studied and documented and this implied collection and *ex situ* conservation of a particular set of material which could be provided for study and subsequently for use in breeding programmes. Although many plant breeders did collect their own material from the wild, it was not realistic for all germ-plasm to be collected *de nouveau* by every breeder every time it was to be used.

Vida had said that conservation in gene-banks was not entirely satisfactory. How true! *Ex situ* conservation of germ-plasm was a science in its infancy and there was a long way to go before the scientific principles behind the management of germ-plasm collections in gene-banks would be fully understood. After massive efforts to collect threatened germ-plasm over the last 10 to 15 years for conservation in gene-banks we were now faced with three major challenges:

To analyse what had been collected;

To prevent genetic erosion in the gene-bank by improving conservation technology; and

To encourage governments to provide continued funding for maintenance of the material in their care.

This last point was perhaps best addressed by demonstrating that genetic resources were useful assets for the future by showing that they were useful assets for the present. IBPGR was, therefore, beginning to give greater attention to encouraging the documentation, evaluation and use of the genetic resources conserved *ex situ*, rather than continuing to put overwhelming emphasis on further collection and storage.

In situ and *ex situ* conservation had much in common in terms of the background studies that were required to undertake them effectively. It was necessary in both cases, for example, to know the distribution of diversity in the field. This enabled intelligent selection of material for conservation in gene-banks, and informal selection of the best sites to be gazetted as reserved. In each case it was also necessary to know the size and structure of populations of the species of particular interest for conservation and the breeding biology of the species concerned. There were excellent opportunities, therefore, for collaboration in the study of germ-plasm for conservation by either method.

Szabó said that The Biosphere, at a first approach, might be regarded as a biological information generator and condenser system. In this system, chemical information, genetical (biological) information and cultural (social) information were integrated in an ecologically more or less balanced whole, which had advanced today towards a cosmic informational unit.

Accordingly, the theory of survival of The Biosphere and the practice of our survival with(in) The Biosphere was, in essence, a problem for informational scientists, or of philosophy.

Looking on the problem of genetic conservation more pragmatically, there were some model systems in our Biosphere, where, in their sphere of influence, traditional human (ethnic) communities had preserved successfully genetic resources belonging to taxa which had emerged after long natural and/or artificial selection. He was inclined to believe that men, grouped in self-conscious communities, had had from the first some strong instinctive feelings about conservation, and that historically, only the communities which possessed this strange sense of care towards their environment survived on the same territory for a long time.

As a biologist, deeply involved in both the genetic and ethnologic aspects of evolution, he would like to focus, in a consciously simplified manner, on the main point of any *in situ* conservation in order to demonstrate the necessity for a new international approach towards the importance of traditional ethnic communities for the genetic conservation of biological diversity. These statements were based on his former results (Szabó, 1983, 1985, etc.)*.

The first great revolution of Mankind, the *Neolithic Revolution*, was in essence a coevolution process between the information content of wild plant and animal germ-plasm (biological information) and that of the social information accumulated in

*[*see* p. 186. Eds.]

different human (ethnic) groups. In this coevolution process two new categories of information emerged in The Biosphere: that accumulated in the germ-plasm of domesticated plants and animals and the cultural information accumulated in society. *Ethno-cultures* were strongly connected with the emergence of '*agri-culture*'. Hunting and gathering nomads were and are generally more protective but less constructive towards their environment, compared with cultivators.

It was worth mentioning that even the existence and the importance of the conservation of biological information was first formulated in a rather speculative (i.e. instinctive) manner by A. Weismann (1885) in his famous *Germ-plasm Theory*. This concept was first introduced in science to denominate biological information responsible for the hereditary characters of the organisms. Due to various causes the concept soon became rather confused and later practically 'empty'. During the emergence of the conservational attitude, characteristic of the second half of the 20th century, the term had been re-introduced in science as synonymous with genetic resources, meaning genetic information, preserved in the different virus, bacterial, fungal, plant and animal species relevant to the human environment (Witt, 1985).

With recent developments in molecular genetics the germ-plasm concept was shifting slowly toward a new, synthetical science field at the interfaces of genetics, systematics, informatics etc., a field of science which could perhaps best be designated by a new word: *genematics*. In the light of this new development every piece of relevant genetic information preserved in any endangered organism, spontaneous or cultivated, had a steadily increasing value in The Biosphere.

There was a coherent theory regarding the role of wild animal and plant genetic information in the emergence of domesticated gene-pools and, ultimately, in the emergence and integration of the Indo-European culture. The Neolithic revolution could be regarded as the *First Germ-plasm Revolution* (Zohary, 1982).

A warning that should be heeded was that the *Second Germ-plasm Revolution*, during the period of the great geographic discoveries, when a lot of new crop plants were introduced all over the world, was not associated with a cultural integration but rather with the destruction of many traditional cultures. We were not fully aware of our losses. Nobody knew how much, nor how valuable, germ-plasm had been lost during the almost complete destruction of the different Amerindian and African cultures, or ethnic groups.

In the second half of the 20th century we were living in the period of the *Third Germ-plasm Revolution* to affect the human environment. This was a period which forced us to care for, to cultivate and to multiply *in situ, ex situ,* or even *in vitro*, an increasing amount of the germ-plasm available in The Biosphere. If we intended to survive with, and to survive because of, this genetic diversity, we would be forced to change our attitude not only towards the animal and plant gene-pools, but also towards genetic and cultural diversity, i.e. towards the human gene-pool, too. This change of attitude was itself a slow cultural-evolutionary process which had started only about 150 years ago. He recalled some of the key-points of the process: Alphonse de Candolle (1883) was the first to stress the correlation between human cultures and crop evolution; Charles Darwin (1885) first documented convincingly the evolutionary effects of artificial selection on domesticated plants and animals. The gene-pool concept emerged originally from Russian human genetic theory and was generalized first in science by the famous American immigrant Theodosius Dobzhansky (Adams, 1979). This concept, also based on strong ethnological reasoning, was introduced to genetic resource studies by N. I. Vavilov (1926–1932, cf. Vavilov, 1950). (NB: In its original sense the gene-pool concept included the genes and alleles preserved in different human populations adapted to different ecological conditions and, consequently, forming different cultural [ethnic] groups.)

It is worth noting that, except for Darwin, all these scientists were themselves deeply and personally affected by social intolerance towards ethnic and/or cultural diversity, and by inter-cultural conflicts rooted deeply in miseducation and lack of social knowledge regarding the importance of biological and socio-cultural diversity.

There was no doubt that new principles would have to be outlined at the international level in order to support the more effective preservation of ecologically important information in The Biosphere. Firstly, more scientific attention should be paid towards traditional, human (ethnic) communities which had preserved a long-lasting ecological equilibrium with their environment, thereby preserving valuable genetic resources. Today, characterized among others by the most ambitious biological research project in the history of science (that of the *mapping of the human genome*), the conscious care of human genetic diversity was not only a scientific, but also a moral, necessity for Mankind. If traditional ethnic communities were drastically and methodically destroyed, or even casually and accidentally lost (for political or economic reasons) then their balanced environments, the valuable and effective sites of *in situ* conservation, would be inevitably destroyed as well.

He cited a single example, that of Einkorn (*Triticum monococcum*, an important and still cultivated ancestor of modern wheats). A comprehensive field survey had been carried out in his former home land, Transylvania, badly hurt in recent decades by the exodus of different ethnic groups: Jews, Germans and more recently Hungarians.

Transylvania, which now belonged to Romania, was an area somewhat similar to Switzerland: traditional human values had been carefully preserved there until recently. In order to evaluate the relations between Man and the plant world from a multidisciplinary point of view, he had started some intercultural research projects (not supported officially by the authorities of the former regime) in the early '70s. A larger team had collected data from the whole territory of Transylvania and a team of two, he and an ethnologist-linguist, worked systematically on a more restricted sample territory, registering thousands of basic facts regarding the interactions observed in about 100 ethnic communities (both Hungarian and Romanian) with about 250 characteristic plant communities (coenotaxa), about 2,300 plant species (phytotaxa) and numerous local cultivated plant varieties. Interactions had been recorded at the genetic, taxonomic, community and ecosystem levels, including language and culture.

Beginning with Neolithic and the Roman times and up until recently, wheat ancestors had been preserved there. It could be demonstrated clearly, however, that the autocratic and destructive ideologies which had dominated the region in the second half of the 20th century together with the systematic ethnocultural destruction carried out, had caused a rapid disappearance of genetic stocks preserved for thousands of years. A parallel series of similar side-effects arising from false political reasoning was almost as long as the whole ethnobotanic survey itself!

The time was ripe for the formulation of specific statements and resolutions to protect not only the animal and plant, but also human, ethnic diversity. It seemed to be wise ecologically to pay more attention towards the protection of ethnic communities which had preserved for us, and still preserved effectively, their environment, including insufficiently-known genetic resources of potential, or actual, ecological importance.

The time was ripe not only for Red Books edited for the protection of the plant and animal diversity, but also to survey endangered human communities. A survey of the different endangered languages and cultures which represented, in essence, the common wealth of the whole of Mankind, was important not only from the cultural, but also from the environmentalist, point of view.

The ultimate scope of such pro-active protection was not to conserve the under-development of an area, or of a community, but to integrate its traditional knowledge and values into newly-emerging situations – to promote both the maintenance and propagation of traditional, and ecologically desirable, behaviour of different ethnic groups. *Cited:* Adams, B. M. (1979). From the 'gene-fund' to 'gene pope': On the evolution of evolutionary language. In: Coleman, W. & Limoges, C. (Eds). *Studies in the History of Biology*. U. P.Johns Hopkins, Baltimore, USA: 285 pp.; Darwin, C. (1885). *The Variation of Plants and Animals under Domestication*, I–II 2nd rev. ed., John Murray, London, cited from Hungarian Edition (1959), Akadémiai Kiadó, Budapest, Hungary: 418 pp; Candolle, A. de (1883). *L'origine des plantes cultivées.* Ed. Masson, Paris, France: 341 pp.; Dobzhansky, Th. (1973). *Genetic Diversity and Human Equality. The Facts and Fallacies in the Explosive Genetics and Education Controversy*. Basic Books Inc. Publ., New York: Hungarian Translation by A. Szabó (1985). Kriterion Publ., Budapest, Hungary: 204 pp.; Szabó, A. (1983) *Applied Biology in the Evolution of the Cultivated Plants* (in Hungarian). Ceres Publ. House, Bucaresti, Romania: 276 pp.; Szabó, A. (1985): Introduction: The natural environment; The vegetation. In *Plant Kingdom and Traditional Human Life in the Calata Area* (Kalotaszeg), (Eds Péntek, J. & Szabó, A.) Kriterion Publ. House, Bucaresti, Romania: 368 pp.; Vavilov, N. I. (1950). *The Origin, Variation, Immunity and Breeding of Cultivated Plants.* Selected and translated from the Russian by K. Starr Chester, *Chronica Botanica*, 13, (1/6). The Chronica Botanica Co., Waltham, Mass, USA: 364 pp.; Vida, G. (1992). This volume, chapter 8, p. 174 *et seq.*; Weismann, A. (1885). *Die Kontinuitat des Keimplasmas als Grundlage einer Theorie der Vererbung.* Gustav Fischer, Jena, Germany: 420 pp.; Witt, S. C. (1985). *Biotechnology and Genetic Diversity.* California Agricult. Land Projects, San Francisco, California, USA: 145 pp.; Zohary, M., (1982), *Plants of the Bible.* Harvard University Press, Cambridge, Mass., USA: 223 pp.

Bazzaz noted that the preceding contributions had assumed that there was a relationship between genetic diversity and ecological tolerance, but if that were not true then it could be argued that maintaining diversity was not so important! He thought that this issue should be considered in the light of the great genetic diversity revealed by such techniques as that of electrophoretic profiles, restriction fragment length polymorphisms (RFLPS), or any other method. As the behaviour of plants in Nature might actually be determined by this relationship, it was essential to investigate it. At present he thought it best to leave open the possibility that there might be no relationship between genetic diversity and ecological tolerance, or even – but much less likely – a negative relationship.

A second assumption which he questioned was: Is there a relationship between population stability and genetic diversity? He thought that a variety of other factors were involved so that genetically diverse populations were not always, of necessity, stable. He took as an example a species with an apparently broad response, capable of tolerating a very wide range of environmental fluctuations. It could be made up of a large number of individuals of different, highly specialized, genotypes each differing in its tolerance limits, or it could be composed of fewer flexible genotypes, each more plastic and broader in their response. The similar response could have been achieved in quite different ways, and more importantly, the consequences of selection on the two populations would be quite different. In one case environmental change, or degradation, would result in the total loss of some of the genotypes, *i.e.* loss of genetic diversity, but not in the other because of their plasticity. The former kind of response was more common in weedy species of disturbed habitats, the latter was more common in species which occupied late successional stages, *i.e.* relatively stable communities: that, incidentally, could have implications for conservation. Much more thought should be given to the important issues underlying conservation so that they were quite clear and did not just emerge from the general scientific and social context in which conservation would have to be achieved.

Lastly, he drew attention to the important issue of the release of genetically engineered organisms into the environment and their effects on natural populations.

There were two contrasting views about such organisms. One was that they could be so well engineered that there was nothing which differentiated them from the products of traditionally-bred varieties; the other was that anything which had been tinkered with was terribly dangerous and uncontrollable. The truth probably lay somewhere in between, but ecologists and environmentalists should pay particular attention both to the kinds of organism which were engineered and how that was achieved, while ensuring that molecular biologists were made well aware of their concerns. No transgenic higher organisms had been released so far in the USA but there had been some testing of bacteria. The most notable case was the *ice*-mutant of *Pseudomonas syringae*, a common resident of leaf surfaces. At 0° to 2°C it caused ice crystals to form but the mutant reduced this temperature to –6° to –8°C before frost damage developed. Experimental trials at Davis, California, involving spraying mutants on to plants, had demonstrated that they were both safe and successful. Nevertheless, this was an area which needed to be watched by experimentalists.

Kefeli was delighted that **Vida** had recommended Vavilov's work and he wished to draw attention to that of another of his fellow countrymen, V. I. Vernadsky, who had been concerned with maintaining the environment without disturbing Nature. In that context, he was uncertain that the development of transgenic plants represented an optimal objective for the current environmental conditions. As a plant physiologist concerned with plant hormones, he was working with transgenic plants which had undergone transformation by the use of *Agrobacterium tumefaciens*. Such plants often suffered enormous hormonal imbalances. They were also often more aggressive than normal plants. They rooted more intensively, and grew much longer without a normal quiescent period. Thus they were capable of outgrowing normal plants in ordinary ecological situations. The result was paradoxical.On the one hand, gene-banks were being established to preserve plants selected naturally across the range of ecological situations which they had met in Nature, while on the other, new, aggressive genotypes were being engineered which would compete with, and block the development of, the weaker species which had been conserved!

There was another paradoxical aspect to collections of varieties which he considered important. At the Institute of Photosynthesis and Soil Science (at Pushchino), as elsewhere, they maintained a collection of seeds and meristem cultures, but because of the anthropogenically induced changes in the environment, these were, in effect, species from an environment which was disappearing! He believed, therefore, that it was important to test these collections periodically against the environment to see if they were capable of adapting to the new ecological situations. By the coordinated use of such testing with that of the genetically new forms he believed that progress could be made without creating a worse environmental situation in the future.

Stanton remarked that, when he was in western Malaysia, he had founded a new botanic garden. He was delighted by the cooperation between the gardens of Malaysia, Thailand, Indonesia and the Philippines, over the collection, conservation and exchange of plants. That was good for assessing and maintaining biodiversity. But botanic gardens were only the shop windows of the rain-forest and he had been particularly concerned, in the light of population pressure in those lands, to increase the carrying capacity of the rain-forest without altering its physiognomy. In his view the main limitation was the supply of energy foods rather than of proteins and accessory nutrients. Most of the potential forest sources of carbohydrates and fats required skilled processing to make them safe. Man had, therefore, at the food-gathering stage of his development, harnessed microbes even before he became a 'system enricher', in his efforts to gain energy foods from the forests. We were still at an early stage in recognizing the importance of this concept. Forest resources such as roots, cassava, yams and energy-rich seeds and trunks, all required what had come to be called 'biotechnological processing', but that aspect had been neglected. The fault with the recent past, and still prevalent today, had been the idea that forests should be destroyed in favour of grasses such as sugar-cane, rice and pasture, as necessary precursors of food production. Thus apart from the destruction of forests, research on roots, tubers, and sources of tree-foods, had been neglected and conservation of their genetical diversity would prove more difficult than that of grasses. These energy-rich forest foods were in danger of being lost.

The problem was associated with the wider issue of tropical forest conservation*. He had seen the forest land-cover in western Malaysia shrink from 60% to 10% in 10 years; now Sarawak was going the same way. He was especially concerned for areas such as Mt Kinabalu and its environs where there were lower Dipterocarp, higher Dipterocarp, and oak, forests leading to subalpine zones. There were twenty-five times as many species of oak (*Quercus* spp.) in the moist forest as in the whole of Europe. Moreover, the region supported the world's most dramatic, parasitic climbing plants, *Rafflesia* spp. They would be lost if the trees went. He was against setting, as suggested by **Poore** (Chapter 3), a low, 10% retention target for conservation. In his view, this was a hidden counsel of despair, and should not be voiced. Publicity for any specific figure only gave strength to Forest Action Planners, even when it was interpreted as *total retention* of some areas to be offset by the more massive destruction of others. Quite apart from the question of conserving genetic resources, he could not stress strongly enough that we still had no clear picture either of the mosaic pattern of rain-forests, or of their recycling rates. Both these statistics were harder to obtain than those for the total species variation.

Areas like Mt. Kinabalu had now become popular tourist areas, which increased the risk to them. He, therefore, made a strong plea for such areas, especially montane and swamp forests, to be set aside entirely *free from tourism* – by analogy with the fate of the original cave paintings in Europe. He recognized there was a tourist need but this might be satisfied by providing 'synthetic' afforested areas in the regions visited by tourists, as an extension of the botanic gardens concept!

N. J. H. Smith & **R. E. Schultes** had submitted a note on the importance of the Amazonian rain-forest. This was the largest stretch of tropical rain-forest in the world, spanning 5,000 km from the Andes to the Atlantic, and some 4,000 km from the Guianas and the Upper Orinoco to the scrub *cerrado* of the Brazilian shield and the seasonally-flooded grasslands of the Pantanal. This vast mosaic of forest communities, secondary growth, natural and Man-induced grasslands, and swamps, contained the richest array of plant and animal species in the world, as well as the richest storehouse of genes for crop improvement. Rampant deforestation, fuelled by development schemes and pioneer farmers, now threatened to destroy genetic resources for many economic plants and potential crops, before they could be tapped for the benefit of people throughout the world. Also, loss of tribal cultures was resulting in the disappearance of unique varieties of many of these crops.

Wild or spontaneous populations of more than 47 perennial crops were at risk. They included the Rubber Tree (*Hevea brasiliensis*) and Cacao (*Theobroma cacao*), as well as several crops that were now emerging from relative obscurity. These included Peach Palm (*Bactris gasipaes*), now being extensively planted for palmito production, Annatto (*Bixa orellana)* which produced a natural red dye, and Brazil Nut (*Bertholletia excelsa*). Numerous minor crops such as Cupuacu (*Theobroma grandiflorum*), a relative of Cacao, and many locally-important fruits could become more important if more widely known, provided that enough of the genetic reservoirs of these plants were safeguarded in forest environments.

Oza drew attention to the coincidence, within narrow limits, of tropical forests and sources of economically important crops, north and south of the Equator. If we wanted to improve the quality of our cultivars we would have to go back to these centres of origin, which were now being decimated.

*[A short account of the role of FAO in the conservation of biodiversity in forests is given in Annexe 3, pp. 191, Eds.]

Tisdell thought it likely that there would be general agreement that the conservation of gene-pools would be of economic value. However, although **Vida** had placed an hypothetical value on the preservation of a gene, it was desirable that a real value could be determined. There were retrospective studies of the value of some wild varieties but no hard data on the economic value of germ-plasm banks. It was desirable for economic reasons that such data should be obtained, particularly because it seemed that contemporary, high-yielding varieties were apparently driving out a lot of the traditional varieties which could create tremendous problems for the future.

McCusker said that IBPGR was beginning to try to estimate the cost of conservation of genetic resources *ex situ.* It was very difficult to arrive at a realistic figure because there was not a simple relationship between the cost of the resource and the benefits derived in a variety of breeding programmes. Nevertheless, they would continue to attempt an estimate for their own purposes.

She also took the opportunity of saying that the notion of basing crop improvement on a narrower and narrower genetic base was not popular with developing countries. That was what they had been provided with and they had had some benefits from using such crops. However, they had now come around to thinking that when the crunch came with a bad season and yields fell, that they would prefer to aim for reliable yields under a variety of conditions rather than maximum yields under optimal conditions. That, of course, made different demands on plant genetic resources. Consequently, it was now considered by IBPGR that it might be desirable to plant some of the material which had been collected directly into the field, or simply multiplied for field planting, rather than feeding it into crop-breeding programmes. She thought that IBPGR and probably other programmes had thought too often of serving the plant breeder, and too little about other aspects and possibilities.

Burnett remarked that the wild *H4* rice of Sri Lanka was a good example of this possibility. After collection it had simply been used in a single trial and then put into production. On the matter of costs, he suggested that the case of the wheat cv *Sonalika* was worth examining. Its success over 20 years throughout N. India, Pakistan, and NW Bangladesh had depended on the presence of two rust-resistant genes. The financial return on the wheat grown in those regions over the twenty-years' period divided by two would provide an initial figure for those who wished to assess the economic value of genes and gene-banks!

Poore said that one tended to think of *in situ* conservation as being concerned with the size, shape, and risks of extinction of small reserves, representing particular ecosystems, in a protected area. Historically, however, vegetation and ecosystems had been constantly changing in response to fluctuations in, or larger changes of, climate. This had sometimes led to extinctions but that had been offset by migrations to adjacent regions. Natural vegetation had now been dissected by Man's activities to leave, at best, anthropogenically changed tracts between areas occupied by natural ecosystems. This pattern reduced the chances of successful migration, so, with climatic change, the probability of extinctions had been increased.

He thought that the present practices for protection were, at best, based on rather general ecological principles. More detailed knowledge was required. He stressed, in particular, the importance of examining with some care the features of the zones which separated recognizable and stable communities – ecotones. These were regions across which migration, in one direction or another, could potentially occur with environmental change, provided that the conditions were right.

He also suggested that two other things were desirable. First, a very large area to include a lot of internal variation within it. For example, changes in altitude, because if there were changes in temperature movement could take place up, or down; or variation in wetness and drainage conditions, so permitting internal migration, brought about perhaps by changes in rainfall patterns. Second, was the provision of reserves within a pattern of land-use which was not too different from that being conserved, e.g. natural forest within managed forest, or the provision of corridors or islands of appropriate vegetation to act as stepping-stones for migration. These could be Man-assisted, or natural.

Burnett observed that the kinds of area that **Poore** had suggested for *in situ* conservation bore a remarkable resemblance, in their characteristics, to the natural centres of variation discovered and described over 75 years ago by Vavilov. One could probably not do better than to try to create such natural conservation areas.

Ramakrishnan said that he had been surprised by the suggestion of **Szabó** that a Red Data Book should be prepared of traditional ethnic communities. He thought it had many dangerous implications and he hoped that it had not been a serious proposition.

Szabó responded by saying that one of his interests had been in determining the role played by *Triticum monococcum, Aegilops* spp. and others in the history and geographical spread of wheat and wheat-like species. Isolated ethnic communities occur which, like the plants, sometimes had their own allelic structure, as well as differing from the surrounding population in language or dialect, culture and habits. Like the plants, they conserved these attributes, even though they were surrounded from time to time by hostile populations. It was because it was important to know of these attributes, which could then be protected, that he had suggested that a Red Data Book should be prepared of endangered ethnic communities.

Ramakrishnan reiterated his concerns. The existence of such a catalogue would lead governments to keep such communities in their original state rather than let them evolve and progress. Society should not be kept in a static state as had been suggested by some anthropologists.

Annexe 3: Forestry and Biodiversity*

CRISTEL PALMBERG

Chief, Forest Resources Development Branch, Forest Resources Division, FAO, Via delle Terme di Caracalla, Rome 00100, Italy

Genetic variation, accumulated in all living organisms on Earth during some three-thousand million years of biological evolution, constitutes the genetic resources of our planet. This diversity provides a buffer against environmental changes and is thus essential for maintaining the stability and biological balance of The Biosphere.

PRINCIPLES AND STRATEGIES IN THE CONSERVATION OF FOREST GENETIC RESOURCES (ECOSYSTEMS, PLANT AND ANIMAL SPECIES)

In addition to buffering species in the wild against changes in climate, soil and other adverse environmental influences (including pests and diseases), genes from the wild are needed to ensure the adaptation to changing environmental conditions and changing socio-economic needs of domesticated plants, including those 16 major crops that today provide the bulk of human food, world-wide. Further domestication and long-term, sustainable use of these species and species yet to be identified, are dependent on the availability of genetic variation from which new varieties can be developed through selection and breeding. Increased genetic knowledge and technologies such as genetic engineering, may provide new possibilities to utilize genetic materials from a great variety of species to improve cultivated plants. Similarly, wild animals provide a gene-pool of potential value to domesticated animals.

The values derived from wild genetic resources are generally associated with the the different levels of organization of diversity that exist in Nature, from ecosystems to species, populations, individuals and genes. In considering the conservation of genetic resources it is necessary to specify clearly the objectives aimed at. This is of the utmost importance, as it is possible to conserve an ecosystem and still lose specific species; and to

*[This contribution was originally prepared within the framework of FAO's Tropical Forestry Action Plan in January 1990 but was submitted as a paper to the Conference in the unavoidable absence of the author. It has unfortunately, not proved possible to update the details since the original submission. Eds.]

conserve a species and lose genetically distinct populations, or genes which may be of value in adaptation and future improvement of the species. Conversely, the loss of an individual species, of distinct populations, of individuals, or even genes, may pose a threat to the continued existence of specific ecosystems.

Thus, when planning a conservation programme, attention must be focused on the level(s) of organization of immediate concern. However, as coexisting organisms interact, and as different levels of organization are to a certain degree interdependent, it is essential that management decisions taken also reflect such interdependencies.

For the conservation of genetic resources of species and diversity within species, there are two basic strategies: (i) *in situ* (on site) conservation, *i.e.* conservation in their natural or original habitat; and (ii) *ex situ* (off site) conservation, *i.e.* conservation in gene-banks (as seed, tissue or pollen for plants; as sperm, ova and embryoes for animals); in plantations and zoos; or in living collections (as plants, in botanic gardens, arboreta and specially established *ex situ* conservation stands, which give due consideration to the representativeness and genetic base of the material thus conserved). These two strategies are complementary, both have a role to play, and they should be used in parallel whenever possible. It should be noted that provided some basic, genetic principles are applied and sustained, utilization-oriented forest management is generally compatible with the *in situ* conservation of genetic resources of priority species, or species under pressure or threat of extinction in parts or all of their ranges. Similarly, plantation forestry and genetic improvement programmes can and should incorporate components of genetic resource conservation; such programmes can even, if well planned, enhance the genetic variation of target species. Harmonizing utilization and conservation of actually or potentially important species is a key to genetic conservation in the long term.

In relation to the need to manage genetic resources, it should be recognized that genetic and ecological variations are very seldom, if ever, in an evolutionarily stable state; the goal for managers of genetic resources is thus not to preserve a static state, but to contain a dynamic system. Ecological or multipurpose management is not necessarily less intense than sound management systems aimed at production of other goods and services; however, it is frequently more extensive. Contrary to popular belief, 'leaving an area alone' can as easily decrease diversity as increase it. Some species will require management intervention in the ecosystem, while others will depend on ecosystem diversity; maintaining such ecosystem diversity may or may not require human manipulation.

FAO'S ROLE AND PROGRAMME IN THE CONSERVATION OF FOREST GENETIC RESOURCES

Conservation of genetic resources cannot be viewed in isolation, but must form part of short, medium or long-term development programmes in the countries concerned. Under the over-all umbrella of the Tropical Forestry Action Plan, and guided by Statutory Bodies such as the FAO Commission on Plant Genetic Resources, the FAO Panel of Experts on Forest Gene Resources, the Committee on Forest Development in the Tropics and the Regional Forestry Commissions, FAO's Forestry Department supports the conservation, collection, evaluation, exchange and wise utilization of germplasm; advises countries in technical, legal and policy-level decisions related to these issues; and disseminates information at policy-making, technical and grassroots levels, aimed at the raising of awareness, training, and coordination of efforts at global and regional levels.

Foresters are in a key position to help ensure the conservation and wise use of natural renewable resources. Forests and woodlands contain not only woody species and wild animals, but – especially in the moist and seasonal tropics and in some parts of the subtropics – a wealth of other plant species of actual or potential socio-economic value.

The programme on the Conservation of Tropical Forest Ecosystems of the Tropical Forestry Action Plan, coordinated by FAO includes components of both ecosystem and genetic resource conservation. It is closely related to the other four component programmes, and particularly so to the programme on Forestry in Land Use. Supporting FAO activities include programmes on the conservation of genetic resources of arid/semi-arid zone arboreal species, the improvement of rural living and the establishment of pilot *in situ* conservation areas and related research, in collaboration with national institutes in all three major tropical regions. They also include collaboration with institutes working in the genetic resources field in the collection, handling, storage and distribution of seed and other reproductive materials; technical assistance in the conservation and management of wildlife resources; assistance to countries in the planning and management of national parks and other protected areas and in the multipurpose management of forest reserves and gazetted forest areas, in which compatibility of genetic resource conservation on the one hand, and the production of goods and services on the other, is stressed. FAO's field programme related to the conservation and sustainable use of forest genetic resources (*senso lato*), has greatly expanded over the past few years, and lays emphasis on subregional or ecological networking and the promotion of TCDC between countries with similar environmental conditions and socio-economic needs.

Elaboration of national and regional/subregional Tropical Forestry Action Plans, helps to establish national priorities and to develop ideas and programmes in this vast and fundamentally important field.

FAO's Inter-Departmental Sub-Group on Biological Diversity of the IDWG on Environment and Energy, was established in 1988. Present activities include the elaboration of an FAO Issues Paper on the conservation of animal, fish and plant genetic resources; and an FAO policy in the conservation of biological diversity.

Within the framework of the Ecosystem Conservation Group (FAO,

UNESCO, UNEP, IUCN), an extraordinary meeting is scheduled to be held at FAO Headquarters in March 1990, with the specific aim of reviewing legal instruments related to the conservation of biodiversity, with special reference to the draft UNEP/IUCN convention on the subject, and its relationship to the FAO Global System on Plant Genetic Resources (the Commission, International Undertaking and International Fund on Plant Genetic Resources), and other already operational or planned conventions or agreements.*

The mandate of the *ad hoc* Working Group on *in situ* conservation of plants of the ECG, chaired by FAO/FOD, has recently been expanded to cover questions related to biodiversity in general. Its membership has been expanded to encompass, in addition to the earlier members (FAO, UNESCO, UNEP, IUCN, IBPGR), other international agencies on an *ad hoc* basis (depending on the subject under discussion at each meeting), such as the Worldwide Fund for Nature (WWF) and the World Bank. All these agencies also collaborate in the implementation of the Tropical Forestry Action Plan, thus linking and coordinating activities between the groups.*

*[See footnote on p. 191. Eds.]

9. Shortcomings of Ecology, with Special Reference to Diversity Patterns and Processes

PÁL JUHÁSZ-NAGY

Professor, Department of Plant Taxonomy & Ecology, Eötvös Loránd Science University, Kun B. ter. 2, H-1083 Budapest, Hungary

INTRODUCTION

Let us start with a naïve but pertinent question: is there a proper, reliable and widely accepted, foundation of ecology? The only objective answer is: *No*, unfortunately, *there is none, as yet.* This gloomy reply may seem highly surprising to those who realize that, including the very important Humboldtian period of its history, ecology is rapidly approaching its *bi*centennial.

Taking a single example out of the many, the authors of *Theoretical Ecology* (May, 1981), holding an 'ultra-atomistic view', seem to be quite insensitive to good old observational facts and simple rules of Nature such as plant and animal community successions, zonality relations, etc. This is why their presentation, though elegant and relevant, becomes somewhat inconclusive and even 'unbalanced'. One may remark rather sadly that, say, 'Harper-ology' in itself may be as one-sided as 'Braun-Blanquetism' or Tuxenism' are in themselves.

Indeed, I believe that, contrary to some common belief, a reliable foundation is by no means a 'theoretical luxury'. Instead, it is an *everyday need* in order to clarify properly the demarcation lines between the fields of our actual knowledge and the huge regions of ignorance. Such a foundation should be imperative for a much better understanding than we currently possess of almost all problems of the environmental future – which problems, *in majoribus*, are rather fuzzy or even obscure, despite all the efforts of the organizer and participants of this unique series of conferences (Polunin, 1972, 1980; Polunin & Burnett, 1990).

The goal of this paper, of course, is very limited. It cannot do more than elucidate some problems – and, indeed, *only some* problems – which seem to be relevant to the critical meeting-points of an improved foundation of ecology and an enlightened understanding of our environmental future. In order to avoid an 'Odyssaean roam', the majority of present arguments are centred on the 'state of the art' of the present-day diversity research (including some criticism, modelling propositions, etc.).

Unfortunately enough, even the history of this reduced area of study is rather confused; despite the efforts of MAB and IUBS programme 'Biodiversity Crisis', even recent research shows a rather slow or zig-zag development. Fortunately, however, the overall importance of 'diversity reasoning' has become sufficiently well-known today, and a fairly large body of valuable publications makes it quite superfluous to introduce here a too-general kind of argumentation. It is to be noted, however, that some historical points and some literary sources will be discussed briefly nearer the end of this paper.

Some Primary Problems

One cannot, of course, get away from the vexing question: what is the main job of ecology? As a primary answer or starting-point, we can accept here the opening sentence of D. Tilman's excellent book (1988): '*The central goal of ecology is to understand the causes of patterns (which we observe in [the] natural world)*'.

This sentence, without doubt, is attractive because of its shortness and simplicity. If necessary, Tilman's sentence can be made much more precise, even by raising a few additional points; for instance, by recognizing such points as the following:

- that there are too many kinds of patterns in Nature;
- that ecology is concerned with a very important family of these patterns, which may be termed '*bio-filtered patterns*';
- that there are several levels, ways, and means, of understanding (depending partly on the different *scaling* properties of bio-filtered patterns); and
- that there frequently exists multi-causality and circular causality as well, etc., etc.

Without bothering much here with the semantic precision of Tilman's intuitive notion, let us concentrate now on some further implications of his quoted statement.

First of all, the *phenetic* patterns and the *latent* ones are to be clearly distinguished, even if we know that there are several degrees or intermediate states of being 'phenetic' or 'latent'. In a very simple case, the former category may refer to a vegetation map of some kind, and the latter category may imply a number of meteorological, pedological, etc., representations.

Secondly, it is to be noted that even 'phenetic' patterns are not necessarily 'observable', *i.e.* 'physiognomical'; indeed, unfortunately enough, the vast majority of relevant primary patterns are non-physiognomic ones.

Thirdly, the natural question arises as to whether our ability to represent properly phenetic patterns is sufficient to make an acceptable interpretation of what 'causes of patterns' may mean.

The only answer is again: No! One of the most troublesome shortcomings of ecology as a whole, whether as an environmental component or otherwise, has always been and still is, an essential weakness in developing adequate pattern-modelling, and especially in selecting 'relevant patterns' in a given context, or in representing pattern transformations of several dy-

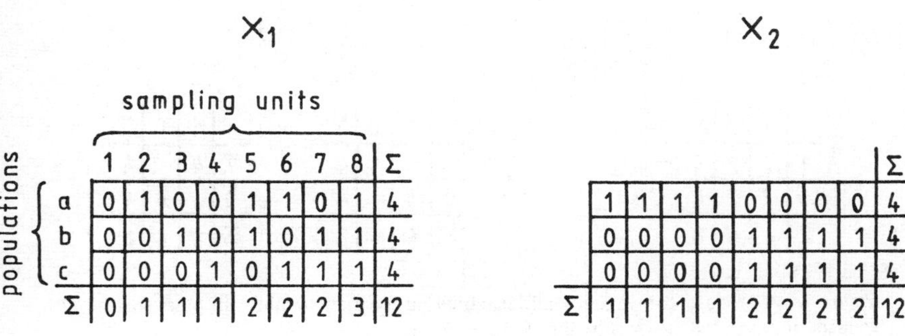

X_1

sampling units (populations a, b, c)

	1	2	3	4	5	6	7	8	Σ
a	0	1	0	0	1	1	0	1	4
b	0	0	1	0	1	0	1	1	4
c	0	0	0	1	0	1	1	1	4
Σ	0	1	1	1	2	2	2	3	12

X_2

									Σ
	1	1	1	1	0	0	0	0	4
	0	0	0	0	1	1	1	1	4
	0	0	0	0	1	1	1	1	4
Σ	1	1	1	1	2	2	2	2	12

Figure 9.1. Two over-simplistic binary tables, showing the simplest type of compositional representations.

namic contexts. In order to see why this weakness is by no means unconnected with a bunch of *'l'art pour l'art'* problems, let us turn now to some major questions regarding the environmental future.

What is the most frightening danger of today and tomorrow? In brief, it is the *degradation of The Biosphere*, in other words, the growing proliferation of degraded patterns all over the globe. The shortcomings of ecology are shown clearly by the facts that our knowledge of the degradational processes is still very limited, and that, surprisingly enough, we have very few models representing degradation as a special kind of pattern transformation. For this reason, a very urgent need of ecology is to find efficient ways of *comparing 'original'* (relatively intact) *patterns with degraded patterns*. This comparison should be sensitive enough to make detectable not only 'semi-drastic changes' but even *in statu nascendi* degradation as well.

It is clear enough that diversity, *inter alia*, should have an important role in such a comparison. It is, however, much more obscure as to what this proper role may be; how to relate diversity properties to other attributes of our object (say, a community of some kind); and how to use a pattern-and-process approach in a study of degradation. These difficulties lead us to the present arguments for some particular reconsideration, as we'll now indicate.

Diversity: Primary Criticism and Postulation

Such a reconsideration can be made concise enough for practical use by introducing a few conceptual and methodological 'oppositions'.

(1) *Scalars* versus *Vectors*:– For too many people, diversity still means the 'number of species' (and/or other taxa), the 'only' tangible danger being extinction. But, using the *ex verbis* terms of the late Professor A. Rényi, a 'number of components' (alias 'richness') refers always to *zero-ordered diversity*, and extinction of one or more elements must be considered to be a tragic 'finality' (end-result) of a more or less long stochastic process. If we want to study the process itself, then, to say the least, we need always vectorial representations, instead of scalars.

(2) *Static* versus *Dynamic Representations*:– If the need is for a dynamic approach to be accepted, then a number of very difficult problems should be

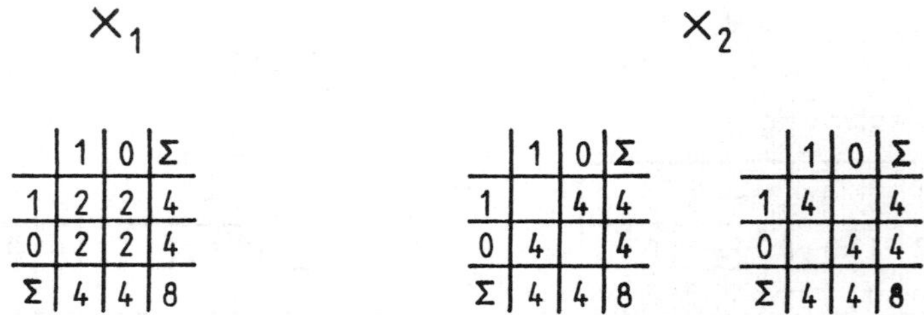

X_1

	1	0	Σ
1	2	2	4
0	2	2	4
Σ	4	4	8

X_2

	1	0	Σ
1		4	4
0	4		4
Σ	4	4	8

	1	0	Σ
1	4		4
0		4	4
Σ	4	4	8

Figure 9.2. Fourfold contingency tables, showing frequency values in pairs for the raw vectors of X_1 and X_2 shown in Fig. *9.1*.

faced. As is well known, the adjective 'dynamic' is used most frequently for temporal processes only (as in the vast majority of differential or difference equations in theoretical physics or theoretical biology). In such representations of temporal processes, space is still 'static' (by considering space as 'homogeneous', spatial heterogeneity as 'negligible', etc.). Contrary to this common and convenient view, the most important biological processes (such as evolution, ecological succession, etc.) are clearly *spatio-temporal processes*, where both space and time should be taken into account (even if the methodology involved is frequently troublesome). If we want to study diversity changes (say, during the process of a degradative succession*), then we must find ways and means of comparing spatio-temporal coordinates, and with special reference to the very neglected spatial processes.

(3) *Single Vectors* versus *Classes of Vectors*:– In returning to argument (1), it is fairly important to consider at least the first part of a 'singularistic' versus 'pluralistic' dilemma. Figure 9.1 shows how deceiving it can be to say that X_1 and X_2, two binary compositions (where '1' and '0' mean presence and absence, respectively), are practically the same, because their raw marginals as vectors are identical and column marginals have only a small degree of difference. (In actual fact, using weighted Shannon's estimates [*cf.* Rényi, 1962], raw marginals have 19.02 bits in both cases: column marginals of X_1 and X_2 can be characterized by 32.27 and 35.02 bits, respectively). In order to get a better insight, it is relevant to see the difference between 'inner configurations' (composed of raw and column vectors, respectively); in some way, X_1 can be regarded as a '*random* composition', whereas X_2 is a '*regular* composition'. In a more explicit way, let us consider first set Q, Q = [a, b, c]; secondly, Q", the power set of Q[set of all possible subsets], Q" = [Ø, a, b, c, ab, ac, bc, abc]; thirdly, the fact that X_1 uses *all* elements of Q" *at once*, whereas X_2 uses only two elements (a, bc) *four* times. In consequence, $m\hat{H}_1 = 24$ and $m\hat{H}_2 = 8$ bits, where both quantities are weighted entropy estimates for X_1 and X_2, respectively, and where m = 8 (number of sampling units). In a more detailed way,

$$m\hat{H}_1 = 8 \log_2 8 - 8(1 \log_2 1) = 24$$
$$m\hat{H}_2 = 8 \log_2 8 - 2(4 \log_2 4) = 8.$$

*[*i.e.* retrogression – compared with relatively undisturbed previous states – as when a habitat becomes less and less habitable by 'desirable' dominants or co-dominants. Eds.]

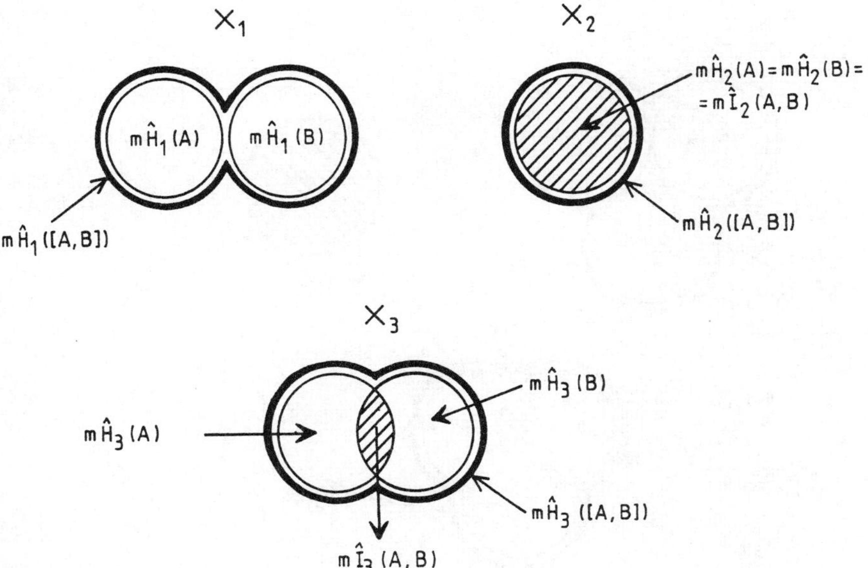

Figure 9.3. Venn-diagrams for a pairwise comparison of data in Fig. *9.2.*

(4) *Diversity* versus *Dependence*:– Even this over-simplistic example and the main properties of this type of diversity (say, '*combinatorial diversity*') show properly that diversity in most cases should not be used *per se* but interconnected (coupled) with other types of phenomena, namely processes. Let us consider now the simple (pairwise) comparison of raw vectors of Figure 9.1 (say, as an *interlocal* comparison), using the well-known 2 x 2 contingency tables as shown by Figure 9.2. It is easy to see that all contingency tables for all pairs of Q elements of X_1 represent clear-cut cases of stochastic *independence*, whereas all 2 x 2 tables for X_2 represent *maximum degrees of dependence* (i.e. 'non-independence'). In a more explicit way, and concentrating on the (a,b)-pair, the values of *joint entropy functions* are:

$$m\hat{H}_1([A, B]) = 16 \text{ bits}, \quad m\hat{H}_2([A, B]) = 8 \text{ bits};$$

the values of *simple*/marginal/*entropy functions* are clearly identical:

$$m\hat{H}_1(A) = m\hat{H}_1(B) = m\hat{H}_2(A) = m\hat{H}_2(B) = 8 \text{ bits};$$

in consequence, the values of *contingency information* (say, association) *functions* are:

$$M\hat{I}_1(A,B) = m\hat{H}_1(A) + m\hat{H}_1(B) - m\hat{H}_1([A,B]) = \\ = 8 + 8 - 16 = 0,$$

$$m\hat{I}_2(A,B) = m\hat{H}_2(A) + m\hat{H}_2(B) - m\hat{H}_2([A,B]) = \\ = 8 + 8 - 8 = 8 \text{ bits}.$$

This relevant difference can be described by the usual notation of set theory as well, i.e. $m\hat{H}_1(A)\ m\hat{H}_1(B) = \emptyset$; $m\hat{H}_2(A)\ m\hat{H}_2(B) = m\hat{H}_2(A) = m\hat{H}_2(B) = m\hat{I}_2(A,B)$, or, it can be depicted by means of the usual Venn-diagrams (*see* Figure 9.3). In addition, Figure 9.3 shows a third case, X_3, where the intersection of $m\hat{H}_3(A)$ and $m\hat{H}_3(B)$ is neither empty, nor 'full' (as before-

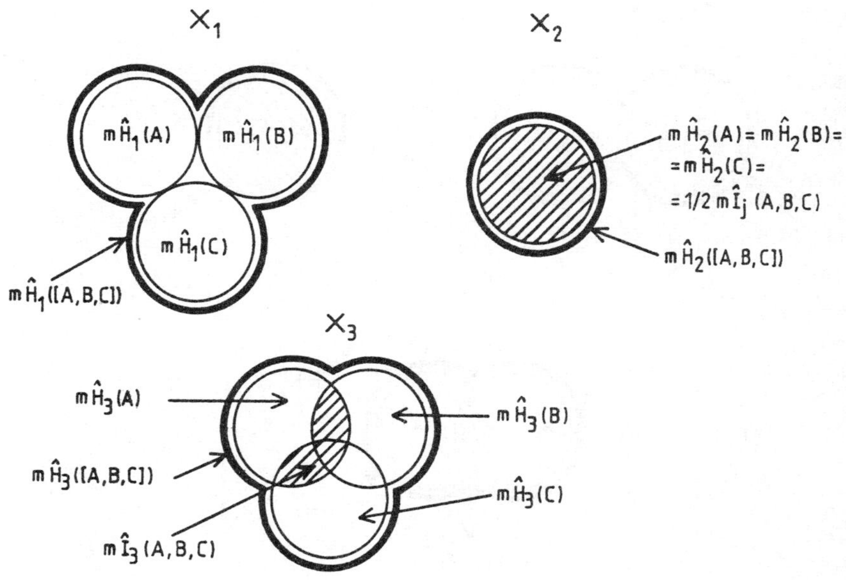

Figure 9.4. Venn-diagrams for a three-ways' comparison.

hand) as is the case in the vast majority of empirical situations. Note that in all Venn-diagrams of this type, diversity appears always as some 'envelope' and dependence as some non-empty *intersection*.

(5) *Entropy* versus *Information*:– The last statement remains true even if we do not consider pairwise comparisons. In the case of our primitive example, it is straightforward to introduce triple comparisons, for all elements of Q (by means of 2 x 2 x 2 contingency tables). Such a comparison is depicted in Figure 9.4, quite analogous to Figure 9.3. It is easy now to identify the quantities at the end of <3> ; namely $m\hat{H}_1 = m\hat{H}_1([A,B,C]) = 24$, $m\hat{H}_2 = m\hat{H}_2([A,B,C]) = 8$ bits, where the actual valuation is due to the fact that, for X_1, all the possible intersections are empty (and, therefore, the value of joint entropy is of *maximum*), while for X_2, as all intersections are 'full' (with *maximum* values), therefore the value of joint entropy is of *minimum*. This contrast shows why it can be very dangerous to interpret entropy *versus* information relations in a vulgar way ('the less the uncertainty, the greater our information'). Instead, it is much better to think in terms of *osse–esse-relations* (i.e. relations between possibilities and 'realizations' as even Duns Scotus guessed rightly, in the dawning period of our age) *cf.* Muralt (1991). In a more explicit way, we may introduce here the function $m\hat{I}_j(A,B,C)$,

$$m\hat{I}_j(A,B,C) = m\hat{H}_j(L) - m\hat{H}_j([A,B,C])$$

where $mH_j(L)$ is an entropy estimate for all raw marginal values.

As in the binary compositions of Figure 9.1 raw marginals are identical, and $m\hat{H}_2(L) = m\hat{H}_2(L) = 24$ bits, therefore, $m\hat{I}_1(A,B,C) = 0$ and $m\hat{I}_2(A,B,C) = 16$ bits (as a special kind of *max*-value). Again, if some

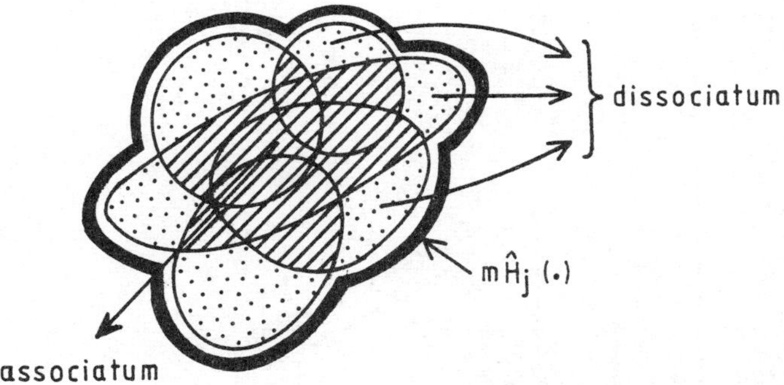

Figure 9.5. A non-elementary Venn-complex.

intermediate third case, X_3, is introduced, then it is expected to have 'intermediate values' between the extreme values indicated above. This shows clearly why it is so important to know the proper *extrema*; how to manipulate with relevant relations (*e.g.* <, ≥) in estimating empirical values; how to use the so-called 'sigma-conditions', proposed by the best authors in the field of information theory (*e.g.* Khincsin, 1957; Kullback, 1959; Luce, 1960; Rényi, 1962; etc.).

(6) *Diversity* versus *Complexity*:– We may call the configurations of Figure 9.3, and even the more advanced Figure 9.4, elementary Venn-complexes (where the adjective 'elementary' refers to a pairwise or three-ways, respectively, comparison of same sort). If we have *s* kinds of populations (where *s* is a natural number, the number of elements in a set Q), and if *s* is more than 2, then we get some *non-elementary Venn-complex* (*see* Figure 9.5). Although it is much beyond the scope of this paper to show the details of a proper methodology (*e.g.* how to manipulate with 2 x 2 x ... x 2 tables), we may guess at least some simple properties of such a complex. First, it has, in the manner of Figures 9.2 and 9.3, an overall diversity envelope. (*Nota bene*: as such a complex has or may have as well several 'inner subenvelopes' of different orders, therefore a relevant but still missing concept, *subdiversity*, should be taken into account.) Secondly, supposing that the intersections involved are neither empty nor full, such a complex can be subdivided roughly into an 'overall intersectional region' and an 'overall outer region'; the former can be represented as a multivariate association (called *associatum*), the latter as a multivariate dissociation (called *dissociatum*), referring to the 'common' and 'uncommon' (particular) behaviour of the populations involved.

It is easy to recognize that the quantity, in <5> , $m\hat{I}_j$(A,B,C), is a very simple illustration of associatum. (*Nota bene*: due to the 'inner structure', sub-diversity relations of such a complex, a number of subassociata and subdissociata of different orders are to be distinguished.) Thirdly, the ordering relations mentioned above can make hopefully tractable at least some main features of *complexity* of our object (including the seemingly simple question: how to relate 2 x 2 properties to 2 x 2 x ... x 2 properties).

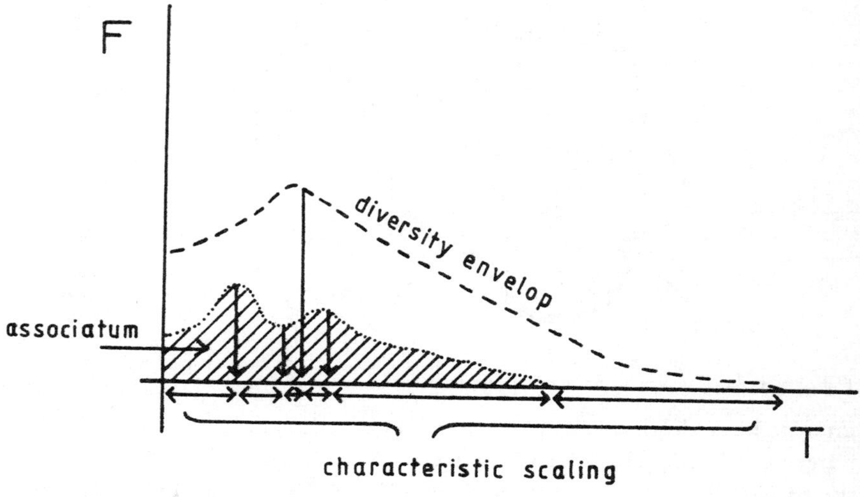

Figure 9.6. Some illustrations of elementary scaling.

Fourthly, it is important to study a non-elementary Venn-complex 'in motion', when a number of its characters (such as the sizes of intersections) may change in a spatio-temporal referential system.

Without listing any further possible 'oppositions', some tentative conclusions derived from the preceding ones suggest:

– that there are many types of diversity, which accordingly include a highly 'diverse body' of phenomena;

– that a particularly relevant and fairly inclusive type of diversity may be termed *compositional diversity*; and

– that compositional diversity makes good sense only if it is related to other attributes (such as dependence) of our objective.

Surprisingly enough, the simplest binary compositions (such as that shown in Figure 9.1) can generate more consistent, and much richer, families of models than the non-binary (so-called 'numerical') compositions (*cf.* Juhász-Nagy, 1976, 1984; Juhásy-Nagy & Poldani, 1983).

Towards Improved Understanding

The former sections have made allusions to a pattern and process approach. However, it remains to be seen how such an approach can be applied, pointing only to ways and means.

First of all, some simple means of a *spatial processing* needs to be considered. Such a processing is realized in many circumstances by *T*, a topographical vector whose points represent the geometrical sizes of sampling units of growing order (processing means here primarily a series of increasing plot-sizes). For the sake of a proper evaluation, it is always supposed that $s<m$, and $m>\infty$ (*i.e.* we have very many plots of each size; the number of plots must be greater than the number of populations). If we put *T* versus *F* (where *F* is a function, or a set of functions, of section 3), then we get some

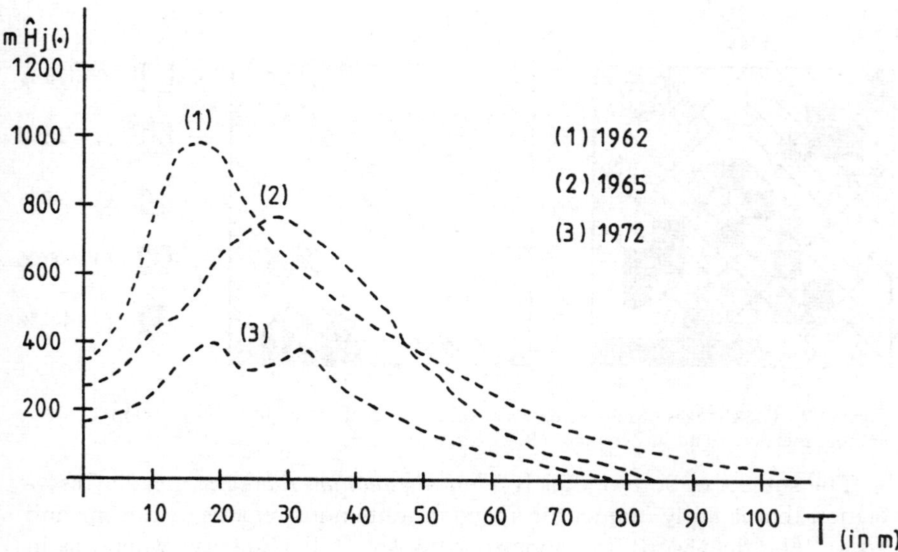

Figure 9.7. Diversity changes of a meadow (Nardo-Festucetum rubrae) in Zemplén, NE Hungary.

curves or graphicons like those shown in Figure 9.6. Although, for the sake of simplicity, Figure 9.6 shows only two graphicons, we may have here quite a number of similar representations – *e.g.* according to the reasoning of (6) in the previous section.

Let us introduce now the simple convention of *elementary scaling*, where only max–min properties (such as peaks) of such graphicons are used. As is shown in Figure 9.6, this convention generates, in the axis *T*, sets of *characteristic points, intervals, orderings*, etc., which sets can be regarded as essential attributes of a particular object. Needless to say, the more functions we have – the property 'being characteristic' – the more reliable our observations become. In a somewhat metaphorical sense of the word, we can say that this type of *characteristic scaling* shows the way in which a community of some sort can 'sense for itself' different degrees of spatial heterogeneity. (It is to be noted that the most classical feature of characterizing scaling has been the emergence of the concept 'minimum area' in the first decades of this century. In the present line of interpretation, 'minimum area' is a point or interval of T, where the value of $m\hat{H}_j(.)$ becomes zero, at least asymptotically. In addition, we can introduce the concept of 'maximum area', i.e. as the actual 'peak', as in Figure 9.7(1), or an interval of $m\hat{H}_j(.)$.)

The next question is *how to change characteristic scaling in time* – in particular if some degradational process is going on in a particular spot. Naturally, this kind of change can be of very different types, depending partly on the properties of *T* and *F* and what functions are to be used, and partly on the 'scaling sensitivity' of a particular object. Some simple features of such changes are illustrated in Figures 9.7 and 9.8.

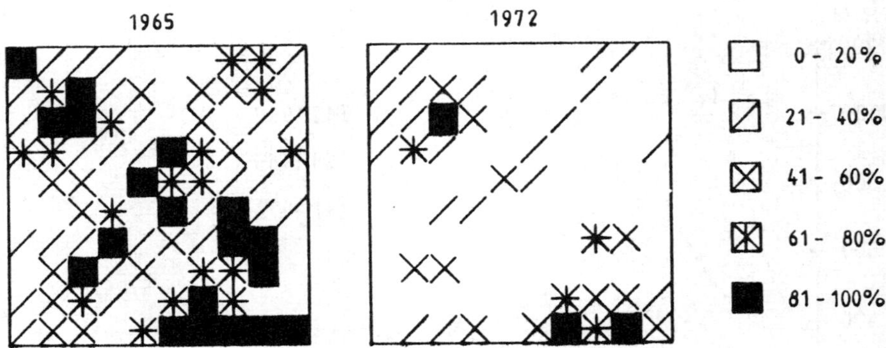

Figure 9.8. The changes of relative diversity values in an allocated form for a 100 m^2 of another meadow stand in Zemplén, NE Hungary.

The subject of both figures is *Nardo-Festucetum rubrae* (a meadow association that is fairly frequent in many mountainous regions of Europe and Asia). The field-work was done in Zemplén (NE Hungary) where, as in many other parts of Europe, these meadows became more and more degraded from year to year. The most drastic changes are *the decrease in the number of populations* (*e.g.* in the stand of Figure 9.8, in which the number of plant species was reduced from 84 to 69 during 7 years), the *permanent substitutions* (*e.g.* the aggressive *Nardus stricta* invades spots that were dominated by *Festuca rubra* beforehand), a *trend towards homogeneity* (which makes these masterpieces of landscape, originally rich in structures and colours, now rather monotonous).

Both these figures represent *diversity pattern transformations* of different kinds. For the sake of simplicity, Figure 9.7 shows only 'the step-by-step movement' of an overall diversity envelope (regretfully omitting many other interesting functions). The primary explanation is due to the reduction of Q", power set of Q (i.e. if the combinatorial possibilities are reduced, then diversity peaks decrease also, and both maximum and minimum areas will have other positions in *T*). Figure 9.8 shows pattern transformation in a more concrete form, in an allocated way for a 100 m^2 sampling area of a meadow. The novelty here is the application of a new concept, *relative diversity*, where the %-participations of an $m\hat{H}(.)$ have been estimated. The sad conclusion is clear enough; this figure illustrates the frightening danger of dissipating diversity.

Short Discussion and Outlook

If we reconsider historically some of the main features of 'diversity reasoning' in ecology (or, otherwise, in science in a much wider context), then we can see at once the strange oscillations of advantages and disadvantages, fashions and counter-fashions, 'forbacks and drawbacks', of *ratio scientifica*.

It is true that the brilliant modern guess is due to Charles Darwin's genius in recognizing that some 'threshold diversity' is always needed for any successful development. But, as E. Mayr's book (1984) clearly shows, very few people were able to follow these kinds of arguments for a surprisingly long time. It is true also that Sir Ronald Fisher called attention again to the importance of diversity in taxonomy, genetics, etc., in the early 1940s (*see also* C. B.Williams, 1964), without any meritable echo during his lifetime, and despite his remarkable *Genetical Theory of Natural Selection* (Fisher, 1930). The present author remembers well the striking confusion in the Presidency of the British Ecological Society (the oldest body of its kind in the world), when they received the manuscripts of some pupils of R. H. MacArthur (*cf.* 1965) on diversity patterns for the jubilee meeting of the Society in 1964; very learned and clever people were unable to comprehend the simple message of those manuscripts.

If he or she tries to follow carefully the most relevant publications in this field of study (to mention but a few, *e.g.* Margalef, 1958; MacArthur, 1965; Woodwell & Smith, 1969; Peet, 1974; Pielou, 1975; Washington, 1984; Magurran, 1988; Wilson, 1988), then the critical reader can trace again the remarkable oscillation of appearance and disappearance of good ideas and ideals in a very strange sequence. For instance, since the activity of R. H. MacArthur's group, it is quite clear (almost a scientific commonplace) that, without patterning, almost all diversity data are meaningless. But this simple truth does not appear in the recent book edited by E. O. Wilson (1988); as far as I can see, very few scientists are aware of it – even in the programme entitled 'Biodiversity Crisis'. One may well ask: why is it so?

For a better or at least more comprehensive answer, let us turn now to the elegant wording of Aldous Huxley (1937) on the 'Nature of Explanation':

> 'The human mind has an invincible tendency to reduce the diverse to the identical. That which is given us, immediately, by our senses, is multitudinous and diverse. Our intellect, which hungers and thirsts after explanation, attempts to reduce this diversity to identity. Any proposition stipulating the existence of an identity underlying diverse phenomena, or persisting through time and change, seems to us intrinsically plausible. We derive a deep satisfaction from any doctrine which reduces irrational multiplicity to rational and comprehensible unity ...
>
> 'The effort to reduce diversity to identity can be, and generally is, carried too far. This is particularly true in regard to thinkers who are working in fields [that are] not subjected to the disciplines of one of the well-organized natural sciences. Natural science recognizes the fact that there is a residue of irrational diversity which cannot be reduced to the identical and the rational. For example, it admits the existence of irreversible changes in time. When an irreversible change takes place, there is not an underlying identity between the state before and the state after the change. Science is not only the effort to reduce diversity to identity; it is also, among other things, the study of the irrational brute fact of becoming ...'

One of the most important points is to detect (or at least to 'feel') the very old roots of the same problem. The 'identity and diversity' dilemma of Aldous Huxley is similar to the *idem et diversum* argument of Duns Scotus (1270?-1308) who, while attacking the doctrines of Thomas Aquinas, had used a block of dialectical phrases (such as *posse et osse, singulare et plurale*, etc.) that we might consider the 'scholastical kernel' of the matter. Without going into the exciting controversy of the above 'two great doctors' '*subtilis*' and '*angelicus*', or into details of many further controversies, it may be satisfactory here to mention some more recent and more explicit aspects of the same dilemma.

For instance, to a biologist, *idem et diversum* may appear at first to be a dilemma between the very different approaches of Carolus Linnaeus and Charles Darwin. Indeed, I hazard a shrewd guess that Aldous Huxley, a grandson of 'Darwin's bulldog', the celebrated Thomas Henry Huxley, had always had this polarity in mind. If we go from the ingenuous guess towards the more 'sophisticated' reasonings of our days, we can trace always *idem et diversum* (in the sense of 'oneness and plurality') in a chronically renovated form, *e.g.* how The Biosphere, a marvellously *unique* construction in its 'Gaia-character', consists of a vast *multiplicity* of components. Due to the irreversibility argument of Aldous Huxley, it is not too difficult to see why modelling is needed in most cases for a better understanding of the 'brutal fact of becoming' even an infinitesimally small part of it.

This is is the point where our chief topic, namely *limitations of science*, can perhaps be faced with some remote hope of some degree of success. One feels that science has – very roughly – two main requirements: one is due to the 'matter of [our human] inner affairs', while the other is connected with the 'matter of outer affairs'. The first condition requires an *organic development* of science itself, where all actual ideals are to be based carefully on all previous ideals; the second condition requires that scientific results, relevant thoughts, etc., be made *public and useful* at a certain level and tempo (*i.e.* science should be *communicative* enough to be widely used and useful.

The very severe limitations of science are shown all-too-clearly by the painful fact that either scientific development is frequently far from being 'organic' (indeed it is widely very 'inorganic' in our case), or else that science is not capable of being 'successfully communicative', as the *public awareness* concerning 'diversity problems' is still extremely weak.

One can say that this is so because scientists, as an 'exclusive gang of strange people', often cannot explain relevant things in proper language, even to the 'man in the street'. But we must always realize that communication of any such kind needs at least two partners – a donor and an acceptor – and that if either partner is 'too passive', then the result is always doubtful. Many people in our part of the world (*vulgo*: 'Eastern Europeans') know only too well that totalitarian systems (*i.e.* 'counterdiversity systems') are not based only on guns and drastic methods; they are also based substantially on the mental laziness of those who, having no courage to face the risks and difficulties of a 'spiritual diversity', accept willingly 'singularistic dogmas' as a matter of simple convenience.

It is true, of course, that some sorts of reasoning (say, models) can be fairly difficult. It is true also that such concepts as 'patterning' or 'pattern and process' may sound rather doubtful even within 'the campus of natural sciences', at least at the first hearing. In such cases there is no other course than proper explanation and successive approximation, pointing for instance to:

- the simplicity of good-old observational facts (*e.g.* tropical regions usually have a greater degree of diversity than regions of moderate or severe climates);
- the need of a better interconnection of 'globality – regionality – locality'; and
- the fact that, in a sense, 'patterning' is nothing but a necessary 'refinement' of the well-known regional reasoning of our naturalist ancestors.

We must believe, indeed, that somehow or other, even the most 'difficult' circumstances, phenomena, and relationships, can be explained to some extent or to a certain point. We must believe that some kind of mental 'meliorization' is always possible by confronting and solving the severe limitations of science. We hope that such meetings, in the manner of the Budapest conference, can help the mutual understanding of very different kinds of people. *Ergo: vivat et floreat diversitas.*

References

Fisher, R. A. (1930). *The Genetical Theory of Natural Selection.* Clarendon Press, Oxford, England, UK: xiv + 272 pp., illustr.

Huxley, A. (1937). *Ends and Means (An Inquiry into the Nature of Ideals and into the Methods Employed for their Realization).* Chatto & Windus, London, England, UK: 336 pp.

Juhász-Nagy, P. (1976). Spatial dependence of plant populations: Part I. *Acta Botanica Hungarica*, 22, pp. 61–78.

Juhász-Nagy, P. (1984). Spatial dependence of plant populations: Part II. *Acta Botanica Hungarica*, 30, pp. 363–402.

Juhász-Nagy, P. & Poldani, J. (1983). Information theory methods for the study of spatial processes and succession. *Vegetatio*, 51, pp. 129–40, illustr.

Khincsin, A. I. (1957). *Mathematical Foundations of Information Theory.* Dover, New York, NY, USA.

Kullback, S. (1959). *Information Theory and Statistics.* Wiley, New York, NY, USA.

Luce, D. (1960). The theory of selective information and some of its behavioural applications. Pp. 5–119 in *Developments in Mathematical Psychology* (Ed. D. Lucas). The Free Press, Glencoe, Minnesota, USA.

MacArthur, R. H. (1965). Patterns of species diversity. *Biol. Rev.,* 40, pp.510–53, illustr.

Magurran, A. E. (1988). *Ecological Diversity and its Measurement.* Princeton University Press, Princeton, New Jersey, USA: x + 179 pp., illustr.

Margalef, R. (1958). Information theory in ecology. *General Systems*, **3**, pp. 36–71, illustr.

May, R. M. (Ed.) (1981). *Theoretical Ecology: Principles and Applications*, 2nd edn. Blackwell, Oxford, England, UK: ix + 489 pp., illustr..

Mayr, E. (1984). *Die Entwicklung der biologischen Gedankenwelt (Vielfalt, Evolution und Vererbung).* Springer, Berlin, Germany: ix + 974 pp.

Muralt, A. de (1991). *L'Enjou de la Philosophie Médiévale* (Études Thomistes, Scotistes, Occamiennes et Gregoriennes). Brill, Leiden, The Netherlands.

Peet, R. K. (1974). The measurement of species diversity. *Ann. Rev. Ecol. Syst.*, 5, pp. 285–307, illustr.

Pielou, E. C. (1975). *Ecological Diversity*. Wiley, New York, NY, USA: ix + 165 pp., illustr.

Polunin, N. (Ed.) (1972). *The Environmental Future: Proceedings of the first International Conference on Environmental Future*, held in Finland from 27 June to 3 July 1971. Macmillan, London & Basingstoke, England, UK, and Barnes & Noble, New York, NY, USA: xiv + 660 pp., illustr

Polunin, N. (Ed.) (1980). *Growth Without Ecodisasters? Proceedinqs of the Second International Conference on Environmental Future (2nd ICEF)*, held in Reykjavik, Iceland, 5–11 June 1977. Macmillan, London & Basingstoke, England, UK: xxvi + 675 pp., illustr.

Polunin, N. & Burnett, J. H. (Eds) (1990). *Maintenance of The Biosphere: Proceedings of the Third International Conference on Environmental Future* (3rd *ICEF*). Edinburgh University Press, 22 George Square, Edinburgh, Scotland, UK: xvi + 228 pp., illustr.

Rényi, A. (1962). *Wahrscheinlichkeitsrechnung, mit einem Anhang Über Informationstheorie*. VEB Deutscher Verlag der Wissenschaften, Berlin, Germany.

Tilman, D. (1988). *Plant Strategies and the Dynamics and Structure of Plant Communities*. Princeton University Press, Princeton, New Jersey, USA: xi + 360 pp., illustr.

Washington, H. O. (1984). Diversity, biotic and similarity indices: A review with special reference to aquatic ecosystems. *Water Res.*, 18, pp. 653–94, illustr.

Williams, C. B. (1964). *Patterns in the Balance of Nature*. Academic Press, London, England, UK.

Wilson, E. O. (1988). *Biodiversity*. National Academy of Sciences, Washington, DC, USA.

Woodwell, G. M. & Smith, H. H. (Eds) (1969). *Diversity and Stability in Ecological Systems*. Brockhaven Symp. Biol., No. 22.

10. Science and its Shortcomings in Environmental Management: The Social Angle

P. S. RAMAKRISHNAN

Professor of Ecology, School of Environmental Sciences, Jawaharlal Nehru University, New Delhi 110067, India

INTRODUCTION

Environmental degradation, initially visualized as a problem facing the industrialized west, has in fact become an issue threatening the very survival of some of the less-developed nations, though the issues involved in them are qualitatively different from those in the west. The problems of both the less-developed and developed nations have global ramifications, even if the issues may be of a local or regional nature. An example of this is the problem of large-scale deforestation and desertification, with obvious implications of a local nature but with possibilities of effecting a world-wide climate change. It is this concern for tackling environmental issues at local, regional, and global, scales that led to the urgent call, by the General Assembly of the United Nations, to the World Commission on Environment and Development, to propose long-term strategies for sustainable development (World Commission on Environment and Development, 1987).

At a purely academic level, ecological theories have undergone rapid changes. Human-engendered ecosystems are being seen more and more as consumer-controlled, rather than as being predominantly determined by the producer in the manner that was visualized a few decades ago when the International Biological Programme (IBP) was in operation. This has brought about a major shift in ecological theories, with recognition of disturbance as the major driving-force for changes in ecological processes and outcomes. Consequently, ecosystem management needs to have adequate emphasis placed upon Mankind as a focal point (UNESCO, 1986). It is not enough merely to talk about ecological science and eco-technology as the basis for environmental management, but is important also to have a close tie-up between natural (including ecological) and social sciences, because environmental management to a large extent is also an issue of human-judged values.

This paper considers the question of environment and sustainable development through some selected case-studies, indicating how science has often failed to solve environmental issues in the absence of an adequate appreciation of the social dimensions. Many environmental management-related projects in India and elsewhere have failed largely because of ignoring the social angle, though a few of the case-studies considered here are in fact success stories, because of the link-up that was worked out between ecological and social components before the projects were undertaken.

The Problem of Nomad Economy in Africa

In an interesting essay on nomads of Africa, Widstrand (1975) has discussed the rationale behind their economy and evaluated the implications of this for the future. Nomads usually have a variety of livestock – cattle, sheep and goats. They are able to tide over a variety of ecological and other difficult situations. For the pastoralists, livestock is the capital and their offspring constitutes the surplus. Where fodder is not available for cattle, goats can manage under stress situations. Livestock being owned by individuals, but not so the natural resources such as grazing-land and water, nomads cannot afford to decrease the pressure on the grazing land. Livestock management based on herd diversification and movement of herds to newer pastures, and complementary activities such as agriculture, hunting, and gathering and storage of food for the lean season, are some of the ways in which risk management is pursued by the nomads.

What has been widely lacking hitherto is an appreciation of how the whole system functions, including social attitudes. Widstrand (1975) rightly emphasizes the herd and family as being mutually-defining concepts in many pastoral societies of Africa. If one is tampered with, the other is obviously affected. Dividing grazing-lands into blocks for pastoralists to rotate between themselves conflicted with the nomads' concept of grazing-land management and often made the people more nomadic than before.

There is urgent need to integrate nomadic people into national economies, whether it be in Sahelian Africa or elsewhere – such as in arid regions of India. For already they have the knowledge and skill to manipulate marginal land effectively, and consequently any resettlement strategy should be based upon an understanding of their perception of environmental management. There are a few scattered and isolated examples of such an integrated approach to development. The village of Seed, near Udaipur in India, provides such an example where the village committee manages common lands in a controlled way for meeting the fodder needs of the villagers (Agarwal & Narain, 1989). For such an integrated approach to development of nomads, input of knowledge from social, economic, and biological, sciences is essential.

Forestry and Community Participation in India

Forestry in India since independence has been an exclusive activity of the national Forest Department. Until recently, it has largely been restricted to management of natural forests for timber extraction, and plantation forestry has been of very limited scope. The limited plantation forestry activities are largely centred around the more important timber species such as Sal (*Shorea robusta*), Teak (*Tectona grandis*), and a few others. Because forest replanting activities have not kept pace with the rate at which forests have been cleared, the problem of deforestation has become a major environmental issue and has attracted maximum public attention.

Social Forestry Programme

It is being increasingly realized now that peoples' participation in the forestry sector is becoming more and more critical if our rapidly declining natural forests are not only to be conserved, but the area under forest is to be extended effectively in order to maintain due ecological balance in the Indian subcontinent. It was in this context that the social forestry programme was initiated in India more than a decade ago.

Social forestry was conceived in 1976, by the National Commission on Agriculture, to involve people in forestry activities and to meet their basic needs of fuel-wood, fodder, and timber. This programme was essentially envisaged with three components: (a) farm forestry, to encourage farmers to plant trees on their own farmlands; (b) woodlots, to be planted and maintained by the Forest Department along roadsides and in vacant public lands and; (c) community woodlots, planted and maintained by the village communities on degraded common lands in the villages.

Of these three components, farm forestry has caught on because it is more cost-effective than the others (Agarwal & Narain, 1985). As private farmlands are with the big farmers, the beneficiary has largely been that section of society. Furthermore, because the choice of species has been largely restricted to *Eucalyptus* and to *Populus* spp. that have industrial value, the programme has largely catered to urban rather than to rural needs. The weaker sections of the rural society often got left out of the social forestry system, because of the weak social linkages in an otherwise well-conceived forestry activity.

Forest Restoration through Community Participation

West Bengal is one State in India where community woodlots under the social forestry programme have, perhaps, been more successful than elsewhere (Agarwal & Narain, 1985). Their success was because of the greater interest that could be generated amongst the rural communities than elsewhere, and their consequent active involvement in the programme. It is in this context of cooperation between villagers and the local office of the national Forest Department in West Bengal, that another experiment in forest protection and regeneration has to be viewed (Malhotra & Poffenberger, 1989).

In 1972, a Sal (*Shorea robusta*) regeneration project had been initiated by the Forest Department in the Arabari area in south-west Bengal. Sal is an important timber-tree species that is highly productive and economically attractive compared with other tree plantations. The Forest Department had two options – either to police the forest reserve or to involve the villagers

Table 10.1. Forest Protection Committee (FPC) Coverage (1,000 ha) in West Bengal, India (after Malhotra & Poffenberger, 1989). A total of more than 1,250 villages were involved.

District	Geographical Area (*ha*)	Forest Area (*ha*)	FPC Protected Area (*ha*)
Midnapore	408.1	170.2	67.3
Bankura	688.2	139.7	49.3
Purulia	625.9	92.3	38.5

in the management programme. The latter option demands that the villagers' needs have to be adequately met. One way would have been to employ the villagers for protection. This, however, has a catch: if they are paid for everything, the moment the monetary incentive is withdrawn, the forest would be the first casualty.

The Forest Department therefore adopted another approach, namely of meeting the lost income of the community due to forest protection by giving them rights to all minor forest products, a 25% share on timber extracted, and supplemental employment programmes. The people, in return, were to give free labour and guaranteed protection of the forest. When the villagers surrounding the Arabari forest tract were given exclusive rights to restrict access and to use minor forest products with the bonus of a share in the timber, they effectively protected over 600 ha of Sal forests. Even where degraded, the forests regenerated very rapidly.

The experience of Arabari caught on. During the last 8 years, throughout India, more than 1,250 villages have formed their own Forest Protection Committees (FPCS), covering in all about 152,000 ha of degraded forests (Table 10.1). As protection had been through community participation and consequently generally effective, the cost of restoration of the degraded forest ecosystem was minimal. Village communities were allowed unlimited collection of dead twigs for firewood, dried leaves of specific tree species for tableware (plate-making) and for rolling 'Bidis' (a local cigarette), and medicinal plants. In mutually agreed and identified sites, they were even permitted to practise agriculture to a limited extent. The approach for forest management was truly an integrated one, and provided annual income of up to about Rs 3,500 per family, based on access to one specified and marked hectare of forest per family (Table 10.2).

General Considerations

What lessons do we learn from the above experiment? Obviously, we need good technology and the selection of the right kinds of species. Social forestry has to be for the people, and therefore the tree species selected should be appropriate to meet the needs of the local community even before considering cash income through export to urban centres. Only then are

Table 10.2. Estimated Income from One Hectare of Regenerated Sal Forest (after Malhotra & Poffenberger, 1989).

Product	Annual income (Rupees) per family
Sal Pole and fire-wood (25% share)	500
Sal seeds	92
Silk cocoons	640
Sal-leaf plates	1,040
'Bidi' leaves	1,000
Total	3,272

equity and social justice possible.

In a programme initiated in a cluster of villages in the Kapkot region of the Kumaon hills in the Himalayas, the local community wanted to use two species of bamboo (locally called 'Ringal'), namely *Thamnocalamus helfri* and *Chimnobambusa falcata*. These two species were declining rapidly because of overexploitation by villagers and the disinterest of the foresters in them. Species selection had been based on close interaction with the local community, scientific input being only one component.

When we initiated this work under the Himalayan ecodevelopment programme, our major problem was to develop appropriate technology for propagation and ensuring maximum survival-rate for the propagules. While considering propagation in different ecological conditions, possible mixtures with other trees, shrubs, and herbs, became important. Here again species selection had to be based on a combination of social and scientific considerations. If the plantations are to be grown on village commons, then the social angle for management becomes important. For the use of bamboos as part of agriculture, an understanding of traditional agro-forestry systems become significant. If fodder species are under consideration, the linkage with the animal husbandry sub-system is important, and if the choice is of fuel-wood species, then linkage with the domestic sub-system becomes critical. An integrated approach with the village as the unit for development (Ramakrishnan, 1984, 1985) is critical. This alone will ensure participatory technology transfer and reconcile the enormous diversity in Indian villages (Agarwal & Narain, 1989).

In this management plan, what seems best called the 'village social ecocomplex' or socio-ecocomplex (formerly in my writings 'village'), with all its different sub-systems such as agriculture, animal husbandry, and the domestic, have to be considered as an integrated whole, the natural resource-base being enhanced to meet basic needs and thereafter to improve quality of life through export outside the village boundary. In this scheme social considerations, such as equity in utilization of resources by the different sections of society, are important.

Forest Management through Village-level Forest Protection Committees provides a basic framework for building up village-level institutions that could effectively manage the common resources and resolve disputes. Social forestry programmes carried out in China in the recent past, through collective units such as peoples' communes and production brigades on their own land, are quite impressive (FAO, 1979). Similarly massive fuelwood plantations were established, and degraded forested lands were rehabilitated, in Korea through village forestry associations. Profit-sharing arrangements were well laid out between the owner, the planter, the manager, and the harvester (FAO, 1982).

Agarwal & Narain (1989) have discussed a few case-studies in India where village-level institutions have succeeded. These institutions could then act as interacting focal points with nongovernmental voluntary agencies and Governmental agencies such as the Forest Department. However, conflicts can arise, for example within the village itself where levels of inequality and social stratification are very high. Village-level institutions that allow everyone to take part in decision-making along with the advisory inputs available from the Governmental and nongovernmental experts, could help in avoiding or resolving disputes and ensure appropriate science and technology inputs under the given social structure. In this, it is essential to ensure effective participation by the womenfolk, because they have always played a crucial role in Indian village social functioning.

The case of the tribal women of Bankura in West Bengal who organized themselves to undertake and implement ecosystem restoration activity through the initiative of the International Labour Organization (ILO) New Delhi office, and nongovernmental womens' societies who did much the same thing elsewhere, altogether constitutes a success story which offers many lessons. The tribal women were organized into a 3-tiers' structure. Ordinary members of the villages send up executive members to a society which is formed between each 2–3 villages. Each society then nominates 2–3 members to the apex body (ILO, 1988). The plantations of Sal (*Shorea robusta*) and Arjun (*Terminalia arjuna*), organized by each society in a cluster of 2–3 villages, have come up in little over seven years. Many hectares of land are covered, involving 36 villages and 1,500 families. The supplementary income to the women came not only through the use of leaves of trees for making tableware (plate) and 'Bidis' (local cigarettes), but also through rearing of Silkworm Moths (*Bombyx mori*) and collection of a variety of non-timber forest products. Agricultural activities in identified sites are also encouraged. The forest-based income generated is over Rs 7,000 per hectare.

SHIFTING AGRICULTURE IN THE HUMID TROPICS OF THE WORLD: IS THERE A WAY OUT?

The increasing agricultural yields of the last few decades were made possible through industrialization of agriculture involving heavy subsidies. Though these agro-ecosystems are efficient in terms of human time and labour, they are highly inefficient in ecological terms. The obvious limitations of such systems as models of development in an energy-limited world (Steinhart & Steinhart, 1974) has led to a renewed scientific interest in traditional systems of agriculture, as they are considered to be models for ecological efficiency.

Shifting cultivation agriculture, variously termed 'slash-and-burn' agriculture and 'swidden', or through a variety of local names such as 'jhum' in India, 'chitememme' in Zambia, and 'chinampa' in tropical America, have been held as models of production efficiencies with up to 50 units of food-energy harvested for each unit of energy input into the system (Rappaport, 1971; Mishra & Ramakrishnan, 1981, Ewel, 1986). The forest farmer in the humid tropics has managed his shifting agriculture for centuries, with optimum yield on a long-term basis – rather than trying to maximize production on short-term considerations (Spencer, 1966; Ruthenberg, 1976; Toky & Ramakrishnan, 1982; Ramakrishnan, 1984). It has been considered possible to increase production without departing too much from this traditional system.

The ecology of shifting agriculture, and its viability in the humid tropics as a land-use system is considered here with emphasis on the shifting agriculture ('jhum') in northeastern India, as this is one system that has been intensively studied from both the ecological and the social aspects. During the following discussion, however, I have also considered some of the scattered studies on similar systems elsewhere.

In north-east India, shifting agriculture is practised in the humid sub-tropical low (*c.*100 m) and high altitudes (*c.*1,500 m) having more than 200 cm of annual rainfall. The soil is either an oxisol (at lower altitudes) or a podsol (at higher altitudes). At lower altitudes the climax vegetation is a semi-evergreen, broad-leafed forest whereas high up in the mountains the vegetation is evergreen broad-leafed climax forest or, more commonly, degraded stages of secondary successional pine forests.

The 'Jhum' Cropping Pattern

The land-use system, as in northeastern India, involves slashing the vegetation, burning the dried slash before the onset of the monsoon, raising a mixture of crops on a temporarily nutrient-enriched soil for a year or two, fallowing the plot for regrowth of natural vegetation, and eventually returning to the same plot for another cropping phase after a few years. This time- interval between two successive croppings constitutes one shifting-agriculture cycle. Until a few decades ago, this cycle took 20 years or more to complete; but now it has come down to a shorter cycle-length of 5 years or less in northeastern India and in many other parts of the world (Nye & Greenland, 1960), due to increased population pressure and reduced land availability.

Historically, Man has long been part of the tropical forest ecobiome and has used this resource to support major civilizations. Pollen evidence from

Table 10.3. Sequential Harvesting of Crops on 'Jhum' Plots on 30-years' Cycle at Lower Elevations of Meghalaya in north-east India (after Toky & Ramakrishnan, 1981*b*).

Species	Harvesting time
Setaria italica	Mid-July
Zea mays	Mid-July
Oryza sativa	Early September
Lagenaria spp.	Early September
Cucumis sativa	Early September
Zingiber officinalis	Early October
Sesamum indicum	Early October
Phaseolus mungo	Early October
Cucurbita spp.	Early November
Manihot esculenta	Early November
Colocasia antiquorum	Early November
Hibiscus sabdariffa	Early December
Ricinus communis	(Perennial Crop)

All the seeds were sown in April.

Sumatra is suggestive of agricultural disturbance 5,000 years before the present time, whereas similar studies from New Guinea highlands consider earlier civilizations going back to 9,000 years BP (Flenley, 1979). Shortening of the agricultural cycle below 5 years has often resulted in a fallow system (FAO/SIDA, 1974) where burning has been dispensed with. Higher population densities with marketing facilities around large-city centres often lead to sedentary agriculture. These transitions have occurred in north-east India (Gangawar, 1987), which agrees with the hypothesis (Boserup, 1965) that limited land-area would eventually lead to settled, intensive agriculture as is also the case with Maya tribes in Central America (Lambert, 1985).

A standard feature of almost all shifting-agriculture systems is the practice of mixed cropping, which was formerly considered as primitive by agronomists and soil scientists (Sanchez & Buol, 1976), but is now suggested as a means of increasing world food production (Andrews & Kasam, 1976). Thus, during the cropping phase, the farmer in northeastern India may raise 8–35 crop species on a small plot of 2–2.5 ha, practising simultaneous sowing and sequential harvesting (Table 10.3) (Ramakrishnan, 1984). The Hanunoo farmer in the Philippines may raise up to 40 cultivars at the same time (Conklin, 1957).

Such an organization of crop mixtures provides good crop-cover for preventing losses of nutrients through runoff and leaching by water (Toky & Ramakrishnan, 1981*a*), helps in optimizing resource-use from a plot of land by providing more space for the next crop that is ready to mature after the harvest of the earlier one (Toky & Ramakrishnan, 1981*b*), concurrently

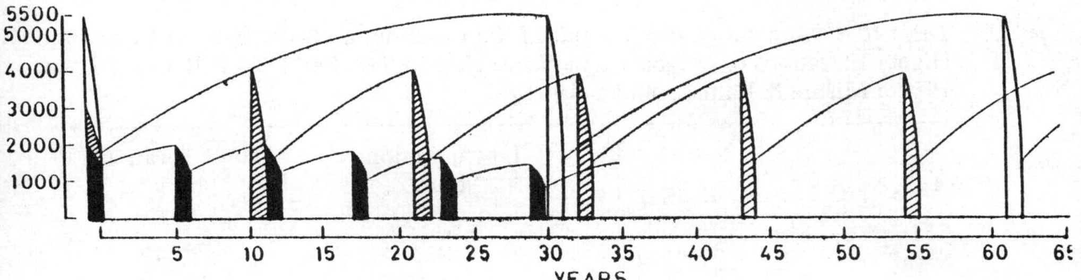

Figure 10.1. Crop yield patterns under 'jhum' cycles of 30 (open), 10 (hatched), and 5 (solid), years at lower elevations in north-east India (after Toky & Ramakrishnan, 1981*b*).

recycling biomass and nutrients through crop and weed residues (Mishra & Ramakrishnan, 1984; Swamy & Ramakrishnan, 1988), and improving soil characteristics through surface mulching (Ramakrishnan, 1984).

The economic efficiency of the agricultural system to a large extent is related to the length of the agriculture cycle. A comparison of the yield-pattern under 30-, 10-, and 5-, years' agriculture cycles at lower elevations in Meghalaya in northeastern India showed that the yield declined with the shortening of the cycle (Figure 10.1), but with optimum economic (Toky & Ramakrishnan, 1981*b*) and energy (Mishra & Ramakrishnan, 1981; Toky & Ramakrishnan, 1982) efficiencies under a 10-years' cycle.

A variety of patterns in shifting agriculture are to be found, based on ecologic, economic, and social, differences between one area and another. In northeastern India, the lower-altitude 'jhum' system (Toky & Ramakrishnan, 1981*b*), with clear-cutting of forest followed by total slash-and-burn and subsequent mixed cropping on unprepared steep slopes with emphasis on rice, is different from the higher-altitude system. This higher-altitude (*c*.1,500 m) system (Mishra & Ramakrishnan, 1981) involves elaborate preparation of the land into ridges and furrows, partial slash-and-burn of the sparsely-distributed pine trees and emphasis on tuber (particularly Potato, *Solanum tuberosum*) and vegetable crops on the ridges, with the furrows acting as water channels running down the slope. These modifications are made because of the lower nutrient status of the acidic soil of podsolic origin and slower regeneration of the sub-humid montane forest with pine as the dominant element (Mishra & Ramakrishnan, 1981) than prevails lower down. In spite of this, the economic returns from the high-altitude system can be more than three times as much as those from the low-altitude type described above (*see* Table 10.4). Based upon the organization of the crop mixture, the economic returns could vary considerably over short distances (Gangawar, 1987). Mere manipulation of the crop mixture, therefore, offers possibilities of improved returns to the farmer.

Weeds, which normally are considered as undesirable and adversely affecting the crop-yield of agro-ecosystems (Seaavoy, 1973), often play a useful role in traditional agriculture. The traditional forest farmer leaves about 20% of the weed biomass *in situ*. Under such a husbandry practice, the crop yield is unaffected and indeed the weeds help to conserve nutrients

Table 10.4. Monetary Output-Input of 'Jhum' under a 10-yrs' cycle at Lower and Higher Elevations of Meghalaya in North-East India (after Toky & Ramakrishnan, 1981*b*; Mishra & Ramakrishnan, 1981*a*).

	Low elevation 'jhum'	High elevation 'jhum'
Input	1,830	3,842
Output	3,354	14,171
Net gain	1,524	10,329
Output-input	1.8	3.9

through plant cover over the soil. Even the weed biomass that is pulled up but subsequently put back into the agro-ecosystem, contributes to nutrient cycling (Swamy & Ramakrishnan, 1988). Weeds also create an environment that is unfavourable to biological pest invasions (Altierie, 1983). Such a useful role of weeds in traditional agriculture opens up a new area of agro-ecological studies, with possible integrated management capabilities for tropical agriculture (Gliessman *et al.*, 1981; Swamy and Ramakrishnan, 1988).

Weed Potential and Secondary Succession under Shifting Agriculture

When a forest is converted to cultivable land, not only is its original vegetation destroyed but the site is subjected to continuing perturbations due to fire, introduction of crop species, weeding, and crop harvest. These human interferences result in a progressive reduction in species' diversity, so that the early stages of any ensuing ecological succession contain fewer species than would have been involved if the process had remained natural. The number would increase gradually with the passage of time (Whitmore, 1975; Toky & Ramakrishnan, 1983). Under the frequent perturbations of a short agriculture cycle, the early successional weedy species take over (Saxena & Ramakrishnan, 1984). Often, weed species having short life-cycles, high fecundity, and seed-dormancy capabilities, are selected. In south-east Asian countries, the grass *Imperata cylindrica* is frequently a problem (Kartawinata *et al.*, 1981), while pre-adapted aliens may be favoured over local genotypes.

As native species are often unable to survive under increasingly simplified weedy vegetation, further vegetative augmentation may take place through aggressive exotics (Kellman, 1980). These exotic weeds, represented by *Eupatorium* spp. and *Mikania micrantha*, form extensive pure stands in northeastern India (Saxena & Ramakrishnan, 1984) and in Africa (Nye & Greenland, 1960), with major consequences on ecosystem properties.

If the agriculture cycle is longer than 5 years, the weeds are suppressed largely by bamboos in northeastern India because of poor light-availability at the ground level. In this suppression, *Dendrocalamus hamiltonii, Bambusa tulda, B. khasiana,* and *Neohouzeaua dulloa,* are some of the important species (Toky & Ramakrishnan, 1983; Rao & Ramakrishnan, 1987). Bamboos in northeastern India are eventually replaced by a variety of broad-leafed trees.

Valley Cultivation

Amongst the various other land-use options discussed below, valley cultivation has some advantages over 'jhum' in that soil and nutrient losses are heavy from the slopes from land under 'jhum' (Ramakrishnan *et al.*, 1981). Valley cultivation moreover constitutes a nutrient sink without any major losses from the system, which makes it more efficient than 'jhum'. From a monetary viewpoint, however, 'jhum' is the more favourably placed, due to diversification of crops and the higher returns from potato compared with rice. Moreover the valley system is restricted in the hilly terrain for want of suitable sites. Yet a distinct advantage that valley cultivation of rice has over 'jhum' is that the land re-use factor for the former is 1 whereas that for the latter is 0.1/0.2 (*i.e.* the land is cropped only once or twice in 10 years). The energy output/input per ha per year for both the systems work out similarly to that discussed by Leach (1976) for pre-industrial farming, with 'jhum' showing up badly due to the low re-use factor.

At the lower elevations of Meghalaya, the Nepalis and the Mikirs* take only one crop during the monsoon, under the valley cultivations, whereas the more laborious Garos* make more effective use of the land by taking two crops – one during the monsoon and another during winter (Maikhuri & Ramakrishnan, 1990). However, their land-use system is less efficient compared with that of the Nepalis, who take only one crop in a year. Nepalis take one crop in a year perhaps because of their preoccupation with a well-developed cattle husbandry system for milking. The net return through valley cultivation, with one cropping as done by the Nepalis, is comparable with the 20-years' 'jhum' cycle of the Garos. The Nepalis, who traditionally practise sedentary agriculture, obtain greater returns from valley cultivation than do the local tribal people.

Unlike most of the other tribal communities of north-east India, the Apatanis have evolved sedentary agriculture chiefly in the form of wet Rice (*Oryza sativa*) cultivation in their extensive valley-lands. Irrigation farming, such as wet cultivation of Rice, requires cooperation of several farmers and also communal work to maintain and improve the water-delivery system (Ruthenberg, 1976). In the absence of a disciplined schedule and scale of water distribution among the beneficiaries, economic returns can very often decline drastically. However the Apatanis, with cooperative management of the water delivery system under the overall supervision of the village headman, have optimized water-use in their ricefields.

*[In answer to our queries Professor Ramakrishnan replied (*in litt.*) that the 'Nepalis [are] an immigrant community and the Mikirs a tribe from Assam', [while the] laborious Garos [are] 'a tribe of Meghalaya'. Eds.]

Table 10.5. Ecological and Economic Efficiencies of Rice Agro-ecosystem of Apathanis in North-east India. Values in parentheses are of net return (after Kumar, 1987).

Production Measures	Monetary Input/output (Rs per ha per yr)		Energy Input/output (MJ per ha per yr)	
	Late Variety	Early Variety	Late Variety	Early Variety
Input total: Rice	2,481	2,677	846.5	904.5
Rice + Millet	2,552	2,798	870.2	946.0
Rice + Millet + Fish	2,753	–	906.6	–
Labour: Rice	2,163	2,337	713	769
Millet	67	116	23	40
Fish	102	–	36	–
Organic manure	250	250	125	125
Seed: Rice	68	90	8.5	10.5
Millet	3	5	0.7	1.5
Fish	100	–	0.4	–
Output total: Rice	8,941 (6,490)	7,603 (4,946)	66,284	56,367
Rice + Millet	9,102 (6,651)	7,817 (5,039)	67,956	58,480
Rice + Millet + Fish	10,062 (7,309)	–	68,182	–
Economic/Energy efficiency				
Rice	3.60	2.84	78.3	62.3
Rice + Millet	3.57	2.79	78.1	61.8
Rice + Millet + Fish	3.65	–	75.2	–

Apatanis make the best use of their irrigated land by planting both early- and late-ripening varieties of Rice. The early-ripening variety is sown the furthest away from the village, where disturbance by animals and poorer irrigation facilities could be major constraints. On the other hand closer to the village, where conditions are more favourable, the late-ripening variety is preferred. The yield from Rice is supplemented by Millet (*Eleusine coracana*) which is cultivated on the elevated bunds between the Rice-plots, and also through pisciculture along with the late variety of Rice, because of its more assured water-supply. With a production of about 50 kg of fish providing additional income of about Rs 1,000 p.a., which is equivalent to the cost of about 450–500 kg of Rice, the system including pisciculture compares favourably with similar systems that are practised in Java and Madagascar (Hickling, 1961).

With human labour as the major input, and with very little use of organic manure, the Apatanis obtain high energy output in crop yield. The efficiency of the system, expressed as output–input ratio, is very high (60–78), compared with a value of about 9 for the traditional Indian agricultural system (Mitchell, 1979), and for traditional Rice cultivation in the Philippines (Nguu

Table 10.6. Monetary and Energy Input and Output under Home Garden of the Mikirs of Meghalaya in North-eastern India (after Maikhuri & Ramakrishnan, 1990).

Production measures	Money (Rupees ha^{-1} yr^{-1})	Energy (MJ ha^{-1} yr^{-1})
Input		
Field Preparation	550	350
Weeding	360	219
Harvesting	280	310
Transportation	160	191
Seed	300	70
Total	1,650 + 110	1,140
Output		
Arecanut	10,489	11,565
Betel leaf	3,240	148
Banana	568	2,766
Black pepper	370	470
Total	14,667+685	14,949
Net return	13,014	–
Output/input ratio	8.8	13.1

& Palis, 1977). Such a high energy-efficiency, with a reasonable monetary efficiency of about 3, makes the Rice system of the Apatanis one of the effective models of traditional agriculture – particularly when it is realized that human labour is a free input, being largely obtained from within the family and (for specific tasks) through cooperative effort (Table 10.5). The high energy efficiency obtained here is even greater than that which has been recorded for shifting agriculture in north-east India (Mishra & Ramakrishnan, 1981; Toky & Ramakrishnan, 1982) and elsewhere (Rappaport, 1971; Steinhart & Steinhart, 1974). With 27–35 MJ (Mega Joule) units of energy output per labour hour, the Apatanis' valley system compares favourably with similar systems in China (Dazhong & Pimentel, 1984) and the even-more-modern agriculture of industrialized societies (Leach, 1976).

Home Garden

The 'home garden', with very effective organization of the crop mixture and effective use of space by stratification in the plant community, is a highly-organized production system (Mitchell, 1979) of the Mikirs. Consequently it is reasonable to find from 3–5 times the economic returns from it as compared with those obtained under 'jhum' (*cf.* Table 10.6). Home garden is an important land-use concept that could well provide additional cash income to the tribal communities and thus reduce their dependence on 'jhum'.

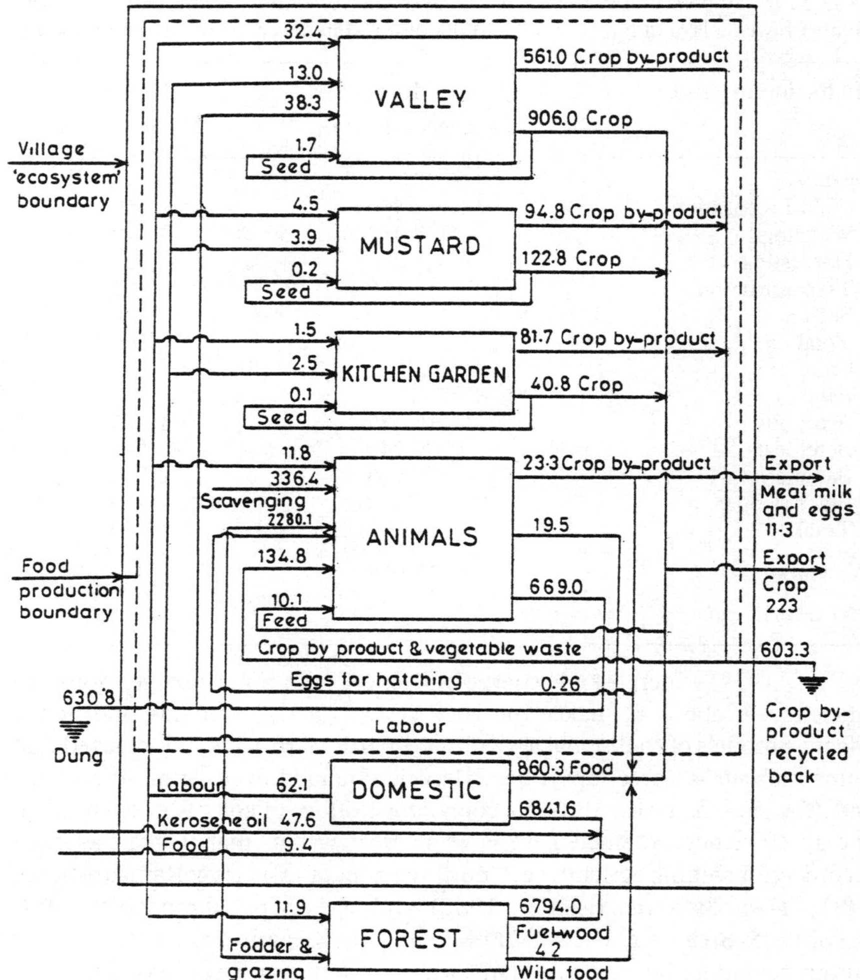

Figure 10.2. Energy flow through a village 'ecosystem' (*i.e.* ecocomplex) of the Karbis in Arunachal Pradesh in North-east India. Unit = MJ x 10^3 (after Maikhuri & Ramakrishnan, 1990).

Terracing

Sedentary farming on the hill-slopes, suggested as an alternative to 'jhum', has time and again been rejected by the farmer for valid reasons such as: (i) terracing is expensive both in terms of labour and monetary input, and is costly to maintain under the high-rainfall conditions prevailing; (ii) heavy input is required of inorganic fertilizers that are costly and in short supply; (iii) weed potential is intensified on the terraces as much as, or even more than, under a short 'jhum' cycle of 4–5 years, the weed potential increasing drastically with any shortening of the 'jhum' cycle; and (iv) even if run-off losses are checked through terracing, leaching of nutrients through percolating water gets accelerated, because of the loose and porous soil.

On the other hand, sedentary agriculture in the form of valley cultivation, which is an important land-use, is both ecologically and economically viable, as the wash-out of the nutrients from the hill-slopes can be exploited on a sustained-yield basis, as is being done by the Karbis (a tribe from Assam) without 'jhum' being utilized as an associated form of land-use (Figure 10.2).

Relatively recently the Indian Council of Agricultural Research (ICAR) has suggested a 3-tiers' system for a given slope as an alternative to 'jhum' (Borthakur *et al.*, 1978). This land-use envisages forest cover for the upper part of the slope, plantation/horticultural crops for the mid-portion, and terraces for the lower part of the slope. Apart from similar problems related to terrace cultivation, the major drawback of the ICAR model is that such a rigid system may come in conflict with the societal organization of the tribal members. Thus, for example, the independence of the family unit would be adversely affected as one family will not be able to maintain this unit. In any case, the ICAR model has not been able to make sufficient impact, so far, to be considered seriously.

Land-use Redevelopment

With many advanced features based on good science, shifting agriculture offers much scope for redevelopment whereby the distortions, in terms of economic yield (Ruthenberg, 1976; Ramakrishnan, 1984) and land degradation (Ramakrishnan, 1985), brought about by shortened agricultural cycles, could be corrected. Agricultural systems must be both sustainable and independent of massive inputs of fossil-fuel derivatives (Ewel, 1986).

The following ecological attributes may accordingly be incorporated into the design of agro-ecosystems for the humid tropics: (i) low requirements for nitrogen and phosphorous as external subsidies, (ii) efficient use of available resources, (iii) protection from biological invasion, and (iv) low risk. These features can be obtained by constructing mixed-species communities that imitate successional vegetation (Hart, 1980).

Agro-forestry

This is the deliberate association of trees and shrubs with crops (often including livestock) and holds great promise for contributing to sustainable land-use systems. Some of the earliest-recognized forms of agro-forestry involved modifying the farming practices of shifting cultivators to include the planting of tree seedlings with food-crops. As the tree canopy developed and began to shade the annual crops, the farmers would move to another area for cropping. Using this system of 'taungya', large areas of teak plantations have been established in Indonesia and Nigeria (Winterbottom & Hazlewood, 1987).

Over the past few years, the International Institute for Tropical Agriculture (IITA) in Nigeria has experimented with the hedgerow technique and developed a new system of alley cropping (Kang & Duguma, 1984). These Authors (*idem.*) also indicated that *Leucaena* sp, a fuel- and fodder-tree intercropped with maize, can fix up to 160 kg per ha per year of nitrogen, or enough to sustain maize yields at a level of 2 t per ha per year.

During a single cropping phase the agricultural system in north-east India may lose about 600 kg per ha of nitrogen without more than one-half of what is lost being recovered during the natural fallow period of five years that follows (Mishra & Ramakrishnan, 1984). An accelerated and directed succession, initiated and supported through introduction of a variety of leguminous plants and the non-leguminous Alder (*Alnus nepalensis*) during the cropping and fallow phases, could restore the soil fertility over a 5-years' cycle period (Ramakrishnan, 1987). Many other early-successional trees that are fast-growing and have a rapid turnover of leaves on the tree, and even fruit trees of economic value, could be useful in those respects; they could also be used as windbreaks around agricultural plots, to prevent blow-off of ashes (*idem.*).

Other strategies could be based on alternative land-use development. Settled terrace-cultivation as tried out in Bangladesh and India (FAO, 1978), intensified valley-cultivation, and a shift to plantation/horticultural crops, are some of the possibilities. Terrace-cultivation in these areas may be of limited value only, because heavy leaching losses of nutrients occur even if run-off losses are checked. Studies from Sarawak suggest that cash-cropping could eventually replace shifting agriculture (Chin, 1982).

The forest village scheme introduced by the Forest Industries Organization (FIO) of Thailand in 1967 emphasized the importance of developing a home garden and establishing and maintaining forest plantations (Boonkird *et al.*, 1984). In these plantations, the people are also encouraged to raise agricultural crops during their first few years. This scheme, which is integrated with health, education, and crop and animal husbandry, is proving to be a successful alternative to shifting agriculture.

Experience in India and elsewhere suggests that rubber plantations, established on a cooperative basis, can ensure peoples' participation (Ramakrishnan, 1985). While short-term strategies may emphasize redevelopment of shifting agriculture itself, long-term strategies may consider alternative land-use systems with peoples' participation. The objective here should be to take the pressure off the land for 'jhum', so that a 10-years' cycle, which was found to be efficient from economic and energy-use viewpoints, as well as for weed reduction and soil fertility recovery (Ramakrishnan, 1984), could be sustained wherever required and found to be feasible.

In Papua New Guinea, apart from traditional fruit trees such as *Pometia pinnata* and *Artocarpus altilis*, Robusta Coffee (*Coffea arabica* agg.) has been introduced as a cash-crop along with Bananas (*Musa sapientum*) and species of *Xanthosoma* and *Leucaena* to give shade. This system, if integrated effectively within the prevailing social framework, is potentially promising.

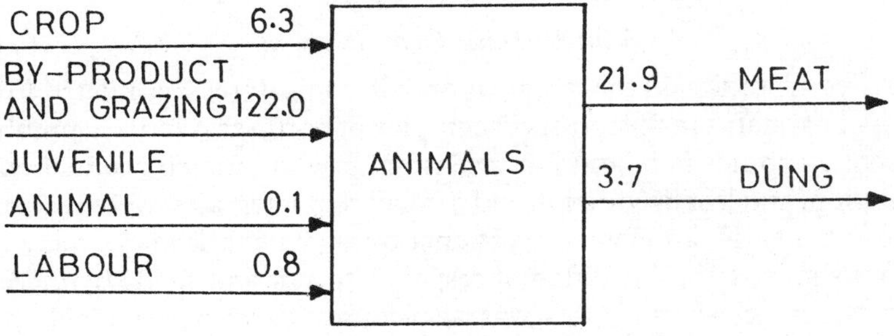

Figure 10.3. Energy input:output pattern and efficiency ratios for animal husbanary sub-system of a Khasi village ecocomplex (after Mishra & Ramakrishnan, 1982).

Integrated Animal Husbandry

Swine husbandry is an integral part of the 'jhum' system in northeastern India (Mishra & Ramakrishnan, 1982), and is also part of it in many other regions of the world. Thus farmers of the Tsembaga tribe raise pigs to be eaten, but such consumption involves religious beliefs and practices (Rappaport, 1971). In fact, the tribal farmer of north-east India consumes pigmeat not only as part of his normal diet but also makes a feast of it during celebrations related to 'jhum' procedures. Again the main reason why swine husbandry is part of the 'jhum' system is because of its inexpensive maintenance costs. It is one animal-husbandry practice, in the traditional sense, that is inter-linked with the 'jhum' agricultural sub-system and yet makes little demand on it.

This animal husbandry practice is based on efficient recycling of resources with a reasonable level of energy efficiency – the principle on which the operation of the 'jhum' system itself is based. In a study of the animal husbandry sub-system of a Khasi village socio-ecocomplex (*see* above) of 20 members (Mishra & Ramakrishnan, 1982), it was found that a major fraction of the total energy input was the feed, crops accounted for about 4.9%, while the rest comprised crop residues *plus* grazing, which are essentially free commodities (Figure 10.3). Animal husbandry is thus an important source of protein for the local people.

General improvement and better management of the animal husbandry available in the region could meet local protein needs and also provide additional income through export. Poultry- and cattle-rearing (where it is ecologically viable and already a part of the traditional animal husbandry practice) also offer scope for development *inter alia* through using improved breeds.

Restoration of Degraded Lands

Various Basic Considerations

In a series of studies on the 'architecture' of tropical trees, as related to their successional status in the forest ecocomplex of north-east India, a number of different attributes in relation to extension-growth, branch production and orientation, and leaf production and display characteristics, were recognized between the early- and late-successional trees (Ramakrishnan, *et al.*, 1982, Ramakrishnan, 1986). A different response to sun and shade by these two seasonal ecological categories was also found: early-successional species were more susceptible to shading than were late-successional ones.

A major outcome of our quantitative morphometric analysis of tree 'architecture' was that we were able to treat it as an adaptative strategy that had apparently evolved through natural selection under a given light-environment. As presumably crown architecture would have evolved for optimization of photosynthesis by appropriate leaf-display, light which becomes limiting over a successional gradient is likely to be a key environmental factor. Thus, this study appears to resolve issues related to strategies of species during forest succession, and so to contribute to an understanding of the process of succession itself. This approach explains to some extent the reasons why certain species hold their ground for a period during succession, being replaced by other species subsequently – a question of continuing interest to ecologists since the concept of succession became a central theme in ecology. Relating tree architecture to photosynthetic efficiency and biomass production is also important for a better appreciation than formerly of this important physiological function and its evolution in both time and space.

Social forestry programmes in a developing country should aim at identification of fast-growing *native* species for meeting the fodder, fuel-wood, and timber, needs of the rural population. Use of early-successional species for social forestry is appropriate because these species tend to have a faster growth-rate and faster accumulation of biomass than later successional ones. The leaf 'population' of early-successionals, having a shorter life-span, should help in promoting faster turnover of nutrients between the plant and the soil, favouring their success in nutrient-deficient wastelands, so that more fertile land should become available for agriculture and other land-uses.

Along with a greater realization that agriculture and forestry can perfectly well form compatible systems on the same site, identification of trees that could optimize production in such a viable agro-forestry system is important. The early successional trees have a crown 'architecture' that permits maximum light penetration to the ground layer, so that the growth and productivity of crop species would not be limited by light. Besides, a fast turnover of nutrients would ensure a constant and steady supply of them in the surface layers of the soil, to be used by the crop species. If the root 'architecture' of the tree species used is such that the roots are more uniformly distributed at different depths, rather than being chiefly restricted

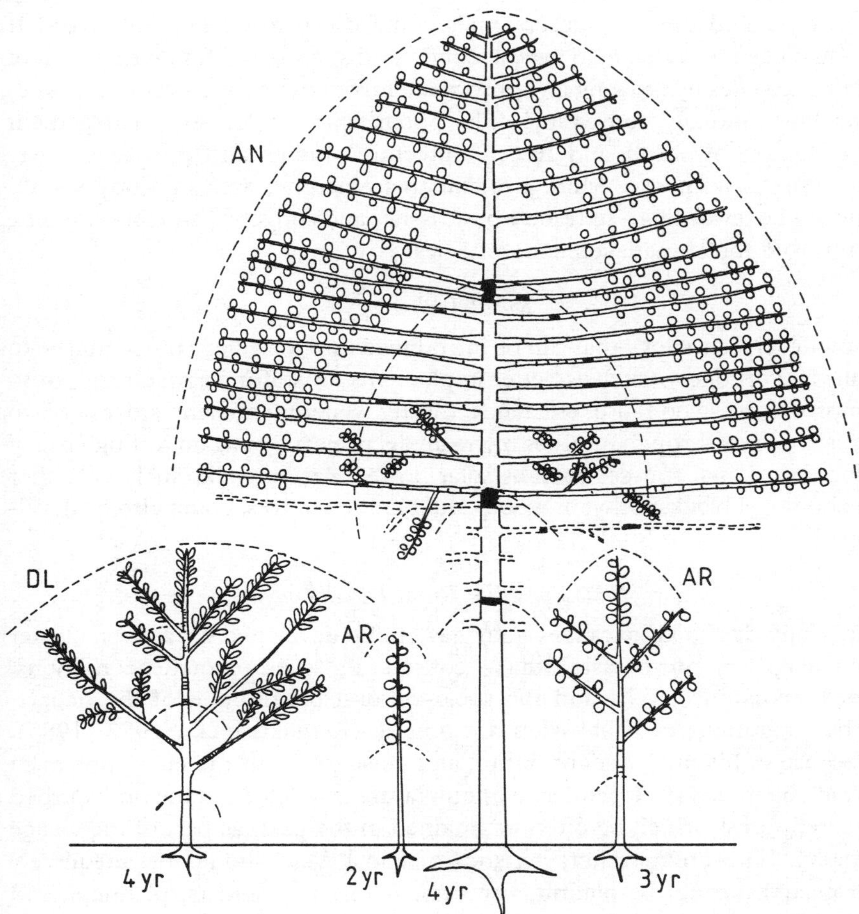

Figure 10.4. Model of a mixed tree plantation involving early successional *Anthocephalus cadamba* (AN) and late successional *Dillenia pentagyna* (DL) and *Artocarpus chaplasha* (AR) (after Ramakrishnan, 1986).

to surface-soil layers, the species-mixture in the agro-forestry system would be more compatible with near-surface crops. Although early-successional trees tend to have a superficially-placed root system, some of the mid-successional trees are better suited in this respect. Identification of suitable root 'architecture' from this continuum should not be difficult.

Mixed plantation programme, having distinct ecological advantages over monoculture forestry, could also be economically as productive as, or even more productive than, the latter if compatible species are used. Fast-growing, light-demanding, early-successional species could form compatible mixtures with shade-tolerant mid- or late-successional species for exploiting the light availability at different canopy-levels, thus optimizing production per unit area (Figure 10.4). Such a mixture would allow optimum use of nutrients from the soil profile, the early-successionals exploiting the surface soil layer and the late-successionals going down to deeper soil layers. It should even be possible to have a 'condensed succession' for revegetation of damaged sites by growing an appropriate mix of different categories of trees that are mutually compatible. The potentiality for forestry management application of shrub and tree architecture studies and growth strategy analysis, is immense. More work in this emerging area of study should indeed be rewarding – to resolve basic issues and concepts, and for designing improved forest management strategies.

Rural Technology

Low-level technology that can be introduced into the village units, maybe to alleviate drudgery (such as better implements for shifting agriculture), or to improve fuel-wood use efficiency through energy-efficient stoves, or to create energy through mini- or micro-hydel projects, or again through use of unconventional sources such as solar power. Artisan skills such as leather technology, blacksmithy, or wood- and bamboo-works, could also be developed.

Philosophy For Rural Development

Specially-designed 'packages' may have to be developed for a given cluster of villages in north-east India – considering micro-climatic conditions, socio-economic levels, and the socio-cultural background of the people. The aspirations of tribal Man are unique (Ramakrishnan, 1983, 1985), because of his independent nature and closeness with the environment in which he lives. His dependence upon Nature is reflected in the undisturbed 'sacred forest' which he often maintained in the past, as part of the village system. The unique characteristics of tribal Man should be adequately protected during the planning process. Involved scientists, planners, and administrators, have often tried to impose, from outside, developmental plans that they consider are good for the people in the region, but without trying to understand the processes that operate in traditional ecological systems. The strategy for development should accordingly be one with which the local people can identify themselves. Such a philosophy in planning would take care of the traditional value-systems, and, therefore, would not only find ready acceptance by the tribal societies but would also ensure their participation in the developmental processes.

The village headman is a key person in each tribal settlement. Village-level institutions operating with the village chief, and involving the District Council set-up for regional planning, could provide the necessary institutional framework for developmental planning, though appropriate adjustments may be required over a period of time, based on experience.

Conclusions

In a collection of edited essays, Johannes (1989) has tried to evaluate the role of traditional knowledge in environmental management. Among items considered were the 'home gardens' of the traditional farmers of Micronesia (Falanruw, 1989). Similar systems exist in the northeastern hill region of India (Maikhuri & Ramakrishnan, 1990). In the same general context one could consider the traditional Rice agro-ecosystem of the farmers of West Africa (Richards, 1985), or that of the traditional Apatani Rice farmers of Arunachal Pradesh in the northeastern hill region of India (Kumar, 1987), or even the traditional rain-water harvesting tanks – now often in disuse – dotting the rural landscape of the Indian subcontinent. An understanding of, and capability for building upon, these locally-available traditional technologies, is important from the point of view of sustainable development. The case-study on shifting agriculture which we discussed above in considerable detail, suggests the dangers of any blind imposition of modern agricultural technology on a society without first evaluating the full implications of ecological and social incompatibilities and the social disruptions that such an imposition may result in.

The case-study on forestry management wherein peoples' participation has to be relied upon for success, suggests the need for a strong tie-up between ecology, sociology, and economics, in order to ensure that any particular technology will work. The Amerindians of Amazonia have always perceived a myriad of ecological zones for resource-needs on a sustained-yield footing (Posey *et al.*, 1984) which could form a satisfactory and lasting basis for forest management . The village of Sukhomajri near Chandigarh in north-west India has developed a resource management plan that is based on the micro-watershed management concept (Agarwal & Narain, 1989); they have effectively managed their water-bodies, increased crop production to nearly three times what it was, protected their forest resources, increased fodder resources, and enhanced milk production. In a short span of just about 5 years, the annual income of a family is said to have increased by an estimated Rs 2,000–3,000. All this was achieved by applying the right scientific input, which was provided by experts through a village-level institution called 'Hill Resources Management Society' that had been specifically created for the purpose, with membership for each family in the village.

The reason why such success stories do not get replicated is partly because of the lack of effective village-level institutions and partly because of the almost total absence of a linkage between scientific rigour and social dynamics. Where such institutional framework was provided and coupled with appropriate technology, as exemplified by the Forest Management operation in West Bengal, or by the Sukhomajri experience or by 'jhum' in northeastern India, the rural ecocomplex redevelopment plan has been a success.

In the recent past, economics has tended to determine the way in which natural resources are managed. If, in future, ecological considerations of how natural systems work can be coupled with economic evaluation, the result will be better technology and project design, with consequent qualitatively improved return. Moreover if, with this, societal perceptions and values can be linked up, by ensuring peoples' participation at all levels of planning of ecocomplex redevelopment, then the outcome is likely to be equitable and sustainable.

References

Agarwal, A. & Narain, S. (1985). *The State of India's Environment 1984–85: The Second Citizens' Report.* Centre for Science and Environment, New Delhi, India: 393 pp., illustr.

Agarwal, A. & Narain, S. (1989). *Toward Green Villages.* Centre for Science and Environment, New Delhi, India: iv + 52 pp., illustr.

Altierie, M. A. (1983). *Agroecology: The Scientific Basis of Alternative Agricultures.* Division of Biological Control, University of California, Berkeley, California, USA: xiii + 162 pp., illustr.

Andrews, D. J. & Kasam, A. H. (1976). The importance of multiple cropping in increasing world food supplies. Pp. 1–10 in *Multiple Cropping* (Eds. R. I. Papendick, T. A. Fenchez & G. B. Triplett). American Society of Agronomy Special Publication No. 27, Madison, Wisconsin, USA: viii + 378 pp., illustr.

Boonkird, S. A., Fernandes, E. C. M. & Nair, P. K. R. (1984). Forest villages: an agroforestry approach to rehabilitating forest land degraded by shifting cultivation in Thailand. *Agroforestry Systems*, 2, pp. 87–102.

Borthakur, D. N., Singh, A., Awasthi, R. P. & Rai, R. M. (1978). Shifting cultivation in the northeastern region. Pp. 330–42 in *Resource Development and Environment in the Himalayan Region.* Department of Science and Technology, Government of India, New Delhi, India: 537 pp.

Boserup, E. (1965). *The Condition of Agricultural Growth.* Aldine, Chicago, Illinois, USA: 123 pp.

Chin, S. C. (1982). The significance of rubber as a cash crop in a Kenyan swidden village in Sarawak. *Federation Museum Journal*, 27, 23 pp.

Conklin, H. C. (1957). *Hanunoo Agriculture: A Report on an Integral System of Shifting Cultivation in the Philippines.* (FAO Forestry Development Paper No. 12.) FAO, Rome, Italy: 109 pp.

Dazhong, W. & Pimentel, D. (1984). Energy inputs in agricultural systems of China. *Agric. Ecosys. Environ.*, 11, pp. 29–35.

Ewel, J. J. (1986). Designing agricultural ecosystems for the humid tropics. *Ann. Rev. Ecol. Syst.*, 17, pp. 245–71.

Falanruw, M. V. C. (1989). Nature intensive agriculture: the food production system of Yap Islands. Pp. 35–40 in *Traditional Ecological Knowledge: A Collection of Essays* (Ed. R. E. Johannes). IUCN, Gland, Switzerland: 64 pp.

FAO (1978). *Forest News for Asia and the Pacific.* FAO, Rome, Italy: pp. 1–26.

FAO (1979). *China: Mass Mobilization of Rural Communities for Reforestation.* FAO, Rome, Italy: 122 pp.

FAO (1982). *Village Forestry Development in the Republic of Korea: A Case Study.* (FAO/SIDA Forestry for Local Community Development Programme.) FAO, Rome, Italy.

FAO/SIDA (1974). *Report on Regional Seminar on Shifting Cultivation and Soil Conservation in Africa.* FAO, Rome, Italy: 248 pp.

Flenley, J. (1979). *The Equatorial Rain Forest: A Geological History.* Butterworth, London, England, UK: viii + 162 pp.

Gangawar, A. K. (1987). *Cropping and Yield Patterns under Slash and Burn Agriculture (Jhum) in North-east India and Related Ethnobiological Studies*,. PhD dissertation, North-Eastern Hill University, Shillong, India: 248 pp. (typescript), illustr.

Gliessman, S. R., Garcia, E. R. & Amador, A. M. (1981). The ecological basis for the application of Traditional Agricultural Technology in the Management of Tropical Agroecosystems. *Agro-Ecosystems*, 7, pp. 173–85.

Hart, R. D. (1980). A natural ecosystem analog approach to the design of a successional crop system for tropical environments. *Biotropics*, 12 (Suppl.), pp. 73–82.

Hickling, C. F. (1961). *Tropical Inland Fisheries*. Longman, London, England, UK: xvi + 287 pp., illustr.,

International Labour Organization (cited as ILO) (1988). *The Bankura Story: Rural Women Organize for Change*. ILO Office, New Delhi, India: 20 pp.

Johannes, R. E. (1989). *Traditional Ecological Knowledge: A Collection of Essays*. IUCN, Gland, Switzerland: 64 pp.

Kang, B. T. & Duguma, B. (1984). Nitrogen management in alley cropping systems. Pp. in *International Symposium on Nitrogen Management in Farming Systems in the Tropics*. International Institute of Tropical Agriculture, Ibadan, Nigeria.

Kartawinata, K., Adisoemarto, S., Riswan, S. & Vayda, A. P. (1981). The impact of man on tropical forest in Indonesia. *Ambio,* 10, pp.115–19, illustr.

Kellman, M. (1980). Geographic patterning in tropical weed communities and early secondary successions. *Biotropica,* 12 (Suppl.), pp. 34–39.

Kumar, A. (1987). *Studies on Ecological Implications of varied Land-use Patterns in the North-Eastern Hill Region of India*. PhD dissertation, North-Eastern Hill University, Shillong, India:

Lambert, J. D. H. (1985). The ecological consequences of ancient Maya agricultural practices in Belize. In *Prehistoric Intensive Agriculture in the Tropics* (Ed. I. S. Farrington). (BAR International Series 232).

Leach, G. (1976). *Energy and Food Production*. IPC Science and Technology Press, Guildford, England, UK: 137 pp.

Maikhuri, R. K. & Ramakrishnan, P. S. (1990). Ecological analysis of a cluster of villages emphasizing on land use of different tribes in Meghalaya in north-east India. *Agric. Ecosys. Environ.*, 31, pp. 17–37.

Malhotra, K. C. & Poffenberger, M. (1989). *Forest Regeneration through Community Protection*. West Bengal Forest Department, Calcutta, India: 47 pp.

Mishra, B. K. & Ramakrishnan, P. S. (1981). The economic yield and energy efficiency at higher elevations of Meghalaya in northeastern India. *Acta Oecol. – Oecol. Applic.*, 2, pp. 269–89.

Mishra, B. K. & Ramakrishnan, P. S. (1982). Energy flow through a village ecosystem with slash and burn agriculture in northeastern India. *Agric. Systems*, 9, pp. 57–72.

Mishra, B. K. & Ramakrishnan, P. S. (1984). Nitrogen budget under rotational bush fallow agriculture (jhum) at higher elevations of Meghalaya in north-eastern India. *Plant & Soil*, 80, pp. 237–46.

Mitchell, R. (1979). *The Analysis of Indian Agroecosystems*. Interprint, New Delhi, India: 180 pp., illustr.

Nguu, N. V. & Palis, R. K. (1977). Energy input and output of a modern and traditional cultivation system in lowland rice culture. *Philippines Journal of Biology*, 6, pp. 1–8.

Nye, P. H. & Greenland, D. J. (1960). *The Soil Under Shifting Cultivation*. Commonwealth Bureau of Soil Science Technical Communication No. 51, Harpenden, England, UK: 1522 pp., illustr.

Posey, D. A., Frechione, J., Eddins, J., Silva, L. F. D., Myers, D. & Macbeth, P. (1984). Ethnoecology as applied anthropology in Amazonian development *Human Organization*, 43, pp. 95–107.

Ramakrishnan, P. S. (1983). Socio-economic and cultural aspects of Jhum in the north-east and options for eco-development of tribal areas. Pp. 12–30 in *Tribal Techniques, Social Organization and Development: Disruption and Alternates* (Ed. N. D. Chaubey). Indian Academy of Social Science, Allahabad, India: 204 pp.

Ramakrishnan, P. S. (1984). The science behind Rotational Bush Fallow Agriculture system (jhum). *Proc. Indian Acad. Sci. (Plant Sci.)*, 93(3), pp. 379–400.

Ramakrishnan, P. S. (1985). Tribal man in the humid tropics of the north-east. *Man in India*, 65, 1–32.

Ramakrishnan, P. S. (1986). Morphometric Analysis of Growth and Architecture of Tropical Trees and their Ecological Significance. Pp. 209–22 in *Naturalia Monspeliensia – Colloque International sur l'Arbre 1986.*

Ramakrishnan, P. S., (1987). Role of tree architecture in agroforestry. Pp. 112–

31 in *Agroforestry and Rural Needs* (Eds P. K. Khosla & D. K. Khurana). Indian Society of Tree Scientists, Solan, India: 363pp., illustr.

Ramkrishnan, P. S., Toky, O. P., Mishra, B. J. & Saxena, K. G. (1981). Slash and burn agriculture in North-eastern India. Pp. 570–87 in *Fire Regimes and Ecosystem Properties* (Eds H. Mooney, J. M. Bonnicksen, N. L. Christensen, J. R. Lotan & W. A. Reiners). USDA For. Ser. Gen. Tech. Report, Washington, DC, USA: 593 pp., illustr.

Ramakrishnan, P. S., Shukla, R. P. & Booth, R. (1982). Growth strategies of trees and their application to Forest Management. *Curr. Sci.*, 51, pp. 448–55.

Rao, K. S. & Ramakrishnan, P. S. (1987). Comparative analysis of the population dynamics of two bamboo species, *Dendrocalamus hamiltonii* and *Neohouza dulloa,* in a successional environment. *Forest Ecol. Manage.*, 21, pp. 177–89.

Rappaport, R. A. (1971). The flow of energy in an agricultural Society. *Sci. Am.*, 225, 117–32.

Richards, P. (1985). *Indigenous Agricultural Revolution, Ecology and Food Production in West Africa.* Hutchinson, London, England, UK: 192 pp.

Ruthenberg, H. (1976). *Farming Systems in the Tropics,* 2nd edn. Clarendon Press, Oxford, England, UK: xvi + 366 pp., illustr.

Sanchez, P. A. & Buol, S. W. (1976). Soils of the tropics and the world food crisis. *Science,* 188, pp. 598–9.

Saxena, K. G. & Ramakrishnan, P. S. (1984). Herbaceous vegetation development and weed potential in slash and burn agriculture (jhum) in N. E. India. *Weed. Res.*, 24, pp. 135–42.

Seaavoy, R. E. (1973). The transition to continuous rice cultivation in Kalimantan. *Annals Assoc. Amer. Geographers,* 63, pp. 522–8.

Spencer, R. E. (1966). *Shifting Cultivation in South-eastern Asia.* (Publication in Geography, No. 1999.) University of California, Berkeley, California, USA: 247 pp.

Steinhart, J. S. & Steinhart, C. E. (1974). Energy use in the U.S. food system. *Science,* 184, pp. 307–16.

Swamy, P. S. & Ramakrishnan, P. S. (1988). Ecological implications of traditional weeding regimes under slash and burn agriculture (jhum) in north-eastern India. *Weed. Res.*, 28, pp. 127–36.

Toky, O. P. & Ramakrishnan, P. S. (1981*a*). Run-off and infiltration losses related to shifting agriculture (jhum) in northeastern India. *Environmental Conservation,* 8(4), pp. 313–2, 10 figs & 6 tables.

Toky, O. P. & Ramakrishnan, P. S. (1981*b*). Cropping and yields in agricultural systems of the north-eastern hill region of India. *Agro-ecosystems,* 7, pp. 11–25.

Toky, O. P. & Ramakrishnan, P. S. (1982). A comparative study of the energy budget of hill agro-ecosystems with emphasis on slash and burn system (jhum) at lower elevations of north-eastern India. *Agro-ecosystem*s, 9, pp. 143–54.

Toky, O. P. & Ramakrishnan, P. S. (1983). Secondary succession following slash and burn agriculture in north-eastern India, I: Biomass, litterfall, and productivity. *J. Ecol.*, 71, pp. 735–45.

UNESCO (1986). *Programme on Man and the Biosphere (MAB) – General Advisory Panel Final Report* (MAB Report Ser. No. 59). UNESCO, Paris, France: 58 pp.

Widstrand, G. C. (1975). The rationale of the nomad economy. *Ambio,* 4, pp. 147–53.

Whitmore, T. C. (1975). *Tropical Rain Forests of the Far-east.* Oxford University Press, London, England, UK: 278 pp.

Winterbottom, R. & Hazlewood, T. (1987). Agroforestry and sustainable development: Making the connection. *Ambio,* 16, 100–10.

World Commission on Environment and Development (1987). *Our Common Future.* Oxford University Press, Oxford, England, UK: xviii + 383 pp.

Commentary on Chapters 9 and 10

CHAIRMAN: Professor István Láng
PANELLISTS AND OTHER CONTRIBUTORS:
La Rivière, McMichael, Koch, Holdgate, La Rivière, Purcell, Kuenen, Marshall, Juhász-Nagy, Juel-Jensen, Koch, Juel-Jensen, Koch, Juel-Jensen, Koch, Fosberg, McMichael, Pellew

La Rivière said that **Ramakrishnan** had illustrated how the incorporation of socio-economic considerations had been successful in the field of environmental management, whereas **Juhász-Nagy** had focused on deficiencies in the theoretical basis of ecology and what should be done about it. He, however, wished to add another shortcoming of science, pertinent to the theme of the Conference, in which the role of science as a provider of the basis for environmental action figured so prominently. But before doing so, he would like to say in passing that he was not sure that it was science that could have shortcomings. Science was a goddess who could have no faults; it was her followers, the scientists and the scientific endeavour as such, that could and did have them. He would not dwell extensively on the shortcomings themselves, but rather on the steps that were being taken to overcome them.

Nobody questioned whether or not science was doing its duty so long as it was only a noble pursuit to satisfy Man's curiosity – just a game for gentlemen and gentlewomen; and this did not change fundamentally when the application of science to industry and, as mentioned earlier by **Goldberg**, in defence systems also, gave science its well-established place in society which was both grateful, and eager to accept its products.

In the past decade or two, however, science had been given a new role to play in society. It was seen as one of its shortcomings that it could not respond fast enough to meet its new tasks. He meant here, of course, its response to the global environmental predicament which to a large extent has been created by the application of the results of science itself. The new role of science was in his opinion two-fold:

(1) to investigate how the Earth system worked and to be a watchdog with respect to global changes, and (2) to develop effective means for transmitting its findings – and also its needs – to Society in an unambiguous manner, without distortion. In discussing the developments in these two areas he would draw on his experience in ICSU and in SCOPE, which, of course, did not constitute all of science but were somewhat representative, especially with respect to the present topics.

It had become quite clear that Man had to know quickly how Planet Earth worked, what the fluxes of matter were between compartments, what factors regulated the conditions for life, what Man's impact was on them and the mechanisms involved. At the same time changes in the Earth system had to be observed, modelled and predicted. This fact-finding research work had become necessary because there was proof that Man's activities had reached a magnitude that did influence considerably the Earth system, e.g. from the work of SCOPE in the '70s on biogeochemical cycles.

This scientific research work required an unprecedented interdisciplinary and international effort to make up for lost time and to provide policy-makers with the best possible basis for action by prevention, mitigation or adaptation. It was striking to note that the genesis of the first attempt at managing Planet Earth had coincided with the new developments in genetics that might, in another way, lead to the control of the destiny of all of life, including Man.

Based upon the earlier work of SCOPE, ICSU had set up in 1986 the IGBP, a study of global change which had the following characteristics: first of all it was *non governmental,* which was very important because that meant independent and unbiased; then it was a *long-term* programme that might well have to run for 10 to 30 years and, as **Goldberg** had already stressed, governments had to be educated to accept that as necessary.

The programme had, to an unprecedented degree, to be *interdisciplinary*, the disciplines ranging from meteorology, geochemistry and biology all the way to geography, anthropology and economics. It had to be *international* in two ways: firstly in all relevant fields the best scientists were needed and thus had to be recruited from the world as a whole and, secondly, the object of study, the Earth, was international and it was necessary to gather data from everywhere. Participation in the programme was by *partnership* between as many countries as possible. Already 40–45 countries had joined but, as mentioned later, the participation of developing countries was still a grave problem. Another group of partners were the *intergovernmental organizations* such as WMO, UNEP and others, including the Commission of the European Communities. *Industry and business* could also be very valuable partners and the first contacts that had been made looked promising.

The research work was to be carried out in a focused core-project in which relevant elements of national programmes participated. The execution of these projects was *decentralized* while the initial planning and their coordination had to be done from *one centre*. While the centre and the Special Committee that steered the IGBP must have, to some extent, a *top-down* approach, the execution at the core-project level was mostly *bottom-up,* as the specialists themselves decide how to do their work. Furthermore, there had to be effective and efficient *management* to enable scientists to devote their time to science with a minimum of bureaucracy. Finally, there had to be appropriate *outreach* activities to keep all groups concerned, inside as well as outside the programme, informed of progress and of the results as these were produced.

ICSU possessed a rich experience in international research programmes, such as the International Geophysical Year, the International Biological Programme, the Global Atmospheric Research Programme and, since 1979, the World Climate Research Programme jointly with WMO and was now working closely with the IGBP. Despite all this experience the new IGBP programme was so vast and diverse as to make it ICSU's most ambitious venture, and one that would happily accept all the help the scientific community could provide.

(2) To set up a programme like the IGBP was not enough if science was to fulfil its second role. There was a great need of improved communication between scientists and policy makers. This problem of interfacing had grown rapidly over the past year or two. While in the past it was very difficult for science to get the attention of the press, or the ear of governments, it now found the latter banging on science's doors to receive the latest results on the state of the global environment! For instance, the French Government had recently held a workshop on how policy decisions could be made in the face of scientific uncertainty, or even controversy. More importantly – and probably without precedent – UNEP and WMO had in 1988 set up jointly an Intergovernmental Panel on Climate Change (IPCC) in order to assess the available scientific evidence, the possible impacts and the corresponding policy options. It seemed to him to be an historic novelty that government delegations had come together to take stock of present scientific knowledge and to invite scientists to present their findings. Many countries had participated enthusiastically and the

panel was presenting its promised report in the summer of 1990 after less than 2 years of intensive work.

Such developments reinforced the continuing mandate of science to acquire new insights, produce new data, improve its models, and maintain monitoring operations for crucial parameters so as to increase understanding of the mechanisms which operated in the Earth system. But, in addition to the interface between science and governments, the relationships between science and intergovernmental organizations and with industry and business had to be improved, and, of course, also with the public-at-large. ICSU had been aware of these problems for some time and had organized a workshop on 'International Science and its Partners' to be held in Visegrad (Hungary) next week.

This year, the IGBP was making the transition from its planning phase to the operational phase. As it was a pioneering adventure, it would certainly have to struggle with unexpected problems. One of these could be: how to obtain the required massive funding without sacrificing independence. It was very important for potential sponsors – governments, the UN, and industry, to realize that it was in their own interest not to attach strings and that the value of the programme would be diminished if it were no longer to be independent. Secondly, in the tradition of ICSU, the programme would present the results of its fact-finding research in a dispassionate manner without attaching judgements, so as to remain distinct from pressure groups and other organizations that had an important mandate to influence decision-makers. Perhaps the largest problem of all would turn out to be: how to finance the participation of the Third World. As all countries – each in its own way – would have to participate in the solution of global environmental problems, it was essential that they were also partners in the research leading to the diagnosis of these problems. Other problems were in the area of access to data, which, depending on their source, might be subject to some restrictions whereas the programme demanded that all scientists had access to them. Then there might well be a shortage of scientists capable of doing interdisciplinary work, but fortunately in some university departments, global change studies were already emerging where highly integrated science teaching was taking place.

In conclusion, he was sure that some of these problems, as well as examples of the misuse of scientific results, would come up elsewhere in the Conference.

McMichael said that the title chosen for this session* had caused him some concern. As a scientist, he could not accept that science, as such, had any shortcomings. It was, after all, no more than a way of discovering and organizing information about natural phenomena. While it could not provide us with all the answers, it did give us the best, indeed, the only information we had about The Biosphere and its workings, and it was up to us to make use of that information in the best way we could. In the long run, we would only survive with The Biosphere if people made decisions to do some things differently. Science could not make these decisions for us, it only showed us the way.

Nevertheless, there was such a thing as bad science, or at least material presented as scientifically sound that was far from being so, and it was this aspect of science that he wanted to address. The extraordinary increase in environmental awareness among people generally, and almost everywhere, was undoubtedly due to the publicity given by the media, particularly television, to some of the conclusions reached by scientists about the environment and human impacts on it. Consequently it was important that the information which was disseminated in this way was correct and based on good science.

It was, of course, difficult to specify exactly what was good science. Just because a scientist arrived at a conclusion, subsequently shown to be wrong, it did not mean

*[Science and its Shortcomings. Eds.]

he or she was guilty of practising bad science and, because other scientists criticized the work of a particular scientist, that did not mean it was bad science. On the contrary, as **Hare** had reminded us yesterday, valid criticism of scientific results by others was, and had always been, an important element in scientific progress.

However, we should not assume that all the science on which the increased public awareness of environmental issues was based was necessarily good science – just because it had been widely accepted. One area that had concerned him for some time was the question of the rate of species-loss, or extinction, due to human interference in biospheral processes. We were all familiar with headlines that proclaimed, 'Scientists estimate that by the turn of the century one million species (or some other very large number) of animals will have become extinct'. Similar statements appeared quite often in the writings of scientists who were generally regarded as highly competent in their fields. But rarely did anyone ask what was the scientific basis for such assertions?

A few years ago he had decided to examine the literature on extinction of species to try to establish just what was known about species extinction-rates. He had found that, in fact, very little was known.

In the first place, it was important to remember that extinction was a natural process. Extinction of local populations of virtually every species was happening all the time, while new populations of the same species were establishing themselves elsewhere. Species extinction began to take place when the rate of population extinction exceeded the rate of population establishment, and happened when the last individual of the last local population died (assuming that no dormant eggs, spores or seeds survived to resume growth when conditions were once again appropriate).

At any one time, many species were rare and approaching the end of their time on Earth while others were actively expanding and, indeed, some were in the process of giving rise to new species. Thus species extinction was to be expected in Nature and the disappearance of a species was not always the direct result of human activity (though no doubt in many cases we had 'helped' species to become extinct more quickly than they otherwise would have done!). Charles Darwin had recognized the naturalness of extinction. In *The Origin of Species* he had written (1859, p. 130):

> 'to admit that species generally become rare before they become extinct – to feel no surprise at the rarity of a species, and yet to marvel greatly when the species ceases to exist, is much the same as to admit that sickness in the individual is the forerunner of death – to feel no surprise at sickness, but, when the sick man dies, to wonder, and to suspect that he died by some unknown deed of violence.'.

So, the ultimate disappearance of a rare species during our lifetime, was not something we should be surprised about nor did it follow that we should always try to prevent it from happening. Nevertheless, we could and should deplore the unnecessary loss of any species, especially of those with which human beings had strong cultural attachments.

Information about historical extinction rates was not very good and, in fact, relatively few known species of animals were with certainty known to be extinct. Some that had not been seen for many years were in all probability now extinct but others continued to reappear with surprising regularity. However, some known species certainly had become extinct as a direct result of human activity, and we should not doubt our capacity to send many other species down the same road if we acted in ways indifferent to their interests. The difficulty was in knowing which species were vulnerable, or under threat, and in knowing when a species was at danger point. It was also very difficult to assess what was happening to the many species that were not 'known' to science but which certainly existed, especially among the less well-studied groups of organisms and those from remote, or uncollected, habitats.

Estimates of current and future extinction rates used in most contemporary literature were generally extrapolations of estimates of historical rates, multiplied by a factor derived from some vague generalization. For example, someone might assert that 'extinction rates are now ten (or a hundred) times as great as they once were, therefore we can expect that X species will become extinct in Y years' – which then became the kind of headline referred to earlier.

Another basis for the assertion that extinction rates were likely to be high was the rate of loss of particular kinds of habitat. It is clear that some kinds of habitat were being destroyed at a very fast rate, for example the Amazonian rain-forests, and we also knew that there was usually an extraordinarily large number of invertebrate species, mostly insects, associated with each rain-forest higher-plant species. Consequently it was relatively easy to jump to the conclusion that massive loss of habitat must result in the extinction of large numbers of species of dependent animals.

Holdgate, in his opening address, had drawn attention to the fact that this link was really the basis for most of our extinction-rate estimates but asserted, correctly, that they were the best we have. He commented that 'They lead on to assertions that 15–33% of all species *may* become extinct in 10–20 years, including 15% of the world's vascular plants and 12% of bird species' (*see* p. 10). These numbers, however, tended to be used in much more precise ways than could be justified from their derivation. Too many conservationists, some distinguished scientists among them, regularly quoted these kinds of numbers as if they were exact estimates. Furthermore, they argued that every piece of natural habitat that was destroyed was adding to the number of extinct species. In fact this might not be so at all.

In Chapter 3, **Poore** had indicated that, in his opinion, it might be necessary to retain only 20% of the tropical rain-forests to conserve 95% or more of the species. **McMichael** was sure that **Poore** would be the first to admit that his figure was as much a 'guesstimate' as was **Holdgate**'s figures for extinction rates, but he suspected, nevertheless, that **Poore** was right. Certainly most nature conservation agencies operated on the assumption that a sampling of some small percentage of natural habitats in a country (perhaps as little as 5% or 10%) would probably capture a substantial percentage of the original species-diversity.

In making these observations he would not want to suggest that we should not fight to save as much natural habitat as was possible, nor should we cease to be concerned about the survival of the greatest range of species possible. He simply wanted to draw attention to the limitations of our scientific knowledge about these matters, and to urge caution in presenting and interpreting such data as were available. If our arguments were not founded on good science then we could scarcely expect to be taken seriously.

Koch thought it was always nice to make clear what we were talking about! He would attempt to define what science was, and what the reasons for its possible shortcomings might be. He therefore considered the features of pure science, parallel to its shortcomings, which stemmed mainly from various weaknesses of human nature.

First he asked whether science was to be regarded as a hobby or a profession. He agreed with F. Jacob, who had once said in a personal conversation '*La science est la chose qui m'amuse*'. In other words, pure science was the pleasure found in the consistent effort 'to know more about things'. True science had no 'practical' objectives at all, being typically '*l'art pour l'art*': an activity that had satisfied the intrinsic curiosity which compelled the first ape to leave the rich yet dull paradise of the tropical jungles just to see 'What was round the corner?' That idea was indeed uncommon, as the forests provided optimal conditions for a simple, comfortable, careless, and apparently safe, life as long as no one wanted to change anything. It was no wonder that the first comprehensive mythological explanation of human presence and troubles on Earth was the biblical tale about the original sin of curiosity and disobedience! That, he thought, was exactly the attitude of simple-minded conservative people towards scientists; as M. McLaughlin had said, 'All societies honour their live conformists and their dead troublemakers'. It was, therefore, easy to explain why so many scientists had been keen to make science a 'useful profession' from the very beginning. Useful, in the trivial sense of the word, was only applied science. Abstract knowledge was the treasure of those who possessed it, and had no value for the average citizen busy at making his living. In this context he cited Plato's (427–347 BC) story about Thales (624–546 BC), who fell into a well while watching the stars and was told by the old woman who rescued him 'Here is a man, who gazes at the stars, but cannot see what lies at his feet'. It was an historical fact that the first real scientists in the history of human knowledge, the ancient Greeks, contented themselves for the most part with theoretical speculations and showed little interest, if any, in experimental proof and practical application. Archimedes (287–212 BC) who, unlike Plato and Aristotle (384–322 BC), was intrigued and amused by practical tasks, such as estimating the grade of purity of gold in King Hieron's crown, seemed to have been an exception; yet, at the same time, he was an example of the absent-minded mathematician who, like – more recently – Newton and Wiener, could not tell whether he had had his meal or not! Nor did Archimedes apparently value his own ingenious technical inventions, for he perpetuated only his mathematical ideas by writing them down for posterity.

Twentieth-century developments in science and technology had walked hand-in-hand with the commercialization of science. Today, science had, unequivocally, become a profession and only a very few, exceptionally talented, heads still pursued science as a hobby, heedless of the money, honour, or position promised by a strictly professional approach. Professionals obliged to 'sell' their science were confronted by the laws of demand and supply. Things intended for sale required publicity, and the professional scientist had not only to do the work, but also to convince people – especially those in charge of grants – of a real demand for what he or she had to sell, *e.g.* for sequencing of the human genome. Thus a modern professional scientist needed a King Hieron who was ready to buy his brain, and he must forget about the fate of the goldsmith who had been executed for cheating. The discovery of the principles that led to the construction of the atomic bomb was a unique scientific achievement, but the horrors of Hiroshima had nothing to do with science. In A.Szent-Györgyi's words: 'The death of an individual is a tragedy; the death of thousands is but statistics' (*The Crazy Ape*, 1970), and he added, after Le Rochefoucauld, that 'we are all very good at tolerating the sufferings of others'.

Ever-increasing competition for the market speeded up the production of newer and newer items. Speed and contemplation were, however, incompatible, *e.g.* the Thalidomide accident, excessive chemicalization in agriculture, and the intolerable levels of environmental pollution in over-industrialized and -urbanized regions. Thus, while science remained sacred as a hobby, it could become a plague as a tool of profit.

Secondly, he addressed the question of the humility and self-confidence of scientists. So long as Man did not know everything, he knew as good as nothing, yet absolute knowledge of everything was a divine attribute far beyond human reach. In other words, science remained humble in the face of Nature and cocksure self-confidence was usually a sign of ignorance.

This did not, naturally, imply that the Encyclopaedists were ignorant; far from that: they were knowledgeable, but prejudiced, and prejudice in that sense meant a special sort of ignorance. The seemingly absolute laws of classical (causal and deterministic) physics caused Laplace (1749–1827) to state that, if a demon knew all parameters and variables of the universe at a given time, he would be able to tell both its future and past precisely. The Encyclopaedists had denied that chance events could happen and regarded such events as delusions associated with some 'hidden parameters' of the system.

The contributions of Einstein, Planck, Heisenberg, Hawking and many other outstanding scientists had, in due course, changed fundamentally the paradigm of classical physics. Contemporary theoretical physicists and cosmologists had learned – and taught – the lesson of why and how humility should be preferred to self-confidence when approaching the secrets of Nature. Humility was, of course, not to be understood as humiliation, but as acceptance of the fact that Nature was – and would remain – wiser than her product the human being, whom Linnaeus had so boldly named '*Homo sapiens*'. Rather than commenting on this pseudonym himself, he referred participants to K. Lorenz's book, on '*The Eight Capital Sins of the Civilized Human*' (1973) and A. Szent-Györgyi's considerations in '*The Crazy Ape*'.

Thirdly, should Man be at peace or fight with Nature? Man needed to be aware of his being both a product, and an object, of Nature. Any human effort to subjugate or conquer Nature was therefore, *ab origine*, in vain. The sole approach to a reconciliation between Mankind and its injured planet was to learn as much as possible about the laws governing Nature and to use that knowledge with due modesty and reason, unfortunately no longer to maintain but rather to restore the conditions of life on Earth, which had been spoiled so dangerously by various recent activities of the so-called 'civilized human' (the *Homo 'insipiens'* or crazy ape). The main capital sin of humans seemed to be undue pride and self-confidence. Spacecrafts and nuclear power-plants were but playing with spillikins compared with the immense potentials still held by Nature in energy-matter and space-time. When viewed from these dimensions, human knowledge was but seagull droppings on the top of an iceberg that represented the Unknown! In this light, human attempts at 'improving' the iceberg of Nature's works seemed ridiculous beyond description. That did not, of course, mean that developments in food production, infectious disease control and in the overall conditions of life for a substantial part of Mankind, should not receive due appreciation and support. It was, indeed, a breathtaking achievement of the human intelligence to have extended the scope of knowledge from the borders of universe to the quanta.

However, before he was stoned as an anti-scientist, he wanted to emphasize again that in his view science *was* sacred as being the greatest achievement of the human brain, and that exactly was the reason why he was always horrified whenever he saw science raped for vile purposes. The headlong conversion of advances in knowledge to improve the war industry, or to change the face of the Earth, scared him to death whenever he considered how much time and care Nature had spent on producing the home of humanity, our green planet. The biological systems of life were so delicately balanced that upsetting their equilibrium was usually realized only at the stage of imminent catastrophe. Human shortsightedness reminded us of the narrow scope of the average politician, who thought only about the forthcoming election. Rather, a scientist should act as a statesman who was able to foresee the impact of his actions on posterity. That was the real responsibility of science.

However, not even full responsibility could avert the hazard of occasional shortcomings due to the inherent limitations of human knowledge. The proof of the pudding was in the eating, but nobody was willing to taste a pudding prepared hastily from inadequate ingredients, as happened not infrequently in 'commercial' science. Genetic engineering was a fantastic achievement of human intelligence yet, at the same time, an extremely dangerous tool in thoughtless hands, not so much for fear of producing chimeras and monsters (which Nature would, anyhow, eliminate as artefacts), as for the intellectual misinterpretations involved. The technocrats might think that the time had come to produce human automatons 'tailored' to specific tasks and would push research in that direction. In fact, very few products of genetic engineering, if any, could function under natural conditions. About two decades had proved to be too short for retailoring the recombinant *E. coli* bacterium, able to produce human insulin, to do their job cheaply on an industrial scale. The evolving science of biotechnology, however fascinating, was helping scientists, for the time being, to discover *new* problems rather than to solve *old* ones.

In these evolving branches of science the 'hobby scientist' saw an opportunity for a deeper insight into the secret workshops of Nature, but the professional scientist saw a possible tool for obtaining more profit. K. Lorenz had reflected bitterly on the blind race after profit and power before a heart failure, or a car accident, put an end to all ambitions. The main objective of science should be, in fact, to find ways of living in peace with Nature instead of making over and again usually vain attempts, in the long run, to conquer her in order to reach short-term goals. The advantages of motor cars seemed doubtful in the light of the damage done to living beings, in general, by the lead, tar and CO-contents of the exhaust gases, and to humans, in particular, by traffic accidents. Urbanization was a blessing so long as it did not involve serious health-hazards at both the physical and mental levels. All the stresses of 'civilization' stemmed from the 'unnatural' conditions of life, for which humans, being products of Nature, seem to be definitely 'unfit'. Thus, while Nature operated by selecting the fittest, civilization seemed to select for 'unfitness' for survival in Nature.

These still-emerging 'unnatural' forms of life had seemed to him like a '*danse macabre*' of Mankind towards self-extinction. The alarm signals of Nature were already there at the lower levels of life; the, often unnoticed, almost daily extinction of a plant or animal species walked hand-in-hand with Man's 'achievements' in conquering Nature.

Let us, he appealed, realize the fatal trap built by ourselves before it is too late.

Holdgate wished to share with the previous speakers the thought that, although humanity had science at the helm of Spaceship Earth, he was a little concerned about the link between steerage and navigation! Generally agreed characteristics of science were the capacity to quantify, to reduce complex systems to order, and to advance by framing testable hypotheses. Yet, thinking back to the speakers on marine issues, he had been worried then, and was still worried, by our evident lack of capacity to relate the magnitude of our social concerns and the setting of our social priorities to the quantification which, as scientists, we would claim to deliver. For example, the relative significance of inputs to marine pollution had been discussed. But those which were enhanced fluxes of natural substances, such as metals, representing substances new to organisms and to which they might be less tolerant, had not been examined. There had been imprecision about the need to relate input to dispersion, dilution, and, eventually, to impact. **Clark** had capped that concern by observing that even when, by scientific logic, one had arrived at the best practicable environmental option, the community did its best to evade that outcome when presented with it! Another example of the difficulty of communication had been given by **La Rivière**. IGBP was both a marvellous concept and an extremely well-planned, collaborative, scientific effort on a 10–20 years' time-scale. When it had been planned that was fine, but now governments' were demanding more urgent efforts and interim inputs were clearly expected by the world community before the full scientific study was complete. Scientists were, therefore, being forced into making interim assessments to provide the best practicable interim policy, so to speak.

His conclusion from such considerations was very simple and, in a way, obvious. We could not let science stand aside from the wider social context, but it should always be remembered that science made two kinds of input into that context. One was problem-solving, which scientists were often called upon to do: that was vital. The other, he suggested more important, contribution was that of a particular kind of intellectual thought-process. It included intellectual rigour – in which science was not alone – to ensure the approximations which scientists provided were as near the truth as was possible, avoidance of the errors of excessive extrapolation, or criticism (as **McMichael** had observed in his contribution), and an attempt to get across to the public through the media (which both scientists and economists were happy to criticize) some idea of the universality of serious scientific commentaries. Indeed, he thought that getting across the intellectual rigour and honesty of science might be more important than the role of science in problem-solving, not least because an understanding of the scientific philosophy, or at least some comprehension of scientific method, by recipients of scientific findings in the community was essential, as indeed it was to those practising science. Unhappily, he feared that in Europe, and he suspected in North America, modern education had not ensured that the receptors were attuned to the signal.

La Rivière responded by asking, would not **Holdgate** agree that the first task of science was to find out facts in order to unravel mechanisms that exist in the world around us and then to report them faithfully? That, he thought, went before problem-solving. Then, he reminded participants, there were very good scientists who were very clumsy with public relations, or who could not write a comprehensible report destined for non-scientists. He thought the task of a science writer was very important and that the scientific community should give thought to how it could find ways and means to achieve such writing effectively. He did not anticipate that scientists themselves should become good science writers; they were neither trained for it, nor could many ever learn.

Purcell remarked that today society was looking ever more to science and, in turn, science was looking at sociological and socio-economic factors. Amongst the public and decision-makers, no issue was more under scrutiny than that of the 'greenhouse' global warming effect. He was deeply concerned that the 'capital E', Environmental community (and he recalled **Vallentyne**'s distinction between the ecological and *environmental* approach and that of the ecological and the *ego*logical approach) was assuming, almost in the role of a doctor, that there *was* a 'greenhouse' effect, that global warming *was* a significant feature, and that it *was* happening, despite the evidence for uncertainty, and that a number of factors might have been missed. He thought it absolutely crucial that scientific credibility on this issue, above all others, should be maintained, as it was going to affect many major developments. For example, in the area of energy, a number of people were already using the issue to promote nuclear energy or, as discussed earlier, dams for hydroelectric power, as sources of 'clean, non-greenhouse-gas' energy. He did not think that anyone was yet in a position to make definitive pronouncements, but that should not prevent scientists from speaking as eloquently as possible on what needed to be looked at or discussed, and what needed to be done before an informed judgement could be made on these issues. There was always tension between rushing into a judgement and taking too long. He thought, in terms of contemporary issues involving science and the environment, that tension was especially critical for the world.

Kuenen said that, of course, there were true scientists but amongst those who practised science there were many who were more interested in getting some sociological/political matter accepted than in real science, *e.g.* in Holland there were one or two more interested in nature conservation than in the necessary research! Such individuals were not reliable and it was important to distinguish between their pronouncements under the guise of their title, university or other, and that of scientists proper. Some of the former were not to be trusted and it was important to recognize that a scientist in name is not necessarily a scientist.

Marshall mused on the meaning of the word 'problems' which had been used. He recalled that a former English Prime Minister, Harold Macmillan, had remarked that he understood problems in Euclid, they were soluble; but political problems were different and he was sceptical about their solution. Scientific problems were more of that nature. It was sometimes possible to get near solutions but, more often, it was only possible to make problems easier or simpler of solution, although any good scientist was going to try to solve them.

Juhász-Nagy reflected on **Holdgate**'s remarks. He agreed with the notion of a dual contribution by science – problem-solving and an intellectual framework – yet these ideas were less than a century old. He remarked that one important contributor had been Sir Ronald Fisher, a 'biometrics man', who was neither an eloquent speaker nor writer, who did not write popular essays, and who even confused people because he was not always understood! He had made major contributions to the measurement, analysis and study of diversity, so that its importance could be properly recognized and assessed. Yet, as recently as 1964, at a jubilee meeting of the British Ecological Society attended by many Americans and Europeans prominent in the field of diversity studies, it became clear that the statistical aspects of diversity were not properly understood. That illustrated what a long way a really important concept with its roots in the scientific community, or in society, had to travel before acceptance and recognition.

Juel-Jensen queried the implication of **Koch**'s remark that the sequencing of the human genome was a useless enterprise. He recalled very clearly the day a young worker told him of the identification of the gene for a kidney defect, and he was associated with a group, led by Sir David Weatherall, which was investigating the genetic make-up of haemoglobins. Surely such work was important and had valuable practical consequences; he could not believe that **Koch** had meant what he had said.

Koch replied that he had meant it, although in a different sense. What he had meant to make clear was that to sequence one human genome was of no help. Several human beings had to be sequenced and the results compared before a standard could be established and then only 10% of the known functional genes could be used: the remaining 'silent' 90% appeared at present functionless. As a theoretician rather than a practising doctor he appreciated the beauty and pleasure to be derived from sequencing the human genome, but he was concerned that there were some very brutally urgent ecological and medical problems facing the world which should take priority over sequencing.

Juel-Jensen asked whether it was not important to be able to tell a mother that she carried the wrong kinds of genes, and whether she should be told that she could have a defective child, or not?

Koch agreed. However, he observed that Nature was much cleverer than Mankind and the vast majority of pregnancies ended up in a normal healthy birth.

Juel-Jensen pursued his argument by remarking that he could match statistics! About 4 million children die of malaria each year in Africa, south of the Sahara. A statistician would say that that was more important than the death of 10,000 from rabies; it was not a problem. But such a statistic was not of much use to a person with rabies whereas a patient with malaria could be cured. That was a constant and unavoidable dilemma.

Koch agreed entirely that the death of any one person was a tragedy and that of a couple of millions was but a statistic. He knew that, but it simply had to be faced as unavoidable.

Fosberg turned to **McMichael**'s comments to ask whether he would grant that the rate of extinction had accelerated during the past century.

McMichael was sure that it had but his concern was that it was very dangerous to put figures on the rate, with the degree of uncertainty prevailing, because people attached too much significance to them.

Pellew said it was only a fool who would come between two physicians locking horns, especially when one was a pathologist, but he wanted to pursue **McMichael**'s comment about misinformation. It was right not to peddle 'junk data' in the guise of science, and **McMichael**'s warning was especially timely when new technical tools were becoming available, such as geographical information systems (GIS), which greatly increased the potential for applying bogus respectability to rubbishy data! Many of the GIS outputs were little better than fancy wallpaper because of the poor quality of the input.

However, he saw the problem as that of inadequate science policy analysis, rather than of bogus data masquerading as genuine. Science generated data and its analysis provided information. That, blended with experience, provided knowledge: the scarce commodity which all sought. We lived at present in a world which threatened to become swamped with data, yet there was too little *information*, let alone knowledge, upon which to make enlightened decisions. He observed that the geostationary satellites used in the IGBP programme referred to by **La Rivière** were going to produce such vast quantities of data that a new word had entered the English language. We now lived in the age of the terrabyte which fed on the gigabyte, that preyed on the humble megabyte! What seemed to be lacking was a basic analysis of why we were collecting data and what we were then going to do with them. That was not the fault of the scientist but of the policy analyst. Politicians were asking more and more for the outputs of global models and, inevitably, those models required global monitoring systems; but there was yet to be, in his opinion, a rational analysis of what parameters should be modelled. Data were being collected as if data collection was a function in its own right. But data collection was only part of the process; therefore, to kick the scientist for the inadequacies of the system was to kick the wrong person. Instead we should kick those who were involved in deciding what sort of data should be collected, what parameters needed to be modelled, and what the priorities should be: in other words, those involved in policy analysis.

11. Learning to Survive with The Biosphere

JOHN C. SMYTH

President, Scottish Environmental Education Council, University of Stirling, Stirling FK9 4LA, Scotland, UK

&

WILLIAM B. STAPP

School of Natural Resources, University of Michigan, Ann Arbor, Michigan 48109, USA

INTRODUCTION

No thinking person can well be in doubt any longer about the global threats to the survival of our human species in its one-and-only current or foreseeable habitat. Nor can there be doubt as to the basic causes of these threats being our own population growth, 'development', and profligate behaviour (Polunin 1972, 1980; Polunin & Burnett, 1990). The speed and complexity of these changes have drawn together the world's rich assemblage of cultures and peoples into a single global community, increasingly dependent on each other and increasingly exposed to the effects of each other's behaviour. Neither any nation nor any human individual can break away from the complex web of relationships comprising development, in the process of which some cultures have become dominant whereas others have been destroyed. Now, however, the survival of all is threatened by the effects of these changes – often unforeseen – on the human habitat and the resources on which the whole system depends.

Unfortunately, understanding of these changes and realization of their implications lags far behind their taking place. Although phrases such as 'the shrinking planet' have become familiar enough to alert people to the idea, the concept of one superecocomplex for Humankind, and the holistic thinking that is necessary to understand and respond to it, remain unfamiliar and even threatening. One might have hoped that the human mental abilities which were responsible for the speed and complexity of these changes would also be equal to the developments in understanding that are needed to address their effects. The educational development of these mental abilities is, however, something which tends to change more slowly than the effects of their use. Most people acquire little idea either of the global dimensions of the activities in which they are engaged from day to day, or of the degree of their dependence on the continuing activity of other people and ecosystems far away.

Here, then, is a major task for educators, and an urgent one – to expand the education of people to encompass the whole human family, to care for that family, to care for the global habitat on which it depends, and to build global perspectives into the daily patterns of all humans' lives. These concerns were the subject of a notable paper by Stapp & Polunin (1991) which provides the foundation and much of the structure of what follows.

Education for a Shrinking Planet

Professor Donald J. Kuenen, during the conference for which the present paper was originally prepared, referred to the need to change human behaviour if survival with The Biosphere is to be possible. Education must have some responsibility for this, but does not yet seem to have been notably successful. What can be done to make education an effective contributor to our survival?

Education is usually thought of as the process by which the development of behaviour is guided, through cultivation of awareness, acquisition of knowledge and understanding, growth of skills, and fostering of attitudes appropriate to success in life as perceived by the culture within which it is carried out (*cf.* Smyth, 1988). We are apt to define education in terms of formal education, that certainly sets the standards by which people nowadays tend to judge the quality of education and achievement. The formal sector is, however, often guided by influences which are notorious for inertia and resistance to change. In many countries environmental education, as a new element in the process, has been a particular victim of these constraints.

Another approach is through informal education, which is variously offered by both statutory and non-statutory organizations, and is still guided but often more adventurous (although less influential and more uneven) than most formal education. But much education in the widest sense is brought about by influences which are quite unguided, which start before birth and continue throughout life, and which may be quite unrecognized as educational influences. They contribute to the environment of education, to which we shall return. If we are to be successful in remoulding education, we shall have to try to bring these influences also within the scope of our guidance systems.

When we try to redesign education, therefore, we must now try to think about something much more comprehensive than classrooms, courses, and qualifications; we have to create for people what has been described as a *sustained learning experience* – a process in which no significant influence can be ignored.

Unfortunately education, as we have already seen, tends to adapt slowly even where the circumstances in which it operates are changing fast. Our shrinking world, with its too-rapid population growth, often wasteful population movement, changing economics, and changing life-styles, imposes severe stresses on its people and their life-support. It is tempting to compare the effects of stress on human societies with effects of stress on ecological systems (Smyth, 1983). We see declining diversity in the former represented by growing uniformity of life-styles, emphasis on behavioural conformity, and intolerance of minority characteristics. There is a challenge here: how do we foster the sense of 'one human family' and at the same time value and preserve the diversity which is one of its strengths?

We see loss of information content in the suppression of traditional cultures (although, at the 'eleventh hour', conservationists are beginning to recognize the dangers of such suppression).

We see loss of behavioural control systems in the decline of traditions and customs, and loss of a sense of continuity in time, when lack of awareness or respect for the past leads to lack of concern for the future. Short-term and material objectives replace long-term ones with, as a grave consequence, a lack of interest in the care of resources. People who are concerned to maintain a sustainable society for the future are all too apt to be replaced by opportunists whose aims are selfish and of limited time-span: such undesirables are to be found everywhere in modern society, from the street corner to the corridors of power.

These are broad generalizations, but they present an uncomfortable parallel with the features of impoverished ecological systems, and they are also familiar to many present-day educators. In many places, but especially in our cities, the generalizations represent very influential features of the social environment in which education is happening. Parents and teachers are being replaced as 'role models' by commercial creations, and the cherished homeland is giving way to an inferior landscape governed by market forces. But how often do those who design education take that all-important environment into account?

The task for educators, therefore, in adapting education to prepare people for a global society, involves the use of a broad spectrum of approaches – formal, informal, and non-formal. It requires an appreciation of the effects of rapid change on human behaviour, and a sympathetic understanding of how to respond to them. In societies in which the culture has fostered widely egocentric thinking and acting, it also requires many ways of demonstrating how the everyday actions of individuals impinge on the lives of others even far away, and how this works in both directions. The linkages between such global issues as acidic precipitation, rain-forest destruction, thinning of the stratospheric ozone shield, climatic change, and the local everyday actions of 'ordinary people', frequently go unnoticed, and are rarely comprehended with any proper degree of concern (Stapp, 1985). And yet, if The Biosphere is to survive as an at-all-suitable home for Man and Nature, it is of crucial importance that they be very widely known and understood.

Impediments To Understanding and Action

Bringing to people's attention the ways in which their own daily activities set off reactions around the world, can motivate them to understand and act upon issues in a more positive way. It would be far better if such a sense of shared responsibility were ignited through education than forced upon people by ecological and social disaster. But educators are working against all kinds of difficulties, some of them already suggested above. Barriers are many to integrating global environmental issues into education and concomitant action.

Stapp & Polunin (1991), discussing these barriers, quoted data from the United States, drawn from an address by Dr Stephen K. Bailey in 1975. The data are now rather old, and current percentages are not available, but even a 200% growth-rate would still leave serious deficiencies. Briefly, Bailey reported that:

- Only 3% of undergraduate students in the United States were enrolled in courses dealing with international events or discussing other cultures;
- Average newspaper coverage of international events is no more than one-half of one column of newsprint per day (with corresponding staff deficiencies);
- From a list of languages spoken by over 100 million people in the world, the number of Americans who are proficient in speaking many of them is negligible (figures are quoted); and
- Television coverage of world affairs is largely episodic, dramatic, and transient (although impressive exceptions are acknowledged).

Stapp & Polunin note also the tendency of educators at all levels to stress the differences between cultures rather than connections and commonalities, promoting a feeling of separation rather than connectedness. The same might be said of the media for which differences make better copy than similarities. Indeed one suspects that these attitudes are buried very deeply in political, social, cultural, and even linguistic, traditions, so that overcoming them may require much more comprehensive action than has been anticipated. Nevertheless to recognize where the problems lie is the first step to solving them, and there is growing concern world-wide for mutual understanding and joint action (witness the success of international aid programmes, for example). A holistic approach to the global system may never have had a better time than nowadays to gain support, and the time also seems ripe to have it incorporate tolerance and respect for cultural diversity.

Choice of the United States as an example might be considered unfair to the rest of the world, but is reasonable if one considers its dominant place in the global economy. Other countries vary to differing degrees – in many of them the display of linguistic skills, for example, puts both America and Britain to shame. National attitudes to global issues will be influenced by many factors, including physical location (*e.g.* separation by water, especially a wide expanse of ocean), economic status, provision of natural resources, political stability, the personal ambitions of autocratic leaders, and many more. Their amenability to change in outlook will also be affected by institutional structures which reduce or enhance the inevitable conceptual lag in transmission of new ideas.

Everywhere, problems may be due to the separation which often exists in both time and space between causes and effects of environmental and social changes. Educators in each and every country have a duty to assess the impediments to understanding in their particular circumstances, and then to adapt education over a broad front – by appropriate methods such as including a major and practical environmental component in school curricula, and in extension courses to suit different identified target audiences.*

What of the people and agencies who might be expected to help to guide education towards global perspectives? They, too, seem to be suffering from the effects of the stressed society. In an address to the British Association for the Advancement of Science, Colin Blakemore (1990) referred to the current attitudes to science – the public benefiting from scientific achievements yet suspicious of motives, the media largely interested in sensations, and the politicians supporting only programmes with a quick financial return or political advantage. Scientists all too often turn instead to the more sympathetic activities with which they are familiar, and at which they can work undisturbed in their own specialized laboratories. There are some notable exceptions, of course; but the task of outward communication, of interpreting scientific findings on environmental issues, is not well enough done to give public education the support – or even the materials – which it needs.

The holistic approach to understanding The Biosphere, which is becoming more and more widely commended by environmental educators and which might be expected to underpin a global perspective, is treated by many scientists with suspicion and even distaste. The idea of holism was of course originally contrasted with reductionism – the view that a system can be understood in terms of its parts – on which science has long depended far too widely and trustingly. But it is now plain that we need both, and that The Biosphere requires integrated study as much as (for example) the cell. What is necessary to make a holistic approach to The Biosphere less suspect and more rewarding for scientists?

The conservation bodies should also be giving guidance. While generally stressing the need for education, however, they tend to hope that other people will deal with it, and do not exert on curriculum developers the pressure needed to influence the course of formal education, *inter alia* in competition with many other pressure-groups (which are often better organized and, especially, funded). The relatively small number of people who attempt to re-design education to meet the challenges of a stressed society are often faced with a task akin to the labours of Hercules.

Need of Funding and Media Support

The magnitude of the educational task, constantly belittled by the media with their emphasis on problems and crises and hence on the immediate, is

*[Collation and dissemination throughout the world of basic facts and desirable trends in environmental education being one of the main reasons for a group of us founding the International Society for Environmental Education (ISEE) alongside the World Council For The Biosphere some years ago, it is sad that, so far, ISEE has not been nearly as active, or become as widely influential, as it should be for the world's good. Eds.]

one of the most potent influences in discouraging educators, both formal and informal, from giving their support to the necessary educational reforms. How can we break into this lamentable situation and persuade people that it is not futile to embark on education concerning and for the environment? On the contrary, it is one of the main imperatives of this age, and in the view of some the most fundamental of all.

An impediment to any educational process will be inadequacy in the quality of information on which it is based. People have to understand and accept the uncertainty principle in science, and make allowance for it in their judgments. They must also appreciate that the transmission of information involves passage through two sets of filters – selection and interpretation by the transmitter, and selection and interpretation by the receiver – in the course of which the message may be changed either unwittingly or deliberately. The less familiar the information is, the more widely may what is received diverge from the original sense. Training people to think, to be sceptical of received information, and to interpret in a balanced manner, may be an aspect of education that is still under-developed (*see* Blakemore, 1990).

There is also still uncertainty about the value systems against which changes in the global system may be judged. More help is needed from ecologists and other environmentalists to establish what the desirable qualities are of a Man-environment system in a state of relative health – presumably a pro-active view of our objectives for the future rather than a reactive one to mismanagement in the past.

Overcoming the barriers

Much is now being done internationally to promote global environmental education, although never enough (*cf.* Stapp, 1985). There have been several important conferences under the auspices of the International Environmental Education Programme (IEEP), jointly undertaken by UNESCO and UN Environment Programme (UNEP), and including an International Congress on Environmental Education and Training in Moscow in 1987 from which an International Strategy was produced (IEEP, 1988). This programme has also been responsible for regional and national meetings and workshops, and for many valuable publications. Important conferences have been organized by the International Council of Scientific Unions (ICSU, *see* Lewis & Kelly, 1987) and by the World Conservation Union (IUCN). Notable among many other events is the present series of International Conferences on Environmental Future (Polunin, 1972, 1980; Polunin & Burnett, 1990), and of course there have been many others at varying levels on more specialized aspects of environmental concern.

These, together with a gathering flood of books, television programmes, and efforts of popular journalism of varying accuracy, have successfully aroused the public in industrialized countries to a sense of crisis reflected, among other things, in the appearance of Green political parties. In many less-developed countries the problems are more starkly perceived in terms of famine, salinization of soils, desertification, and flooding, as well as in disasters due to inadequately-designed or -managed industries.

These programmes and perceptions are making invaluable contributions in paving the way to environmental education, and leading ultimately to significant changes in the quality of the world environment. Our earlier comments on the continuing impediments to global environmental education indicate where more changes are needed, especially in emphasizing the common needs of Mankind – physical, social, mental, and spiritual – however differently they may be expressed.

There remain gaps to be bridged, however, in scaling these ideas to levels that are tangible to the great mass of world citizens. The themes and gaps are apt to be on such a grand scale as to appear inaccessible or irrelevant to the people in the fields or the streets: only by linking global issues to people's own daily lives, and to the things which they consider directly important, can we hope to make our education really effective.

Dr Holdgate has warned us already at this Conference of the incidence of 'thinking green but acting dirty'. Here people sometimes fail simply because they are overwhelmed by the size of the global issues and crushed by the thought of their own inadequacy to influence them. Perhaps we can learn a lesson from industry. Using approaches that are familiar in business and industrial training, a school programme called Education for Achievement has recently been having some success in the UK (Bates, 1989). It is positive rather than negative in its approach to its chosen topics, and adopts a number of stages which groups and individuals are taught to go through, as follows:

(i) Select an objective (an achievable one) and decide on the criteria for success: the question is not 'What shall we do?' but 'What shall we make happen?'

(ii) Generate different ways of pursuing four chosen objectives and select the most promising one of them;

(iii) Put the chosen plan into operation against a time-table and control the process; and

(iv) Continually review operations and results.

This programme has been successful in schools which have tried it out and, clearly, is readily adaptable to environmental management projects. A very similar programme is now being developed by John Baines in the UK (pers. comm.) as a personal, individual activity for improved environmental management which he uses in training courses for teachers and others. The procedure is to adopt a 'mission statement' for a limited objective of environmental improvement or management, break it up into achievable sections, and then proceed in much the same way as already described. Baines (pers. comm.) finds the results very encouraging, the sense of actually achieving something being a great motivator for further effort.

For those in industrialized countries, the 'green' literature is now full of suggestions for specific contributions to environmental improvement which are both real and realizable. They cover diet, energy conservation, transport, water conservation, waste disposal, recycling, recreation, and many more aspects. They call for answers to searching questions – for example, 'What do I really need?' rather than 'What do I want?'. Individual action may not seem like a great contribution to a global problem, but even small achievements encourage, and the more people who can be cajoled into joining in, the greater the effect will be. Nor must we forget the educative influence that children can have on their parents and families. As Chancellor Hare has already reminded us here, these individuals are collectively (or will become) the voters, tax payers, and customers, whose influence ultimately determines what will work.

In many of the less-developed countries, resource materials of a similar kind are also becoming available, not only through schools but through adult literacy programmes. Here also other channels of communication may be important – street theatre, story-telling, song and dance – with similar capacities for presenting small, achievable packets of relevant action.

Acting locally still has to be connected to thinking globally, but this is easier where there are specific issues to connect. Food-choices, for example, have implications of universal concern: fresh produce saves processing and packaging, locally-grown produce saves transport, vegetarian sources of protein save energy expenditure on raising and marketing meat, and so on. The greatest advantage of these approaches is perhaps that they are positive, based on a concept of environmental health that is continuous with one's own health, or recognizing environmental 'problems' such as sicknesses or injuries needing to be treated. Just as we now recognize primary health-care as a critical approach to improving the human condition, so also should we recognise primary environmental care.

In this context especially we must also abandon any idea that education is something to be left entirely to the educators. Much learning takes place from a study of 'role models' who are often unconscious of their influence. The environment which we create for young people to grow up in is taken as evidence of the environment in which we really believe, and sets the standards whatever we may say about them. The creation of a sustained learning experience for people, from which a healthy behaviour towards The Biosphere will grow, is a task for everyone, in which none of the influences which determine education can be ignored.

Conclusion

We no longer have the option of foregoing a global perspective. Educators in schools and out of schools must prepare a world citizenry that can understand, live with, and act responsibly upon, the critical truth that we have only one world and it is indivisible. The 'haves' must learn that there are 'have-nots', and realize that if these exhibit resentment and bitterness it may well be justified.

Where radical changes are needed it is better that they be guided so as to avoid the most damaging effects of stress. We must all learn that world security thus depends literally on building respect for all nations and cultures – a process that starts at home and in our own communities. We must also learn that people all over the world have rights to reasonable complements of acceptance, peace, and human dignity, and that the self-respect which this implies is a necessary foundation on which to build respect for others.

Had the importance of an evolving pattern of education been more fully realized, and educators (of all sorts) been better versed, the continuing escalation of environmental threats might have been prepared for, or even prevented. There is still time for educators, by precept and example, to help Humankind to survive as an integral part of a healthy Biosphere.

References

Bailey, S. K. (1975). Personal address to the US National Academy for Student Affairs by Dr Stephen K. Bailey, Vice-President and Director, International Education Project, American Council of Education, delivered in Washington, DC, as cited in Stapp & Polunin, 1991.

Bates, E. (1989). Education for achievement. *Roy. Soc. Arts J.*, CXXXVII (5398) (September 1989), pp. 611–7.

Blakemore, C. (1990). Who cares about science? *Sci. Publ. Affairs*, pp. 97–119.

IEEP (1988). *International Srategy for Action in the Field of Environmental Education and Training for the 1990s.* UNESCO–UNEP, Paris & Nairobi: 21 pp.

Lewis, J. L. & Kelly, P. J. (1987). *Science and Technology Education and Future Human Needs.* Pergamon Press, Oxford, England, UK : xii + 185 pp.

Polunin, N. (Ed.) (1972). *The Environmental Future: Proceedings of the first International Conference on Environmental Future,* held in Finland from 27 June to 3 July 1971. Macmillan, London & Basingstoke, England, UK, and Barnes & Noble, New York, NY, USA: xiv + 660 pp., illustr.

Polunin, N. (Ed.) (1980). *Growth Without Ecodisasters? Proceedings of the Second International Conference on Environmental Future (2nd ICEF)* held in Reykjavik, Iceland, 5–11 June 1977. Macmillan, London & Basingstoke, England, UK, and Halsted Press Division of John Wiley & Sons, New York, NY, USA: xxvi + 675 pp , illustr.

Polunin, N. & Burnett, J. H. (Eds.) (1990). *Maintenance of The Biosphere: Proceedings of the Third International Conference on Environmental Future (3rd ICEF).* Edinburgh University Press, Edinburgh EH8 9LF, Scotland, UK: xvi + 228 pp., illustr.

Smyth, J. C. (1983). Education for the antediluvians. *Inst. Env. Sci. Proc.*, iv (i), pp. 1–14.

Smyth, J. C. (1988). What makes education environmental? Pp. 33–56 in *New Ideas in Environmental Education* (Eds. S. Briceño & D. C. Pitt). Croom Helm, London, England, UK: xiv + 219 pp.

Stapp, W. B. (1985). Guest Editorial: Some overall imperatives of the Environmental Education Movement. *Environmental Conservation*, 12(2), pp. 103–4.

Stapp, W. B. & Polunin, N. (1991). Global environmental education: towards a way of thinking and acting. *Environmental Conservation*, 18 (1), pp. 13–18.

12. The Management of Environmental Information

Robin A. Pellew
Director, World Conservation Monitoring Centre, 219c Huntingdon Road, Cambridge CB3 0DL, England, UK

Introduction

We live in a world of rapid change and uncertainty. The inexorable growth in human populations together with the rising expectations and aspirations of people, are placing steadily-increasing pressures upon natural resources to support human development. As a result, species are being driven to extinction at an accelerating speed, whilst hanging over the planet is the spectre of climate change caused by the legacy of past indifference to the global ecological systems upon which Mankind depends.

In these circumstances, the one solid basis upon which we must build our collective response is information. Information is derived from the analysis of data: blended with experience, information is also the source of knowledge, which is a precious commodity especially in times of uncertainty. The practice, as opposed to the concept, of sustainable development – a term that is already becoming widely abused – necessitates the quantification of both the productive capacity of biological systems and the demand for the resources that such systems can provide. For these and innumerable other reasons the demand for reliable, up-to-date information has never been greater than it is today.

Despite our life in a world that is rapidly becoming submerged in data, we seem to have rather little information on which to base enlightened decisions. The new geostationary satellites that form part of the International Geosphere-Biosphere Project will generate such vast quantities of raw data that a new set of words has entered the English language to take care of it. We now say that we live in the age of the *terrabyte* which feeds upon the *gigabyte* that preys upon the humble *megabyte*! What we lack is information presented in a comprehensible and usable form that can be applied directly for improved, ever-more-enlightened management of our resources.

Scenarios of Environmental Change Needed

So what should be done about it? Politicians are now pressing the scientific community to come up with realistic scenarios of the degree and scale of future environmental change, which in turn generates an increasing dependence upon global models. Yet the rates and linkages inherent in such models are so uncertain that even relatively small changes in the input values can have profound effects upon the output predictions. This has been seen recently in the uncertainties of the role of marine phytoplankton as both a source and a sink of atmospheric gases resulting in widely divergent scenarios of possible climate change.

Global models necessitate a network of stations for global monitoring to provide the input data. A network of World Data Centres has been established to monitor geophysical, solar, oceanographic, and glaciological, processes (Allen, 1988). Yet, curiously, comparable centres to record *terrestrial* change have not been established, although there is interest in using some of the existing Biosphere Reserves to form a network of monitoring stations. But what parameters should be recorded if we are to monitor effectively the state of the planet's health? This in turn begs the question of what types of outputs should the models be providing, particularly to support political decision-making?

Information management in the environment sector must be user-led; first define your problem, next determine the outputs needed to elucidate it, and then identify the data requirements to provide these outputs. The data collection and data management system must be the servant, not the master, of the end-user. But the advances in technology are outpacing the science policy analysis. The prodigious quantities of data that we are now producing, and the sophisticated computerized mechanisms for their storage, cataloguing, and assessing, have overtaken the rationale for their collection. We risk accumulating data for which we have identified, and seem likely to identify, no purpose or end-user whatsoever. There is an urgent need for the policy analysts and modellers to articulate their data needs more selectively than at present – particularly in the living-substance component of the IGBP. The days of 'I'll-have-a-little-bit-of-everything' are surely over.

Heighten Research Status of Monitoring

Why is monitoring given such a low priority in the field of scientific research? Why do many scientists regard it in such a disparaging light? I suspect that monitoring is not regarded as a credible research activity because it is not intellectually challenging: it does not lend itself to experimentation and hypothesis-testing, and it is not the stuff of PhDs. Yet long-term monitoring and assessment is fundamental to provide the reliable model-predictions for which decision-makers are now shouting. The logical conclusion must be that science policy, at least in the environmental monitoring sector, is too important to leave to self-indulgent scientists – we need an independent review body with a mandate for science policy analysis to identify what knowledge we need, and thus what data we require to generate this knowledge. The logical institution to provide this body, which must have an intergovernmental role, must be the United Nations and its agencies.

Responsibility for monitoring the environmental status of the planet lies with the Global Environment Monitoring System (GEMS), which is the core of UNEP's (The United Nations Environment Programme's) Earthwatch Programme. Despite its very limited resources in relation to the size of the task, by and large GEMS works pretty effectively at data-collection, using its small coordinating secretariat in Nairobi to catalyse other national and international monitoring initiatives. The key to its effectiveness lies in the development of information networks. There are currently some 25 GEMS global monitoring and assessment networks, which fall into five interlinked groups (Gwynne, 1988). Thus there are:

1. Networks that monitor environmental pollutants – particularly urban-air quality, water quality, and food contamination by pesticides, heavy-metals, and other toxins;
2. Networks that monitor the long-range transport of pollutants and their effects, particularly acidic deposition. This includes EMEP (the European Monitoring and Evaluating Programme) but is now expanding its coverage to include less-developed countries which are beginning to realize that rapid industrialization has an environmental price to pay, including acidic deposition;
3. Networks that monitor marine pollutants and living marine resources – particularly marine mammals – which have focused on regional seas and coastal areas through the development of Regional Seas Protocols coordinated by the UNEP Oceans and Coastal Areas Programme Activity Centre;
4. Networks that monitor climate and atmosphere, including the WHO (World Health Organization) Background Air Pollution Monitoring Network, Climate System Modelling, the World Glacier Inventory, the World Climate Impact programme, and the WHO/UNEP Intergovernmental Panel on Climate Change; and
5. Networks that monitor renewable natural resources, ranging from the monitoring of forests and other types of land cover, soil degradation, and desertification, to the loss of habitats and the status of endangered species.

Processing of Incoming Data

The above is an impressive list of networks, but how are the incoming data to be processed to provide usable information? Realizing that far more efficient use could be made of the incoming data than is currently practised, some five years ago GEMS established a Geographic Information System (GIS) data management system called GRID – the Global Resource Information Database. Its objective was to integrate the various data-sets to address specific environmental problems, thus providing a practical assessment or management tool when focused on a specific area or resource (UNEP, 1985).

GRID has now carried out a number of pilot projects, including land-use analyses of such countries as Uganda, an assessment of the factors influencing the status and distribution of African Elephant (*Loxodonta africana*) populations, and analysis of the effects of climate warming and sea-level rise upon river-delta countries such as Egypt and Bangladesh. These have shown the power and application of GIS techniques for environmental management, but, because of its restricted application and the substantial appetite for cash resources by GRID, the bottleneck of the under-use of the GEMS data remains. What is now needed is to download the analysis function to other organizations in the same way that the data capture has been downloaded through networking.

Aware of this bottleneck, GEMS is responding by setting up a network of GRID nodes for regional data management – Nairobi, Geneva, and Bangkok, have already been established, and further nodes are planned for West Africa, Latin America, and the South Pacific (Gwynne, in press). Such networking of the analysis capability is very encouraging, but more could be done through collaboration with existing governmental and nongovernmental agencies – particularly universities – rather than setting up expensive new centres.

There are an increasing number of institutions in both the developed and developing worlds which are using GIS for environmental planning, and which could link into a broad-based GRID network for the regional or sectoral analysis of GEMS data. The first national GRID node, which was inaugurated in 1989 in Norway (GRID-Arendal), is an example of this sort of collaboration: the network of national monitoring and assessment centres that is currently being set up in Africa is another example, while a third is the link that GRID has for biodiversity data with my own organization, the World Conservation Monitoring Centre (WCMC) (Pellew, 1990).

The World Conservation Monitoring Centre

WCMC provides a model, within the conservation sector, of the use of networks for both the collection of data and the dissemination of information. It is a joint venture between the three partners in the World Conservation Strategy: IUCN (The World Conservation Union), UNEP (United Nations Environment Programme), and WWF (World Wide Fund for Nature). Its mission is to support conservation and sustainable development through the provision of quantitative information on the status and distribution of the world's biological diversity (Pellew & Harrison, 1988). Its activities involve the collection, analysis, and distribution, of biodiversity data with three main themes:

- Species data, including plant and animal species of conservation concern and of potential value for sustainable development: the database comprises some 60,000 plant and 25,000 animal species, including their conservation status and distribution data.
- Areas-related data – particularly habitats, critical sites for the conservation of biological diversity, and protected areas: the database includes some 17,000 national parks and protected areas down to the level of local

forest reserves; extensive data files on tropical forests, wetlands, and coral reefs; also digitized data files on tropical forest distributions for GIS analysis.

- Trade-related data covering the utilization of threatened species and their derivative products: the database has been compiled mainly in support of CITES (Convention on International Trade in Endangered Species of Wild Fauna and Flora) and includes some 1.5 million trade transactions with a particular focus on elephant ivory: a more recent activity has been the monitoring of the trade in tropical hardwoods and its impact upon the *in situ* conservation of tropical forests in support of ITTO (International Tropical Timber Organization).

The above are substantial data-holdings, yet WCMC is currently doing little more than scratching the surface of the problem as a whole. To monitor the status and change in the world's biological diversity is a massive task. Recent estimates, based on rates of tropical-forest loss (Wilcox, 1988), suggest that some 15–30% of all species may be driven to extinction within the next three to five decades, representing several hundred vertebrates, many thousands of plant species, and more than a million species of insects (*cf.* M.W. Holdgate, this volume).

Distributed Information Networks

Faced with the wellnigh impossible task of monitoring the accelerating rate of species and habitat loss with the relatively limited resources at its disposal, WCMC recognizes the fundamental necessity to develop networks of distributed information to facilitate the capture, organization, and distribution, of data. What is now needed is a global conservation network for the free flow of information, with WCMC serving as the central repository of data, which must be catalogued and indexed for remote access through telecommunication links.

For this data management system to succeed, four fundamental requirements must be satisfied:

1. The data must be regarded as a public resource, available for free distribution through the network. Networking for data capture will work only if the two-ways flow of free data, providing reciprocal benefits, can be ensured. An agency will willingly contribute data *only if* through access to the central information repository it *gets back more* than it puts in: equally, the central repository is expanded through each new data accession that it receives. This type of reciprocal exchange is morally acceptable only if both parties agree that the data must be placed in the public domain. The tendency of some research institutions and scientists to regard their data as financial assets, or to tie their use up in complex copyright agreements, I regard with considerable suspicion;
2. Data transfer necessitates the development and acceptance of standard nomenclatures, terminologies, and classification systems, if dissipation of crippling amounts of staff time is to be avoided in re-formatting incoming data. WCMC is actively involved in the preparation of such standards,

particularly for the classification of habitats. In collaboration with IUCN, it has developed a classification system for protected areas throughout the world (IUCN, 1984), while the classification of threats to species is currently being reviewed (Mace & Lande, in press). Such standards must be incorporated in transfer formats to facilitate the electronic incorporation of incoming data-sets with minimum staff involvement;

3. The database must be structured to facilitate remote access and data availability. Each new accession must be source-stamped, date-stamped, and allocated a reliability rating based upon the method of its origination – ranging from merely anecdotal to derived from stratified sample-based survey. The data holdings must be catalogued and indexed to simplify remote access. Data entries should, whenever possible, be supported by key references, and the supporting bibliography should also be on-line; and
4. Data verification is transferred to an output, rather than an input, function. Some superficial checking of new accessions will be made, but raw data outputs incorporating multiple sources are likely to incorporate data conflicts. The database management system should highlight such inconsistencies, but their resolution is an end-user responsibility. If required, WCMC can undertake such verification, analysis, or assessment; but this service, which involves the investment of added-value to the data, will incur a charge.

Conclusion

The technology and expertise to develop such networks already exist: what we continue to lack are the resources and political commitment for their establishment, full operation, and effective use. Yet unless such information management systems are set up, and not just in the conservation sector but across the entire environmental spectrum, short-sighted decisions will continue to be made because of the lack of adequate information.

UNEP, building upon an initiative begun by IUCN, is currently preparing a new legal instrument for the conservation of biological diversity. The drafting of this international convention is likely to have a 'bumpy ride', not least because of the problems of the exploitation by the developed world of the genetic resources of the less-developed world, and the consequent need for innovative fund-raising mechanisms to make good many glaring lacunae. But even if a protocol can be agreed upon, it will be no more than a 'paper tiger' unless the mechanisms are developed to provide the supporting information without which the convention will not work.

We live in a world not only of change and uncertainty but also of technological advancement and information skills. The opportunities which these latter provide must be effectively mobilized if we are to increase the knowledge-base for the duly-enlightened management of the global environment.

References

Allen, J. H. (1988). The World Data Centre System, international data exchange and global change. Pp. 138–52 in *Building Databases for Global Science* (Eds H. Mounsey & R. Tomlinson).Taylor and Francis, London, England, UK: xv + 419 pp., illustr.

Gwynne, M. D. (1988). The Global Environment Monitoring System (GEMS): some recent developments. *Environmental Monitoring and Assessment*, 11(3), pp. 219–23.

Gwynne, M. D. (in press). Global monitoring, data management and assessment within GEMS and GRID. In *Global Natural Resource Monitoring and Assessment: Preparing for the 21st Century* (Eds. H. G. Lund & D. F. Hemenway). American Society for Photogrammetry and Remote Sensing, Falls Church, Virginia, USA.

IUCN (1984). Categories, objectives and criteria for protected areas. Pp. 47–53 in *National Parks, Conservation and Development: The Role of Protected Areas in Sustaining Society* (Eds. J. A. McNeely & K. R. Miller). Smithsonian Institution Press, Washington, DC, USA: xiii + 825 pp., figs & tables.

Mace, G. M. & Lande, R. (in press). Assessing extinction rates: towards a re-evaluation of IUCN threatened species categories. *Conservation Biology.*

Pellew, R. A. (1990). The World Conservation Monitoring Centre (WCMC): What it is and what it does. *Environmental Conservation*, 17 (2), pp. 179–80.

Pellew, R. A. & Harrison, J. D. (1988). A global database on the status of biological diversity. Pp. 330–7 in *Building Databases for Global Science* (Eds. H. Mounsey & R. Tomlinson) Taylor and Francis, London, England, UK: xv + 419 pp., illustr.

UNEP (1985). *Global Resource Information Database* (GRID). Global Environment Monitoring System, United Nations Environment Programme, Nairobi, Kenya: 16 pp.

Wilcox, B. A. (1988). Tropical deforestation and extinction. Pp. v–x in *IUCN Red List of Threatened Animals.* International Union for Conservation of Nature and Natural Resources, Cambridge, and Gland, Switzerland: 172 pp.

Commentary on Chapters 11 and 12

CHAIRMAN: Ambassador Harun Ur Rashid
PANELLISTS AND OTHER CONTRIBUTORS:

Daoudy, Sacks, Furedy, Thorndike, Sacks, Pellew, Craig Davis, Tisdell, Craig Davis, Kefeli, Purcell, J. Petts, Ramakrishnan, Wasawo, Mische, C. Westing

Daoudy spoke of the lack of uniform approach amongst UN agencies both in the provision of information, and towards education. This was a serious matter for UNEP with its small budget of $US50 millions, only 5 offices world-wide, and the task of informing the world about environmental matters. He would certainly be reporting on this lack of coordination to the next General Assembly of UN in the hope that matters could be improved: this should help UNEP, whose work he greatly admired.

Sacks wished to focus on a selected set of 'good news' and 'bad news' items concerning the state of environmental information and education. He began with some points of 'good news'.

First, the efforts of the environmental community, most recently evidenced through activities associated with the twentieth anniversary of Earth Day, had been remarkably successful in raising the general state of environmental awareness among the public at large. Although we had a very long way to go indeed to achieve any serious scientific and environmental sophistication and balance in the general public, nevertheless, through the combination of formal and non-formal educational modes, as well as through the print and electronic media, we had done much to sensitize the public to the environmental *problematique*. We had also done much to help individuals to recognize their responsibilities to act in environmentally sound ways. In great measure, the public was supportive of environmental controls and willing to be taxed – within limits – to support those controls. Good news indeed!

Secondly, a world-wide effort had been initiated both to introduce environmental subjects and environmental thinking into school systems at all levels, and to insure that environmental education was a lifelong process.

Thirdly, the future of environmental education was brighter today than it had been since the first explosion of public and professional concern in the late '60s and early '70s. Within the next five years or so, we could reasonably expect that there would be better communication among environmental education NGOs and more action in the environmental education and information arena. There would be more resources poured into the development and delivery of environmental information to multiple audiences. We had a reasonable expectation that greater efforts would be made to 'internationalize' and 'globalize' the curriculum in the public schools and in higher educational institutions as the notions of environmental, economic, and social, interdependence and global environmental security took firmer hold.

The place of environmental education would become increasingly secure as both governments and non-governmental organizations began to recognize environmental education as one of the key, critical steps in altering human perceptions and changing human behaviour.

He then turned to the 'bad news'. Firstly, society 'writ large', and environmental educators in particular, had a very difficult time, both in keeping abreast of the vast volumes of data that society collected and stored, and in translating these data into useful information to assist society in selecting sound public policies, or to assist individuals in making sound personal environmental decisions. As a Wisconsin colleague of his had described it, the flow of data, of information if you will, was so great that it was like trying to take a drink from a fire hose. We were overpowered by the sheer quantity of data and information.

This veritable flood of data had made it especially difficult for environmental educators to keep up-to-date in their teaching. When educators attempted to keep up-to-date, this often meant that they paid attention to specific features of the environment – such as the status of a particular species – rather than to ecosystem and biospheral processes. Such a lack of holistic, integrative thinking worked to the detriment of the education provided.

The second piece of 'bad news' related to higher education. We had seen little progress, if any, in cracking the continuing barriers to integrating knowledge across disciplines, or in leaping the high walls that existed among departments in universities. Although more lip-service was paid today to integration and interdisciplinarity than had been paid since 1970, we, the educational and scientific communities, had done a pitifully poor job over the past thirty years in preparing enlightened students, who now occupy important positions as members of faculty, departmental chairmen, deans, provosts, and even chancellors.

If anything, the narrowness of vision, and resistance to interdisciplinarity, had increased over the past quarter-century, not diminished. As those interdisciplinary programmes which began in universities in the late '60s had matured and grown, they still remained on the fringes of academic acceptability; they had not been drawn into the mainstream of higher educational institutions, and they continued to be regarded with hostility, as mechanisms that threatened to undermine scientific rigour; they continued to be seen as working competitively to undermine the primacy of the disciplines, parasitically draining essential resources from the host.

At least some of these obstacles to interdisciplinarity had their origin, he believed, in feelings similar to the prejudice and bigotry that affected race relations. The interdisciplinarian was perceived as a 'slacker,' an 'interloper,' if not an 'invader,' someone whose motivations were suspect and whose mind was just not as quick as that of the disciplinarian. From the disciplinarian's perspective, the interdisciplinarian simply did not share 'our' language, 'our' paradigms, indeed, 'our' values.

This attitude, expressed more genteelly or at times less so, continued to pervade higher education. This attitude would damn our science, and it would certainly continue to damn our ability as a society, as a culture, in its efforts to respond meaningfully to the environmental problems which were overtaking us.

All this was not only bad news, it was old news, and it was old news made worse precisely because it was so old. Largely, we in higher education had failed miserably to overcome such barriers and attitudes, and he saw little to suggest that we would succeed in doing so in the future, at least in the near term.

Lastly, there was one final piece of 'bad news.' As a community, educators, and higher educators especially, had also failed miserably to instil an environmental, or conservation, ethic. We were able to talk about environmental problems; we were able to define their parameters; we were even able to devise technical solutions to many simple and even some complex environmental problems; but, we had been largely unable to crack the dominant themes and values of our society and our world which underpinned environmental degradation. Perhaps the problem was that the formal educational enterprise was simply ill-suited to venturing into areas as nebulous and unsteady as the formulation of ethics and values, at least ethics and values that fell outside the professions themselves, about which educators did feel it was their responsibility to educate students. Or perhaps it was a matter of not being able to educate our young because we were not able to educate ourselves, their teachers.

In any case, he saw little hope for major behavioural change unless, somewhere in our society, we were able to educate our young successfully towards an environmental ethic. As environmental, economic, and social conditions worsened, and as population continued to increase exponentially, the prospect in this area dimmed: it did not brighten. Again, this was old news.

Therefore, he asked, do we give up? No! At this point he joined the famous anarchist, Peter Kropotkin, who rejected his princely standing in Czarist Russia after witnessing countless abuses of absolute power. In his *Memoirs of a Revolutionist*, Kropotkin had said, 'It is hope, not despair, which makes successful revolutions.'

In his humble opinion, what we needed for a truly successful future for environmental education, for our only Biosphere, was a revolution in our educational system, in our own thinking about what constituted that system and about what *really* mattered. Unless we were prepared to take some bold and forceful steps we would not be able to fulfil our responsibilities as educators, or as global citizens, and we would continue to replicate in our students the faults of their teachers.

Furedy spoke about some of the needs in developing countries for environmental education. She said that *informal* education would be called upon to meet urgent needs for environmental education in developing countries for several decades at least because most of the people whose actions would affect local environments would not stay in schools long enough (if they entered at all) to acquire formal environmental education, and voluntary agencies, eager to address environmental issues, would work mainly outside formal institutions. The informal educators would often be ahead of their trained colleagues in environmental awareness (as **Sacks** had pointed out), but their resources would be slim. Thus resources had to be used creatively to achieve 'sound communication' (**Holdgate**), basic information-sharing (**Pellew**) and 'sustained learning experiences' (**Smyth**).

Environmental educators with few resources had to select strategies for optimal impact, evaluate and adapt learning projects, and balance advocacy with open-minded enquiry.

It might be time to rethink the great emphasis given to children in environmental education. When one asked why almost all the environmental education efforts of voluntary organizations were designed for children, the first answer one received was that children were the future decision-makers. If one then asked whether society could afford to wait ten or fifteen years for environmental change to be mediated by that generation, one was told that parental values and behaviour were changed by children carrying ideas into the homes. Although educational campaigns were rarely evaluated, there was some evidence for such intergenerational influences. Some case-studies in Japan suggested that children could 'shame' adults into action; and education on recycling in Ontario schools was said to be affecting the willingness of households to cooperate with source-separation of household wastes. But, in general, children's environmental campaigns were not built upon an understanding of whether, and to what extent, children influenced adult behaviour; they followed this course from convenience and because it was a pleasure to work with children.

At least some of the resources currently given to children's education in developing countries might more usefully have been allocated to environmental awareness for women. Women taught basic hygiene and housekeeping to children, besides making important decisions in agriculture and domestic resource-management. Thus they had the potential for substantial impact on local environments. That was her first suggestion.

Her second suggestion for effective use of scarce resources was that environmental education in developing countries should make more effort to tap local sources of knowledge about ecosystems. In this way we could preserve what was sound in traditional knowledge systems, as illustrated by Burnett (p. 85) and Ramakrishnan (pp. 209 *et seq.*), while developing practical ideas about how humans could fit their basic needs and life-styles into ecologically sustainable practices.

An example of how education and knowledge documentation could be brought together was seen in the work of the Mazingira Institute in Nairobi (Figure 12.1).

Figure 12.1. The Mazingira Institute of Nairobi, Kenya, uses its children's competitions to engage young people in environmental issues and also to gather information about environmental awareness in Africa.

Figure 12.2. The logo of the 'Magic Eyes' anti-litter campaign of television, Thai Community Development Association, Bangkok. Through television cartoons, school and neighbourhood drives, and booklets and fliers, this campaign has become the best-known public message in Thailand.

Their environmental competitions designed for children and youth had a research purpose as well: the entrants supplied information that was used to start information bases which were later developed by more systematic research. For instance, competitions had asked entrants to document, illustrate and comment on local flora and fauna, and on practices in home gardening. Such an approach could be the way in which to build up the 'ecological histories' that Craig Davis thought to be important.

Besides empirical observation, environmental values were a component of these important knowledge systems. This dimension was often neglected by educational and improvement projects. For instance, a great deal of effort and money had gone into installing water, sanitation and waste-disposal systems in developing countries; yet there was not one good study, not even a case-study, on people's concepts of wastes! Engineers and public health experts saw wastes as hazards; poor people saw wastes as resources: the perspective of the poor was not considered worth investigating, although it confounded many an infrastructure, or hygiene education project.

Holistic approaches to environmental work were, as **Vallentyne** had urged, crucial. Furthermore, promoters of environmental projects needed to balance enthusiasm (which was the mainspring for informal environmental education) with critical thinking. A cautionary tale came from the 'Magic Eyes' campaign of Thailand (*see* Figure 12.2). Magic Eyes was started as an anti-litter campaign by a community development association in Bangkok. The concept was based on the omniscient eyes of spirits in a Thai folk-tale. People were urged, in jingles and television cartoons, not to litter because the magic eyes would see them. The Thai Community Development

Association had been very successful in getting corporate support for their cause, which was promoted through a logo (Figure 12.2) on products, T-shirts, buttons, posters, and so on. The theme was currently considered the most popular public-service message in Thailand, and the organization had received a UNEP Global 500 citation.

But a broader, 'ecosystem' perspective on Thailand's urban waste problems would have been unlikely to produce an anti-litter campaign, for such campaigns did *not* address the *causes* of waste problems, nor did they raise awareness of the need to reduce wastes and to recycle. They focused on 'out-of-sight, out-of-mind' values. The campaign's promotion of plastic bags for the disposal of refuse had undermined the city's attempts to make compost for farming. The campaign organizers had failed to ask: 'What are the causes of waste problems in our cities?'

In summary, she said that informal education would be the main vehicle for influencing the everyday environmental behaviour of poor people in developing countries.

Efforts should reach beyond children, to adults, preferably to women if a target group had to be chosen. The educational effort should respect people's empirical understanding of their local ecosystems and enhance our knowledge of them. Without a commitment to holistic thinking, without critical thinking and evaluation of education, enthusiastic volunteers could not make the best use of scarce resources.

Thorndike wanted to clarify two issues. Firstly, she asked how best could universal environmental education – the creation of environmental awareness in every human being as **Tolba** had described it – be achieved? Would it be through computerized databases and the other opportunities which modern technology now offered, or would it continue to be necessary to provide experiential situations – discussions on waste education, 'nature-centre' experiences especially for children, or adult training in the field?

Secondly, what kind of mechanism should be promoted to deal with the problem raised by **Pellew** of too many data and too little information? Two problems were involved in this. One solution might be to have somebody, possibly interdisciplinary, that would sift the data to decide what was useful and what was not. But there would still be the necessity to ensure that decision-makers received information in a form both useful to, and usable for, making policy decisions.

Sacks responded that universal environmental education, like universal education, should be delivered through the various existing educational systems. He did not think that anyone present would consider computerized information systems to be the primary vehicle for transmitting environmental education to the general public, or for making them more aware of, and more involved in, environmental matters. Such methods were expensive but could play a part in a targeted, decision-making context. He saw no other effective route than existing public education, non-formal education, through zoos for example, and the popular media.

Pellew also responded to **Thorndike**. In his opinion, databases would be a rather blunt instrument with which to try to engender an environmental ethic in the public. While they could supply supporting material, they were no substitute for the 'experiential' element. On the matter of the infinitude of data, he said that means had to be devised, with only limited resources available, to determine, in a coherent and rational way, which were the essential parameters, the rates and determinants that should be monitored to enable decisions to be made in an enlightened manner. He believed that, before data collection began, it would be necessary to carry out an analysis so that it was known precisely what information was to be collected, to ensure that the end-user became the driving force rather than the nature of the technology involved, which was the present situation. This implied that the users had to be known before the process began. He was not a supporter of the notion of generating databases on the off-chance that they might prove useful. He believed the use and user should be defined first, and that would predetermine how data were collected and processed.

Craig Davis returned to the first of **Thorndike**'s questions. He agreed with the views of **Sacks** but added two further points. Firstly, education had to be local, by which he meant, for example, that in an Indian village it should be done by Indians, preferably by those conversant with the local culture, but using all the skills and techniques available. Environmental education, in this sense, could be global although expounded locally.

Secondly, and here he spoke as a US citizen, existing formal education had taught him a great deal about US history in the context of western civilization – his heritage – but virtually nothing about the universal biological heritage. People were regarded as civilized people, not as 'ecological' people; he thought that criminal. In his view, nobody should graduate from any college in the USA, or elsewhere, without a thorough understanding of the biospheral base upon which all the worlds' civilizations ultimately rested. If that were the norm, then much environmental education would be achieved through osmosis in the home between parents and children.

Tisdell thought he had detected some disappointment in some speakers that environmental education had not advanced as rapidly as they would have wished. He could only comment that, since 1970, he was sure that there had been tremendous progress. He had watched the environmental activities of his children and believed that they would grow up with quite a different perception of the world from that of his generation. Then, in the '70s, universities had developed interdisciplinary activities and several had created Schools of Environmental Studies which were still growing strongly. Moreover, more traditional departments, such as his own, were finding a new lease of life in relation to environmental questions. For example, a Master's degree in Environmental Management had been introduced at Queensland. In Australia, and elsewhere, there had also been an enormous growth in coverage of environmental questions by the media, contrary to the views he had heard expressed by some; one needed to sit down and assess that contribution objectively. He had a definite impression that it was both far greater and better now than ever before.

Craig Davis agreed with **Tisdell** on the great increase in university involvement in environmental courses, although the scale needed to be increased, *e.g.* at Ohio State University there were 57,000 students but only 325 in the environmental programme.

Kefeli mentioned that at Pushchino they had developed an American/Soviet project in international environmental education. A variety of educational approaches were employed, that of the Americans basically orientated to sustainable agriculture, that of the Soviets to understanding The Biosphere in a manner related to the research of his institute. Each year 25 students plus 5 teachers exchanged on this course which they were finding extremely fruitful.

Purcell commented, in connection with remarks on the importance of interdisciplinary programmes, that they still had a long way to go. He recalled being invited to assist in one where the Professor of Chemical Engineering commented that anything which was not chemical engineering was 'good neighbour stuff'! There was much still to be done.

J. Petts explained that she was responsible for training professional people in industry, an area of some importance environmentally. She thought that universities had a very important role in providing environmental subjects for specialists such as chemists, chemical engineers, civil engineers, accountants and lawyers, to name a few. They were needed at that stage if they were to be carried forward into their careers, when they could be built on by ongoing education and training. It was particularly important to provide such training for *everyone* in industry, not merely the regulatory authorities or those with environmental responsibilities. There was still a long way to go with the accountants and lawyers although some progress had been made with engineers and technicians. It was also essential that the training should begin with entrants at the bottom of the company ladder because they provided, in due course, the managing directors of companies.

Ramakrishnan remarked that he was not convinced of the existence of genuine interdisciplinary research but of the importance of bringing things together! Putting different disciplines in the same School or Department did nothing for interdisciplinary research. Motivation of different people in a specific problem in which each is interested was the only way, and that interest had to come from within: it could not be imposed from without.

Wasawo said he had become worried in listening to the discussion that we were in danger of making environmental education yet another subject; that, he thought, would be a mistake. The aim should be to educate a person about his surroundings and relationships. That involved all subjects and was essentially the promotion of an attitude.

Mische commented on **Wasawo**'s concern that another discipline was going to be presented to educators. He believed it important to educate for an understanding of global interdependence which, inevitably, encompassed every discipline and interest. That notion provided an entry for environmental issues to those interested in other kinds of interdependence, such as economic interdependence, but who would be interested in, and prepared to consider, all kinds of interdependence.

C. Westing said that to launch a viable environmental programme, people had to be made aware of how it affected them as individuals; then they were willing go along with it, as **Tolba** had demonstrated with the ozone programme. She believed that this was the reason why environmental education was so successful in Norway. There, everyone was deeply fond of Nature, everyone had a cottage in the mountains or by the sea, so that they were well aware of Nature and naturally concerned for the good of the environment. It was necessary, as **Furedy** had said, to discover what the people's perceptions were and what they valued. Then, if it could be shown that what they valued might be lost, they would respond from personal conviction.

Annexe 4: Education and Environment: A View from Least-developed Countries*

HARUN UR RASHID

Ambassador, Permanent Representative of Bangladesh to the United Nations in Geneva

The global impact of certain environmental changes on the long-term growth and development of the developed and developing countries alike – and indeed on the future of Mankind – is being increasingly recognized. Experts believe that structure and method of production, and the pattern of consumption in the developed countries over decades, have contributed immensely to degradation of the global environment. The developing countries, on the other hand, are confronted with environmental decline which is inextricably linked with poverty, illiteracy, unemployment and increasing population pressure. These compel the poor, in their efforts to cope with subsistence needs, to adopt unsound farming, grazing and other agricultural methods, or to depend heavily on ecologically fragile systems. Their impact is often compounded by adverse effects of certain natural phenomena such as natural disasters, soil erosion, drought, desertification, etc. A recent report of the UNCTAD correctly states that:

> 'It is no exaggeration to say that halting environmental degradation is one of the most formidable, difficult yet inescapable challenges that least-developed countries will face in the 1990s. Because of its ramifications and implications, environmental degradation in LDCS, should it continue, can in fact not only compromise and even nullify the LDCS' development efforts, but also affects the whole equilibrium of the Earth's natural patrimony.'

Education and motivation can most certainly play a crucial role in halting environmental degradation in developing countries. If properly planned and administered, they can greatly contribute to raising public awareness concerning environmental issues and encourage people to adopt production patterns that are consistent with the preservation and improvement of the environment. But can poor countries, left to themselves, undertake and implement such programmes? Can education alone provide an adequate remedy to halt environmental degradation in developing countries? What role can the international community play in this regard?

To answer these questions, one should take a close look at the nature of

*[This contribution was circulated to members of the Conference in connection with the discussion of Chapters 10 and 11. Eds.]

environmental problems in developing countries. As has just been indicated, poverty is the root cause of their environmental degradation. However, poverty is but a manifestation of the development crisis faced by these countries. There, low *per caput* income is responsible for scanty savings, very little of which is investable. In the absence of an adequate external flow of resources, and in most cases straining under crushing debt burdens, most of the developing countries are unable to generate sufficient growth. Whatever growth there is, gets eaten up by ever-growing population. Poverty, therefore, remains invincible, accentuating, in turn, environmental degradation.

Education, indeed, can play a crucial role in promoting environmental awareness in developing countries. However, it must be noted that the education sector, in spite of the high priority accorded to it by those countries, is beset with a series of problems. Overall resource constraint manifests itself in this sector as well. Prevailing low literacy militates against a quicker expansion of the percentage of literate people. Low incomes and unemployment discourage education and encourage drop-outs. Further, in view of stringent structural adjustment efforts, particularly in the Least Developed Countries (LDCs), investment in the social sector, including the education sector, has suffered heavily.

Any serious effort to halt and reverse degradation of the environment in developing countries, particularly in those classified as LDCs, must focus primarily on the alleviation and eradication of poverty. These economically disadvantaged countries must be provided with adequate financial and technical resources for development. In overall, as well as sectoral terms, the international community must afford them sufficient means in order to reactivate their development process and to put them back on the path of sustained growth and development. In this general context, an added emphasis on environmental education would be feasible and meaningful only if, and when, it was supported by adequate resources. The international community must bear a special responsibility particularly towards the LDCs in providing such resources. Based on such support, those countries should strive to place a high priority on environmental education, which should be aimed at enhancing awareness of the economic and social benefits of environmental protection for the improvement of their lives. Actions in this regard could take the form of both formal and informal instruction and motivation. Existing institutional arrangements and the media could be utilized for this purpose. To this end, it is imperative that the international community should provide LDCs with the necessary technical and related support.

We frequently hear the concepts and notions of 'one world' or 'global village'. These expressions assume full meaning in the context of the protection and preservation of the environment. However, confronted with tremendous resource constraints, least-developed countries can look only to their development partners for support. For developed countries, it is an important responsibility to assist adequately the LDCs in this uphill task, as the environment of least-developed countries is also everyone else's environment!

13. The Human Population Problem: As Explosive As Ever?*

NORMAN MYERS

Consultant in Environment and Development, Upper Meadow, Old Road, Headington, Oxford OX3 8SZ, England, UK

PAUL R. EHRLICH

Bing Professor of Population Studies, Department of Biological Sciences, Stanford University, Stanford, California 94305, USA

&

ANNE H. EHRLICH

Department of Biological Sciences, Stanford University, Stanford, California 94305, USA

INTRODUCTION

The population explosion is not over. Indeed it is probably now entering its most explosive phase. Though the past sixty years have seen a far greater growth in human numbers – from 2 thousand millions to well over 5 thousand millions – than in all our human history, that is little compared with what could come. During the next forty years we may well see another 5 thousand millions added; that is, 60% as much time could witness a 166% greater increase in human numbers. So, because of the lack of attention to the issue by political leaders and the general public, the greatest-ever growth in human population could still lie ahead, and surely does lie ahead unless there are massive calamities such as broadscale starvation episodes. As somebody once said, 'We ain't seen nuthin' yet.'

Note, moreover, that the anticipated increase of 5 thousand millions in the next 40 years is a *medium* projection (United Nations, 1988; Zachariah & Vu, 1988; United Nations Fund for Population Activities, 1989). If most of the less-developed countries were to follow the regrettable example of the Philippines and most countries of sub-Saharan Africa, of allowing their population growth-rate to keep on rising, the ultimate total for Humankind's numbers is projected to reach well over 14 thousand million people. But if, by contrast, most countries were to follow the splendid examples of China, South Korea, Taiwan, Indonesia, Thailand, Kerala State in India, Sri Lanka, Tunisia, Cuba, Colombia, and Mexico, we would end up with a total of little more than 8 thousand million people.

*[This paper was given at the conference by Dr R. Paul Shaw in the unavoidable absence of the authors. Eds.]

That difference between the high and low projections, namely of 6 thousand millions, is far more people than exist on Earth today. In other words, the population prospect is both a profound problem and a magnificent opportunity. There is still time for us to get on top of the problem before it gets on top of us – though its effective tackling will require a far greater planning effort, and a vastly greater sense of urgency, than humanity has demonstrated to date.

Population Growth: Why a Problem?

Before looking further at our population predicament, let us consider why it is a predicament at all. Essentially it has to do with the ecological concept of carrying capacity, which in this context may be defined as the number of people that the planet can support without irreversibly reducing its ability to support people in the future (Ehrlich *et al.*, 1989; *see also* Ehrlich & Ehrlich, 1990; Myers, 1990). This simple-seeming definition actually reflects an entire complex of interrelated factors such as human activities' impacts on food-producing systems (soils, water stocks, etc.), on natural ecosystems, and on the atmosphere and climate. So we can summarize the population-growth problem by saying it is a function of three interacting factors: the number of people, their *per caput* affluence (reflecting their consumption of life-sustaining resources), and the environmental injury imposed by technologies deployed to supply each unit of increased affluence (Ehrlich, 1989).

To be a trifle technical, consider the equation I = PAT, where I is impact, P is population, A is *per caput* affluence and consumption, and T is technology of an environmentally malign sort. Note that the three factors PAT interact in a multiplicative fashion, *i.e.* they compound each other's impacts (Ehrlich & Holdren, 1971; Myers, 1987, 1989*a*, 1989*c*; Ehrlich *et al.*, 1989; Ehrlich & Ehrlich, 1990). The equation immediately makes clear why there are population problems not only in developing countries, but also in developed countries where the A and T multipliers for each person are unusually large.

The equation I = PAT also makes clear why 'developing' nations with big populations, albeit with little economic advancement, can generate an enormous impact on The Biosphere: because the P multiplier on the A and T factors is so large. Consider, for instance, the vast repercussions that stem from small *per caput* amounts of coal-burning or chlorofluorocarbon (CFC) manufacture in China or India. This means that control of 'greenhouse' gases from industrial expansion in China or India can be obtained only with difficulty – even though the 'greenhouse' effect will wreak havoc on the agriculture of those two nations, soon projected to comprise almost two-fifths of all humanity. This is because their contributions of 'greenhouse' gases are tightly tied to population size (for further analysis of this question, *see* below). Consequently, there can be no realistic strategy to reduce their 'greenhouse' emissions without, among other measures, population planning of a scale and urgency far greater than any we have witnessed to date.

To illustrate how the equation's interactions work, suppose that, by dint of great effort, Humankind managed to reduce the average *per caput* consumption of planetary resources (A in the equation) by 5%, and to improve its technologies (T) so that they caused 5% less environmental injury on average. This would reduce the total impact (I) of humanity by roughly 10%. But unless population growth (P) were restrained at the same time, it would bring the total impact back to the previous level within less than six years.

True, certain economists (e.g. Simon, 1981) may protest that this analysis grievously underestimates the scope for technological expertise to keep on expanding the Earth's carrying capacity. But note that if, by new technological breakthroughs, we were able to do better than we have ever done before, notably by reducing the adverse impacts of humanity's activities on the environmental resource-base by 10%, and if at the same time we were to forego all advance in *per caput* affluence, population growth could eliminate all the benefits within just six years.

To illustrate through a specific example, consider the Earth's carrying capacity with respect to food production for humans. According to the world Hunger Project (Kates *et al.*, 1988), the planetary superecocomplex could, with present agro-technologies and with equal distribution of food supplies, support 6 thousand million people *if they all ate a vegetarian diet*. If they derived 15% of their calories from animal products, as do many people in South America today, the total would slump to 4 thousand millions, but if they gained 25% of their calories from animal protein, as is the case with most people in North America today, then the Earth could support only 2.5 thousand million people.

Certainly, these calculations reflect no more (and no less) than today's food-production technologies. Theoretically we can hope that sundry advances in agro-technologies will still be available to come on stream. But world grain output, after increasing a phenomenal 2.6-fold between 1950 and 1984, has levelled off from 1985 onwards. The 1989 harvest of 1.67 thousand million tonnes of grain was only 1% above that of 1984 – even though the world's farmers had invested thousands of millions of dollars to expand output, and fertilizer-use had increased by 14% (Brown *et al.*, 1990; US Department of Agriculture, 1990). Yet by 1989 there were 440 million more people on Earth to feed. This means that between 1985 and 1989, world population increased by almost 8.5%, whereas *per caput* food output declined by more than 7%!

To put it another way, each day sees an additional 250,000 more people seeking to share the Earth and its resources. This is the equivalent of adding a Los Angeles-sized city to the world population every two weeks, an entire Mexico every year, and a United States *plus* Canada every three years. In turn, this means that every 14 seconds the world's population grows by 40 people, whereas the planet's stock of arable land declines by one hectare. Furthermore, some of the worst agricultural problems are now starting to manifest themselves in the form of a fall-off in production in many areas.

To continue, irrigation has been the most productive sector of expanding agriculture since 1950, but has now run up against its most basic constraint: the per-person amount of irrigation water is now declining (Brown *et al.*, 1990). Erosion of topsoil is estimated to have reached 24 thousand million tonnes world-wide yearly in recent years, which is practically equivalent to all the wheatland topsoil in Australia (Pimentel & Hall, 1989; Brown *et al.* 1990). In addition, low-level ozone and acidic precipitation among other forms of pollutants are taking an ever-heavier toll of cropland productivity. And if rapid climate change is added to the mix, the outlook could be grim indeed (Daily & Ehrlich, 1990).

For an extreme case of the carrying-capacity dilemma, consider Kenya. The country could conceivably support rather more than 50 million people in the distant future, if ever it had the chance to build up its manufacturing base and develop export markets enough to enable it to buy sufficient food overseas – supposing that extra food were available. But Kenya, with its population growth-rate still at 4% per year, is projected to expand from its present 23 million people to 50 millions in less than 20 years – a prospect that would be daunting to even the most richly endowed and best governed nation. Moreover, Kenya's ultimate population size, *i.e.* the total when population growth is supposed to reach zero at the start of the 22nd century, is projected to top 110 millions!

We must surely anticipate that Kenya will shortly feature growing throngs of impoverished peasants, who will pursue whatever means are available to gain their subsistence livelihoods – and if that means cutting down forests, ploughing up grasslands, and cultivating steep slopes (the dangerous phenomenon of marginal people in marginal environments), they will wreak exceptional havoc on Kenya's agricultural resource-base in short order. This will mean of course that they will further reduce Kenya's carrying capacity in terms of food production alone.

A similar prospect must surely await Ethiopia, with its present population of just 50 millions projected to reach 128 millions within another 30 years. Much the same must apply, too, to Nigeria, where the present 119-millions population is projected to soar to 274 millions by the year 2020; to Bangladesh, where the present total of 118 millions is projected to grow to 230 millions; to Pakistan (currently 113 millions but projected to grow to 242 millions) (projections based on Population Reference Bureau, 1989; United Nations Fund for Population Activities, 1989; *see also* Zachariah & Vu, 1988). (For details of some likely repercussions and the estimated carrying capacity in each of these countries, *see* Goliber, 1989; Myers, 1989*a*, 1989*b*, 1989*c*.)

Realization of these projections depend of course on large numbers of people finding the wherewithal to sustain themselves throughout a normal life-span. To this extent, strictly demographic projections are made in something of a carrying-capacity vacuum. (Of course they are mere projections, not predictions, and still less are they forecasts; but they tend to be perceived by the media and disseminated to the world public as assertions of likely fact.)

Moreover, we must remember that an entire nexus of associated factors is relevant, notably technology types, economic systems, energy inputs, political persuasions, trade relations, and a host of other factors that either reduce or aggravate the impact of population growth *per se* (McNamara, 1984; Repetto, 1987; Benedick, 1988; Conservation Foundation, 1988; Hinrichsen, 1988; Sadik, 1988; Davis *et al.*, 1989; Keyfitz, 1989; Shaw, 1989). But as we shall see below, these associated factors generally serve to compound the impact of population growth, *i.e.* they have a multiplicative rather than a merely additive effect. The present paper will demonstrate this, below, with regard to the energy sector and global warming as concerns China and India.

We must also bear in mind, as the 'Brundtland Report' (World Commission on Environment and Development, 1987) reminds us, that within another 35 to 40 years there are projected to be twice as many people on Earth, consuming three times as much food and fibre, seeking four times as much energy, and engaging in at least five times as much economic activity, as currently. We should bear in mind, however, that this prospect is not predetermined: it will occur only if we allow ourselves to be controlled by events, instead of deciding to control events. Projections are not destiny. We can still choose. But first we must do what we show few signs of undertaking, and that is to *choose to choose*. The year 2000 AD will be only about 2,800 days away when this book is published, and we lose one per cent of our planning time every few weeks.

Population, Industrialization, and the Environment: Two Illustrative Cases

To point out some specific linkages between population growth, emergent industrialization, and environmental impacts, and to highlight the specific role of population growth, let us consider the case of China and India. China's energy-use has been growing for much of the 1980s at the rate of 5% per year, and the country entertains even more ambitious plans for the future (Zhao & Sun, 1986), in part because it possesses almost one-third of the world's stocks of coal. Already China has surpassed the United States as the number one coal-burner, accounting for 10% of global carbon emissions. While its *per caput* carbon emissions are only one-tenth those of Americans, its sheer human numbers make the difference. By 2000 AD the country hopes to burn almost twice as much coal annually; and by the year 2030, if plans for industrialization work out, China may well be contributing 20% of global carbon emissions – roughly equal to its share of global population if the populace has then reached its projected total of 1.5 thousand millions (Oppenheimer & Boyle, 1990).

In addition, China plans greatly to increase its national stock of refrigerators, utilizing CFCs that are both a prime component of global warming and the major source of stratospheric ozone depletion. To date only one Chinese household in ten possesses a refrigerator – though the proportion in the capital, Beijing, has risen during the 1980s from less than 3% to more than 60%. The country has already built 12 CFC-production plants in order to accommodate the refrigerator needs of many more of its today's 250 million households. By the year 2000, the Government plans to expand CFC production 10-fold, which will still leave *per caput* output at only one-fifth that of the United States until recently – but China's vast human numbers will make the CFC impact, in terms of both the 'greenhouse' effect and ozone-layer depletion, a critically determining factor.

As for India, with a *per caput* income of still only $290 per year, *i.e.* about one-sixtieth of that of the United States, electricity capacity today is only 55,000 megawatts, about twice that of New York State. Although the country possesses meagre coal reserves compared with China, it is burning them so fast that it now ranks as the world's fourth-largest coal-burner. In 1950 India's coal use was of only 33 million tonnes, but by 1989 it had soared to 191 million tonnes, while production of crude oil (another fossil fuel) rose from 0.3 to 30.4 million tonnes, and total power-generation increased from 5 to 217 thousand million kwh) (Dave, 1988). The Government plans that every family should have a television set, in order to foster development processes through a basic communications network reaching into every home in the land. But this goal alone would require the production of an additional 80,000 megawatts of power. Another major development goal is to supply electricity for lighting to half the homes in the country, totalling another 80,000 megawatts. Attaining these two goals alone would more than double India's present carbon emissions (Dave, 1988; Oppenheimer & Boyle, 1990).

We must bear in mind, moreover, that India's present population of more than 850 million people is projected to reach 1,043 millions by the year 2000. But even with its low *per caput* income (the 'A' factor of the I = PAT equation), and its less-than-advanced technological capacity (the 'T' factor), its huge population (the 'P' factor) makes for a disproportionately large potential contribution to global warming.

Suppose India managed to reduce its fertility rate to replacement level (about half the present number of children per completed family) within the next three to four decades; and suppose, too, that at the same time it did no more than double its *per caput* use of commercial energy (roughly matching that of China today), using coal. This increase, in conjunction with population growth, would result in India's emitting enough carbon dioxide into the global atmosphere to more than cancel out the benefits of a putatively extreme step on the part of the United States, namely the termination forthwith of all coal-burning without replacing coal with any other carbon-containing fuel (Ehrlich & Ehrlich, 1989).

These two graphic examples serve to demonstrate the overwhelming consequences of sheer human numbers, *plus* the continuing growth of these same numbers, when tied in with plans for even a modest amount of industrialization.

The Great Challenge

To reduce population growth in developing countries, we should tackle two predominant sets of needs. We should supply family-planning services to those couples who possess the motivation but lack the contraceptive hardware; and for those who still lack the motivation, we must attempt the much more complex and costly task of supplying incentives to reduce 'fertility'. Of course there are many other related factors, notably the alleviation of pervasive poverty in most of the less-developed countries, together with improvements in the status of women. But due to the limited length of this paper, we shall concentrate on the two challenges cited.

Unmet Needs

With regard to the first and more straightforward of the two challenges, there is some heartening news. As already noted, a considerable list of countries have made remarkable progress in reducing their population growth-rates. No doubt this has partly been due to a degree of socio-economic advancement. But it has been due in major measure to nation-wide availability of family-planning services. In Indonesia there were only 400,000 couples practising contraception in 1972, but 18.6 millions in 1989; during that same period the total fertility rate fell from 5.6 to 3.4 children (and infant mortality fell by 40% – a critical factor that we shall return to below) (Keyfitz, 1989). In Zimbabwe, contraceptive use rose from 14% to 43% between 1982 and 1988, while population growth declined from 3.6% to 2.9% (Awle *et al.*, 1988). In Mexico, contraceptive use increased from 23% to 45% between 1973 and 1986, while the total 'fertility rate' fell from 6.3 children to 3.8 per woman of reproductive age (Sherbinin, 1990). In Colombia, following a widespread increase in the provision of birth-control services, the 'fertility rate' declined by almost 50% between 1960 and 1985; comparably in Chile by 51% and in Cuba by almost 54% (World Bank, 1987).

This brings us to the question of unmet needs, being the needs of those people who want to defer the start of child-bearing, want no more children, or prefer to wait a longer period before having another child – but do not have access to the means to do so (for some background reviews, *see* Alan Guttmacher Institute, 1987; Jacobson, 1987; World Bank, 1987). These unmet needs appear to be great, though there are no precise figures of the scale of the problem. They range from an estimated 75% of women in Latin America, to 43% in Asia, and 27% in Africa (United Nations Department of International Economic and Social Affairs, 1987; *see also* Boerma, 1987; Fortney, 1987; Jacobson, 1987; Winikoff & Sullivan, 1987). If unwanted births could be eliminated – surely a measure to be implemented on grounds of human rights, whatever its impact on population growth and the future of our world – 'fertility rates' could be reduced by at least 30%.

The cost of supplying family-planning facilities need not be great, being roughly an average of $15 per couple per year – a tiny fraction of the cost of the basic human needs that would otherwise have to be met in terms of health care for mother and child, additional food supply, housing, schooling, etc. Indeed in comparison with the benefits derived, it is a massively worth-while, *i.e.* an extremely cost-effective, investment. A fairly recent analysis (Nortman *et al.*, 1986) of the Mexican Government's family planning programme has estimated that, for every peso spent on family planning for its urban population during 1972–84, there was a saving, in terms of infrastructure services not required (schools, health facilities, housing, employment, etc.), of four to five pesos.

A similar study in Thailand from 1972 to 1980 shows a saving ratio of 7 to 1 (Chao & Allen, 1984; Knodel *et al.*, 1987; Ismartono, 1989). If we suppose there are 800 million women of reproductive age in developing countries, and 30% of them, namely 240 millions, experience unmet needs, the cost of supplying them would be $3.6 thousand millions a year, or not much more than a doubling of today's total expenditures on population programmes.

Motivation for Smaller Families

This is a much bigger challenge – far more complex and expensive – than the first, and it appears to apply to roughly two-thirds of all developing-country couples 'at risk'. It relates to a host of socio-economic factors (*e.g.* degree of economic advancement and social equity), and especially to the overall status of women (including their levels of education and employment, and their cultural role generally). This is not the place to discuss them all: volumes have been written on the multiple variables in question. Rather let us focus on just one: infant mortality. Without assurance for parents that their offspring will survive, there will remain a seemingly dominant compulsion to keep on producing children in order to ensure that 'enough' will reach adulthood. But this will not be achieved as long as developing-world communities experience the demographic haemorrhage of 40,000 children dying each day (United Nations Children's Fund, 1990) – the equivalent of a jumbo-jet full of children hitting the ground every ten minutes throughout the year. If recent trends persist, we can expect to witness the deaths of 100 million children during the 1990s – with a grossly reduced prospect of persuading parents to engage in birth-control!

The challenge of preventing most of these deaths is ultra-relevant to population planning overall, not least because it is technologically feasible and remarkably cheap. Thanks to recent breakthroughs in health-care, three main strategies are available (United Nations Children's Fund, 1989).

- First is immunization against measles, whooping cough, tetanus, and other great childhood killers. Although we now save 2 million children each year through this means alone, we could save another 3 millions if immunization were made universal. The cost is trifling, no more than $1.50 per child.
- A second solution lies with Oral Dehydration Therapy (ORT). For the equivalent of a mere 10 US cents, a packet of salts staves off the rigours of

diarrhoea. Already ORT saves 1 million children each year. But two out of three developing-country families do not avail themselves of the remedy; we could readily save another 1 to 2 million children per year by this means alone.

- Third, we should do much more to encourage breast feeding. Bottle-fed infants contract many more illnesses and are 25 times more likely to die than breast-fed infants. The cost of a comprehensive educational campaign to promote breast feeding is 'peanuts' compared with the benefits – which coincidentally include a contraceptive effect.

The total cost of such a three-parts programme should not amount to more than $2.5 thousand millions a year by the late 1990s (United Nations Childrens Fund, 1990). This is roughly equivalent to what the United States spends on cigarette advertising per year, what the Soviet Union spends on vodka each month, or what the European Commission spends every five weeks to subsidize its farmers to produce beef mountains, milk lakes, etc. – and what developing countries pay each week to service their foreign debts. It is also equivalent to less than one day's military expenditures world-wide in each year, or the same as developing countries devote to armaments and the like every three days.

So motivation for smaller families is not a matter of cost, it is a question of political will and active promotion. There are few instances in the whole field of development where so much could be accomplished for so many at so little cost. Accordingly it is not a case of 'Can we afford to do it eventually?' Rather is it a case of 'How can we afford not to do it right away?' Yet our lackadaisical response means that we are implicitly saying it is still too costly. How low does the cost have to fall before we accept it as a sound investment?

The Policy Perspective

Fortunately, much of the world has officially come around at last to the view that population planning is a 'must' – in principle at least. In 1970 only a minority of developing-world governments accepted the need for planning of any sort, but today the great bulk of governments are committed to the cause, albeit with often a mere fraction of the energy and urgency required in practice. How about developed-world governments?

Curiously, while the developed world is all too ready to urge the virtues of population planning on the developing world, not a single developed nation has established a population policy of its own. How many Americans, and at what level of resource-consuming living, are *really* good for the United States, let alone for the world as a whole? How many Britons can be sustainably supported within a purely British context, let alone the broader context of their nation's impact on the planetary system? When the two governments involved, together with all other developed-world governments, have formulated an answer to key questions such as these, they will be better able to determine how many citizens they should aim for by, say, the years 2025 and 2050 AD, and hence the measures they should implement to get from here to there – bearing in mind that demographic momentum places a premium on long-term planning.

To put the problem in perspective, recall that each year the global family is expanded by 1.7 million new-born Americans and 18.4 million new-born Indians, or almost 11 times as many Indians as Americans. But an average American consumes 55 barrels of oil per year which, multiplied by 1.7 million, means 93.5 million barrels of oil – with all that means for the 'greenhouse' effect (to consider just a single repercussion of consumerism). An average Indian consumes only 5 barrels of oil per year, which means an annual increase of only 92 million barrels of oil burned. So who also needs a population policy, in order to take explicit and systematic account of their nation's impact on Earth's carrying capacity for themselves as well as for the rest of Humankind?

There is thus quite clearly a case for developed nations – and especially for over-developed nations – to consider the rationale for a steady reduction of their populations over the long haul, as a measure to be broached with all due despatch.

Plainly we all need to look to our population problems. To point a critical finger at developing-world 'laggards' is akin to proclaiming that their end of the boat is sinking.

Conclusion: The Decisive Decade

We predict that the 1990s will be the most decisive decade in human history, at least to date. During these coming ten years, we face a final window of opportunity. We can still plan our planet's future, especially in terms of the population issue, to positive effect. Or we can abandon it, as has been largely the case to date, to a *laissez-faire* outcome. Either way, our 1990s 'planning' will leave its mark for centuries and indeed far beyond – when we consider, for example, the effects of mass extinctions of species. Through our actions, or through our inaction, we shall determine the planet's future more thoroughly than it has been possible for any human community to do in the past. By design or by default, the die will be cast.

But however much negative evidence there may be accumulating at ever-faster rates, we should not despair over the population prospect, even if in many ways it is our key and basic problem. On the contrary, we should bear in mind that human communities are capable of the most extraordinary shifts in understanding, foresight, planning, and behaviour. Here we are in April 1990 in Budapest: who would have supposed just a short two years ago that the political landscape of eastern Europe would be transformed to such exemplary purpose today? And who would have supposed only a year ago that the Berlin Wall would be reduced to collectors' items? Let us hope that other 'walls', notably the barriers that still block our path to an acceptable population future, will come tumbling down just as swiftly and conclusively.

References

Alan Guttmacher Institute (1987). *Women at Risk: The Need for Family Planning Services*. Alan Guttmacher Institute, New York, NY, USA.

Awle, E. Van de, Fala-Giakenda, M. & Ohadike, P. (1988). *The State of African Demography*. International Union for Scientific Study of Population, Liege, Belgium.

Benedick, R. E. (1988). Population–environment linkages and sustainable development. *Populi*, 15, pp. 14–21.

Boerma, J. T. (1987). Levels of maternal mortality in developing countries. *Studies in Family Planning*, 18, pp. 213–21.

Brown, L. R., Durning, A., Flavin, C., French, H., Jacobson, J., Postel, S., Renner, M., Shea, C. P. & Starke, L. (1990). *State of the World 1989*. W. W. Norton, New York, NY, USA: xvi + 25 pp., illustr.

Chao, D. & Allen, K. B. (1984). A cost–benefit analysis of Thailand's family planning programme. *International Family Planning Perspectives*, 10, pp. 75–81.

Conservation Foundation (1988). Population and environment. *Conservation Foundation Letter* 3, Conservation Foundation, Washington, DC, USA.

Daily, G. C. & Ehrlich, P. R. (1990). *An Exploratory Model of the Impact of Rapid Climate Change on the World Food Situation*. Division of Biological Sciences, Stanford University, Stanford, California, USA: 25 pp.

Dave, J. M. (1988). *Policy Options for Development in Response to Global Atmospheric Changes: Case Study for India for Greenhouse Effect Gases*. Nehru University, New Delhi, India.

Davis, K., Berstam, M. S. & Sellers, H. M. (1989). *Population and Resources in a Changing World*. Morrison Institute for Population and Resource Studies, Stanford University, Stanford, California, USA.

Ehrlich, P. R. (1989). Facing the habitability crisis. *BioScience*, 39, pp. 480–2.

Ehrlich, P. R. & Ehrlich, A. H. (1989). *How the Rich Can Save the Poor and Themselves: Lessons from the Global Warming*. Stanford Institute for Population and Resource Studies, Stanford, California, USA.

Ehrlich, P. R. & Holdren, J. P. (1971). Impact of population growth. *Science*, 171, pp. 1212–17.

Ehrlich, P. R. & Ehrlich, A. H. (1990). *The Population Explosion*. Simon and Schuster, New York, NY, USA: 320 pp.

Ehrlich, P. R., Daily, G. C., Ehrlich, A. H., Matson, P. & Vitousek, P. (1989). *Global Change and Carrying Capacity: Implications for Life on Earth*. Stanford Institute for Population and Resource Studies, Stanford, California, USA: 17 pp.

Fortney, J. A. (1987). The importance of family planning in reducing maternal mortality. *Studies in Family Planning*, 18, pp. 109–14.

Goliber, T. J. (1989). *Africa's Expanding Population: Old Problems, New Policies*. Population Reference Bureau, Washington DC, USA: 52 pp.

Hinrichsen, D. (1988). *State of the World Population Report 1988*. United Nations Fund for Population Activities, New York, NY, USA.

Holdren, J. & Ehrlich, P. R. (1974). Human population and the global environment. *American Scientist*, 62, pp. 282–92.

Ismartono, Y. (1989). Thailand's approach to a touchy subject. *People*, 16, pp. 12–13.

Jacobson, J. L. (1987). *Planning the Global Family*. Worldwatch Institute, Washington, DC, USA: 54 pp.

Kates, R. W., Chen, R. S., Downing, T. E., Kasperson, J. X., Messer, E., & Millman, S. R. (1988). *The Hunger Report: 1988*. Brown University, Providence, Rhode Island, USA: 40 pp.

Keyfitz, N. (1989). The growing human population. *Scientific American*, 261, pp. 119–26.

Knodel, J., Chamratrithirong, A. & Debavalya, N. (1987). *Thailand's Reproductive Revolution: Rapid Fertility Decline in a Third World Setting*. University of Wisconsin Press, Madison, Wisconsin, USA.

McNamara, R. S. 1984. The population problem: time-bomb or myth. *Foreign Affairs*, **62**, pp. 1107–31.

Myers, N. (1987). Population, environment, and conflict. *Environmental Conservation*, 14(1), pp. 15–22.
Myers, N. (1989*a*). Population growth, environmental decline, and security issues in Sub-Saharan Africa. Pp. 211–31 in *Ecology and Politics: Environmental Stress and Security in Africa* (Eds A. Hjort & M. A. M. Salih). Scandinavian Institute of African Studies, Uppsala, Sweden: 255 pp.
Myers, N. (1989*b*). Environment and security. *Foreign Policy*, 74, pp. 23–41.
Myers, N. (1989*c*). Environmental security: the case of South Asia. *International Environmental Affairs*, 1, pp. 138–54.
Myers, N. (1990). *Future Worlds: The Gaia Atlas of Change and Surprise.* Robertson McCarta, London, England, UK, and Doubleday, New York, NY, USA: 190 pp.
Nortman, D. L., Halvas, J. & Rambago, A. (1986). A cost–benefit analysis of the Mexican social security administration's family planning program. *Studies in Family Planning*, 17, pp. 1–6.
Oppenheimer, M. & Boyle, R. H. (1990). *Dead Heat: The Race Against the Greenhouse Effect.* Basic Books, New York, NY, USA: xii + 268 pp.
Pimentel, D. & Hall, C. W. (Eds) (1989). *Food and Natural Resources.* Academic Press, Inc., San Diego, California, USA: 256 pp.
Population Reference Bureau (1989). *World Population Data Sheet 1990.* Population Reference Bureau, Washington, DC, USA: 1 page.
Repetto, R. (1987). Population, resources, environment: an uncertain future. *Population Bulletin* 42. Population Reference Bureau, Washington, DC, USA: 144 pp.
Sadik, N. (1988). *State of the World Population Report: Safeguarding the Future.* United Nations Fund for Population Activities, New York, NY, USA: 21 pp.
Shaw, R. P. (1989). Rapid population growth and environmental degradation: Ultimate *versus* proximate factors. *Environmental Conservation*, 16 (3), pp. 199–208, 3 figs.
Sherbinin, A. de. (1990). Survey report: Mexico. *Population Today*, 18, p. 5.
Simon, J. (1981). *The Ultimate Resource.* Princeton University Press, Princeton, New Jersey, USA.
United Nations (1988). *World Population Prespects, 1988.* United Nations, New York, NY, USA: *see* Sadik (*q.v.*).
United Nations Children's Fund (UNICEF) (1989). *State of the World's Children 1989.* UNICEF, United Nations, New York, NY, USA.
United Nations Children's Fund (UNICEF) (1990). *State of the World's Children 1990.* UNICEF, United Nations, New York, NY, USA: 102 pp.
United Nations Department of International Economic and Social Affairs (1987). *Fertility Behavior in the Context of Development: Evidence from the World Fertility Survey.* United Nations, New York, NY, USA.
United Nations Fund for Population Activities (1989). *State of the World's Population.* United Nations Fund for Population Activities, New York, NY, USA.
US Department of Agriculture (1990). *World Agricultural Production.* (Circular series, WAP–3–90.) Foreign Agricultural Services, US Department of Agriculture, Washington, DC, USA.
Winikoff, B. & Sullivan, M. (1987). Assessing the role of family planning in reducing maternal mortality. *Studies in Family Planning*, 18, pp. 128–43.
World Bank (1987). *World Development Report 1987.* World Bank, Washington, DC, USA: xii + 285 pp.
World Commission on Environment and Development (1987). *Our Common Future.* Oxford University Press, Oxford, England, UK: xv + 400 pp.
Zachariah, K. C. & Vu, M. T. (1988). *World Population Projections.* Johns Hopkins University Press, Baltimore, Maryland, USA.
Zhao, D. & Sun, B. (1986). Air pollution and acid rain in China. *Ambio*, 15, pp. 2–5.

14. Population and Environment in Perspective: The Mediterranean Picture

MICHEL BATISSE
President of the Blue Plan Regional Activity Centre
Sophia Antipolis, 06560 Valbonne, France

In 1975, at a time when concern about the state of the Mediterranean Sea was rising in public opinion, the Coastal States, gathered under the aegis of UNEP, adopted a 'Mediterranean Action Plan' and signed a convention for the protection of their common Sea against pollution. At that time already, some people, including Mostafa Tolba of UNEP and Serge Antoine of France, realized, however, that most of what was going wrong in the Sea found its origin in what was taking place on land in the surrounding countries. Consequently, they pressed for introducing in the Action Plan some elements concerning 'socio-economic' issues, thus implying that terrestrial activities in those countries and their consequences on the state of the Sea would be considered under the cooperative arrangements that were being developed. In this context, the main idea which emerged was to carry out a comprehensive study of the interactions between population, resources, environment and development in the whole region, and to find out how these interactions could develop in the mid-term and long-term. This is the 'Blue Plan'. In fact it is not a plan in the rigid sense of a binding set of mutually-agreed-upon objectives, but a review of the possible futures which could characterize the region and an ensuing invitation to select those courses of action which will minimize harmful effects on people and the environment in the region. The Blue Plan is thus based on the analysis of the Mediterranean 'system', which is viewed as consisting of the following elements: population, urbanization, agriculture, industry, energy, tourism, transport, soils, forest, inland water, coast and sea. In the exercise the multiple and complex interactions between these elements are reviewed and their possible evolutions are presented according to a number of 'scenarios'.

The time-horizons selected for this prospective study were 2000 AD and 2025, with 1985 taken as reference year. The course of things to come is pretty much set until the end of the century, but evolutions may considerably diverge between now and 2025, an horizon which may seem far away in political and economic terms, but not for environmental action, such as reforestation, erosion control, or pollution abatement, which require long-

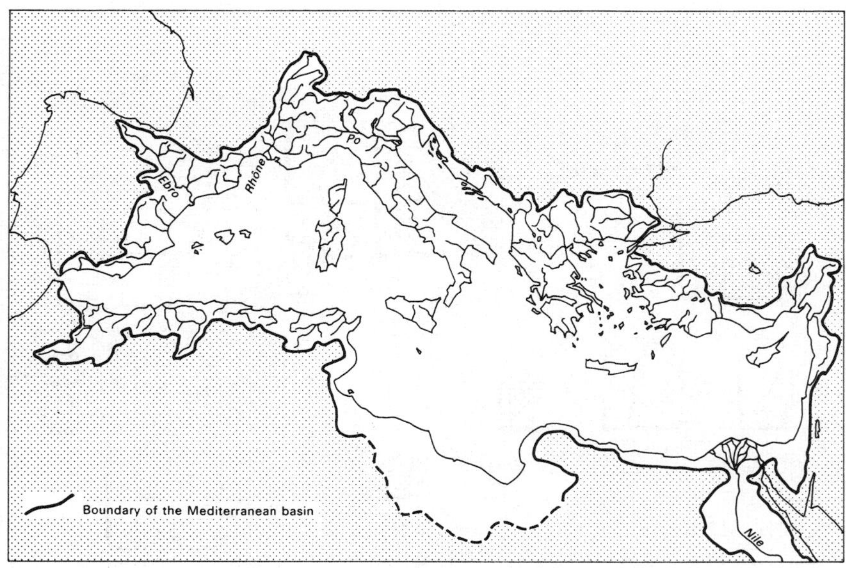

Figure 14.1. The Mediterranean watershed. For the sake of clarity, small rivers are not indicated on this map. Watershed boundaries in very arid areas are approximate. (*Source*: adapted from J. Margat, 1988.)

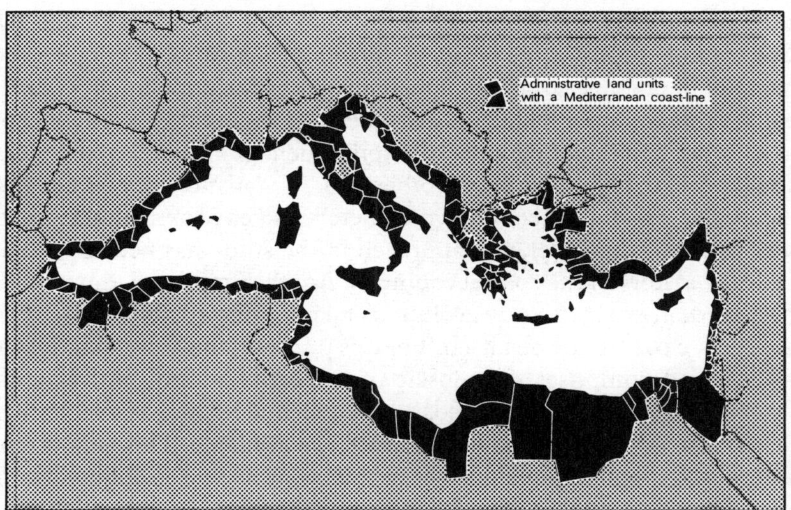

Figure 14.2. The Blue Plan Mediterranean regions. There is no commonly accepted delineation of the Mediterranean region. For most purposes, particularly population, urbanization, tourism, and coastal activities, the Blue Plan regional prospective studies had to be based on a suitable administrative definition, namely the territorial units located on the coast and for which statistical data are available. For some countries, such as Italy and Greece, these cover a very large portion of the country. For others, including Morocco and Syria, only a small part of the country is thus covered. For certain subjects a broader geographical basis was used, namely the hydrographic boundaries of the watershed. (*Source*: UN/Blue Plan.)

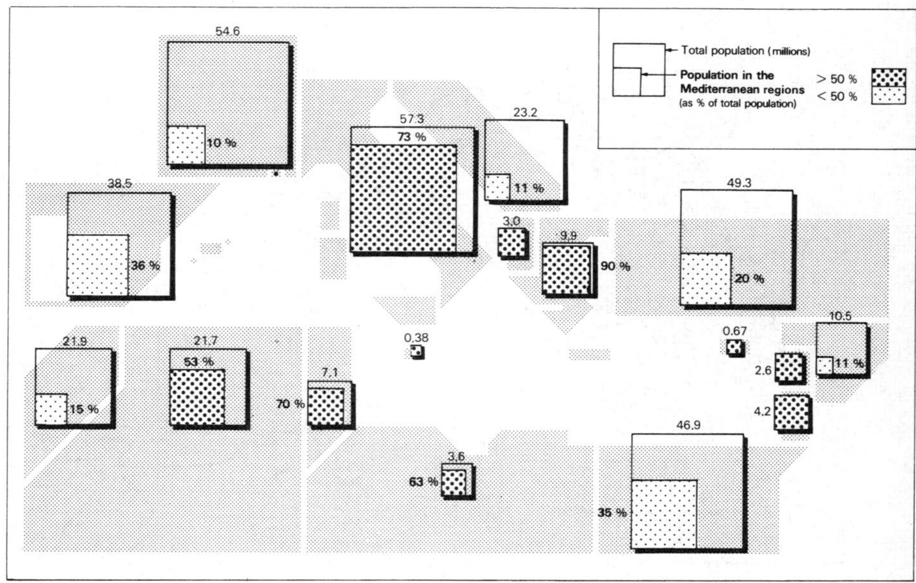

Figure 14.3. Population in the Mediterranean regions of each country, 1985. (*Source*: UN/Blue Plan.)

term efforts. The region considered by the Blue Plan study is the Mediterranean Basin. This is well defined in hydrological terms by the watershed boundaries of the inflowing rivers as shown in Figure 14.1. It is, however, practically impossible to obtain coherent figures for population, economic activity, or environmental factors at the level of these river basins, except of course for water-related parameters. Consequently for most of its work, the Blue Plan had to define the region otherwise. As statistics are easily available at national level, and as economic parameters and environmental legislation mostly relate to that level, one approach in the study has been to consider the entire territory of the coastal countries. Another approach, however, had to be added, in order to study evolutions taking place at a more 'Mediterranean' level, closer to the Sea itself. For this purpose, the area of study which was identified comprises the mosaic of territorial administrative units in each country which touch the coast and for which comparable statistical data, particularly on population dynamics, are available. This area of study, referring to 'coastal regions' in a broad sense, is shown in Figure 14.2. In terms of population, these coastal regions represent varied percentages of the countries' total population and these percentages are given in Figure 14.3.

Five scenarios have been constructed on the basis of coherent sets of assumptions bearing on the macro-economic context, on demographical changes and on national development and environment strategies. Three of these scenarios follow, to a greater or lesser extent, the economic and environmental trends which are currently observed. A *reference trend scenario*

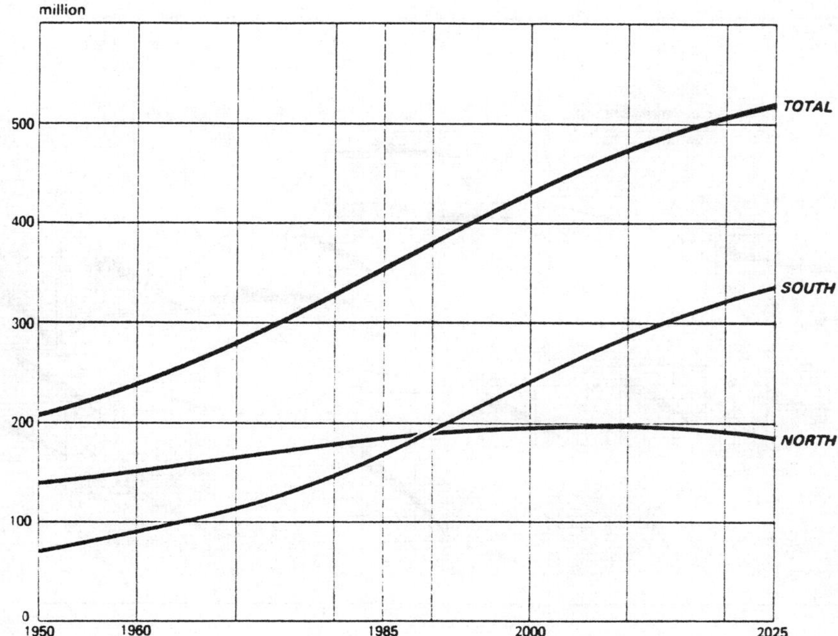

Figure 14.4. Population in the Mediterranean countries: Trends 1950–1985 and average scenario 1985–2025. The rate of population growth in the Mediterranean countries as a whole tends to level off as from the year 2000. The population of the southern and eastern countries exceeds that of the northern countries as from 1990. (*Source*: UN/Blue Plan.)

(Tl) represents the actual continuation of present economic trends. A *worse trend scenario* (T2) represents the situation with a lower economic growth and fierce international competition. Conversely, a *moderate trend scenario* (T3) indicates the consequences of better coordination of economic policies, higher economic growth and adoption of environmental corrective measures.

These trend scenarios lead to more or less unacceptable social and environmental conditions, particularly in the South and East of the Mediterranean Basin. Therefore two 'alternative' scenarios have also been explored. The *reference alternative scenario* (A1) rests upon a closer North-South cooperation among Mediterranean countries for development as well as for environmental protection, including agreements with the EEC on trade and overall migratory flows. The *integration alternative scenario* (A2) implies equally strong North-South cooperation and care for the environment, but also the building up of subregional relations particularly among groups of countries in the South and East (for instance a 'United Maghreb'), with opening of trade and migration within these groups.

It clearly appears that, in all five scenarios, the demographic situation constitutes the dominating factor of evolution in the region. The population figures used in these scenarios at country level are those of the UN Population Division. For the future, the hypotheses chosen are also those given by the

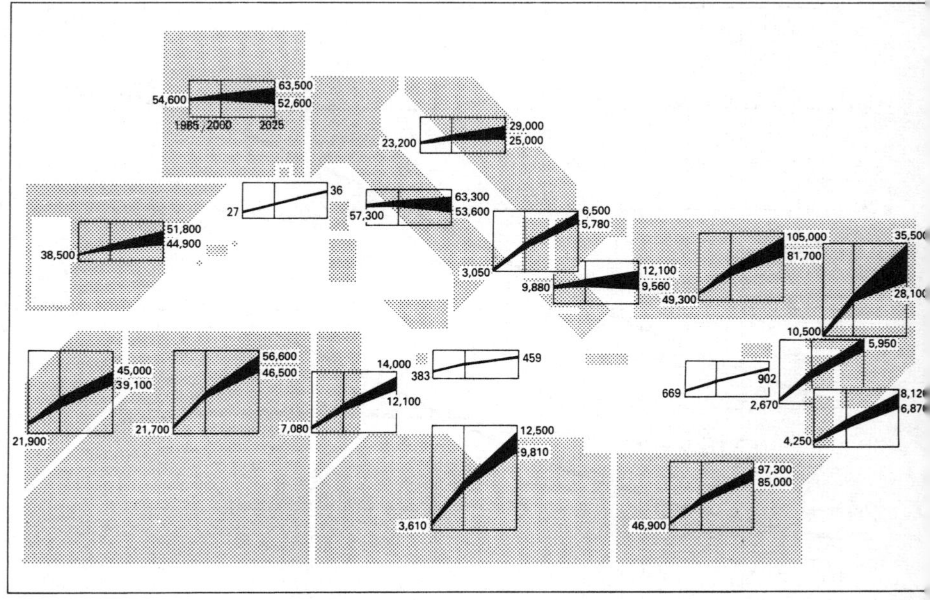

Figure 14.5. Population trends, 1985, 2000, and 2025 AD: Extreme scenarios. (*Source*: UN/Blue Plan.)

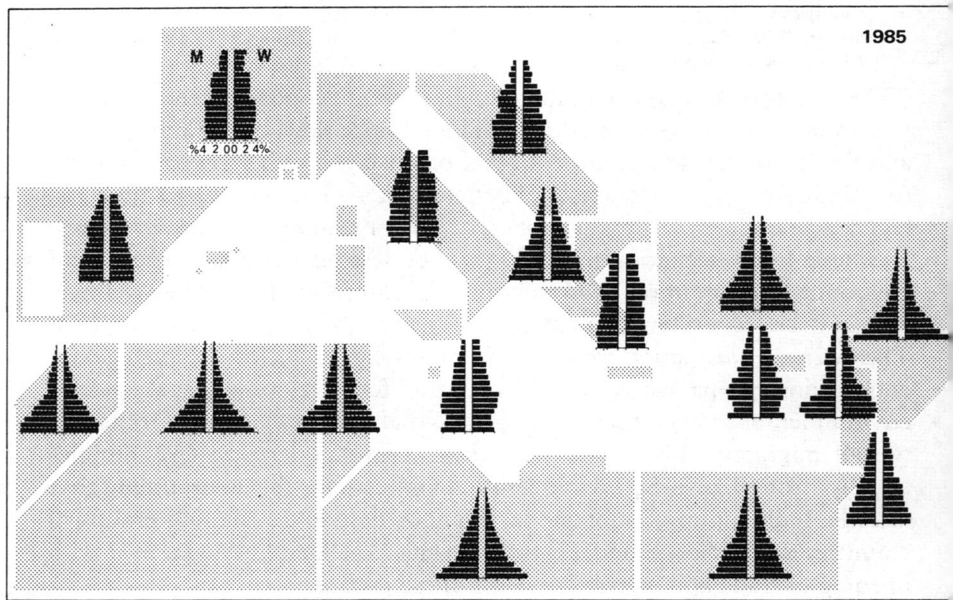

Figure 14.6. Age structure, 1985, as percentage of total population (5-year age-groups). (*Source*: UN/Blue Plan.)

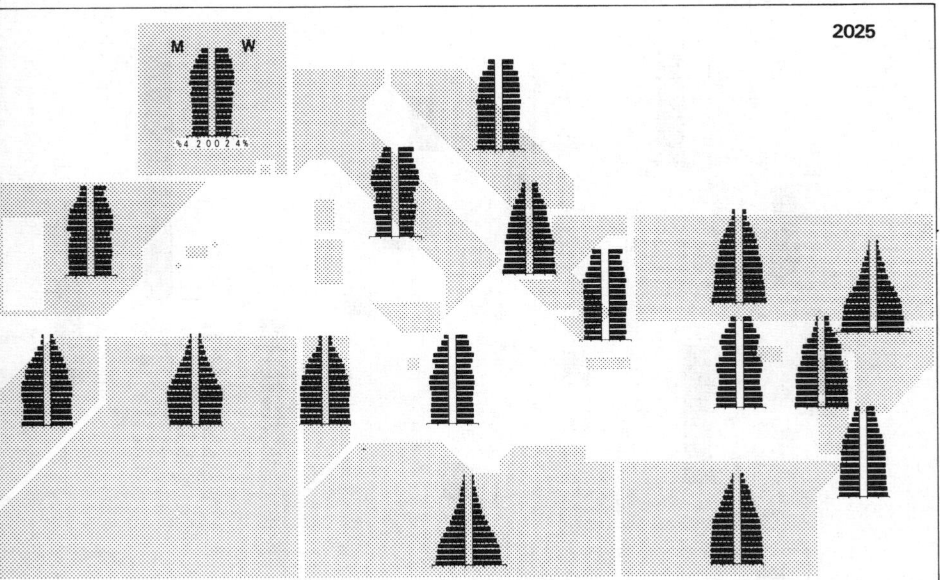

Figure 14.7. Age structure, 2025 AD, as percentage of total population (5-year age-groups). In 2025 the age structure in the southern countries will tend to resemble that of the northern countries. (*Source*: UN/Blue Plan.)

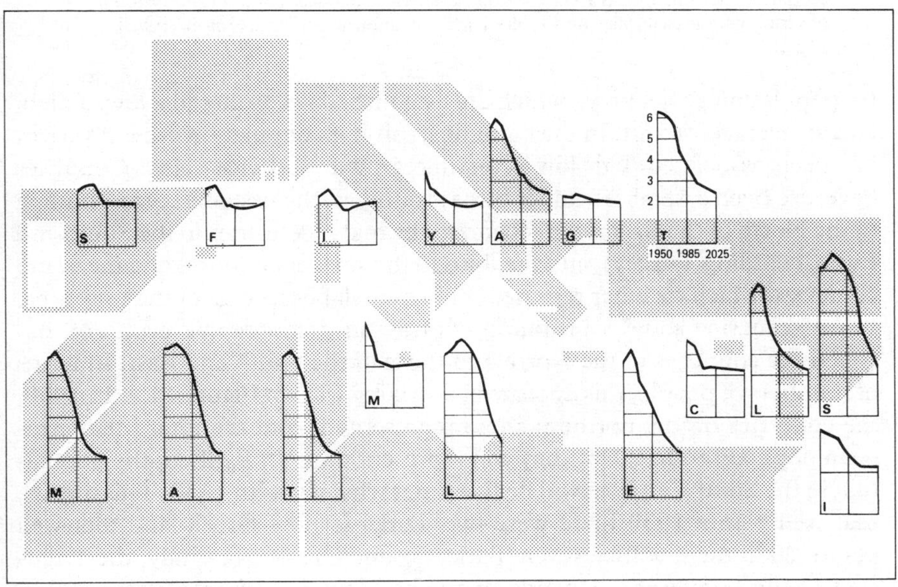

Figure 14.8. Number of children per woman: Trends 1950–1985 and average scenario 1985–2025. (*Source*: UN/Blue Plan.)

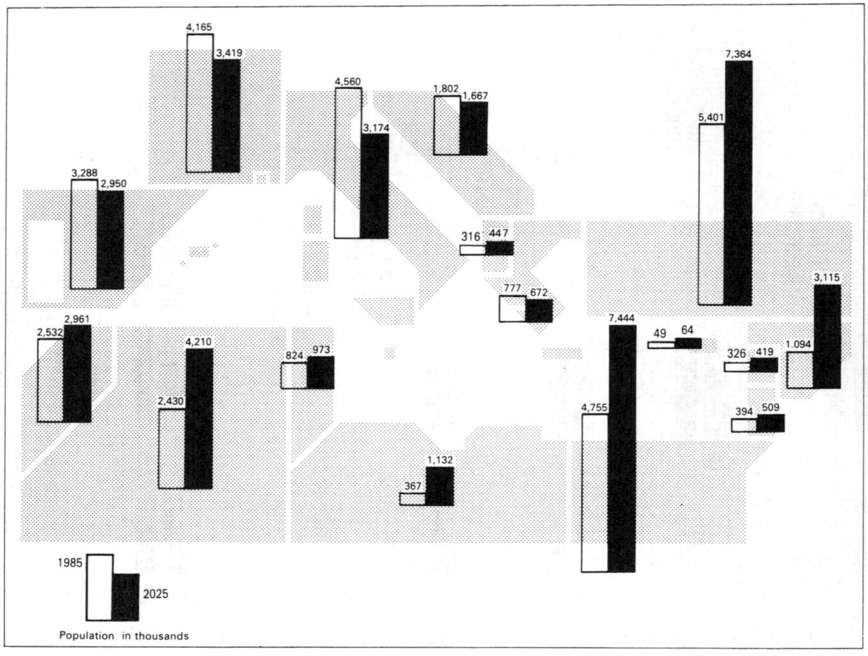

Figure 14.9. Population aged 15–19 in 1985 and 2025AD (000s): Average scenario. Analysis of the 15–19 age-groups helps to assess some of the problems of employment dynamics and related socio-economic aspects. (*Source*: UN/Blue Plan.)

UN population projections, which are established according to a low, a high, and an average variant. In the scenarios, this high population growth variant has been associated with low economic growth and vice-versa. Countries have also been assembled under three groupings, the first one corresponding to the North of the Basin, the second to the major countries in the South and East (including Turkey), and the third to the smaller countries and the island countries, which together represent only a small percentage of the total area.

All scenarios show a profound contrast in demographic evolution between the countries of the North and those of the South and East. In terms of numbers of people, this appears in a striking way on Figure 14.4. In 1950, the countries on the northern shore represented ²/₃ of the total Mediterranean population. In 2025, they will represent only ¹/₃. By then, the population in the South and East will be five times what it was in 1950. This means, that well before 1950 there were two Turks for one Greek, and that well before 2025 there will be seven Turks for one Greek. Naturally, the evolution of the population depends upon the scenarios. From a total of 330 millions in 1980 it could reach 520 (scenario A1) to 570 millions (scenario T2) in 2025, the difference being not small as it represents about the equivalent of one Egypt or one France of today. These evolution ranges are shown on Figure 14.5 for each of the Mediterranean countries.

SPAIN, FRANCE, ITALY, GREECE, YUGOSLAVIA

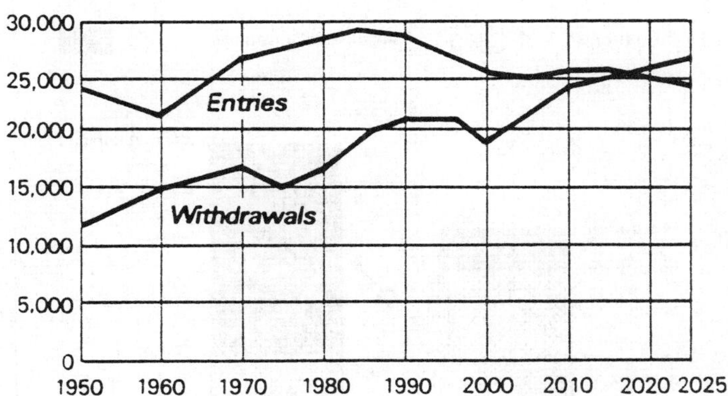

TURKEY, SYRIA, EGYPT, LIBYA, TUNISIA, ALGERIA, MOROCCO

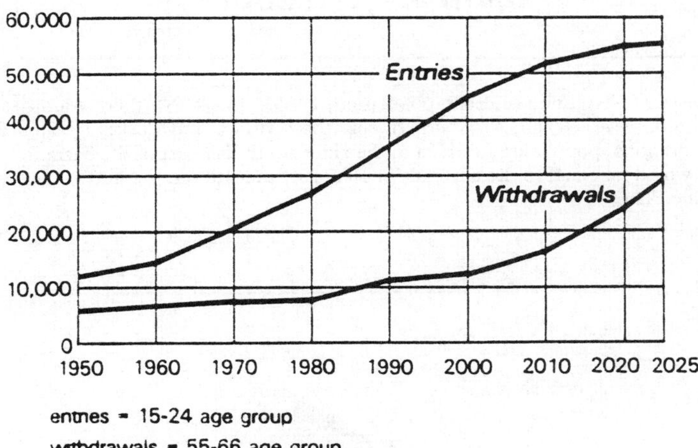

Figure 14.10. Entries into, and withdrawals from, the labour market (000s): Trends 1950–1985 and average scenario 1985–2025. (*Source*: UN/Blue Plan.)

This shift in population numbers between North and South is, however, accompanied by another equally important shift in the age-structure of population as shown in Figures 14.6 and 14.7. While populations in the North with low fertility-rate keep ageing, populations in the South and East remain very young. Only around 2025 the age-structure in the latter countries begins to resemble that of the North, when the decline in fertility begins to show its effects, as indicated by the number of children per woman shown in Figure 14.8. Thus, a dramatic increase in the number of people in the 15–19 years age-group occurs in the South and East in all scenarios. Figure 14.9 shows this contrasted evolution on a country by country average basis. In the South and East this evolution is bound to lead to formidable problems in

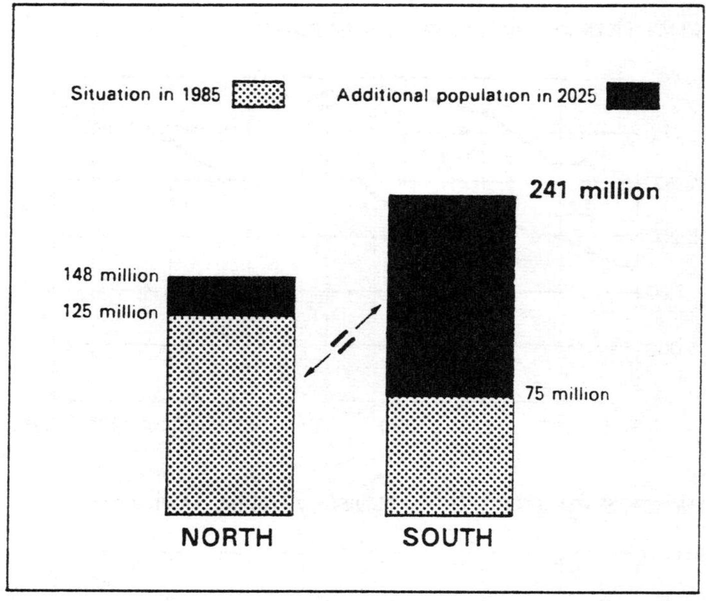

Figure 14.11. Evolution of urban population, 1985–2025AD. Northern countries: Spain to Greece (Region A). Southern countries: Morocco to Turkey (Region B). The additional population foreseen in the cities south and east of the basin in 2025 would be equal to the current population of the cities in the north. (*Source*: UN/Blue Plan.)

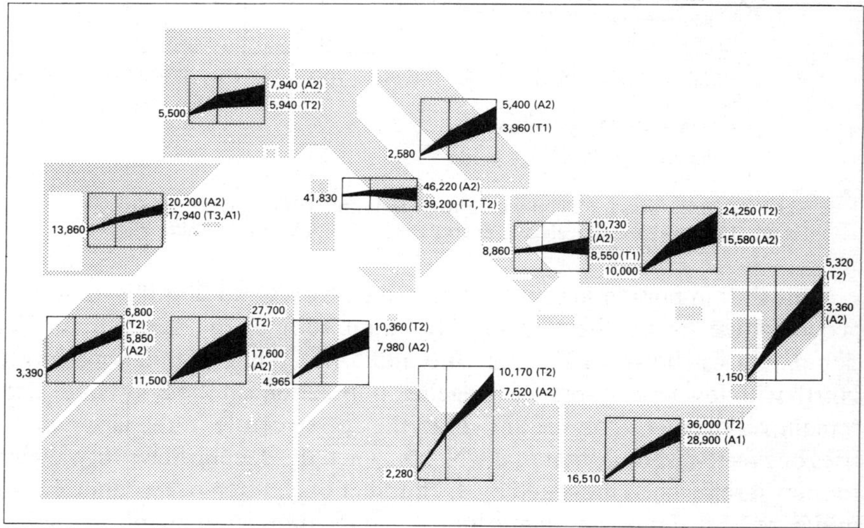

Figure 14.12. Coastal population in the Mediterranean countries (000s): Extreme scenarios 1985–2025AD. The coastal areas considered here correspond to the coastal administrative regions illustrated in figure 14.2. (*Source*: UN/Blue Plan.)

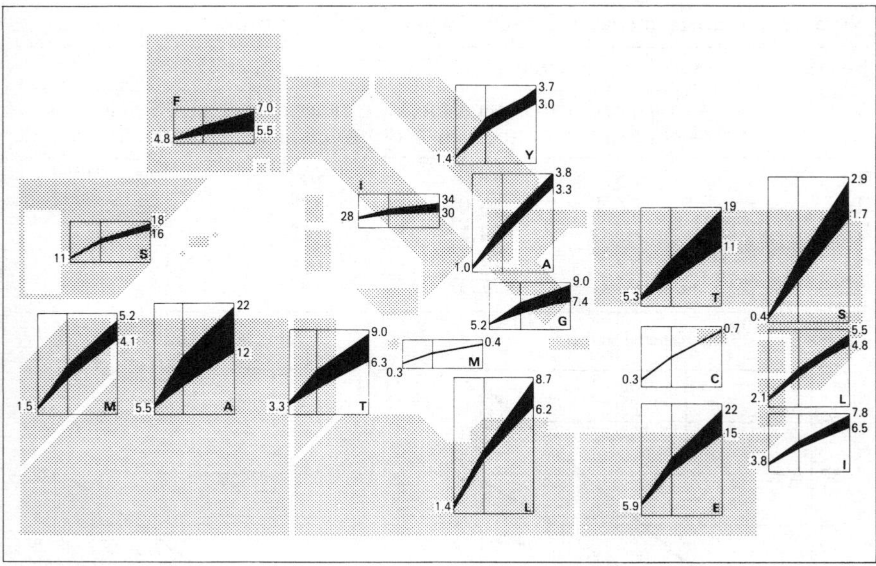

Figure 14.13. Urban population in coastal regions (m.): Extreme scenarios 1985, 2000, and 2025AD. (*Source*: UN/Blue Plan.)

such areas as education and social order, and particularly in the employment situation where entries to the labour market will be far in excess of withdrawals, as shown in Figure14.10.

Another factor of major significance is the trend towards urbanization, which has already taken place in the countries of the North and which is affecting more and more those in the South and East, although their rural population is bound to remain high in density. Figure 14.11 shows, in a striking way, how the average projected increase in the urban areas south and east of the basin would equal the current population of the cities of the north by 2025AD.

An aggravating feature of these population trends is the concentration of people and economic activities in the coastal regions. Figure 14.12 shows the evolution according to extreme scenarios in these regions as defined earlier – and Figure14.13 indicates the particularly high increase in urban population in these same regions. As a matter of fact, much of this concentration takes place in a rather narrow coastal strip, within a few kilometres of the shoreline. This 'littoralization' process, either permanent or temporary in the form of tourism, leads to extremely high human concentrations in this very fragile environment.

The Blue Plan scenarios have paid special attention to tourism as this is a major economic sector in the region, being one of the greatest (and often the greatest) asset of many Mediterranean countries, and also an activity which has considerable impact on the environment. Figure 14.14 shows the formidable increase in numbers for both international and domestic tourists

Number of tourists in coastal regions, 2000 and 2025 AD (m.)

Scenario	2000		2025	
	International tourism	Domestic tourism	International tourism	Domestic tourism
T1	85.4	53.9	147	72
T2	76.4	45.0	125	48
T3	94.0	64.1	162	98
A1	97.7	71.4	168	130
A2	107.0	77.3	193	148

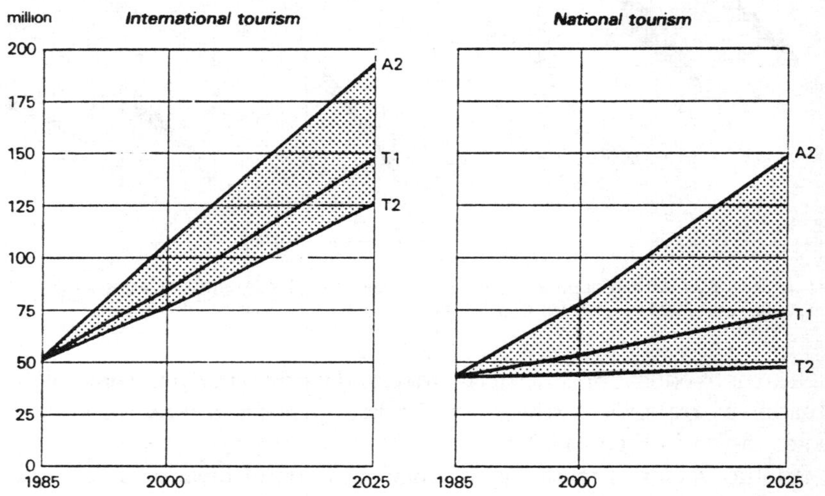

Figure 14.14. National and international tourists in the coastal regions, 1985–2025AD: Scenarios T_1, T_2, and A_2. (*Source*: UN/Blue Plan.)

which could take place between now and 2025AD in the Mediterranean countries as a whole. It is interesting to note on these diagrams that the highest increase corresponds to alternative scenario A2, which is one of the favourable scenarios. This result may appear paradoxical as increased numbers of tourists would inevitably increase space requirements (for hotels, parkings, roads and sports equipment), the water demand in peak periods coinciding with the dry season, the discharge of sewage and solid waste, the automobile and boat traffic, and other impacts along the coastal fringe. The high number of international tourists in scenario A2 reflects the assumptions which it makes on opening of trade and of human migration. The sharp increase in domestic tourism corresponds to the improvement of socio-economic conditions in all countries. This scenario illustrates the major importance of tourism for the economy of the Basin in the future. But it also assumes that the environmental impacts of tourism could be mitigated, *i.e.* that hotels, tourist attractions and equipment will be properly conceived and located, and that tourists will show more respect for the physical and cultural environment which they largely come to experience. This can be

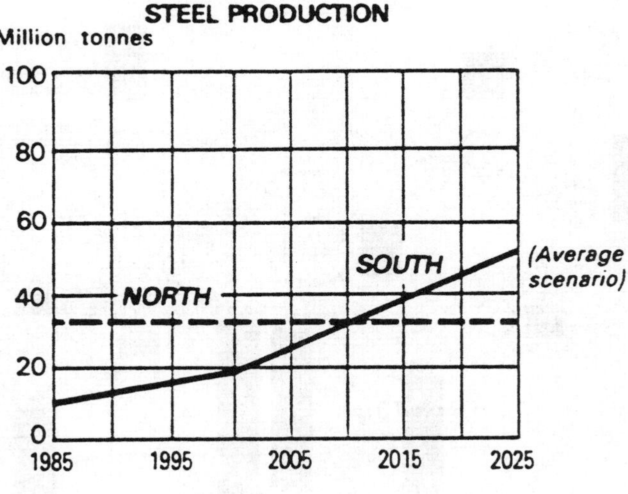

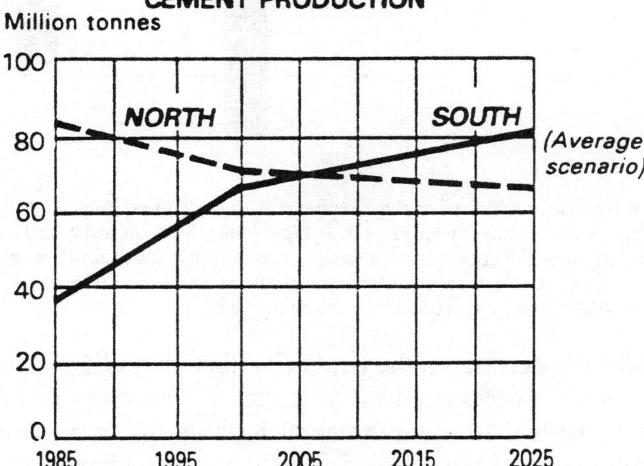

Figure 14.15. Two industrial activities in the Mediterranean to the year 2025 AD. (*Source*: UN/Blue Plan.)

done, although it will require considerable will-power, determination and organization, and the scenarios have attempted to estimate the future demands for space and water associated with the development of tourism, particularly in the alternative scenarios.

Generally speaking, all scenarios have attempted to correlate the anticipated increases in permanent and/or temporary population with the different economic sectors and environmental elements. Although quantification is not always possible, it has been attempted for a number of key indicators. For instance, the evolution in the production of major industrial commodities such as steel and cement, shown in Figure 14.15, is strongly related to the evolution of population. In the North, these heavy industries are expected

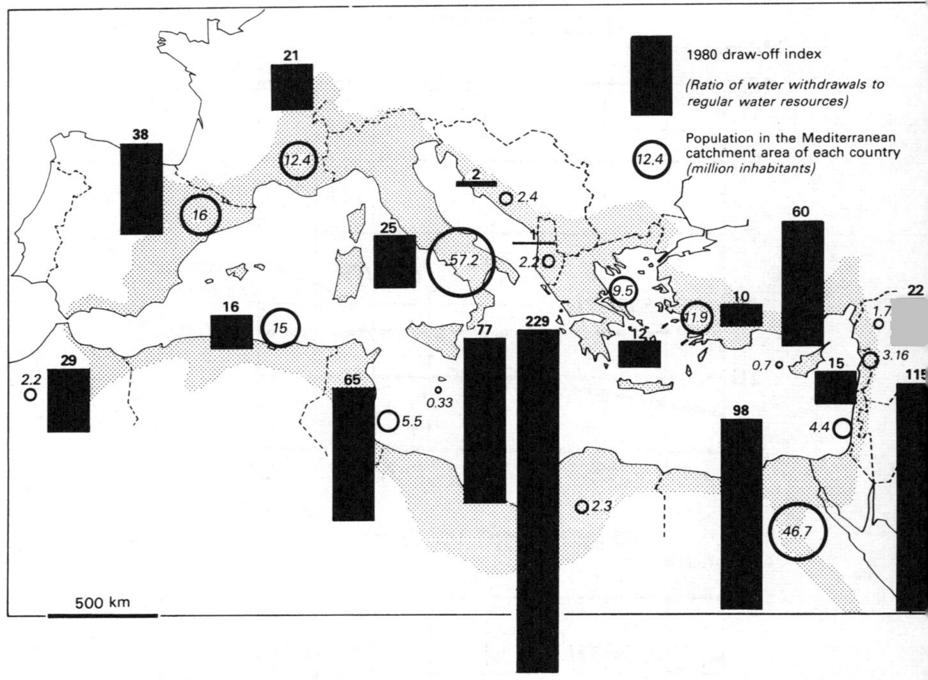

Figure 14.16. Water draw-off in the Mediterranean catchment area, 1985 (annual withdrawal as percentage of resources). A high water draw-off index calls for utilization of non-conventional resources: non-renewable fossil aquifers, recycling of used water, multiple water uses, reduction of losses, desalination, etc. (*Source*: UN/Blue Plan, adapted from J. Margat,1988.)

to peak, and even decline, to the benefit of high-technology industries and the tertiary sector, whereas in the South, they will have to expand considerably, together with their environmental hazards, to meet increasing demands resulting from population growth and associated domestic and commercial building.

Directly related also to population growth is the increase in water demand for domestic use, for industry and energy production (for cooling) and, above all, for agriculture. Irrigation is by far the largest consumer of water, usually supplied at a very cheap price, but all Mediterranean countries in the South and East, with the exception of Turkey, will have to import food massively in the future even if they do not do so already now. Thus, intensification of agriculture through irrigation is a necessity which they cannot bypass. Water resources vary of course from basin to basin in the region and so their assessment in global terms is of little meaning. A rough evaluation of the current situation at the country level is given by the 'exploitation index', which is the ratio of estimated draw-offs to total available water resources. These percentages are given in Figure 14.16, which shows that Israel and Libya, with figures higher than 100%, have to rely already on non-conventional resources through recycling of used water, desalination, or exploitation of non-renewable fossil aquifers. Egypt is close to the same situation, and an index above 60, as in Cyprus, Malta and Tunisia, means that strict water-use planning is already necessary in these countries. By 2000 AD the index is expected to exceed 100% for Egypt, Malta and Tunisia, and by 2025 AD, Algeria, Morocco, and Spain, are also likely to experience difficulties.

These are but a few examples of the paramount importance of the population factor in the dynamics of environment/development interactions in the Mediterranean Basin. The full Blue Plan report, which has recently been published,* of course enters into much greater detail on this matter. As an official exercise carried out in an intergovernmental framework, it limited itself to scenarios which show no drastic rupture with an orderly evolution of behaviour and decision-making. It is well known, however, that history does not necessarily follow such a path, and 'surprise' or 'catastrophic' scenarios could also be considered for this highly contrasted region. Within its self-imposed limitations, and on the assumption that economic growth will help to reduce population growth, the Blue Plan advocates that immediate major attention be paid to the Mediterranean coastal regions through integrated planning and management, and to the promotion of a better understanding of population-resources-environment-development interactions among national and local decision-makers in both the public and private sectors, and among all Mediterranean people.

*Grenon, M. & Batisse, M. (Eds.) (1989). *The Blue Plan: Futures for the Mediterranean Basin.* Oxford University Press, Oxford, England, UK: 280 pp., illustr.

Commentary on Chapters 13 and 14

CHAIRMAN: Dr R. Paul Shaw
PANELLISTS AND OTHER CONTRIBUTORS:
Medawar, Oza, Enyedi, A. Cloudsley, Medawar, Stanton, Medawar, Bazzaz, Shaw, Harun ur Rashid, Bazzaz, Ramakrishnan, Shaw, Simon, Purcell, Ramakrishnan, Barton Worthington

Medawar had not been present at the start of the Conference but hoped that it had already agreed, without exception, that humanity, and The Biosphere, could not survive unless it succeeded in balancing human numbers to human needs and with the needs of the other 4–5 million species which inhabited the Earth. If it had not, she hoped to be of help in converting any opponents of contraception, achieved by all acceptable means. As to her credentials for the task, she had two sons and two daughters, four grandsons, two granddaughters, and three godchildren, and she wanted universal family planning because she loved children and wanted them to have a future. She was also the Director of the Margaret Pyke Trust in London, now 21 years old, which supported the largest and most comprehensive family planning centre in the world.

Years ago her husband [the late Sir Peter Medawar, biologist and Nobel Laureate. Eds] had said that opposition to contraception could be understood if it was realized that the idea of being responsible for the size of one's family was *new*. Until the end of the nineteenth century, men and women had children and *death* reduced their numbers; that was no longer true. Despite the human haemorrhage of the daily death of 40,000 children and, in the Third World, the death of a woman every minute from abortion or childbirth, an increase of a million more children every 5 days was not prevented.

She believed that fear of this new idea – of being responsible for one's family's size – could be reduced to three unnecessary fears, and a fourth which could be overcome.

The first was *tribal*, the notion that we must have more people to be strong. The worst, recent example of this had occurred in a country not far away and not long ago.

The second was *pessimism*. People feared that the problem was insoluble, and too costly. Experience had shown that, where good family planning services had been set up, *e.g.* in Thailand, Indonesia and Colombia, the birth-rate had fallen rapidly, and so had maternal and infant mortality. The cost had not been prohibitive and, anyway, at less than the cost, both human and financial, of not spending the money.

The third fear was *ignorance*. Western opponents feared that family planning was being imposed on the Third World and that women did not want it. They had not learned that, according to the World Fertility Survey, the majority of women *wanted* to plan their families and wished for quality in their children, not quantity.

The last, fourth fear came from *underdeveloped men*, by which she did not mean underendowed or puny men, but those who were morally and intellectually underdeveloped. Such men were unsure of themselves and so disliked and feared the loss of their power over women. They feared that if women could control their fertility they would dominate the men and even threaten their livelihood, and they would neither want nor be able to look after them. As testimony she quoted the Rev. J. W. Burgon (1889): 'Inferior to us God made you and inferior to us to the end of time you shall remain'!

She believed that it was time for the world's really strong men to convert their weaker brothers: to realize that men and women got more from understanding and trust than from the insecurity or fearfulness. The effects of such a crusade would have a great influence, both on balancing human numbers with human needs, and on human happiness.

She ended with two quotations. The first showed that Indian thinking on birth control had long been ahead of English thinking, that managing human numbers was not only a panacea for the world's ills but it was essential if all other welfare activities were to succeed:

> 'We have fought disease ... the expectancy of life has gone up and we have reduced child mortality. But believe me ... all this is like writing in the sand. You write in the sand and the tide of population comes in and washes out all that is written'.
>
> High Commissioner M. C. Chagla (1963)

The second quotation showed that the future King of England had also understood the relation between human numbers and conservation:

> '...the pressure of people is the force above all that makes conservation necessary'.
>
> HRH Prince Charles, Prince of Wales (1984)

She hoped for progress.

Oza said the most tragic aspect of the population explosion was that it was mainly in Asia, Africa, and Latin America (where most of the people were already living at, or near, bare subsistence levels, with inadequate food, housing, education, and medical care) that the rates of growth were so alarmingly high. Even in the underdeveloped countries which had adequate potential resources, excessive population growth was swamping agricultural and economic development.

Asia, with far less *per caput* resources, faced an even more dismal future if it was to support the 4 thousand million inhabitants expected by the end of this century. The need for more food was the most urgent problem facing the world today. More than half of the world's people did not get enough to eat, or at least not enough of suitable quality.

About one-third of Mankind lived in an environment of relative abundance whereas the other two-thirds remained entrapped in a cruel web of circumstances that severely limited their access to the necessities of life. They had not been able to achieve the transition to self-sustaining economic growth. These were the people belonging to the poorer, mainly underdeveloped eastern countries. The gap between the rich and poor nations was no longer merely a gap – it was a chasm. The misery of the underdeveloped world was a dynamic misery which was continuously broadened, and deepened by a population growth that was totally unprecedented in history. This was why the problem of population was an inseparable part of the larger overall problem of development and environmental maintenance.

There was a moral responsibility on the wealthier nations, with low birth-rates, to help the people of underdeveloped lands to exercise responsible parenthood. It was in this context that the developed nations should give every measure of support to those countries which had already established family planning programmes.

It would be tragic if primitive religious taboos, irrational political dogmas, biological illiteracy, and political expediency, should conspire to prevent or delay a rational solution to this problem. Population growth must be controlled either by high death-rates, or low birth-rates. The world would soon have to choose whether future population growth was to be controlled by enlightened but artificial birth-control, accompanied by economic and social advancement, or by the ancient destroyers – pestilence, famine, and war. In the near future, it would not be surprising if the death-toll due to starvation outnumbered all the deaths which had resulted from armed conflicts the world over (Oza & Gaekwad, 1979*b*)*. The willingness of the major nations of the world to reduce armaments was a ray of hope, but the present huge transfer of armaments to the Third World was one of the great tragedies of our time.

What was in store for the younger generations in the Indian Sub-continent by the year 2021 AD? The basic priorities should be to take care of the vital needs of the members of the human society, *i.e.* food, clothing, shelter and the alleviation of suffering.

Over the years, many of the problems facing conservation had changed quite drastically. The 20th century had witnessed the development of two, largely new, traits – appreciation of wildlife and the explosive growth of human population, both of which loomed larger and larger as the second half of the century progressed. The soaring demand for food, timber, and housing, had, for example, led to the destruction of our natural abodes – the virgin forests of the plains and in the hilly regions of India – and of their beautiful wildlife (Oza, 1986).

The three crises which we had to face by the turn of the century were wood-hunger, food-hunger and energy-hunger, all correlated with population pressure. The Report of the Science and Technology Advisory Council of the Prime Minister on an *Approach to a Perspective Plan for the year 2000 AD – the Role of Science and Technology*, had drawn particular attention to the population problem and insisted that India's population must be restricted to not more than 970 millions by the year 2000 AD.

Population control and its associated social and economic aspects was a herculean task. Women needed to be educated on the subject. The 1981 census revealed a population of 685 millions which experts believed had been under-estimated by 43 millions. The Planning Commission had estimated that the population by 2000AD would exceed by 36 millions the 1985 estimates, because the expected decline in birth-rate had not come through as expected. The World Bank had estimated the rate of growth of population for India at 2.1% in the period 1980-87. Thus, India's population might cross the one thousand million mark by 2000 AD. The Operations Research Group study, based on the average annual increase in life expectancy and assumptions about fertility trends and couple 'protection rates', through family planning methods, had made two alternative population projections of 1,224 millions (high couple protection) and 1,402 millions (low couple protection) by 2021AD.

'Family planning' in India had become a 'dirty word'. Even the name of the Ministry of Family Planning had been changed to Health and Family Welfare! India was adding 15 million people a year – over one million people a month to its population; that was the equivalent of Australia, or a seventh of Brazil, annually, and in spite of having the oldest family planning programme in SE Asia! Clearly, if the present growth-rate was not dealt with reasonably, the population would explode into suffering, violence, and inhumanity.

With this were linked problems of slums, urban conglomeration, water resources, malnutrition, housing, unemployment, poverty and the retention of their life-styles by tribal communities. Unless significant and substantial measures to curb population growth were implemented, we should have to bid farewell to any hope of sustainable development. Even the *World Conservation Strategy for the 1990s* would have no option but to stress the need to control the global population explosion. It would also be necessary to impose ecological checks on the planning process in developed, developing and underdeveloped countries.

India led the world in the holistic view of Man and his environment. Paradoxically

*[*See* p. 301. Eds.]

enough, it was today on the brink of an ecological disaster. It had already experienced major hardships due to poor agricultural yields in the last two decades when the impact of population growth was most significant. No one seemed to consider ecological factors as responsible, ultimately, for the heavy loss of life.

In recent years, much fertile soil had been lost as a result of erosion. Deforestation was a tragic episode. People had, foolishly, destroyed much of the forest wealth on the plains and in the Himalayan regions, and had slaughtered their rich heritage of wildlife – mostly for temporary commercial gain alone. The wanton axing of the larger indigenous trees and uprooting of other floristic elements in the hills, and on the lower slopes of mountains, had gradually impoverished them and been followed by erosion that ultimately left them barren. The lower Shiwaliks, for example, had been ruthlessly axed down and devegetated (Oza, 1980).

A primary reason for the loss of fertile soil was that Indian forests were permitted to be widely overgrazed and, subsequently, forestry departments all over the country had become parties to the illegal cutting down of trees and/or silent witnesses of the gradual processes of deforestation.

He had no illusions as to the magnitude of the tasks before India. No one who was a committed conservationist could close their eyes to the alarming threat of desertification. Moreover, successive 'drought' years in a majority of the Indian States had been tragic. To add to their miseries, the loss of vegetational soil-cover in these regions had resulted in frequent, devastating floods. Droughts and floods, combined, had reduced much of the land to a pitiable shadow of its former self.

The prices of agricultural products had not been stabilized and were for ever on the increase. Ecological viability had been destroyed over a wide area and, consequently, human welfare was at stake. There had also been a widespread deterioration of natural resources. India had thus been affected badly by the actions of its own people, who had changed the ecological situation. If the conservation of the forests was not given a high priority, the chances for economic stability would look extremely grim in the years to come.

Farming lands had started to turn into uncultivable lands, and any drought conditions led farming communities towards virtual starvation. Consequently, the movement of the rural community was towards the urban areas – where most of the people were basically without land, employment, and food.

Now was the pregnant moment to involve youth in conservation and the environmental movement, for they had a greater stake in the future. Young people should be in the front-line to overcome biological and environmental degradation. They had the potential to restrict, and even reverse, such major catastrophes. Public education on *'environmental awareness'* (incidentally, the name of INSONA's quarterly journal!), of the need for fertile soil, for limiting family numbers, for good-quality livestock, as well as the importance of vegetational cover on the land through massive planting of broad-leaved trees on the mountain-slopes, valleys, plains, roadsides, and desert areas, and the maintenance of pastures and indigenous floristic elements in and around the Indian cities, must, henceforth, all be vital considerations.

The natural resources of the Rajasthan and Gujarat deserts were too limited to sustain the present vast community of people, while those in need of wood, and the agricultural development of more and more marginal lands, had stripped the Indian plains. Forest trees were vanishing from many areas, and mountain slopes, unfit for fruitful farming, had suffered from ignorance and an urge to take more and more from Nature.

As a result, the people of Jammu and Kashmir State, of Himachal Pradesh, of Uttar Pradesh and, to a larger extent, of Assam, Bihar and Orissa, were becoming victims of environmental deterioration, without a single mountain slope spared from agriculture. Yet there were still pressures from migration, and grazing of livestock on the last remnants of the forest heritage. To allot land to the landless in forested areas could be disastrous, for tribal people did not concentrate on the forest-based industries but relied on farming. When once the soil had lost its fertility, they shifted to another piece of land. Ultimately, the land remained neither useful for forestry, nor for agriculture, and the human misery multiplied (Oza & Gaekwad, 1979*b*).

We were all aware of the Sahel crisis. As a result of the environmental degradation, and the reduction in genetic diversity and variability in the natural habitats, it would not be surprising if the Sahelian latitude, including its grave problems, extended to the Indian Sub-continent. It seemed inevitable that we would witness extreme hunger for cultivable land. The pressures of population growth were so severe in India that it was difficult to maintain any balance between the hungry and the land. Education (to the educated!) in judicious land-use and the need to conserve natural resources might, however, bring some hope for the future.

Dr M. S. Swaminathan, when President of the World Conservation Union (IUCN), delivering a Valedictory Address at a recent conference on 'Communicating Conservation and Sustainable Development', held in Bangalore, India, on 6th April 1990, had pleaded for the establishment of an Environmental Amnesty System to monitor the environmental violation by the present generations at the global level ('*genetic violation*'), for environmental damage went beyond the present generation. Such a system could take care of human damage to the environment. It should serve, it was hoped, the same objective as Amnesty International, which took care of violations of human rights ('*somatic violation*'). For ecological viability and in order to alleviate human suffering, disarmament and economic justice were crucial. We must have the right to clean up the environment with economic security.

It was essential to have indicators of sustainable and non-sustainable development in order to prepare an environmental safety audit for any region. An unequal world just could not sustain conservation. The life-styles of the rich could not support the Earth. We would have to endeavour to eradicate environmental illiteracy. Social unrest was due to economic reasons, generated as a result of unsound environmental planning. We could not dream of a 'better common future' if we did not have a 'better common present'. Conserving today, we could save tomorrow. The 1990s was a critical decade for the future. We still had a chance to reverse the processes of degradation.

In the eyes of Indian culture and civilization we must salute the Divinity of Earth. Divinity of Mother Earth was the beauty of Mother Nature. We should take only for sustenance, only for fulfilling human needs and not for greed. At a time when the atmosphere of Planet Earth had begun to change, we must have the concept of *Vasudhaiva Kutumbakam* – the World as a Community, World Anthem, World Motto (INSONA's motto – *The Lord in his Grace created this beauteous Universe; the bounden duty to preserve it rests heavily on us!*) and World Flag with Planet Earth. Now was a time for a new model of thinking – a pre-nuclear attitude. Spiritual and ecological values in our lives could shape the road towards harmony, peace and understanding of human personality.

Planet Earth was very beautiful and so fragile. Mother Earth had nurtured us as her children. Would we be able to repay our debt? We should have to mobilize our resources to bring about a major breakthrough. Humanity would have to move towards a new level of consciousness to survive in universal harmony, or sink. Dr Karan Singh, President of the Hindu Virat Samaj, delivering the Keynote Address in the IUCN Conference on 'Communicating, Conservation and Sustainable Development', had urged that the spiritual and ecological values of the Indian cultural system should be renewed. The second version of the *World Conservation Strategy* (WCS) should adopt the Atharva Veda's holistic view of the Earth. The Atharva Veda had the magnificent *Hymn to the Earth* which was redolent with ecological and environmental values. Singh urged that the Assisi Declarations on Man and Nature (WWF 25th Anniversary, 29 September 1986) should be implemented, pointing out that all religions desired to sustain and cherish the Earth.

If the present degree of destructive activity by Man continued, as seemed likely, would we not risk suffering, before the turn of the 20th century, the worst ecodisasters and other catastrophes in the history of Mankind? Was the future of Mankind secure with the development of science and technology? The impact of Man was having a detrimental effect on The Biosphere, and the ability to save the global environmental future was largely in human hands.

The first task for Man was to curb population-pressures. The next step was to maintain and protect our Biosphere. Another urgent need was to save the world's tropical rain-forests and their ecosystems which sheltered the greatest diversity of plant and animal species on Earth.

The capability of choosing our destiny thus lay within us. We had totally to reject planning which included deaths of human beings as one of the 'costs'. The future would belong to those who were contemplative enough to give Mankind well-founded hope and a sensible, human way to attain that hope. The only option left to Man was to live in *Harmony with Nature*, not in conflict. It had to be either conserve or perish (Oza, 1983, and cf. 1986)!

We were fighting against time. What had we to lose? He concluded with a quotation from the *Gita*, the Bible of the Hindus: 'If you are slain on the battlefield, you shall indeed go to heaven; if you live through the battle, you shall continue to enjoy the Earth; therefore, get up and decide to fight'.

Cited: Oza, G. M. & Gaekwad, F. P. (1979*a*). Population increase – a global problem and remedy. *Proc. & Jt. Report, World Future Studies Conference & DSE – Preconference, Berlin (West), 4th–10th May 1979, part 2,* pp. 1175–79; Oza, G. M. & Gaekwad, F. P. (1979*b*). Environmental deterioration causing fears of food shortages in India. *Environmental Conservation,* 6(3), pp. 243–44; Oza, G. M. (1980). Potentials and problems of hill areas in relation to conservation of wildlife in India. *Environmental Conservation,* 7(3), pp. 193–200, 5 figs.; Oza, G. M. (1983). An Indian view of Man and Nature. *Environmental Conservation,* 10(4), pp. 331–35; Oza, G. M. (1986). Threats to unique wildlife through Indian habitat destruction. *Environmental Conservation,* 13(2), pp. 131–36, 4 figs.

Enyedi commented that **Myers** and the **Ehrlichs** (Chapter 13) had addressed the population versus environment problem at a macro-regional level. He wished to introduce a micro-regional aspect, *i.e.* the geographical concentration of the population. It was well known that strong population concentrations, especially in metropolitan areas, represented a heavy burden on the environment. In these urban concentrations all forms of pollution were present intensively, and the local destruction of The Biosphere was serious. Thus, global trends in urbanization could also influence environmental deterioration.

The North-South difference was explicit in another respect. The *first stage* of the modern urbanization had been chracterized by *urban explosion* everywhere, which meant the rapid concentration of population in urban areas. This stage was characteristic of the developed, industrial world in the 19th century. Urban growth in Western Europe was fed by rural outmigration, but was constrained by two conditions: a) the birth-rate dropped dramatically with the relocation of the population from rural to urban areas, and b) rural overpopulation was eased by outmigration to overseas.

In *Stage 2* – during the first half of the 20th century – a relative de-concentration had occurred within the urban zones, although the total urban population continued to grow. There was a continuous population de-concentration within metropolitan areas (due to suburbanization); moreover, urban growth was channelled towards medium-sized, and later to small, cities. Thus, population growth became distributed among a great number of urban settlements.

From the 1960s, metropolitan growth had ended (*Stage 3*). The phenomenon of de-urbanization had become almost general in the industrial world (with a few exceptions *e.g.* Japan). The bulk of the population growth was relocated into non-metropolitan areas. By the end of the '80s, we could register once more some metropolitan growth, but on the new, not-yet-overcrowded metropolitan areas.

In sum, urbanization processes in the industrialized countries had been characterized by a continuous population de-concentration in the second half of the 20th century. This fact had somewhat diminished environmental deterioration.

Most of the Third World (or South) countries were in the *first stage* of urbanization. Urban explosion had produced here a frightening growth of megacities. Of the twenty-five largest cities of the world (with over 5 million inhabitants), 16 were located in Third World countries. As far as the *megacities* were concerned (with over 10 million inhabitants), none of them was in Europe, only two were in North America (New York City and Los Angeles), and one was in Japan (Tokyo-Yokohama). All the others were in less-developed states.

Myers and the **Ehrlichs** had stated that in the highly developed countries the weight of *per caput* pollution was 15 times as high as in the developing countries. One should not conclude that environmental problems were concentrated in industrial countries. Megacities in the Third World had an extremely high population density, they concentrated poverty, they had no adequate infrastructure for environmental protection; consequently, they had – poorly monitored – serious pollution. This pollution was concentrated in small parts of the Earth's surface but it touched hundreds of millions of people.

The majority of the Third World megacities' population belonged to the traditional social sector; they did not form a modern urban society, nor did they change their earlier, rural, demographic attitude. Urban population growth had been fed by high natural increase, consequently, a curb on migration would not alter remarkably the population-growth pattern. Different remedies – from integrated rural planning to propagating urban societal values – were well known but their implementation had had limited success so far. The situation seemed hopeless to him, even in the long run if the present (technology and consumption-oriented) economic model became general on the Earth. He assumed that an alternative economic model ('sustainable development') would have to be developed, first of all in the industrialized countries – but he would not risk forecasting the success of this model!

A. Cloudsley was encouraged by the emphasis of **Myers** and the **Ehrlichs** on infant welfare. Increased emphasis on infant welfare was important even though it might result in the short- or medium-term in an increase in population. However, infantile mortality was not the only anxiety that drew women into having large families; their husbands demanded them because it demonstrated their virility. It was important to recognize this and take it into account, as **Medawar** had pointed out. For instance, many women in Sudan and Nigeria had told her that 3–5 children would satisfy them; but their husbands wanted more and the women feared divorce, or a second marriage, if they did not oblige. Furthermore, such uneducated men did not like using contraceptives and were ignorant about the aims of family planning. What kind of education should be provided for such men? Incidentally, if there were less children in these countries, fewer animals would need to be herded !

Medawar responded that she thought that it was for International Planned Parenthood and IUCN to decide the question of educational programmes. As a pointer she believed that President Borguiba had suggested that there had to be a change of heart so that virility meant 'Manhood', not 'Rabbithood', and that proper men looked after their women. She also thought that fear was a good contraceptive. If it could be brought home to men that what was happening to the environment was due to their enormous families which they could not look after, then perhaps they would think again. The consequences of not planning needed to be brought into every school and subject. She hoped that that would not become monotonous but one had to start with the next generation and start with fashion: it should not be fashionable to be a rabbit!

Stanton reminded participants that it had been said that it was necessary to remove taboos and inhibitions in relation to the family planning behaviour of indigenous peoples. Might that statement be more true for those societies where ecosystem-attuned cultures had been partially destroyed or obliterated, rather than to truly autochthonous groups? An extension of that notion was that it might be appropriate to identify, restore, and extend, ancient, inherent, population-control practices such as extended lactation – and he noted that the commercial world had been responsible for the swing away from breast feeding – seasonal reproduction, group segregation, and periods of untouchability. No doubt anthropologically-orientated, family-planning experts were aware of other such practices.

Medawar said that she had not accused indigenous (Third World?) people of having taboos against family planning but she had drawn attention to tribal fear, *i.e.* that a tribe had to keep its numbers well above those of its enemy, as one of the fears which prevented the use of contraception. She was all in favour of encouraging any decent practices which might help to balance human needs with human numbers, while offering commonly-used methods at the same time.

Bazzaz thought it imperative, because of the importance for policy and planning, to explore every angle and wished, therefore, to ask two questions. Firstly, was it not possible that urbanization *per se* would lead to a decline in the population increase? There was a relationship between the change from rural to urban populations and the birth-rate. Secondly, should not the issues between the 'haves' and 'have nots' be considered within countries as well as between countries? He thought the consequences of urbanization to be important.

Shaw [who had presented the substance of Chapter 13 on behalf of **Myers** and the **Ehrlichs** at the Conference] agreed that urbanization might lead to a reduction in the birth-rate, but more because urbanization was a generally modernizing force which exposed people to new ideas such as family planning. On the other hand there were countries so poor that migration to 'bright light' areas became simply migration to different but equally destitute areas. There the pattern of high fertility was maintained. The issue was difficult; some would argue that urbanization should be promoted to bring down the birth-rate but that needed to be reconciled with the view that the urban explosion was associated with other negative effects and that the correct policy was, therefore, to promote decentralized communities and curtail urban migration.

On the second point, he agreed that disparities within countries were important but they were difficult to deal with. The issue was constantly brought up at the UN but, because of the concept of national sovereignty, it meant that the UN had to accept that a country was speaking fairly for all its citizens even when it was obvious that there were inequalities, *e.g.* ethnic dominance, which resulted in UN action not being applied even-handedly, as in family planning matters. There was no obvious solution to this dilemma as such issues could not be honestly and openly discussed.

Harun ur Rashid said that he came from a country where the density of population (at 1900 persons per square mile) was one of the highest, if not the highest, in the world. He had been told that that was equivalent to squeezing the present world population on to the land territory of the USA! Amongst the population problems Bangladesh was facing was the extended family concept, *i.e.* grandparents and other relations lived with the families. As there was no old age pension or social security system, children – of whatever social class – had to ensure the economic security of the older members of the family. Another aspect of family size was related to the nature of the country's agriculture and agro-based industry. This was labour-intensive because of the lack of mechanization and, in any event, for a poor family to have 3, 4, or 5, children working on the land increased income and made the family a viable economic unit.

A second problem was to educate women. This was affected by the strong role which religion played in the status of women in society. Education led to employment and this changed the traditional status of women. He was not sure how effectively this problem was being tackled.

A further societal problem was differential growth within the population. The élite families were mostly educated but represented only 2–3% of the population and usually had far smaller families than the remaining 95%. Over the next 50 years this would lead to an even greater imbalance between the illiterate with many children and the educated with far fewer.

Finally, he pointed out that as over 45% of the population were 15 years or younger, and that family planning, so far, had only affected a small percentage of the population, the predicted population explosion was imminent. It had been recognized that an integrated approach to the problem – pre- and post-natal care, the provision of health centres and family planning services, adequate housing and education – was essential but most of these had still to be implemented and would require aid.

Bazzaz was rather concerned by the simple application of the multiplicative nature of the impact formula (p. 271). In industrialized and developed countries the amount of energy utilized, or impact on the environment was much higher than in less-developed countries. That, however, hid the fact that in some countries, Japan for example, part of the impact on the environment was due to the production of goods to be exported to, and used by, less-developed countries. Hence, it followed that some of the impact should really be assigned to these latter countries. This kind of situation should be allowed for in any models that were made and used.

Ramakrishnan commented, somewhat wryly, that he thought the last statement of **Bazzaz** was very dangerous as it should not be forgotten that Japan was not exporting goods to developing countries out of charity!

Shaw responded that the model was only illustrative and intended to point up differences between developed and less-developed countries. He was surprised that he had not seen many more sophisticated models and would certainly like to see further developments.

Simon thought that one had to be very careful when suggesting more sophisticated models. The point in developing equations in a fairly sophisticated form was that they were a sort of heuristic device to explain some key features, but it could only be a small step in attempting to give unjustified analytical, perhaps even political, rigour to something which could not really stand up to analysis. While the point made by **Bazzaz** was valid, it was also true that the environment in the Third World was being damaged in numerous ways in order to export more to advanced industrial countries; that further complicated the equation. There had, indeed, been several examples at the Conference of this kind of damage, *e.g.* the destructive logging which was depleting the tropical forests. Another example was the highly capital-intensive, highly mechanized, chemically-based agro-production cropping often promoted by institutions such as the World Bank as part of a structural adjustment export orientation in a very open global economy. But most of such production was for export rather than for the domestic economy. Another important complicating issue was that, over the last 10–15 years, one of the dominating influences on global output and the distribution of manufacturing capacity had been the environmental protection developed in North America, Western Europe, and perhaps more recently in Japan, although several other industrialized countries had increased their 'dirty' technologies. Meanwhile, dangerous production processes had been exported to the Third World where protection was less and enforcement mechanisms inadequate, even if in theory laws and regulations existed.

The point he wished to make was that, if international accounting and comparisons were going to be made, he suspected that the balance might be different from what might be suggested at first sight. There was a great need for political, and consciousness-raising, exercises to be simple and, on the other hand, for more substantive policy analyses and formulations to be analytically rigorous. Moreover the two, although distinct, should go hand-in-hand. That certainly applied to population issues.

Purcell liked the equational approach and believed that it should be developed as it enabled account to be taken of the net impact, *i.e.* the transfer back and forth when environmental activities in one sector were going to be for the benefit of another, the quality of environmental degradation, the degrees of control exercised, and so on.

He was deeply concerned to promote population control for all the reasons already adumbrated but, as an American, he found it difficult to push population control to the world with a straight face! At present there was a rapid increase in population in the USA supported by television advertisements and initiated by President Reagan. More recently, in his presidential campaign, Mr Bush had taken about 30 seconds to talk about the environment and *2* minutes about his *12th* grandchild. What kind of a tone had that set for the rest of the world!

Ramakrishnan considered that the existing available mechanisms for handling population problems were very inadequate: India provided an example. There were many instances of success stories of local communities, particularly village communities, which had been effectively mobilized and the people involved in decision-making where population problems had not been considered in isolation but integrated into the whole process of development and environmental management. Instances were known where the population had been reduced as a result. What was required far more widely, therefore, was a rather different kind of institution – one to promote population control based primarily on the village community as the basic building block, rather than the reverse which was more usual.

Barton Worthington, reflecting on the whole discussion in the light of the Conference, commented upon human population dynamics. Problems raised by demands on natural resources consequent on the ever-increasing population of the world had recurred frequently at the Conference. When first mentioned in discussion at an early stage, an attitude which had become fashionable in recent years was expressed (by **Holdgate**), namely that the population problem was so delicate that its consideration must be careful and approached with tact. However he, himself, had insisted that this was the most important ecological problem of all facing the world and that the Conference could do a good service by facing up to it squarely and decisively.

The rapid increase in numbers of people in most developing countries, often doubling in each generation, had been recognized 20 to 30 years ago, but now, with this problem much more acute, it had been superceded by that of atmospheric pollution which might lead to changes in climate. Though speeches by scientific, economic, and political leaders had recognized population pressure as a problem, it had generally been relegated to a kind of footnote. The Conference should bring it once again to the highest priority. Important as climatic changes might be, The Biosphere, including Mankind, had already weathered several since the Ice Age, whereas human population dynamics, if they carried on as at present for another generation or two, were likely to cause a catastrophe once and for all.

During the colonial days in the Third World, up to about the 1930s, a rising population was regarded as essential for economic development, but then A. V. Hill, the British biophysicist and Member of Parliament, returned from a tour of India. In a series of forceful lectures he pointed out that the average expectation of life among

Indian peasants was about 27 years and that the infant mortality rate approached 400 per 1,000. But in spite of this the population was increasing at a truly alarming rate. India was not unique, for vital statistics from countries in Africa and elsewhere, meagre though they were, began to show the effect of health services, which were generally the first public services to be introduced. The expectation of life at birth increased, infant mortality was reduced dramatically, but there was no evidence of any reduction in the number of babies per family. Current data and predictions which were widely known (*see* Oza, pp.297 *et seq.*) could only lead to disasters which were already all too obvious in the famines of Ethiopia, the Sahel and parts of the Far East.

Exceptions to this gloomy foreboding were the very few countries, including China, where the law had been invoked to reduce birth-rate. There was also some hope in countries where leaders had spoken out strongly on the subject and, consequently, family planning had been able to establish firm footholds. Moreover, side by side with the gloomy evidence from the Third World, the recent experience from Europe provided an optimistic note: through a combination of economic progress, social attitudes and family planning, the increase of population which followed the industrial revolution during the 18th century had been brought under control to the extent that France, for example, was now worried about a declining population!

The problem was clear: its solution was less so. The argument widely used that developments, backed by public services already under way, would provide a solution in due time, was flawed; for the time was long overdue. The Conference recognized that a solution depended on many factors and that the problem should be attacked from many directions. As one of these, it was noteworthy that the British Chapter of the Club of Rome, recently reactivated, was focusing attention on the many religions in developing countries and how they might be persuaded to exert pressure to reduce birth-rates.

In general there was little hope of solving this overriding problem unless it received world-wide attention: at least as much as the 'greenhouse' effect and climatic change had at the present time. Leaders in countries of the North as well as the South should provide maximum support; the media should follow, emphasizing the practical rather than the sentimental, with propaganda reaching to villages and fields; and adequate finance should be provided so that both governmental and non-governmental organizations could play their full part in the campaign.

Towards the end of the Conference a question was put to **Tolba**, after his delivery of the Baer-Huxley Memorial Lecture (*see* Chapter 21) in which he had outlined the major environmental problems facing the world today and the organizational improvements needed for their solution. The question was: at what level of priority would he place the population problem? With no hesitation he had answered 'Top'.

[Postscript: The report of the UN Population Fund, *The State of the World's Population, 1990*, which was launched in London on 14 May, stated that the world's population was now increasing faster than ever before, with 250,000 babies born each day. The updated total of 5.5 thousand million people is expected to increase by between 90 million and 100 million ***every year this decade***.]

15. Human Instability and the Release of Dangerous Forces

ARTHUR H. WESTING

Senior Research Fellow, International Peace Research Institute, Oslo: Fuglehauggata 11, N-0260 Oslo 2, Norway, and Adjunct Professor of Ecology, Hampshire College, Amherst, Massachusetts, USA

INTRODUCTION

It is our privilege to be gathered here this week on the initiative of Professor Dr Nicholas Polunin in order once again to be looking into the environmental future – this time to be seeking ways of 'Surviving With The Biosphere'. It is the fourth such occasion, starting with that of 1971, for which Dr Polunin deserves such enormous credit (Polunin, 1972, 1980; Polunin & Burnett, 1990).

On the other hand, it is a tragedy that the necessity for these meetings has not diminished during the intervening two decades, but, rather, has become ever-more-urgent. It is thus my purpose today to alert you to one of the newly-emerging challenges to surviving with The Biosphere – one that provides an ever-larger potential for undermining human stability on this Earth.

In order to set the stage for my presentation, I begin with a few indicators of the increasingly alarming general level of human instability, both in the civil and military sectors of human society. For this purpose I dwell only upon the two decades that have elapsed since the First International Conference on Environmental Future. These preliminary remarks are meant to lay the groundwork for achieving one of the increasingly indispensable components of *environmental security**, to wit, the attainment of security from the

*The attainment of environmental security is the *sine qua non* of our long-term survival and well-being on Earth (Westing, 1989*a*, 1989*b*, 1989*c*, 1989*f*): for on one hand, environmental security is an inextricable component of comprehensive human security (with its further requirement of social security); and on the other, environmental security is a profound obligation that we have to the creatures with which we share this globe. Environmental security has two basic and inseparable components. The first of these – *environmental protection* – has three parts: (1) protection from wartime and similar vandalism; (2) protection from medically unacceptable environmental pollution; and (3) protection, of special areas, from all but temporary human intrusions.

The second basic component of environmental security – *sane resource utilization* (whether non-extractive or extractive) – depends upon exploitation (use or harvesting) at levels and employing procedures that either maintain or restore optimal resource services or stocks. Exploitation of renewable resources must be carried out strictly on the principle of sustained use or sustained discard, and that of non-renewable resources strictly on the principle of frugality. Moreover it should go without saying that both of the two components of environmental security unavoidably derive from – and are thus unavoidably constrained by – the fundamental principles of ecology.

pent-up dangerous forces in the industrialized environment – the release of which is becoming an ever-more-likely concomitant of military hostilities.

Human Instability

The Civil Sector

The demands of the *civil* sector of human society upon Nature ultimately derive from our unconscionably high numbers – a fact that is only recently becoming quite widely recognized in most parts of the world, though not so effectively in the East. Prior to the 1970s and even the 1980s, the demands upon Nature that were occasioned by our growing human numbers and their increasing aspirations, were sufficiently cryptic to escape widespread popular attention. This was owing in part to their alleviation by technological advances, in part to a not-widely-appreciated, 'grabbing' or so-called 'borrowing' from the future (namely *via* non-sustainable resource exploitation on one hand and non-sustainable waste disposal on the other), and in further part to continued material prosperity, especially in the West. Indeed, this was the case despite a clear recognition of both the enormity and poignancy of the problem by such perspicacious conservationists and naturalists as H. H. Fatehsinghrao P. Gaekwad of Baroda (speaking, for example, at the first of these conferences [Gaekwad, 1972]) and Dr F. Raymond Fosberg (speaking at the second of these conferences [Fosberg, 1980]).

But our sins have been catching up with us during the past two decades. In 1970 there were only 24 abjectly poor countries in the world (those that the United Nations General Assembly designates as 'least developed' on the basis of their dismally low levels of *per caput* income, literacy rate, and industrial development); by 1980 their number had risen to 31; and by today (1990) it has reached 42 – a 20-years' increase of 75% in the number of hapless nations (*Ceres*, 1975; UNCTAD 1984, 1989). Yet the trend is more pervasive than even these sad figures suggest. In 1970 only one country (Chad) was experiencing a continuing decline in gross national product *per caput*; by 1980 the number of countries thus 'losing the economic race with their population growth' had risen to about 35; and now today the number has reached an incredible 90 or more (ACDA, 1982 Table I, 1989 Table I).

As those of us gathered here this week know so well, growing human numbers *plus* increasing aspirations are unfortunately taking place in a global environment of fixed size. Thus, in 1970 there were still 3.6 hectares of land surface on Earth per person; by 1980 that area had shrunk to 3.2 hectares per person; and by today (1990) it is down to only 2.5 hectares per person – a 20-years' decline of 30% in global *per caput* land availability (ACDA, 1982 Table I, 1989 Table I). Moreover, expansion of irrigated farmlands can no longer keep up with expanding needs: irrigated area *per caput* was still rising steadily in 1970; but by the late 1970s this value had peaked, and it has been declining steadily since (Postel, 1989 pp. 9–10). Furthermore, world grain production *per caput* was rising in 1970, as it still was in 1980. However – despite continuously rising fertilizer and pesticide inputs – world grain production *per caput* peaked in the mid-1980s, although it is, of course, still too early to tell whether the downward trend since the mid-1980s will prove to be a continuing one (Brown *et al.*, 1989).

It would be shockingly and tragically easy to extend this litany of anthropocentric and ecocentric indicators of the increasing instability of our civil sector, but I will refrain from this here, having already indulged in such extension at our last-preceding conference three years ago (Westing, 1990*b*). I refer you as well to the complementary indications by Drs Martin W. Holdgate and Mostafa K. Tolba at this Conference.

The Military Sector

The demands of the *military* sector of human society upon Nature ultimately derive from our unconscionably bellicose behaviour. Indeed, it may be useful to remind ourselves how much a part of human society its military sector is. Some 142 of the 171 sovereign nations of the world currently maintain armed forces (Westing, 1990*c*). Moreover, merely since World War II, the armed forces of about 100 of those 142 militarized nations have intruded upon the territory of some other sovereign nation for hostile purposes, once every 7 to 8 weeks on average (Tillema, 1989). Moreover, during this time there has been an ever-more-frequent use of those armed forces for domestic repression with deadly force.

The demands of the military sector upon Nature, occasioned by our bellicose behaviour, fall into two categories (a) those deriving from the continuing *peace-time* – that is to say, interwar – training and other military activities; and (b) those deriving from the disruptions of the environment caused by the frequent *war-time* military activities.

The peacetime drain on Nature by the military sector has been a relatively unchanging one in recent decades. The size of the military sector of society – and thus the general magnitude of its consumptive and disruptive environmental impacts – can be roughly estimated in a number of indirect ways, for example (Westing, 1988*b*): (a) the military sector accounts for just over 2% of the global labour force; (b) the military sector accounts for almost 6% of a world-wide summation of gross national products; or (c) the military sector accounts for about 4% of all the ships at sea (as measured in terms of their tonnage).

Having suggested that the peace-time activities of the military sector represent some modest though finite fraction of the totality of human activities, I must hasten to add that a reduction of the military sector – even to the point of *complete* abolition – would be only modestly sparing of the environment. This is so because such reduction would not represent a significant retrenchment of human activities, but rather, primarily, a shift from military to civil activities.

Nonetheless, a reduction in the military sector of society would be an environmentally desirable goal for at least two reasons: (a) such a reduction would diminish the likelihood of wartime devastation (as elaborated below), and (b) such a reduction would provide a source of financial, material, and intellectual, resources for environmental protection that are not otherwise readily available. Especially advantageous would be a shift of intellectual resources from weapons' development to environmental protection, as some 20% to 25% of all scientists and engineers in the world today are engaged in military research and development – that is to say, currently engaged in destructive efforts rather than constructive ones (Westing, 1988*b*).

The occurrence of the spasmodic disruptions of Nature that have been brought about by armed conflict, has also been remarkably frequent and remarkably consistent in recent decades (Westing, 1982; Eckhardt, 1989). Here I dwell only upon the more significant of the dozens of wars that are always liable to be in progress somewhere in the world, and again restrict myself to the 1970s and 1980s. During these past two decades, Humankind has been embroiled in at least 24 wars that each resulted in 33,000 or more direct combat fatalities (counting both military and civilian ones). And, of these 24 'significant' wars – at least 8 might well be referred to as 'major' wars – namely those that each resulted in one-third of a million or more direct fatalities (Westing, 1982; Eckhardt, 1989).*

Although any war has the potential for substantial environment disruption, such disruption becomes especially likely in what I have just defined as major wars (*cf* Westing, 1980). Moreover, there has been a clearly discernible tendency in recent decades for wars to become ever more environmentally disruptive, for a number of technical and policy reasons (Westing, 1980 pp. 2–5).

Despite the foregoing, the greatest contribution to human instability from the military sector derives not so much from what has actually been occurring all along, but rather from what might quite conceivably happen in the future. Two potential threats to human stability from the military sector must be stressed, both of which have progressed to the level of potential cataclysm largely during these past two decades. The first of those cataclysmic threats to the human environment is a major nuclear war, whereas the second is the release of pent-up dangerous forces. The threat of nuclear war is by now well known to most of us (*cf.* Westing, 1987), so that I need not discuss it further here. By contrast, the threat of released dangerous forces is *not* well known, and surely deserves our keenest attention will now be explained.

* The eight wars of the 1970s and 1980s that each resulted in one-third of a million or more fatalities (the cut-off point, on a logarithmic scale, between 10^5 and 10^6 fatalities) are: (1) Second Indochina War of 1961–75; (2) Bangladesh War of Independence of 1971; (3) Ethiopian Civil War of 1974–; (4) Cambodian Insurrection of 1975–77; (5) Angolan Civil War of 1975–; (6) Afghanistan Civil War of 1978–88; (7) the Gulf (Iran–Iraq) War of 1980–88; and (8) the Sudanese Civil War of 1984–. (Westing, 1982; Eckhardt, 1989.)

The Release of Dangerous Forces

Current Status

Many of the nations of the world are becoming ever-more highly developed and heavily industrialized. Of the various potentially dangerous artefacts of our rapid world-wide industrialization, I here single out three: (1) nuclear power-plants; (2) chemical factories; and (3) dams. Damage of huge proportions to the human environment could be accomplished by attacking these nuclear, chemical, or hydrological, facilities, whether such attack were intended or not. Thus, the dangerous forces that have become ever-more likely to be released over wide areas in a future war (whether overtly or through sabotage) now include *radioactive gases or aerosols* from nuclear facilities, *toxic gases or aerosols* from industrial chemical facilities, and *impounded waters* from hydrological facilities, which I accordingly consider in sequence:

Nuclear facilities

Dealing first with the radioactive threat, the human environment now contains about 200 civil nuclear power-plant clusters in 26 countries (altogether totalling about 466 separate plants, of which some 429 are currently in operation) (IAEA, 1989 Tables 1 & 13), *plus* a number of nuclear-fuel reprocessing plants and nuclear-waste storage sites. All of these nuclear facilities have been constructed since World War II – and more than 80% of them during the past two decades.

The possibility exists that a destroyed nuclear facility will contaminate a large surrounding area with iodine-131, caesium-137, strontium-90, and other radioactive debris – an area that would be measurable in hundreds or thousands of hectares. The most heavily-contaminated inner zone would become life-threatening; an outer zone of lesser contamination would become health-threatening; and a still greater zone beyond would become agriculturally unusable. Such a radioactively-polluted area would defy effective decontamination. Its degraded status would recover only slowly, over a period of years or decades, as has been demonstrated by the Pacific test islands. And, of course, the Chernobyl accident of April 1986 suggests very well the extensive disruption to the human environment that could be expected from an even worse occurrence of this kind (Westing, 1989*e*).

Chemical facilities

Turning next to the toxic threat, the human environment now contains many tens of thousands of factories, of which – by way of important example – many thousands (employing of the order of nine million workers) manufacture industrial chemicals (UN, 1985). Indeed, some 72 countries each now have at least 1,000 chemical workers, of which 13 countries each now has fully 100,000 or more such workers. Some factories that manufacture industrial chemicals might well release explosive or toxic substances into the environment if attacked. Similarly dangerous chemicals in storage or in transit add to this category of threat. Two past events will suggest some of the tragic possibilities:

1. An accidental explosion at a chemical (trichlorophenol) factory in Seveso, Italy, in July 1976, contaminated a surrounding zone of about 300 hectares with health-threatening levels of 2,3,7,8–tetrachlorodibenzo-*para*–dioxin, necessitating the long-term evacuation of many hundreds of local inhabitants, the sacrificing of thousands of their domestic animals, the destruction of huge quantities of local produce and crops, and the arduous several-years' process of decontaminating the affected farmlands, homes, and other local artefacts (Westing, 1978).

2. An improper procedure at a chemical (carbaryl insecticide) factory in Bhopal, India, in December 1984, led to the escape of a massive cloud of methyl isocyanate, which killed more than 2,000 local residents and permanently disabled an even larger number; there was also massive livestock mortality (Bowonder *et al.*, 1985).

Hydrological facilities

Turning last to the threat of flooding, the world's human environment now contains some 777 dams, scattered throughout 70 countries, that are at least 15 metres high and impound over 500 million cubic metres of water each; in fact, some 522 of these dams (in 63 countries) each impound over 1,000 million cubic metres (ICOLD, 1984; Mermel, 1988). Most (more than 90%) of these hydrological facilities have been constructed since World War II – and more than 60% of them during the last two decades. It is obvious that a substantial proportion of those many hundreds of huge impoundments now scattered throughout the world would make eminently suitable military targets; and it is additionally obvious how devastating the downstream effects could be on the human environment.

Indeed, the breaching of dams for the purpose of releasing impounded waters has been spectacularly successful in past wars, including both World War II and the Korean War of 1950–53 (Westing, 1984, 1990*a*). I present one specific example from World War II that demonstrates how very tempting such targets can be. During World War II, the Allies in May 1943 destroyed in the same operation two major dams in the Ruhr valley of Germany, the Möhne and the Eder (Quast, 1949; Brickhill, 1951). A vast amount of damage resulted from the breaching of these two containment structures – actions that had each released of the order of 120 million cubic metres. Despite these being not very great quantities: (a) 125 factories were destroyed or badly damaged, 25 bridges vanished, and 21 were badly damaged, while a number of power-stations were destroyed, numerous coal mines were flooded, and various railroad lines were disrupted; (b) some 6,500 cattle and pigs were lost and 3,000 hectares of arable land was ruined; while (c) 1,300 German lives were lost *plus* those of unnumbered slave labourers.

Overall, it becomes clear that the release of dangerous forces from nuclear, hydrological, or chemical facilities – whether by hostile intent or incidentally – would now constitute one of the grave threats to the human environment in any possible major war of the future and, perhaps to a somewhat lesser extent, under any circumstance. It thus becomes appropriate for me to devote the remainder of my presentation to suggesting means of preventing the release of such dangerous forces, or at least of mitigating their impact. I first touch upon technical approaches and then upon legal approaches.

TOWARDS A MORE SECURE FUTURE

Technical approaches

Technical approaches to preventing the release of dangerous forces from the artefacts of an industrializing society, or of mitigating their impact, can take one or another of several forms. A potentially dangerous facility could be made more nearly safe (Westing, 1990*d*): (a) by doing away with the facility altogether and either substituting for it a more benign source of the societal benefits it had been providing, or else doing without them; (b) by adopting safer designs and procedures*; or (c) by surrounding the facility with a buffer zone that is free at least from permanent human habitation, and large enough for any highly-dangerous releases essentially to spend themselves within that zone.I dwell here upon the notion of a buffer zone.

The establishment of buffer zones, free from permanent human habitation, around facilities that contain potentially dangerous forces, appears to be an obvious precaution to adopt (Westing, 1989*e*). Thus, each of the 200 or so nuclear power-plant clusters in the human environment should *invariably* be surrounded by a buffer zone that would coincide with the area from which any rapid and long-term evacuation would be necessary – that is, an area perhaps 500,000 hectares in extent, which would mean having a radius of about 40 kilometres).

The suggested buffer zone has to be free (or made free) from all permanent human habitation, from all major transportation arteries, and from all

*Various possibilities exist with respect to more-nearly-safe design features for facilities containing potentially dangerous forces (Westing, 1990*d*): for example, all present and any future nuclear power-plants should be fitted with massive nuclear-containment structures. The core-cooling systems – both the routine ones and the emergency ones – should be rigorously protected against possible disruption, whether such disruption is on-site or off-site. And, for any future plants, the type of reactor should be chosen or designed with inherent safety as a paramount goal. As an example, gas-cooled reactors (such as those prevalent in the United Kingdom) are inherently less likely to permit catastrophic releases of radioactive contaminants into the human environment than the far-more-widely-employed water-cooled reactors, or than the graphite-cooled reactors (such as the one at Chernobyl). Chemical plants that contain potentially dangerous forces might well be constructed partly underground, and with adequate containment structures. Dams that impound large bodies of water should be built more massively than the present design criteria calls for. As for prudent protocols, war-time contingency plans for nuclear power-plants that are subject to attack could include operation at lower power-levels in order to reduce their radioactive inventory, followed by their being shut down. Similarly, war-time contingency plans for water impoundments that are subject to attack, should include reduction of their water-levels.

militarily attractive targets. Moreover, it would be appropriate not to construct a nuclear plant closer than about 50 kilometres from any territory of another nation. Buffer zones surrounding potentially dangerous chemical plants (but smaller than those just suggested for nuclear facilities) make equally good sense, as would buffer zones in areas that are subject to catastrophic flooding following the breaching of a dam.

Needless to say, had a buffer zone of the sort that is being suggested here been in place around Chernobyl, most of the immediate human agony, long-term human sequelae (psychological, somatic, teratogenic, mutagenic, and genetic), and the monumental post-accident expenses, would have been avoided. Indeed, the numerous peace-time accidents involving nuclear, chemical, and hydrological, facilities combine to emphasize the enormous importance of such buffer zones in association with all artefacts in the human environment that contain dangerous forces, no matter whether releases from them might occur in time of war or peace.

Although the buffer zones that are being proposed here could well be devoted to agriculture, range management, or forestry, in many cases it might be even more desirable to set them aside as Nature reserves, that is, as reserves conforming to one or another of the protection categories I to V of the World Conservation Union (Gland, Switzerland) (IUCN, 1985 pp. 4–11). If such a buffer zone became a Nature reserve, it would add to the currently inadequate global extent of protected Nature (Westing, 1990*b*).

Legal approaches

As for legal approaches to preventing the release of dangerous forces into the human environment, a body of international law does exist that has as its purpose the prevention of utter human and environmental devastation in time of war (Goldblat, 1982; Lupis, 1987 Part II; Westing, 1988*c*). However, it is flawed in a number of fundamental respects, especially because: (a) the use of nuclear weapons is not thereby expressly proscribed (Westing, 1989*d*); and (b) there is no widely accepted form of unconditional compulsory arbitration or adjudication to provide for the non-violent resolution of interstate conflict (Westing, 1990*c*).

The most salient (though nonetheless inadequate) restriction on the release of dangerous forces in time of war derives from a pair of closely-related treaties: these are Bern Protocols I and II of 1977, additional to the Geneva Conventions of 1949 relating to the Protection of Victims of Armed Conflicts. According to both Bern Protocols I and II of 1977, belligerents are prohibited from attacking dams or dikes and nuclear electrical generating stations, if such actions could result in the release of dangerous forces that would cause severe losses among the civilian population (Westing, 1988*c*). As circumscribed as this proscription is, it nevertheless remains necessary for me to have to urge the many still recalcitrant countries – indeed, roughly half the nations of the world – to become party to Bern Protocols I and II of 1977.

Also of relevance with respect to treaty restrictions on the release of dangerous forces, is the (regrettably still nascent) principle that nations have the responsibility to ensure that their activities do not cause damage to the human environment in areas beyond their jurisdiction. This principle has been most clearly enunciated in the, albeit non-binding, Stockholm Declaration of 1972 on the Human Environment (constituting its Principle 21) (UNGA, 1973). Needed in this regard is a multilateral treaty that enshrines the concept and makes it generally applicable to the transboundary release of dangerous forces in both peace-time and war-time.

Conclusion

Thus the two decades since the First International Conference on Environmental Future have witnessed extraordinary advances in technological and industrial capacities in many of the nations of the world. To begin with, the threat of cataclysm as the result of a major nuclear war will be with us for as long as nuclear weapons remain in the arsenals of the world. The further threat of catastrophe from non-nuclear war derives in part, of course, from the possibility that such a war might escalate to a nuclear one, but also because of its own growing potential for devastation – devastation in a global Biosphere that is already being stretched to its very limits and in some respects beyond them.

The advances in industrialization have incorporated into the human environment veritable storehouses of dangerous forces that are likely to be released by future hostilities. Thus, even a future non-nuclear war could readily dwarf World War II in its human and environmental impacts. Moreover, the losses would inevitably include the invaluable cultural heritage of the world that has accumulated and survived over the millennia. It thus becomes amply clear that, as industrialization progresses and spreads ever-more-widely throughout the world, the freedom to wage war shrinks apace.

The very survival of Humankind now hinges upon the widespread development, in all segments of society, of attitudes sufficiently supportive of substantial restrictions on the pursuit of war. This development must be able: (a) to induce the relevant governments to adopt such restraints formally, thereby developing ever-more appropriate legal norms (Westing, 1988*c*, 1989*d*, 1990*c*); and (b) to provide the societal support which would ensure that the restraints would be honoured, thereby establishing ever-more appropriate cultural norms (Westing, 1988*a*, 1989*d*, 1990*e*).

In that last regard, it has been especially interesting to see the key role that is now being played in this process by environmental groups ('green', etc., movements). These concerned groups have been striving during the past two decades, with increasing imagination, forcefulness, and flair, in an attempt to achieve true comprehensive human security – an agenda that includes, as it must, sincere attempts at arms reductions, at social justice, and at ecological balance.

The World Charter for Nature of 1982 proclaims that 'Nature shall be secured against degradation by warfare or other hostile activities'; and further, that 'Special precautions shall be taken to prevent discharge [into natural systems] of radioactive or toxic wastes' (UNGA, 1982). This mutually-supporting pair of admonitions must now be truly taken to heart by all the peoples of the world, in order to ensure for themselves and subsequent generations a habitable, fruitful, and reasonably pleasant, human environment – one, moreover, that can be exploited in a manner which permits its sharing with the other creatures living on Earth.

This chapter suggests that the stability of Humankind, has, during the past two decades, become an ever-more-elusive goal – and, thus, that our 'Surviving With The Biosphere' has acquired an ever-more-precariously maintained status. Offered in primary support of this prognosis is the rapid increase since 1970 of abjectly poor ('least developed') nations; and, in general, of nations the increases in gross national product of which can no longer keep pace with their growth in human numbers. It is further suggested that, although the military drain on The Biosphere (both consumptive and disruptive) has remained relatively constant in recent decades, such impact has in the past two decades become increasingly inappropriate, whether in peace-time or in war-time.

The drain on The Biosphere becomes increasingly inappropriate in part because the financial, material, and intellectual, resources routinely devoted to the military sector are now ever-more-urgently required for the amelioration of the growing environmental and social ills in the civil sector. It has become so in even greater part because future military activities might well escalate to a level at which they would result in cataclysmic disruption of The Biosphere. Such cataclysmic disruption of the human environment – and thereby of human stability – could be brought about in one of two ways: (a) *via* a major nuclear war; or (b) *via* the release of dangerous forces, especially those contained in nuclear, chemical, and hydrological, facilities.

It is pointed out that potential targets in a future war which contain dangerous forces have become part of the human environment largely during the past two decades. There now exist some 200 nuclear power-plant clusters, distributed among 26 nations; large numbers of major chemical factories, distributed among about a dozen nations (*plus* significant additional numbers in a further 60 or so nations); and almost 800 huge artificial water impoundments, distributed among about 70 nations.

It is recommended that all potential sources of dangerous forces that are not decommissioned, be invariably surrounded by a buffer zone free from permanent human occupancy that coincides with the zone of potential necessary rapid and long-term evacuation in the event of a major release. It is further recommended that all of the nations of the world become parties to multilateral treaties proscribing the hostile release of dangerous forces.

In short, three major conclusions emerge from the present analysis: (a) that human instability has been increasing with frightening rapidity during the past two decades; (b) that as human demands on The Biosphere continue to increase, it becomes ever-more-necessary to shift financial, material, and intellectual resources from the military sector to the civil sector; and (c) that as industrialization progresses ever-more-widely, sources of profound danger to The Biosphere multiply, and our freedom to wage war shrinks concomitantly.

References

ACDA (1982). *World Military Expenditures and Arms Transfers 1970–1979.* US Arms Control & Disarmament Agency, Washington, DC, USA: Publication No. 112, iv + 134 pp., illustr.

ACDA (1989). *World Military Expenditures and Arms Transfers 1988.* US Arms Control & Disarmament Agency, Washington, DC, USA: Publications No. 131, v + 137 pp., illustr.

Bowonder, B. Kasperson, J. X., & Kasperson, R. E. (1985). Avoiding future Bhopals. *Environment* (Washington), 27(7), pp. 6–13, 31–7.

Brickhill, P. (1951). *Dam Busters.* Evans Brothers, London, England, UK: 269 pp. + 13 plates.

Brown, L. R., Flavin, C. & Postel, S. (1989). World at risk. Pp. 3–20, 195–200 in *State of the World 1989* (Eds L. R. Brown *et al.*). W. W. Norton, New York, NY, USA: xvi + 256 pp., illustr.

Ceres (1975). Face of poverty. *Ceres* (Rome), 8(3), pp. 18–20.

Eckhardt, W. (1989). Wars and war-related deaths. 1945–1989. Pp. 22–23, 56, in *World Military and Social Expenditures 1989*, 13th edn. (Ed. R. L. Sivard). World Priorities, Washington, DC, USA: 60 pp., illustr.

Fosberg, F. R. (1980). Whither terrestrial ecosystems?: preservation of the habitat of Man. Pp. 63–82 in *Growth Without Ecodisasters?: Proceedings of the Second International Conference on Environmental Future (2nd ICEF), held in Reykjavik, Iceland, 5–11 June 1977* (Ed. N. Polunin). Macmillan, London & Basingstoke, England, UK: xxi + 675 pp., illustr.

Gaekwad, F. (1972). A Viewpoint from a Developing Country. Pp. 445–53. in *The Environmental Future: Proceedings of the first International Conference on Environmental Future, held in Finland from 27 June to 3 July 1971* (Ed. N. Polunin). Macmillan, London & Basingstoke, England, UK: xiv + 660 pp., illustr.

Goldblat, J. (1982). *Agreements for Arms Control: A Critical Survey.* Taylor & Francis, London, England, UK: xvi + 387 pp.

IAEA (1989). *Nuclear Power Reactors in the World*, 9th edn. International Atomic Energy Agency, Reference Data Series No. 2, Vienna, Austria: 60 pp.

ICOLD (1984). *World Register of Dams*, 3rd edn. International Commission on Large Dams, Paris, France: 753 pp.

IUCN (1985). *1985 United Nations List of National Parks and Protected Areas*, 3rd edn. International Union for Conservation of Nature & Natural Resources (IUCN), Gland, Switzerland: 174 pp., illustr.

Lupis, I. Detter de (1987). *Law of War.* Cambridge University Press, Cambridge, England, UK: xx + 411 pp.

Mermel, T. W. (1988). Major Dams of the World: 1988. *International Water Power & Dam Construction* (Sutton, England, UK), 40(6), pp. 52–64.

Polunin, N. (Ed.) (1972). *The Environmental Future: Proceedings of the first International Conference on Environmental Future, held in Finland from 27 June to 3 July 1971.* Macmillan, London & Basingstoke, England, UK: xiv + 660 pp., illustr.

Polunin, N. (Ed.) (1980). *Growth Without Ecodisasters?: Proceedings of the Second International Conference on Environmental Future (2nd ICEF), held in Reykjavik, Iceland, 5–11 June 1977.* Macmillan, London & Basingstoke, England, UK: xxvi + 675 pp., illustr.

Polunin, N. & Burnett, J. H. (Eds.) (1990). *Maintenance of The Biosphere: Proceedings of the Third International Conference on Environmental Future (3rd*

ICEF). Edinburgh University Press, Edinburgh, Scotland, UK: xvi + 228 pp., illustr.
Postel, S. (1989). *Water for Agriculture: Facing the Limits*. Worldwatch Institute, Paper No. 93, Washington, DC, USA: 54 pp., illustr.
Quast, H. (1949). (in German) [Destruction and reconstruction of the Möhne and Eder dams]. *Wasser- & Energiewirtschaft* (now *Wasser, Energie, Luft*) Baden, Switzerland, 41, pp. 135–9, 149–54.
Tillema, H. K. (1989). Foreign overt military intervention in the nuclear age. *Journal of Peace Research* (Oslo), 26, pp. 179–96, 419–20.
UN (1985). *United Nations Industrial Statistics Yearbook*, United Nations, New York, NY, USA: Vol. 19, Part 1, xi + 633 pp.
UNCTAD (1984). *The Least Developed Countries: 1984 Report*. UN Conference on Trade & Development, Geneva, Switzerland: Document No. TD/B/1027, xi + 216 + 76 pp., illustr.
UNCTAD (1989). *The Least Developed Countries: 1988 Report*. UN Conference on Trade & Development, Geneva, Switzerland: Document No. TD/B/1202, xiii + 223 + 84 pp., illustr.
UNGA (1973). *Report of the United Nations Conference on the Human Environment, Stockholm, 5–16 June 1972*. UN General Assembly, New York, NY, USA: Document No. A/CONF. 48/14/Rev.1, vi + 77 pp.
UNGA (1982). *World Charter for Nature*. UN General Assembly, New York, NY, USA: Resolution No. 37/7 (28 Oct 82), 5 pp.
Westing, A. H. (1978). Ecological considerations regarding massive environmental contamination with 2,3,7,8-tetrachlorodibenzo-*para*-dioxin. *Ecological Bulletin* (Stockholm), 1978(27), pp. 285–94.
Westing, A. H. (1980). *Warfare in a Fragile World: Military Impact on the Human Environment*. Taylor & Francis, London, England, UK: xiv + 249 pp.
Westing, A. H. (1982). War as a human endeavor: the high-fatality wars of the twentieth century. *Journal of Peace Research* (Oslo), 19, pp. 261–70.
Westing, A. H. (1984). Environmental warfare: an overview. Pp. 3–13 in *Environmental Warfare: A Technical, Legal and Policy Appraisal* (Ed. A. H. Westing). Taylor & Francis, London, England, UK: xiii + 107 pp.
Westing, A. H. (1987). The ecological dimension of nuclear war. *Environmental Conservation*, 14(4), pp. 295–306.
Westing, A. H. (1988*a*). Cultural constraints on warfare: micro-organisms as weapons. *Medicine & War* (London), 4, pp. 85–95.
Westing, A. H. (1988*b*). Military sector of society *vis-à-vis* the environment. *Journal of Peace Research* (Oslo), 25, pp. 257–64.
Westing, A. H. (1988*c*). Multilateral treaties constraining military disruption of the environment: excerpts. Pp. 163–8 in *Cultural Norms, War and the Environment* (Ed. A. H. Westing). Oxford University Press, Oxford, England, UK: xiv + 177 pp.
Westing, A. H. (1989*a*). Guest Comment: Comprehensive human security and ecological realities. *Environmental Conservation*, 16(4), p. 295.
Westing, A. H. (1989*b*). Environmental approaches to regional security. Pp. 1–14 in *Comprehensive Security for the Baltic: An Environmental Approach* (Ed. A. H. Westing). Sage Publications, London, England, UK: xii + 148 pp.
Westing, A. H. (1989*c*). Environmental security for the Danube basin. *Environmental Conservation*, 16(4), pp. 323–9.
Westing, A. H. (1989*d*). Proposal for an international treaty for protection against nuclear devastation. *Bulletin of Peace Proposals* (Oslo), 20, pp. 435–6.
Westing, A. H. (1989*e*). Reflections on the occasion of the third anniversary of the Chernobyl disaster. *Environmental Conservation*, 16(2), pp. 100–101.
Westing, A. H. (1989*f*). Regional security in a wider context. Pp. 113–21 in *Comprehensive Security for the Baltic: An Environmental Approach* (Ed. A. H. Westing). Sage Publications, London, England, UK: xii + 148 pp.
Westing, A. H. (1990*a*). Environmental hazards of war in an industrializing world. Pp. 1–9 in *Environmental Hazards of War: Releasing Dangerous Forces in an Industrialized World* (Ed. A. H. Westing). Sage Publications, London, England, UK: xvi + 96 pp.

Westing, A. H. (1990*b*). Our place in Nature: reflections on the global carrying-capacity for humans. Pp. 109–20 in Polunin, N. & Burnett, J. H. (Eds). *Maintenance of The Biosphere: Proceedings of the Third International Conference on Environmental Future (3rd ICEF)*. Edinburgh University Press, Edinburgh, Scotland, UK: xvi + 228 pp., illustr.

Westing, A. H. (1990*c*). Towards eliminating war as an instrument of foreign policy. *Bulletin of Peace Proposals* (Oslo), 21, pp. 29–35.

Westing, A. H. (1990*d*). Towards preventing the release of dangerous forces. Pp. 61–76 in *Environmental Hazards of War: Releasing Dangerous Forces in an Industrialized World.* (Ed. A. H. Westing). Sage Publications, London, England, UK: xvi + 96 pp.

Westing, A. H. (1990*e*). War, environment, and cultural norms. *Environmental Awareness* (Baroda), 13, pp. 15–7.

Commentary on Chapter 15

CHAIRMAN: Dr Mostafa K. Tolba
PANELLISTS AND OTHER CONTRIBUTORS:

Fosberg, Juel-Jensen, Cloudsley-Thompson, Stone, Tolba, Stone, Medawar, N.V.C Polunin, Persányi, J. Petts, Juel-Jensen, Samatar, Mische, Miller, Westing

Fosberg paid tribute to **Westing** who had devoted his intellectual energy in the world's major institution dedicated specifically to the promotion of peace towards mitigating the forces that threatened our tenancy of Planet Earth through environmental and other threats to peace.

Westing had dwelt on an aspect of human behaviour that was seldom considered and not at all well understood – the condition or state of instability. If the reality of this condition were questioned, the impressive list of examples, and their consequences which he had given in his talk, were ample evidence. These consequences and potential consequences, should frighten us all. The chronicle of incredible, human self-destructive activities was impressive. It made one thing evident, one which we perhaps did not like to admit – that despite our high opinion of ourselves, and despite the name of our species, *Homo sapiens*, we were, after all, really irrational beings. And, perhaps, being irrational, we should expect instability.

He had long been impressed by evidence and examples of irrational acceptance of things, conditions, and people. Many of these, of course, were sources of satisfaction and beneficial motivation, but many were unquestionably deleterious and even self-destructive. He had not before thought of instability as one of the consequences of this irrationality, but it was a very serious one.

We might ask a question, one undoubtedly on **Westing**'s mind, but not stressed in his talk – what was, or were, the basic causes of human instability? There were probably many causes, working together, to produce this condition, and they were very hard to disentangle and isolate in order to examine. He would only mention and examine, briefly, two.

The factor of stress seemed to him to be one of the principle sources of unstable behaviour, both by individuals, and by aggregations, or communities. Stress had always been with us, as had been war, hunger and disease. But stress had, even in his lifetime, seemed to increase, with the increasing complexity of our industrial and technical life. The things he had to cope with in his youth seemed simple in comparison with the situation now, and the increase in complexity and, with it, stress, was clearly geometrical. This was a dismal prospect, especially if we were to expect a corresponding increase in instability. The frequency of mental breakdown and nervous ailments had become appalling and, with it, the intractable drug-abuse problem.

A further, even more basic and serious problem, doubtless related to the stress problem, was population growth, emphasized by **Westing**, and dwelt upon in some of his writings. This growth also seemed to be geometrical. What was not always appreciated was the correspondingly, but inversely geometrical, decrease in the resources available to us. Such were living space, materials, necessities such as potable water, and now perhaps, as had been indicated, even food, in spite of intensified agriculture and 'green revolutions'. As had been emphasized, and was being stressed every day in the environmental literature, Planet Earth is *finite*. The growing threat of the Ehrlichs' 'Population Bomb' should frighten us all. Nature and Blasco Ibañez's *Four Horsemen of the Apocalypse* might solve this problem for us – and we would not like the solution!

Juel-Jenson disagreed profoundly with **Westing**. There was no such thing as 'civilized war'. It was naïve to think that you could prevent a nation at war from destroying dams because it might hurt people. The example chosen was a particularly unfortunate one. Germany had started the war, and Britain was fighting for its life and for civilization as we knew it. The destruction of two major dams in the Ruhr valley by the Royal Air Force, an example of extraordinary bravery on the part of a few people, had probably helped to turn the course of the war. **Westing** was too young to know what the Second World War was like. Had the dams not been breached, we should probably not have been here today.

He was no psychiatrist, and would, therefore, not embark upon an alienist's explanation of why people behaved as strangely as they do. We all owed a debt of gratitude to **Samatar** (*see Annexe 5.2*) for giving us the naked truth. Much of the chaos in tropical Africa was due, not only to the difference of material resources between the developed and the underdeveloped world, but in many, and alas in an increasing number of countries, to the rapacity and corruption of pocket dictators, or extreme groups, who, in search of power, had usurped their countrymen.

He wished to give one example of how there was nothing wrong with the enterprise in one African country: Ethiopia. People did a lot of talking about the environment and about ecology, but not many did much else about it. When, after the murder of Prince Ras Seyoum Mengesha in the early 1960s, his son Prince Ras Mangashia Seyoum became Governor-General of Tigray, he did all the things that people were now talking about. There was a good deal to be said for a *benign* feudal system. He had built roads, not at third hand, sitting in an office, but he had gone out and driven caterpillar machines himself. He had started reforestation. He had repaired terraces. He was able to post a man with a gun at the new plantations or the new terraces, and that deterred anyone from pulling up the trees. He had started small industries, including environmentally harmless enterprises such as harvesting frankincense and myrrh from *Boswellia* sp. and other gum-producing trees. He had started a clothing factory. He had educated 307 destitute children, and when appropriate, sent them on to colleges and universities, either in Ethiopia or abroad. When there was a severe drought in 1971-72, the Province of Wello, immediately south of Tigray, had suffered disastrously. It was hit climatically far less badly than Tigray, but where tens of thousands died in Wello, very few had come to any harm in Tigray, because Mangashia had built roads and was able to get emergency supplies to everyone, however remote they were. When Mengistu Haile Maryam had seized power through a military coup in 1974, Ras Mangashia had had to flee into the Sudan with his youngest son, for he was afraid that the population of Tigray would rise against Mengistu's forces when they were inadequately equipped with weapons, and there would have been a bloodbath. All his efforts had come to nothing, and the barbarous regime of Mengistu had deliberately destroyed what Mangashia had achieved. It had discouraged the peasants from growing crops by compulsory purchase at a third of the market price. It had withheld supplies, and it had since used hunger as a weapon. This was a very good example of how the greed of one person, or of very few people, affected many. He did not need to go into what happened later, with hundreds of thousands of Ethiopian refugees seeking shelter in the Sudan, and the problems which that had created for one of the poorest countries in the world. It was a pity, perhaps, that no time had been set aside to discuss the whole problem of refugees in the world, for that subject should have been part of our deliberations.

Cloudsley-Thompson said that **Westing** had given ample reason why human beings should feel unstable and insecure; but the nature of their fear should really be discussed by a psychiatrist. Fear was by no means a simple response to stressful situations. Indeed, and perhaps paradoxically, the claim had been made that, apart from deaths and other major disasters, the greatest cause of stress for Western families was their holiday travel. **Tolba** had told us how the risk of skin cancer and eye cataracts had swayed public opinion in favour of action to protect the ozone layer.

What, then, was fear and how did it influence people? Most people were frightened for themselves and for their loved ones: for humanity, their fear was only metaphorical. Fear lay in the imagination. The worst fear was fear of the unknown. Waiting for a race to begin, to go on stage, or into battle was, for example, far more stressful than the actual events that one was waiting for. Samuel Taylor Coleridge had expressed this aspect of fear in *The Rime of the Ancient Mariner*:

'Like one that on a lonesome road
Doth walk in fear and dread.
And having once turned round walks on,
And turns no more his head;
Because he knows, a frightful fiend
Doth close behind him tread.'

Fear was not rational. Some people were so afraid of lung cancer that they just had to smoke yet another cigarette merely to steady their nerves. Others might fear loss of power, loss of security, or ridicule, even more than they feared death itself. In *The Face of Battle* John Keegan (1976) had analysed the reasons that prevented ordinary soldiers from running away from the enemy. Foremost among them was that they do not wish to lose the respect of their peer group. Why, Keegan asked, did the British squares at Waterloo not break when charged by French cavalry and fired at with cannon balls? Had they run away, of course, the soldiers could easily have been speared in the back. But this was unlikely to have been their main reason for not so doing. It was probably far more that they knew that, if wounded, they would be pulled inside the square and not be left alone on the battlefield.

Religious fervour, combined with the respect of their peer groups, might have motivated some of the bravest soldiers of modern times: for example the Zulu warriors of the last century and the Japanese *kamikaze* pilots of World War II. When he had congratulated a friend, who was dying of cancer, on his courage and fortitude, he had replied: 'When there is no alternative to death, the question of courage becomes irrelevant'.

As Shakespeare had pointed out, 'conscience doth make cowards of us all'. Here is a trivial example. Shortly after 'D' day, in June 1944, his Cromwell tank was destroyed and he lost all his kit. Even when issued with another tank, he still lacked a knife and fork. So he decided to stop among the ruins of a Normandy village and look for something to eat with. Whilst so doing, the village was shelled by the Germans. This would not normally have worried him unduly. Tank crews were very blasé about artillery and machine-gun fire – what simply terrified us were 88 mm anti-tank guns. But, on this occasion, because he knew that he was guilty of looting, he had felt quite frightened and was more than glad to reach the safety of the tank and leave his guilt outside. Nobody, however brave, can be completely fearless, but our fears varied. A further question still arose. Was it ethical to impose our own beliefs upon others by manipulating their fears? Perhaps there was an analogy here with compulsory vaccination for smallpox in the old days.

The greatest fear of all, except possibly for the very old and very religious, must surely be the fear of death, the great unknown. And this fear was the same for the starving poor of the Third World as it was for us pampered Westerners. Yet Seneca had claimed that he felt no fear of death because he had already experienced it during the centuries before his birth. On his deathbed Trollope's formidable Duke of Omnium had said that he went expecting nothing and fearing nothing.

He was far from optimistic about the future and thought that the only hope for humanity lay in altruism. This quality could influence critical decisions in the future. At the same time, we should recognise that anyone who had too excessive a fear of his, or her, own death, had not yet learned to live.

Stone disagreed with **Westing**'s emphasis on stability/instability *per se*, which was not quite the focus on which he would have wished to concentrate. Instability in itself was neither bad nor good. It was desirable for some things to be unstable, for instance it was desirable to destabilize, i.e. change, poverty, or the present environmental situation in the Sahel. Rather than change, he thought the important issue which had been raised was the notion of threat, which was where fear came in. Fear was associated with short-term threats. For example, the notion that nuclear power-stations should be closed was so motivated, but the solution proposed, more use of coal, also imposed environmental dangers of a different, much longer-term kind. There was a hope that technology could catch up with some of these longer-term hazards, but there was no way that it could help if a dam collapsed, or a nuclear power-plant blew up: those were cataclysmic effects. The problem therefore resolved into how could people be motivated to respond sensibly to environmental threats? It was easier to motivate people over cataclysmic threats, because of their potential immediacy, than it was to motivate them even in the face of some theoretically equivalent number of casualties, or deaths, over time, which had arisen from a long-term problem. Even a similar number of deaths were simply not perceived as an immediate threat. This posed real problems.

He had noted that **Westing** had treated law and technology separately, but those involved in them both, like himself, considered them together. Law often drove technology. For example, technological solutions to keeping people clear of radiation in nuclear plants, or improved safety inspection, were devised by scientists but were greatly strengthened when adopted by law as part of legal policy.

International agreements were a way forward on some issues but international law was a dicey business and one could only have modest ambitions for it. There was much that could be done in anticipation of Bhopal-, or Chernobyl-types of events. It was possible to conceive of bilateral, or even international, agreements between neighbours before constructing a nuclear, or chemical plant. Early-warning agreements of potentially transboundary types of catastrophes – a nuclear explosion, or the release of a toxic chemical cloud – were also feasible and, indeed, we should be working to establish them. He hoped that there would now be some movement on a nuclear transboundary notification convention with the support of the Soviet Union, although that would have all kinds of implications. Such an agreement had been on the agenda in 1972 but had not been accepted: it was interesting to note that it had taken a catastrophe, eventually, to move the world! These kinds of issues were very important items in the international legal agenda.

He was pessimistic about the value of international agreements when it was a matter of the survival of a state, as it would be in a time of war. It became much more difficult to fashion treaties that could be enforced. Who would be left to enforce them, for example?

Tolba remarked that he was at a loss to understand why governments were dragging their heels over industrial accidents with transnational consequences. He had made proposals immediately after the accidents in Basle and Bhopal that there should be early notification and immediate emergency assistance, that everyone should be informed of what was happening in their midst but, as with nuclear issues, there had been no more than a dragging of heels.

On the matter of the law reinforcing, even driving, technical change he queried the notion of legal action to bring about benign technical innovations to existing plant, as **Stone** had suggested. One of the points that had to be borne in mind was to look into who in the private sector owned the capacity to produce the more benign situation or substitute. There had to be a balance between forcing people to do things by law, or encouraging the private sector, who had the money and owned the technology, to be encouraged to act of their own volition. A careful balance had to be struck between using the stick and the carrot, otherwise the possibility of benign innovations might be completely lost.

Stone explained that, by legal strategies to enforce technological change, he had in mind such devices as tax incentives or other forms of positive award. He did not equate legal action solely with punitive measures!

Medawar made two points about fear. Firstly, she saw nothing immoral in using the element of fear to show people what might happen if we did not change human behaviour, *e.g.* in relation to the environment in regard to our reproduction, but she did think it immoral to enforce action on people because it had not worked. Secondly, she thought it a great thing to be a meliorist (Latin, *melior* = better), by which she meant that one should take account of what could be done and put one's heart into doing it. That might achieve a little good; undue pessimism, optimism or whinging, would not. We should not be dismayed by the size of a problem because we had not yet solved any of them: we were still beginners and so could hope to improve!

N. V. C. Polunin remarked that the connection between stress and an increasing density of people in the world to which **Fosberg** had drawn attention might be due to other things. These might include an unprecedented decline in people's sense of the existence of any kind of moral, technical, or other related spiritual guidelines by which to order their lives. He thought that this reflected the effect on people of the development of industry, science, and other types of advanced human activity in opening up a 'can of worms' to which there seemed to be no answers, leading them to shy away from any kind of authority. Even as a scientist he felt this when asked to respond to questions about environmental issues. He did not wish to stir up people's emotions but rather to give them facts, but often the facts were simply not there. What, therefore, as scientists – people recognized by the public as authorities – was the message, and how was it to be got across to the public?

Persányi, from the Hungarian Ministry of the Environment, remarked that the fact that an American child might consume ten times more than an Indian child only recognized the minimal materialistic difference. Which, he asked, would live in the better conditions, amongst less pollution, enjoying more leisure, better health and culture? Ecological security was an important aspect of human security. Traditionally there had been ecological exploitation but to this had now been added the export of hazardous products and wastes, and of modern technologies that used a lot of raw materials, had high energy requirements and often caused much pollution. How were such imbalances to be regulated? The problem was very evident in Hungary where they had exchanged a socio-economic system built on environmentally unfriendly production and consumption systems which created enormous environmental problems. Even today several foreign firms, sometimes as joint Hungarian ventures, were importing the new forms of ecological exploitation. The ecological 'iron curtain' should be destroyed and there was a need for international cooperation to promote and favour human equality and stability both in the east and in the west.

J. Petts wished to make some comments, arising from what **Westing** had said, about chemical installations. Over the last ten years there had been a massive international effort and agreement over the risks from major installations such as had been referred to already. These included the development of reasonably sophisticated, she used the phrase cautiously, risk assessment techniques which did allow for buffer zones, separation distances, good design of plant to take account of potential sabotage, and mitigating measures, including emergency planning, if a catastrophe should occur. All these procedures had been developed on an international basis;

there had, for example, been a massive OECD programme to implement legislation in order to bring about adequate separation distances. We certainly had to ensure that this approach went beyond Europe and the OECD countries to full international acceptance.

The key was how to transfer these very sophisticated technologies to other countries and to ensure that they were effective. In Bhopal, most of the 2,000 casualties arose because the Indian government was unable to exercise planning controls over land-use for buffer zones against migrating people in shanty-town-type developments close to the installation. That could be done in the west but help was needed to ensure that this was achieved elsewhere.

There was also the question of the risk-benefit equation. Major installations did provide massive benefits to communities and societies where they were located as well as exposing them to potential risks. It was both much better and necessary to provide sufficient, adequate information to be able to assess the risks and benefits and balance them off in a sensible, effective and pragmatic manner rather than simply to forbid such installations, or have fixed – say 50 km – buffer zones around them. The latter certainly could not be done in the UK, or the population would all be living in the sea!

In summary, she believed that the right solution was to transfer existing technology, including the very wide-reaching risk-assessment technologies we already possessed (some of which had been in use for a long time) to other countries, and to assist them to ensure that risks and benefits were properly presented and weighed when political, or other, decisions were made.

Finally, as an aside from the main theme, she commented that she had noticed how **Westing** had said that he liked to promote Nature reserves within protected zones. No doubt it was a matter of ethics that he was apparently unhappy to accept risk to human beings but would accept risks to animals and plants! Had he actually thought in those terms?

[In an aside **Juel-Jensen** suggested that rather than animals and plants, politicians should be put in such protected zones, but Tolba said that he would not dare recommend that himself! Eds.]

Samatar thought that, in describing the near-suicide we were about to commit, **Westing** had put too much emphasis on the role of sovereignty: but what sovereignty were we talking about? For example, Somalis needed to be protected from their own government, but there were many nations which, if it was pointed out that they were causing a catastrophe, would claim it was solely a matter of their own sovereignty. Each situation needed to be assessed in its own right.

Mische reflected that irrationality, with which we all had to wrestle, was not so much an innate aspect of human nature as the result of acting, apparently rationally, but according to the logic of the national state. Leaders were very often forced to act irrationally because they were operating in a lawless situation of the survival of the fittest. If we could achieve international systems in the next generation that could deal with the various security issues, including that of economic insecurity, then there was a basis for hope. Problems which appeared too big for people were really too big because they were, in fact, problems that nations were at present unable to deal with rationally.

Miller said that he thought the root of the problem was that we were brought up as competitors. We had regional competitions in schools, national sports, and then international sports. He had nothing against Olympics, they demonstrated the excellence that we all admired, but they were not based, as were other kinds of regional, national, or international, competition, on the threat of dangerous forces. It would be important in the early years, in training in our neighbourhoods, to try to see our contemporaries and neighbours as citizens of the world – committed to working for the world above all else – and to try to inculcate that attitude in others.

Westing said that he had not been certain how best to approach the topic of human instability but, if there was one point he would like to stress, it was that somehow or other we must be able to reinforce the principal, both in our attitudes and in law, that nations had a responsibility to ensure that their activities did not cause damage beyond their own areas of national jurisdiction.

He had no desire to enter into a discussion with **Juel-Jensen** about either the correctness or propriety of World War II, or of how it has been pursued. He had merely mentioned it to demonstrate how militarily attractive was an attack on a dam, and how for a minimum of energy input one could achieve a maximum of damage, much of it disproportionate damage as it went far beyond military necessity.

On the matter of catastrophes, he recalled it being said that the real problem was not the individual catastrophe but the continuing, cumulative inroads into the environment. He was concerned with catastrophes in the context of creating a more humanitarian norm for war which would be, both socially and environmentally, less damaging. He applauded the actions described by **J. Petts** which the OECD had taken over risk reduction. He had agreed with her entirely that the two important steps were to accept risk-prevention as the norm and then to transfer the benefits, financial and technical, to the rest of the world. There was, of course, a responsibility of the industrialized world to different regions within itself and to the entire world, as several speakers had pointed out, and we needed to promote both technology transfer and the transfer of these conceptually quite simple notions of prevention by every possible means.

Annexe 5.1: Environmental Security for the Horn of Africa*

H. E. Mikael Imru**
PO Box 2828, Addis Ababa, Ethiopa

Extent of the Region

Today, with the economic development problems featuring high on the global agenda, and questions of environmental restitution pressing for urgent action, the links between development and environment on one hand and, on the other, those of grinding poverty, political instability, and armed conflict, are becoming increasingly manifest. Violence and repression are almost ubiquitous, and the staple diet of many communities is expressing themselves intensely and protractedly in some such way, or doing so only sporadically in others.

It would not be possible to understand the root causes of this deplorable violence without an appreciation of the difficulties of modernizing societies and a deep awareness of the complicated interrelationship of the above-mentioned factors. In this regard our perception of the Horn of Africa is especially critical. This geographical expression is not of local designation. It is outsiders who, for their own reasons, have applied this nomenclature to our region. However, it can serve well if it does not remain too restricted and static in the extent of the area it covers and the affiliations it demarcates. Indeed it should acquire a dynamism of its own, borne out of the necessities and urgent concerns of its own people.

As it stands at present, the concept of the Horn of Africa is a truncated one, betraying its origins as a purely strategic designation to serve interests that are 'other-regional', and not dissimilar from such appellations as the 'Middle East', the 'Near East', or the 'Incense Coast'. They were elaborated in the formulation and establishment of enclaves – zones of influence for the purposes of incursions and expansion that still haunt the nostalgia of those who dream of the past. They fail to comprehend the imperatives of a rapidly-changing world, bewildered but otherwise untouched by the healing passage

*[The papers in Annexe 5 were presented at a short session of the Conference under this title. They are, however, so relevant to the theme of Chapter 15 that they are included here. Eds.]
**[Research worker mainly on Ethiopian history and culture; sometime Prime Minister of Ethiopia and Chief Political Adviser to the Ethiopian Government, Addis Ababa, Ethiopia. Eds.]

of time, which they could not arrest or even slow down. The definition of extent as it stands is, therefore, an obsolete consideration – especially in the possibility of bringing the peoples of the region closer together. Indeed, it served to reinforce the physical, cultural, and psychological, barriers rather than the oneness, unity, and interdependence, of the region as a whole. It appealed to the taste for adventure, rather than responding to the imaginative conceptions of bold spirits, touching the heartstrings of the people and responding to their modern needs.

What approach can we develop that firmly establishes the eco-geographical complex irrespective of political boundaries, but that proves to be a manageable unit which is conducive to sustainable and equitable socio-economic development? Modern communications and technology, age-old affiliations, socio-economic orientations, and political alliances whether temporary or of a more enduring sort, have, after all, contracted our planet – not only physically but also on an ever-widening level of human relations – thereby rendering the concept of neighbourhoods somewhat ambiguous.

Distances are relative and definitions of 'us' and 'others' have perforce to undergo revision. With greater advances in technology and communications, even mental distances between peoples become affected, and spiritual awareness of the meaning of the 'stranger' deepens.

More and more widely the world is being organized on regional lines, as is evident in the European unification process planned for 1992, the European Free-trade area, the recently-concluded US–Canada Free Trade agreement, and many movements in the Third World that promote cooperation and integration, although with varying degrees of success.

As far as our region is concerned, too-wide a definition of extent would bring together far-flung elements and render futile any efforts at cooperation, let alone foster firm steps towards integration. On the other hand, too-narrow an image may falter on the side that excludes more than it includes, and hence result in irreparable disjunction.

What is the right balance that may lend a measure of intimacy to the exercise and not prove too restrictive at the same time? Separate but related treatments, and considerations of varying geographical areas and subject-matters, may prove salutary and lend a measure of reality to the exercise. They may, hopefully, facilitate trade-offs and weave themselves together in the final outcome, if and when research, practical programmes, and diplomatic endeavours, lead to hard negotiations.

This may offer a reasonable approach that avoids inflexible polarities from developing. Furthermore, there are objective criteria that would have to be followed – the nature of the geography and layout of the land, the demographic imperative, the shape of river basins, the communication networks, the land- and sea-resources, the maximum exploitation of comparative advantage, etc. – will provide links that will sharpen the exigencies of cooperation and highlight the advantages that would ensue.

A very broad definition of the Greater Horn of Africa would encompass parts of North-east, East, and Central, Africa – a wide area that stretches from the Sudan in the north to Tanzania in the south, and north-east Zaire in the west. This *greater* region is underlined, and is circumscribed by the Nile Basin, the Rift Valley, the Red Sea, the Gulf of Aden, and the Indian Ocean. It is thus composed of parts of interlocking highlands and lowlands, coastlands and arid plains, savannas and rangelands, chains of lakes and hot-water springs, and life-giving water systems as well as drylands.

If, on the other hand, we define the Horn of Africa narrowly, limiting it to the three countries of Ethiopia, Somalia, and Djibouti, we will be subject to a self-denying ordinance that limits our scope and dwarfs our vision. There are constraints in both approaches, and the advantage may lie in a third approach that is flexible and capable of serving the needs of fairness to all partner-countries, facilitates a balance of gains and exchange of concessions, and would be informed by the demands of real politics rather than imagined rights and unattainable advantages.

Regarding the limits of the region, problems are likely to arise at the periphery. The Sudan, for example, is closely linked to the rest of the region on both ecological grounds and political considerations. But here we encounter a difficulty; for whilst northern Sudan is linguistically, ethnically, and in religious orientation (staid or fundamentalist), an 'effervescence' of the eastern Arab world, the southern half is decidedly Black African. On the other hand, Ethiopia is a credible, self-contained bridge between the vast Arab world and the emerging Black African giant. It has no ethnic or linguistic affiliations with the Arab world, and can effectively play, in the future, the role of the honest broker between the Arab world and Black Africa. At the other, southern end of the spectrum, it is possible that the 'liberation' in entirety of Southern Africa may tend to exert a decisive pull on Tanzania and alter its East African orientation.

Historical and Physical Setting

The region has excited the interest of outsiders throughout the ages, and travellers have penetrated from the torrid coast and arid lowlands – in some parts narrow, being not more than a mere one-hundred kilometres wide – to the escarpment and the equable plateau, there to learn of a thriving civilization and a powerful, expanding kingdom as long ago as the early centuries of the Christian era. In other places the coastlands merge with the barren scrublands or where there was a former arm of the sea (and today lying well below its level), with salt rocks and exploitable mineral layers to remind us of its past and warn us of its and our vulnerable future.

Beyond the coastlands, the lowlands widen, encompassing the extensive rangelands as if jealous of the fertility of the plateaux and the lofty, rugged highlands that tower above them. To the highlander, the lowlands appear to encroach on, surround, and envelope, his high strongholds, which, however, are threateningly poised in a challenging posture, and prepared to ravish. They vividly remind us of Robert Frost's poem 'Once in the Pacific'; for they widen and descend to the arid, semi-deserts of the Sudan in the North and of northern Kenya in the South. The Sudan, on whose expanses the generosity of the Nile breathes life in the greenery of the lands adjoining its banks, was once the centre of the civilization of ancient Napata. Aksum on the high plateau and Napata in the lowlands, on the bend of the Nile, north of the confluence of Atbara/Tekeze, dominated in turn for more than two millennia the region of the Greater Horn of Africa (Toynbee, 1965).

The rangelands – that surround the fertile and well-watered uplands – with their oasis, well-heads, and grazing expanses, used to be the paradise of the pastoralists. Furthermore, the heavy precipitation and the great rivers that accumulate and flow from the uplands and also feed the underground currents, are the sources of the water-towers of the region. They inextricably link the Nile Basin in the west and the lowlands in the east with the rugged uplands – the dry with the wet lands – making them interdependent.

The extensive Rift Valley, emerging at the Red Sea in the north, divides the region, skirts the escarpments, and in its southern portions extends into Kenya – branching towards the sources of the White Nile in Uganda and the neighbourhoods of Lakes Victoria, Kyogo, and Mobutu (Albert). The concept of the Horn of Africa without the basin of the Nile, is therefore inadequate – indeed somewhat amputated – and would become increasingly untenable if regional cooperation, especially in environmental spheres, is to make a decisive contribution to the stability, peace, and sanity, of the region.

The entire geographical area may perhaps more realistically, and with greater relevance, be termed the 'Greater Horn of Africa'. It includes the basin of the Upper Nile and parts of the vast Rift Valley with its chains of lakes and hot-water springs – also the highland masses which today are sparsely covered by remnants of temperate forests and alpine vegetation, and are marked by torrents which, in their due season, foam when negotiating the deep ravines, whilst the lower undulating ridges are garbed by thinning tropical forests.

The uplands enjoy a rainfall of 1,500 to 2,000 mm per annum and support the bulk of the human population. The slopes descend, some abruptly and others in gently-undulating paces, to the vast, harsh arid lowlands that encircle and envelope the highlands. The contrasting eco-geographical complex is sharply marked by the interdependent divisions that yawn between the sown and the desert areas, between plantations and pastures, and between watered areas and the stark drylands. The region as above depicted is one of sturdy cultivators, pastoralists of great endurance, and hard-working artisans and craftsmen – people who look, in their formidable ruggedness, as if hewn from rocks. Generations of warriors, cultural agents, organizers, and traders, stopping at caravanserais, have roamed the length and breadth of its spectacular and variegated scenery; and of its spiritual habitations, lulling with poetic dreams and stirring to great heights of endeavour, its rustic people.

This 'jigsaw' land of contrasts and puzzles, that has been for millennia under the furrow of the plough, the edge of the axe, the bite of the hoe, and the treadmill of innumerable hooves, was, in its youth, fertile and 'dripping in fat'. But today it is ageing, appearing tired and in need of restoration – retaining nevertheless in its naked memory the glories of its past, and pointing to the promise of the future in its undefeated spirit. It is a land that urgently needs the healing hand of rejuvenation.

This situation emphasizes the growing realization that the lifelines of countries which are isolated tend to be attenuated, and would even become distorted in the future if irrational and monopolistic trends were to be the predominant characteristic of the region – as population expands, environmental degradation deepens, and competition for land and water becomes ever-more acute and bitter. Rivalries and unwarranted encroachment would instigate open hostilities, and render the lives of its peoples even bleaker in the future than they are at present.

Limitations of the Nation State

Confident that the area demarcated must of necessity be manageable, the image that it invokes should be revitalized, so that it can undergo qualitative change and be made to serve a larger and more meaningful purpose than hitherto. It has the potential to emancipate itself – from the myopic, self-limiting and self-immolating image of a narrowing and dismembered Horn, crushed by the debris of cumbersome empires turned bureaucratic. Accordingly one trusts that environmental cooperation, invoking the vision of a comprehensive region, may repair and compensate, and, on the failings of the old notions – freed from their original and narrow associations and limitations – be made to respond to the needs of the times.

Eco-geographical regions rarely coincide with national territorial limits, either falling short of, or spilling over, their borders. In the case of the Greater Horn of Africa, which has the potential of surviving and prospering by becoming closely knit, the resolution of difficult environmental questions lends itself to regional cooperation. The countries that are now dispersed over the entire region, face similar challenges, share common resources, enjoy much the same opportunities, suffer related hazards, and would greatly benefit if these could be viewed, and their solutions perused, in a wide but manageable *regional* context. With modern advances in technology and communications, the extent of manageability is being constantly expanded, bringing, however, into bold relief, the sensitive task of reconciling the twin contrasting needs of concentration and dispersal of decision-making authority – of regional and local concerns, of effective popular participation despite bureaucratic rigidities, plurality of consultations, and straight-jacket centralism.

Without a *regional* overview and grasp it will not be easy to transcend these asymmetries and develop comprehensive strategies that effectively resolve the problems of the area. It is becoming increasingly evident that the world, organized as at present on the basis of sovereign states on one hand

and remote, far-flung bureaucracies on the other, is becoming outdated – suffering from serious limitations in coping with the needs of the times and having to make-do with improvizations to solve many vital problems that confront humanity as a whole. A middle-way will have to be found – perhaps in more realistic regional and sub-regional entities.

The present array of sovereign states suffers from both internal weakness of inadequate integration – giving rise to ethnic tensions, guerrilla wars, and chronic and damaging power-struggles – and also from eternal pressures of the more powerful on the weaker states. Many states are domestically overstretched and externally vulnerable. World-wide international organizations that govern relationships between states, regional organizations that promote cooperation between neighbouring countries, and nongovernmental structures that cultivate and moderate relations between peoples, are all phenomena that demarcate a growing area of contacts and cooperative activities which disguise the dismantling of the nation-states.

In this situation, the necessity of recognizing the inadequacy of the nation-state becomes imperative. Many nation-states even today are forced by economic circumstances, security needs, or famine conditions, to look beyond their frontiers. The question therefore arises of actively seeking the best and most effective way that takes us beyond the nation-state to higher stages of meaningful cooperation. The last fifty years have enabled us to accumulate valuable experience that has made us aware of both the pitfalls and opportunities in this process. Perhaps our best starting-points are institutions that promote regional cooperation, whence we can advance better and become more effectively organized by stages, to the delegation of sovereignty to organizations of world-wide import. But sympathetic relationship, inspired by close acquaintance and intimate cooperation between groups, and that which is aroused by virtue of common humanity, will have to be actively interlinked. The right balance will have to be struck if we are to overcome the problems of magnitude and impersonality of operations on global and regional scales.

The participants in basic political units must enjoy sufficient elbow-room if they are to appreciate that they are not just clogs in a system which descends from the higher to the lower levels of interaction – eclipsing altogether, in the process, the influence of those levels. We can successfully make the transition beyond the nation-state only if we can understand and appreciate basic human needs and aspirations. These developments would enable us to establish the Greater Horn of Africa as an operational unit of an interlinked and interdependent eco-geographical complex – irrespective of physical barriers and national boundaries, and undeterred by contrasting historical heritages, cultural and linguistic differences, and other disparities especially in extant heterogeneous customs and outlooks, with which we have lived for such a long time – aloof and even hostile to one another.

Links Between Underdevelopment and Violence

The difficulties with which the people of the Greater Horn of Africa have to contend, and the problems they have to resolve, are twofold. One problem, of which the causes are manifold, is that, associated with the underdevelopment of the resources of the region and the inability of societies to utilize them, let alone exploit them fully, is the potentialities of their people. The constraints in the way of negotiating successfully the historical stages of economic development; the cultural causes affecting beliefs, attitudes, and conduct, that tend to be deeply entrenched at certain stretches of a people's history, making for either dynamism or inertia, creativity or stagnation; the state of material resources and the means of their exploitation; the processes of modernization and the challenges which they present for those in the spiral grind of catching up with the relatively more advanced countries, are only some of the main questions whose resolution would have a salutary impact upon the effort at sustained and accelerated development. In the economic crisis of recent years, these problems have been gravely accentuated and now seem likely to persist – unless massive and uninterrupted effort is undertaken, over more than one generation, to ease and eventually overcome them.

The second problem which besets many parts of the region – some more intensely than others – is that associated with the proliferation and persistence of armed conflicts of both internal and external origins. Some of these conflicts have been allowed, through stubbornness, lack of imagination, and absence of foresight, to harden to such an extent that today they defy negotiated solutions and vitiate fraternal relations, undermining bonds that had been historically forged over the centuries. Inflexibilities and rigidities, born of vested interests in the conflicts, have set in and made the path of peace a thorny one – sapping the will to reverse the unhappy trend. They have made the parties to the conflict numb and insensitive to the suffering of the people, so that they have called for ever-larger sacrifices. Some of the combatants find it difficult to adapt to new conditions and to attend to urgent national needs. They have become indifferent to the deepening, dire needs of their underdeveloped societies and impoverished peoples, having acquired the taste for war and feeding on growing appetites for violence.

Selfless struggles were heroic when they were inspired by liberation themes and were dedicated to the survival of societies and the emancipation of peoples; although in human affairs an ever-present danger, the corrupting spread of violence, was always there to dilute values and ideals. There is of course always the forlorn hope, indulged in by the enraged – themselves often tossed about and ravished by its heavy waves – that violence can be exercised in justifiable and measured terms. It is true that worthy ideals, strongly adhered to, can have a dampening effect on tendencies towards violence and prevent it from exerting an overwhelmingly negative and spiritually corrosive influence. Violence has, however, its instruments, and when once indulged, its human agents are invariably entangled in them.

When once a struggle for liberation has been successfully accomplished, acquired power has to be consolidated, and progress made in the intricate and difficult tasks of insuring national integration – particularly in heterogeneous, plural societies. Pressed – as many of the economically backward societies are – with the compulsions of sustained and accelerated development, the need for strong government and the setting aside of democratic niceties becomes greatly attractive. With the considerable effort and sacrifice which, perforce, the process demands, it is not without temptations and excuses that many political regimes neglect the vital matter of safeguarding human rights, insuring democratic participation, and protecting the structures and balances that promote due progress. Their exigencies would interfere too glaringly with the attempts of governments at squeezing surpluses from their peoples for the promotion of economic development and the maintenance of ever-growing, insatiable bureaucracies.

The seeds of the above dilemma were sown in the conflict between the demand for strong government and the essential need to be attuned to the heartbeats of the people. The difficulty lay in the struggle of governments to preserve the respect and influence which their peoples may have for them, as they persist in demanding increasingly greater sacrifices in pursuit of hard economic objectives that do not result in immediate benefits. This situation is further compounded by the ever-present tendency for cumbersome bureaucracy to become corrupt, especially where popular vigilance is attenuated by intimidation and force. The dangers make our societies lose their way. Instead of the original purpose of integrating society, a gulf develops between rulers and elites on one side and the mass of the people on the other.

There is no heroism in bitter conflicts between fraternal peoples who have extensive historical affinities and share related cultures and religions. The insidious spread of violence in the Horn of Africa, if it ever had any justification, has now reached an unacceptable stage that is greatly destructive of society. In certain of its aspects it shows signs of aimlessly degenerating into a state of lawlessness, plunder, gangsterism, and pure militarism that appears to be manipulating fighting for its own sake and shedding fraternal blood on an unforgivable scale. It mirrors in great measure the economic predicament into which the region has sunk, being subject to underdevelopment, environmental degradation, and deepening impoverishment – all punctuated by periodic famines. Societies that feed, get intoxicated, and seem to thrive, on violence at this advanced stage, become essentially moral and physical bankrupts.

It would indeed be escapist and oblivious of the realities of the day if we were to discuss, and endeavour to resolve, the outstanding problems that confront the Greater Horn without tackling the question of violence and its proliferation in the social and political life of the region. This would be behaving like the mindless ostrich that keeps its head well buried in the sand. The human degeneration becomes complete as it enthrals in inflicting destruction of great magnitude upon human achievements, cultural accumulations, and fraternal relations, hence jeopardizing the psychological stability of society. As indicated above, it mirrors the dilemma of deepening poverty that is linked with the perpetuation of violence, while in the context of the Greater Horn of Africa, violence and underdevelopment are the two sides of the same coin. The twin problems could remain, sustaining each other in their flagrant embrace, until a successful breakthrough is achieved in resolving the paradox of their relationship.

We should be fully awake to the exigency that it is in an overall regional settlement which promotes peace and economic development, that we can consolidate the task of restoration and rehabilitation. It is in the larger perspective that we should pose the problems of the Greater Horn and seek their definitive solution. The transcending of this impasse should be a principal task for the future, but the struggle to attain a new dimension of freedom from societal decline and the suffocating atmosphere of violence, to be effective must be undertaken on all related fronts in a region-wide context.

The Regional Dimension and Cooperation

There is a need for a deeper appreciation of the enormous problems of the Greater Horn of Africa by the leaders of the International Community. It is a sensitive and strategically-located area that is fraught with troubles and easily ignitable. The wider world can neither remain indifferent to its fate nor regard its problems with equanimity. The region has experienced a long history of strife and natural adversity, including colonial, guerrilla, and civil, wars in the last half-century, and involving the use of highly sophisticated weapons. It is also subject to exensive environmental degradation and periodic, devastating famines.

It is necessary as a first step, therefore, to stabilize conditions in this notoriously tense region that is quite capable of instigating a general conflagration, situated as it is on the cross-roads of international tensions astride the Red Sea and the Persian Gulf, the vortex of the oil routes, the Arab and Israeli sphere of conflict, the Black African and Arab divides, and the extensive area of inflammable religious fundamentalism that outreaches from the southern borders of the Soviet Union in the north to the Indian sub-continent in the south. Once before, over fifty years ago, it was the scene of armed conflict between Ethiopia and Fascist Italy, in which modern weapons – including tanks, bomber aircraft, artillery, and poison gas – were tested as a prelude to the second world war. Freeing such a region from perennial conflict would make a decisive contribution to the peace and prosperity of the world at large.

The notion that neighbouring countries need to cooperate for their mutual benefit is a novel one to many people – a 'delicate plant that needs to be carefully nurtured and cultivated'. Regional cooperation on environmental issues, underlying the exigency of sustainable development, promises to transform old attitudes and ways of thinking, and to reverse entrenched postures. If persisted upon and enriched by varied and intensive activities that can reduce tensions and strengthen the bonds between the peoples of the region, it can heal the wounds of the past and promote general prosperity to the advantage of all partner countries. New ideas and perspectives, at the base of regional endeavours, should open alternative fruitful ways of looking at old problems, and enable us to discard the sterile responses of the past, which have caused so much suffering to the peoples of the region. Our attitude to socio-economic systems, concepts of neighbourhoods, and regional interdependence, would have to undergo radical transformation.

The weighty changes in the trends of cultural, material, and spiritual, dimensions of life, that the region is undergoing, should be given adequate appreciation, so that our understanding would broaden and gain insights into the great concerns of the peoples of the region. It is necessary to clear away the irrationalities and hatreds that continue to clog present-day relationships and vitiate future efforts at composing and establishing new affinities on strong and enduring foundations. By its very nature, success at this hardened front is bound to be gradual, demanding uninterrupted effort at state, expertise, and popular, levels. Intellectuals, spiritual guides, leaders of opinion, and the people at large, must all be involved in developing forward-looking concepts that define the new relationships and popularize them.

Such a commitment notwithstanding, the effort is bound to be subject to vicissitudes – but can be bolstered and made to gather momentum by solid step-by-step achievements that can clear away distrust. It would be unrealistic to assume that long-entrenched outlooks and internalized mental constraints could be easily replaced by more positive attitudes and healthy beliefs. It would require persistent educative work that would need more than one generation to take root. The secrets of eventual success will be to avoid serious breakdown in relationships on the way, and to assure all partners that, in the balance of gains and sacrifices, there will be advantages for all – indeed that the whole would yield greater benefits than the sum of the parts.

Success would nourish the sinews of confidence, and the difficult challenge that would test the determination and staying-power of leaders is their ability to build up a network of relationships that goes beyond tribal, ethnic, religious, and the national, community in the process of cultivating a regional consciousness that would in time constitute a force capable of narrowing and dissipating the adverse effects of existing disparities. Here, environmental cooperation, underlined by region-wide economic interest, and supported by modern means of communications duly revitalized by the fine arts, literature, and related creative endeavours, could play a decisive role. Expanding economic links and trade are especially important in that they can provide practical 'ballast' to the idealistic stirrings.

Need to Culture a Regional Conscience

The challenge and the choice before the peoples of the Greater Horn of Africa is between destructive hostilities or meaningful cooperation. There is no middle way or any lukewarm alternative, for the question that faces our societies is one of sheer survival. If we start to develop a regional consciousness and think in terms of what would most benefit the region as a whole, then a wide vista of cooperation would open up in which rewards and sacrifices could be balanced in overall terms. If we choose life instead of death – creation instead of destruction – we can gradually dissipate the layers of suspicion and the hold of prejudice that accumulated over the centuries and stultified the clarity of our thinking processes. Thus the resources of the region, transboundary and otherwise, instead of being sources of temptation and sterile conflict, would become opportunities for cooperation and sharing for the economic, social, and environmental, benefit of all partner countries – especially if the region-wide factors are introduced into the diplomatic game.

It is not the weapons of destruction that need to be accumulated and rendered more and more potent and greatly damaging to the environment, but the skills and machinery of diplomacy that have become dull and need to be sharpened and the machinery oiled. It is not hostile propaganda and guerrilla warfare that should be sustained by sanctuaries in each other's territories, but regional development activities of benefit to all that need to be revived and extensively implemented. If only environmental cooperation, in some strategically important sphere that definitively resolves perennial problems could be achieved, it would constitute a breakthrough and provide the driving momentum that would in time increase in magnitude and improve the prospects for the future.

Any success that could be scored in this area would justify any amount of effort and expense by the countries of the region and the international community as a whole, for it would help to stabilize a notoriously sensitive region. The Greater Horn of Africa has rich untapped resources and development potentials that remain unexploited. It is high time that its people got a chance to benefit from Nature's bounty that is on offer. Peace and stability can be established and maintained only if measured and steady development can be achieved.

It depends upon the peoples of the region to respond to the challenge of turning their vision into reality by promoting effective cooperation in the spirit of fraternity and equality born of the exigencies and urgencies of the times. To save our societies from bankruptcy, both moral and material, we should exert all in our power to cultivate solid relations that accord with the imperative trends of our shrinking world in the demands of its technological, communications, and developmental, revolutions and the drive towards regionalization that is evident in many parts of both the developed and developing world, where old enmities have been superseded and new, timely and relevant affinities and friendships consolidated.

The situation that prevails today in the Greater Horn of Africa is a far cry from this vision that at present appears impossibly idealistic. However, our efforts must be equal to the challenge that animates us. The peoples of the Greater Horn of Africa *can* meet the challenge: for it needs but little historical imagination to appreciate their potential. Unless intra-regional relations in matters of economic development, environmental restitution, and eventually in political cooperation, are viewed in this new perspective, we will continue to find our societies 'shipwrecked on the rocks of the traditional animosities'. Inevitably and perforce our region, in its wide-sweeping stretch, will have to be pluralistic and sufficiently adaptive and capable of accommodating all interests – not monolithic, politically or otherwise. The viability of the region as a whole should be assured by being sufficiently integrated and resilient, making it conducive to sustainable and equitable socio-economic development that is duly capable of supporting a disciplined system of political consultations insuring effective policies and actions.

The tasks and struggles ahead will be hard and the called-for sacrifices considerable but well worth while – indeed of decisive import. Taking up the challenge of promoting the resolution of the complex problems of the region will require identifying the linkages that exist between environmental stress, the clash of economic interest, and the tensions inherent in political conflicts that escalate to armed confrontations. The objective of the exercise needs to be to develop the case for meaningful practical cooperation on broad, socio-economic, diplomatic, and political, fronts. No single country in the region can solve its economic, environmental, and above all its ethnic, problems all on its own; it will need the effective cooperation of all the others, and the material and moral support of the international community.

One can discern a 'silver lining' behind the threatening 'dark clouds'. At the beginning of the 1990s we can witness, in some modest measure: cumbersome sprawling bureaucracies, that stifle the initiative of the people, being destabilized and losing their grip; nation-states becoming better organized, integrated, and streamlined; and regional organizations that promote cooperation gaining experience and confidence. Within the context of effective regional cooperation and firm, reliable international guarantees that safeguard vital interests of nations and peoples, political associations, including federations, could be experimented with which allow countries to seek and find their natural orbit free from external impositions. It is only through a deeper awareness of the hugeness of the problems that face the peoples of the region, trapped as they are in the coils of war and environmental degradation, that one can experience something of the dilemma in which they find themselves.

'A Little Closer Together'

Environmental cooperation in the Greater Horn of Africa is still in the early stages, despite the fact that countries of some regions have been cooperating in respect of locust control for some forty years. The East African Common Services Organization has not survived the trials of separate, independent statehoods; but this notwithstanding, it gave us a legacy of experience from which we should draw lessons and greatly profit by careful weighing of the difficulties it had faced while ourselves avoiding the mistakes that it may have made. The time has arrived when comprehensive, organized cooperation on an effective scale is becoming an absolute necessity. Meanwhile countries of the region have found it more rewarding to establish relations – rather in the way of becoming client states – with powerful countries far from their borders, than with neighbouring countries that are as poor as they are and in much the same predicament.

Recent developments indicate, however, that the hazards to which they are subject have brought the countries of the Greater Horn of Africa a little closer together. In 1986, they toned down their hostilities and organized the Inter-Governmental Authority on Drought and Development (IGADD). This organization is still in its developmental stage, but is finding its bearings, testing the ground, and gaining experience for the challenging and difficult tasks that lie ahead of it. Its membership is sufficiently comprehensive, including as it does all the countries of the Greater Horn of Africa; consequently it can play an important role in smoothing differences and promoting peace, having established structures in which periodic consultations are held and joint decisions taken at the highest political levels. A permanent secretariat is in place at the organization's headquarters in Djibouti, to initiate activities of cooperation and follow-up on their implementation.

This organization achieved an important success at its outset by bringing together Ethiopia and Somalia, countries between which a state of war had existed chronically, in a forum that enabled them to exchange views in a constructive atmosphere. To date its achievements are modest but already under its auspices a modern early-warning system of impending droughts is being established for the entire region. IGADD's other projects remain in the gestation stage, and finding the necessary finance for them absorbs much of the time of the leadership of the secretariat. Raising its sights to the future, the organization should be careful not to develop a fixation on the problem of drought rather than overall development that enhances the capacity of its member countries to deal effectively with all aspects of environmental degradation.

Thus IGADD should consider the Greater Horn of Africa as a single eco-geographical complex, and contribute effectively to cultivating a regional consciousness that promotes peace and development – on a sustainable and equitable basis, and on the region as a whole. The problems of environmental degradation can be confronted effectively only if the potential of the region is realized and its economic capacity is greatly enhanced. Otherwise, successes scored on the drought and famine front would suffer relapse and remain perpetually piecemeal. The Authority should therefore grow into a wider and more comprehensive organization.

The failure to reinforce the carrying capacity of the land through sustainable development has resulted in mounting difficulties of absorbing and securing high rates of population growth. This is telling badly, having resulted in marked deterioration in the quality of life and increasing the hardships that arouse dormant atavistic excesses in society. Thus the question of food security looms large. Environmental cooperation can play an effective role in helping to solve this problem. Poor land-management, deforestation, and soil erosion, can be counteracted through joint efforts. Schemes for sharing fresh water and aquifers would prove decisive in alleviating tension and promoting cooperation. Solving problems of nomadic movements would enhance peaceful coexistence. This can be balanced by schemes that guarantee access to ports, continental shelves, and fishing resources, as a basis for negotiations.

The establishment of IGADD is an indication that the meaning and understanding of security may be changing in the Greater Horn of Africa – towards the recognition that there is more to security than military strength and confrontation. Improvement of relations between the superpowers also gives rise to the hope that their interests in the region may be less contentious in the future than they have been in the past. Their involvement could become more productive than at present as soon as rivalry gives way to cooperation and a genuine interest in searching for enduring solutions that promote peaceful conditions which accord with their real interest. There is hope that partisan thinking may look beyond traditional strategic interest in the Horn, and begin to address the root causes of instability. Another encouraging sign has been the recent success of the United Nations in the roles it has played in regional conflicts, notably in Namibia and Afghanistan. This has greatly enhanced the image of the organization as a viable instrument for promoting peaceful relations between sovereign states.

Reference

Toynbee, Arnold J. (1965). *Between Niger and the Nile.* Oxford University Press, London–New York–Toronto: 133 pp.

Annexe 5.2: Threats to Environmental Security in the Horn of Africa: The Somali Case

SAID S. SAMATAR
Professor of History, Rutgers University, Conklin Hall, 175 University Avenue, Newark, New Jersey 07102, USA

INTRODUCTION

The recent history of this region, I think, points to two prevailing developments as principal threats to environmental security for the Horn of Africa. This applies especially to that part of the Horn which is inhabited by the Somali people. One of these two threats relates to the colonial imposition of artificial boundaries on the Somalis – political boundaries that all but undermined the integrity of the fragile Somali super-ecocomplex. The other stems from political blunders and economic mismanagement that threaten to bring about a complete breakdown of civilian society in the country.

Regarding the colonial political boundaries, I wish to offer a minor caveat: there is no intention in this paper to present a partisan position in the matter of boundary disputes between Ethiopia and Somalia on the one hand, and Somalia and Kenya on the other – merely a desire to show the boundaries not only as lines of demarcation between sovereign states, but also as barriers to resource-use maximization in a rather uneconomic zone of Africa, where Man and beast, equally, face an interminable, often daily, struggle for survival at the most basic level at least of human existence.

According to a satirical Somali-creation myth:

> 'God first created the family of the Prophet Muhammad and he was very pleased with the nobility of his handiwork; then he created the rest of mankind and was modestly pleased; then he created the Somalis and he laughed.' (Laitin & Samatar, 1987 p. 29).

To judge by the sentiment of this tale, the Almighty had to undergo wild mood-swings in creating humanity – from ecstatic pleasure at the brilliance of his creative act with respect to other members of Humankind, especially the family of the Prophet, to laughter at the ridiculousness of the human parody that goes by the name of Somalis which he blundered into making. The Somalis thus resulted from a creative mistake!

Aside from the cheerful self-mockery which the Somalis enjoy, this tale powerfully signifies the Somalis' consciousness of their corporate unity – a people created separately and distinctly from others, as it were, even though they may be destined to remain a laughable sort of creation. This fervent sense of belonging to a distinct national community, with a common heritage and a common destiny, is rooted in a widespread Somali belief that all Somalis descend from a common founding father, the mythical Samaale, to whom the overwhelming majority of Somalis trace their genealogical origin. (Even those clan-families, such as the Digil and Rahanwayn in southern Somalia, many members of whom do not trace their genealogy directly to Samaale, readily identify themselves as Somalis, thereby accepting, at least in a symbolic sense, the primacy of Samaale as the fore-bearer of the Somali people.) Genealogy therefore constitutes the heart of the Somali social system and is the basis of their collective predilection to internal fissions and internecine sectionary conflicts as well as of the unity of thought and action among Somalis – a unity that borders on xenophobia.

I have addressed elsewhere the formidable challenges that genealogical segmentation places in the way of creating a stable, centralized state in Somalia (*cf.* Laitin & Samatar, 1987). The Somali collective sense of national uniqueness is, in addition to genealogy, largely shaped by the interplay of nomadic pastoralism and oral poetry, the latter being the passion and pursuit of the Somali nomad's life. To facilitate the reader's appreciation of the contours of the Somali 'national character' (a phrase loaded with conceptual shortcomings), it may nonetheless be helpful to summarize certain features of these two cultural traits, namely of nomadic pastoralism and oral poetry.

What gives the Somalis a unifying centre – and fosters in them a sense of common values and ethics that cut across ethnic, class, and regional, contradictions – is their overwhelmingly pastoral base. The Somalis have been called a nation of nomads whose world is defined by a weary cycle of transhumant migrations between campsites, water wells, and grazing grounds. The label is simplistic – as labels often are – but useful in that it places a deserved emphasis on an institution that has been central to the shaping of Somali society. Somalia is the only sub-Saharan African state in which an estimated two-thirds of the population earn their livelihood from animal husbandry and related enterprises. Consequently, the pastoral attitude and outlook wield a pervasive influence in national life, impinging on plans for economic development as well as on almost every case of social interaction.

If geography is the 'mother of history', it could be said that Somali ecology gave birth to Somali pastoralism. Much of the Somali peninsula is an arid semi-desert that is unfit for reliable cultivation and is only precariously suited for raising livestock. This is especially true of the northeastern and central portions of the country, which together make up roughly 80% of Somalia's landmass (*see* Figure 15*a*.1). In this land of hardy vegetation, steaming sand-dunes, and sun-baked coastal plains, the Somalis have evolved, during the centuries, a way of life that is peculiarly suited to their demanding environment: on one hand, a transhumant pastoralism designed

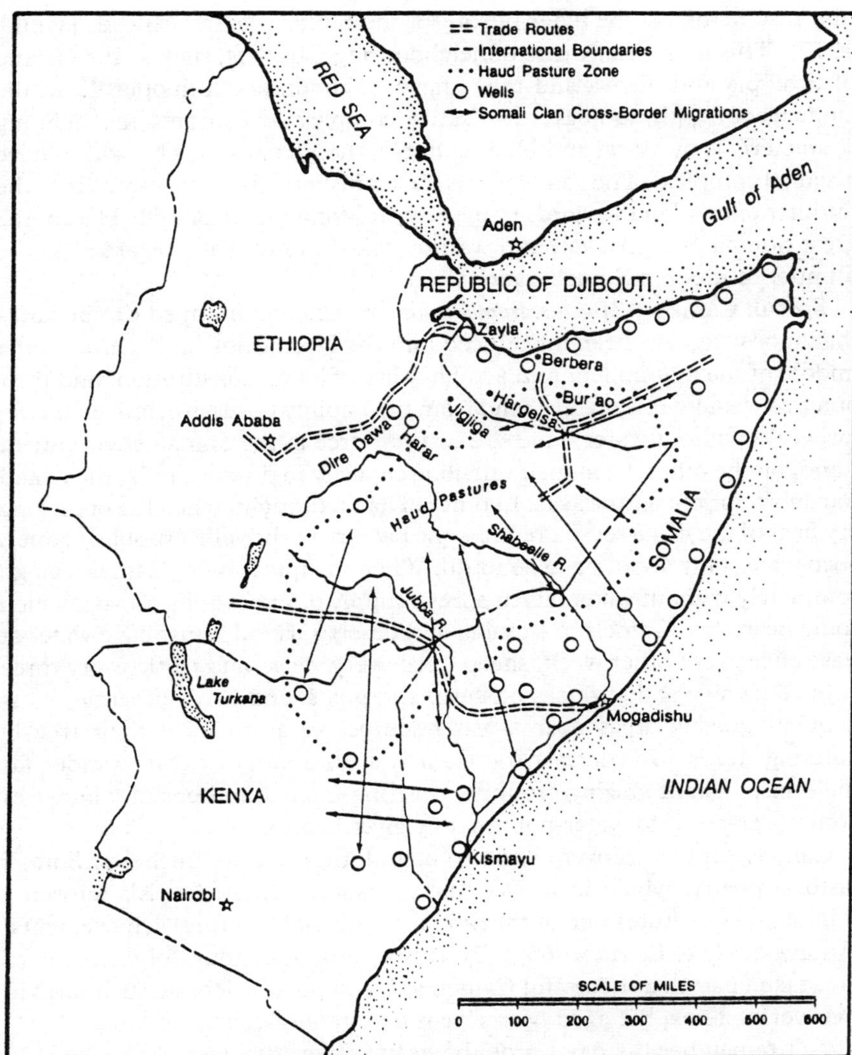

Figure 15a.1. Trade Routes, International Boundaries, Haud Pasture Zones, Wells, and Clan Cross-border Migrations, in the Somali Peninsula, of which some 80% is occupied by arid semi-desert. 1 mile = 1.6 km. (From Laitin & Samatar, 1987.)

to maximize the meagre resources of water wells and pasture-lands for the pastoralists and their herds and, on the other, a social organization that encourages collective action and mutual aid.

Leading Dependence on Domestic Animals

The Somalis raise a variety of animals, including cattle, sheep, goats, and camels. Although cattle-, goat-, and sheep-herding, play a considerable part in the Somali economy, the camel forms the centre of Somali material life –

'the true hero', as one elder put it, 'of the Somali desert' (Sheikh Jaama', 1977). This is especially true of neighbouring Cushites, such as the Oromo of Ethiopia and Kenya and the 'Afar of Djibouti and Ethiopia. If, in the language of anthropology, the 'cattle complex' constitutes the defining characteristic of Masai and Nuer cultures, the Somalis may be said to have a 'camel complex'. The camel forms the mainstay of Somali pastoralism, the 'mother of man' in the words of the Sayyid Mahammad 'Abdille Hasan, the great Somali poet, patriot, and warrior, who is universally regarded as the founder of modern Somali nationalism.

Somali Camels (*Camelus dromedarius*) are the one-humped dromedaries that appear to have been introduced into the Horn from Arabia around the middle of the first millennium BC. They are of hardy constitution, and their practical value stems on one hand from their ability to survive, indeed thrive, for weeks without water – the scarcest resource in the Somali environment – and, on the other, from their unfailing capacity to produce milk, meat, and transport, for the pastoralists. In times of harsh drought, when the punishing dry heat of the *jiilaal* season reduces the pasture to shrivelled rubble, Camels need water only about once a month. When the rains come, Camels can go completely without water, given a fresh supply of moist grass. Goats, which come nearest to Camels in stamina and desert survival, must have water at least once every other week, sheep once every week, and cattle every three days. This means that flocks of sheep and goats, and herds of cattle, must frequent grazing areas near a water-source, so as to meet their weekly watering needs. By contrast the Camel, a born survivor, can wander far afield into distant grazing grounds, covering many hundreds of kilometres from water-point to water-point in any given season.

Camels, and feuds over or about them, form the chief themes of Somali pastoral poetry, which in B. W. Andrzejewski & I. M. Lewis's informed judgment, constitutes one of the Somalis' 'principle cultural achievements' (Andrzejewski & Lewis, 1964 p. 3). It is a mark of honour, for example, to have taken part in a successful Camel raid, and poets celebrate such raids in their verse. Thus, the aged Mareehaan poet reminisced:

'I remember the day I rode the heavy-sinewed horse
And I made the plan of the battle entirely mine
And in the dark I made out the beauty of Idil
And at a sign from me the boys surged forward rushing the herds.
Oh! fine-nippled beast! how she galloped before us.
Then we brought her to a well-watered valley where the he-camel serviced her.
Oh, how I planned skillful schemes! But now have I not become weak?
Lord, of thee I implore the holy water of heaven' (Samatar, 1982 p. 19).

Another poet compares his amorist pains to a periodic affliction (called *dhukaan* in Somali) that devastates Camels in the dry season:

'Like a Camel sick to the bone,
Weakened and withering in strength,

So I, from love of you,
Oh Dudi, grow wasted and gaunt' (Lawrence, 1954 p. 32).

Foreign students of Somali language and culture also refer to the Somalis as a 'nation of poets', whose poetic heritage is intimately connected to the peoples' daily lives. 'Every chief in the country', wrote the romantic, globe-trotting Sir Richard Burton, who frequented the Somali coast in the 1850s disguised as a Muslim holy man with the alias of al-Hajj 'Abdullah, 'must have a panegyric to be sung by his clan, and the great patronize light literature by keeping a poet' (Burton, 1894 p. 82), and continuing, wrote that, 'from time to time usually during the rainy seasons, the clans assemble to engage in poetic contests, and it is here that each clan presents its best poet(s), to do battle with rival poets of opposing clans, their poetic duelling overseen and evaluated by a hoary panel of elder arbiters and literary connoisseurs called *heerbeegti.'* The clan assembly is therefore a focal point not only for the exchange of ideas but also for an intense inter-clan interaction, hence serving to integrate the nation.

Desert, Camels, feuds, and poetry: this aggregation forms the humpty-dumpty of Somali pastoral society, and it may be said in passing that Somali pastoral society today – especially in its passion for poetry and flair for inter-tribal feuds – looks much like Homer's Greek tribal confederacies who engaged, alternately, in bouts of poetry, feuds, and sports. To their subsequent regret years later, the Somalis woke up, at the end of the nineteenth century, to discover sharp colonial-imposed lines of demarcation amongst themselves – lines that not only cut members of ethnic groups apart, but also cut entire tribes from their traditional pasture-lands and water-sources.

Colonial Subdivision Damaging

With the advent of European colonialism, the Somali peninsula, formerly one of Africa's culturally homogeneous regions, got subdivided into mini-lands – a British Somaliland, a French Somaliland, an Italian Somaliland, and Ethiopian Somaliland, and what came to be known as the Northern Frontier District of Kenya (Figure 15*a*.2). When independence came to them in 1960 the Somalis, having successfully unified the British and Italian Somalilands into the unitary Republic of Somalia, got into their heads the ill-advised objective of re-unifying all Somalis into the national fold. The concept of 'Greater Somalia' – a policy whose pursuit all-but wrecked Somalia as a state – was born, as was the elusive doctrine of 'Somalia Irridenta', which was reminiscent of Giuseppe Mazzini's 'Italia Irridenta' (unredeemed Italy), whose revolutionary search for redemption (*i.e.* re-unification) bedevilled European political stability for a century (the nineteenth). Somalia Irridenta has, similarly, threatened the stability of the Horn now for more than a generation – from the 1960s to the present.

Soon after independence, the Somalis in the Somali Republic, quite understandably, wanted their kinsmen across the border in Ethiopia and Kenya to join them in the national fold. Equally understandably, the

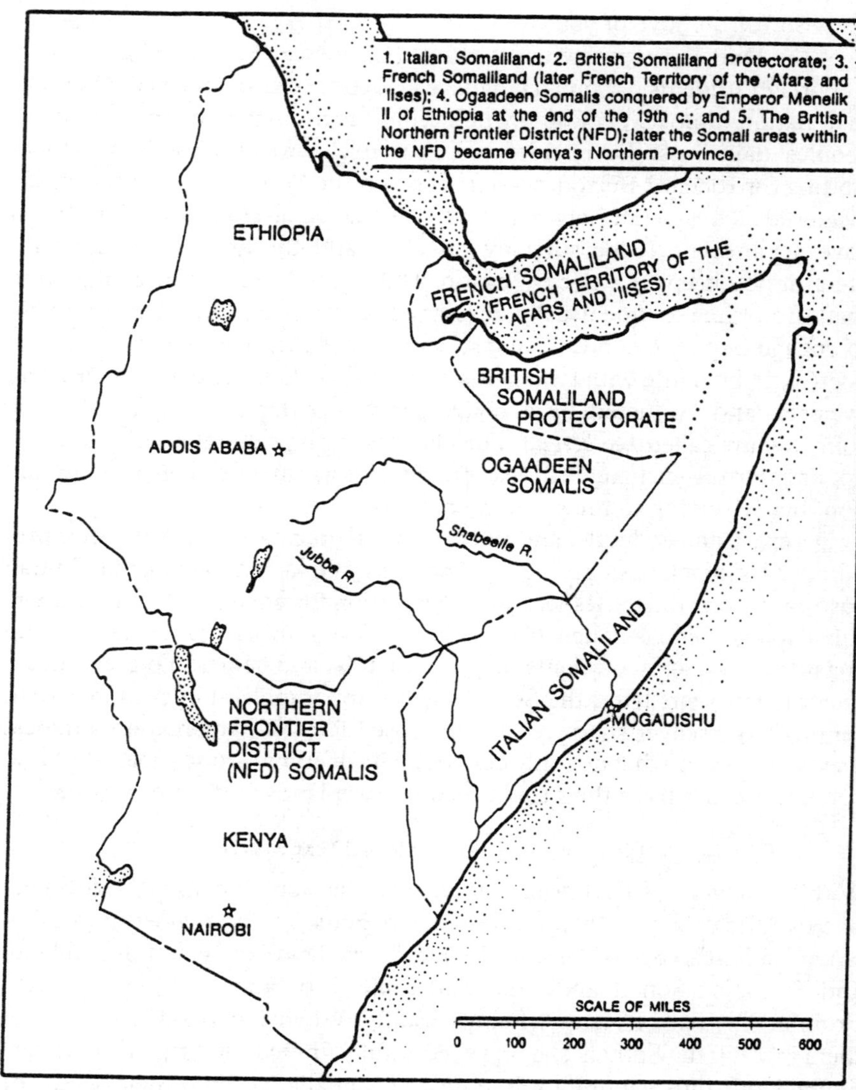

Figure 15a.2. Partition of the Somalis into Five Imperial Zones. 1 mile = 1.6 km. (From Laitin & Samatar, 1987.)

Ethiopians and Kenyans were alarmed by Somali 'irridentism' whose objective, as they saw it, was to take away large chunks of what they regarded as integral parts of their respective countries. The stage was thus set for a localized international conflict that, as often elsewhere, would risk drawing superpower rivals into the Horn, and meanwhile entrenched a festering dispute of thirty years between Somalia and her neighbours that engulfed the whole region in cataclysmic outbreaks of violence.

In what ways have these political problems contributed significantly to severe stresses on the Somali super-ecocomplex? They have done so, simply, by cutting off (through arbitrary political boundary-barriers) those clans on the Somali side from their traditional grazing grounds, whilst others (on the Ethiopian and Kenyan side) have been excluded from traditional sources of water. For example:

> 'In southern Somalia at the tri-junction point of the three frontiers, the Mareehaan clan was cut by the new frontiers into three parts, one part under the British in the Northern Frontier District of Kenya, another under the Italians in the south, and a third in the south-west under Ethiopia; so were the Bah-Gari and Muhammad Subeer Ogaadeen clans, as were a host of smaller clans. The Mareehaan and Bah-Gari sections, who were lopped off to the Italian side of the frontier, were lucky enough to have access to water sources in the Shabeelle (Leopard) River, but they lost valuable grazing grounds on the Ethiopian side. On the other hand, their kinsmen in Ethiopia retained the pastures but were cut off from indispensable water sources on the coast. In the north, the Dulbahante and Isaaq became British subjects, which enabled them to retain the coastal water-points; but they lost the grazing grounds in the Haud (which fell on the Ethiopian side of the frontier), where they traditionally spent six months of the year. The Ogaadeen clans on the Ethiopian side were fortunate to have access to both water and pasture in the Doollo-Harodigeet plains; but they were cut off from coastal trade, especially from the port of Berbera (on the British side), through which 90 percent of their commerce passed' (Laitin & Samatar, 1987 p. 176).

In recounting these events, there is no intent in this discussion to take sides in the political quarrels of the three states of Ethiopia, Kenya, and Somalia, but merely a desire to demonstrate the dangerous overcrowding and overgrazing, with a corresponding abuse of the environment, resulting in a 'serious disruption of the fragile ecological balance between forage plants, animals, and people' (Samatar, 1986 p. 176).

The Somalis' recent past demonstrates the point. The years, for example, of outbreaks of drought, famine, and attendant refugeeism and starvation among the Somalis – 1964, 1967, and 1979 – happen, not coincidentally, to be years of serious border-wars between Somalia and Ethiopia on the one side, and Somalia and Kenya on the other. (I would suspect, too, that the papers by my able colleagues on this panel* should point to a direct correlation between endemic political conflict – *i.e.* nationalities *versus* central government – and rapid ecological deterioration leading to mass starvation in Ethiopia.) Western journalists and newsmen therefore do us disservice when they concentrate solely on the Ethiopian starvation as *sui generis*, without explaining that 28 years of civil war lie at the bottom of the human tragedy that we are witnessing in that country today. Moreover

*[*See* Imru, pp. 327–40; Westing, pp. 354–7. Eds.]

Ethiopia, if managed well and remaining at peace, possesses enough agricultural resources to feed, by one estimate, the entire population of the African continent (*cf.* Befekadu, 1990).

When a border incident occurs, as it often does when Somalis and Ethiopians come in contact, Ethiopia often responds by clamping down the border crossing-lines to the clans on the Somali side, and this produces immediate crisis inside Somalia, or vice versa. The Somalis' sense of loss and calamity as a result of the fracturing of their corporate existence by those colonial dividing-lines is well expressed by the following lament, composed by a pastoral poet shortly after the Ethiopian–Somali war of 1977–8. The poet was denied permission by an Ethiopian garrison near the town of Jigjiga to cross the border to join his family on the other side:

> 'My brother is there (on the other side of the border);
> I can hear the bells of his camels
> When they graze down in the valley,
> And the leaves of the bushes they browse at
> Have the same sweetness as the bushes near my place,
> Because the rain which
> Makes them grow comes from the same sky.
> When I pray, he prays,
> And my Allah is his Allah.
> My brother is there
> And he cannot come to me' (*cf.* Samatar, 1986 p. 186).

The continuing ecological crisis in the Somali habitat is thus directly rooted in political conflict – not just in the Somalis' external disputes with their neighbours in Ethiopia and Kenya, but in intra-Somali conflicts as well – which brings one back to the political nature of the principal threat to environmental security in the Horn, referred to earlier.

Predicament Largely Political

A brief review of recent Somali history will, in my view, demonstrate the political nature of the Somalis' present predicament. In 1969 a military clique under the leadership of General Mohammad Siyaad Barre seized power in Somalia from a discredited civilian government in a bloodless coup. The new regime, then riding high on the crest of popular support, galvanized the nation into a purposive direction. The soldiers 'cleaned up the Augean stables' of the previous civilian regimes, transforming the administration into an efficient working system, undertaking a vast building programme of roads, bridges, and slum-clearance, and energizing the private sector, by word and example, into massive self-help programmes. The effect of all this on state and society alike was dramatic, as the statistics in Table 15a.1 would seem to indicate.

The statistics tell a dramatic story. Within two years (*i.e.* by 1971), the chronically deficit-ridden Somali budget began to report a surplus. From 1961 to 1970 the Somali state budget had been in deficit by, on the average,

Table 15a.1. Ordinary Budget of Somalia: 1961-79 (Ali, 1983) (in millions of Somali shillings*).

	1961	1962	1963	1964	1965	1966	1967	1968	1969	1970
Revenue	109.80	119.40	143.80	151.90	187.90	234.30	245.60	159.60	281.50	288.80
Current Expenditure:	172.30	166.90	189.20	200.80	203.10	257.30	272.10	181.60	301.50	304.10
Foreign Contributions:	62.50	47.50	45.50	48.90	15.20	23.00	26.50	22.00	20.00	20.00

	1971	1972	1973	1974	1975	1976	1977	1978	1979
Revenue:	336.70	406.00	446.40	565.20	628.40	670.70	890.00	1,419.60	1,526.00
Current Expenditure:	298.10	352.70	418.70	530.80	559.30	649.30	801.80	1,361.50	1,572.90
Surplus/Deficit:	38.60	53.30	27.70	34.40	69.10	21.40	88.20	58.10	-46.90

*[One US dollar was equivalent, at the exchange rate prevailing in April, 1990, to roughly six Somali shillings. In response to our request for an updating covering more recent years, Professor Samatar replied that this had proved 'unobtainable'. Eds.]

a whopping one-third that had to be met from foreign aid and grants; in 1972, it showed a surplus of So. Shillings 53.3 millions, equivalent at the prevailing exchange rate of the time to almost US$9 millions.

Moreover, the worst drought (known locally as Daba-dheer, or Long-tailed drought) in Somalia's history struck the nation in 1973–74; yet the budgetary surpluses continued. With a modicum of technical help from the Soviet Union, the Somali regime mobilized the nation to cope with the emergency. They succeeded remarkably, and Long-tail passed without any serious devastation to man or beast.

Considerable – certainly considerable for Somalia – development programmes were also launched, and many of them were successfully implemented, including the introduction of a national script for the writing of Somali, so that the national language took its place among the family of written languages for the first time in 1972. A literacy campaign in that year eventually raised the literacy level from 20% to 50% – a not insignificant degree of progress if judged by African standards. The road to Merka (Merca), one of a dozen paved roads in the country and part of the capital's main artery on the south-west and hence to the 'breadbasket province' of Lower Juba (Giuba), was saved from the encroaching sand-dunes blowing from the shores of the Indian Ocean. A revegetation programme carried out by thousands of volunteers and self-help cadres with no better equipment than garden utensils and their bare hands, succeeded in beating back the sand-dunes lastingly. In addition, dozens of bore-wells were dug and opened for use to improve the herdsmen's water-supply, and thus to make robust the important livestock sector of the economy.

This progress was remarkable. Somalia, considered by the World Bank to be one of the 25 poorest nations on the globe, and long ridiculed by journalists and others as the 'graveyard of foreign aid', was becoming a proud, self-sufficient nation! Neighbouring African states began to take note of the 'Somali miracle'. The Mwalimu Julius K. Nyerere, apostle of Ujamaa, or African socialist self-sufficiency, came over to have a look, and was duly impressed, publishing the opinion that: 'the Somalis are practising to fruition what we have merely preached in Tanzania.' (Nyerere, 1974). Robert McNamara of the World Bank, too, came over to check, and was at once smitten by the 'fever of Somalophilia', to judge from his frequent scurryings from one African capital to another, haranguing African leaders when they agreed to grant him audience, to become 'more like the Somalis' – a phrase that has lately re-surfaced in Washington political parlance as 'the Americans should become more like the Japanese' (McNeil, 1990). But then, just as suddenly, the 'bottom fell out of Somalia'.

From 1977 onwards, a succession of disasters struck, leaving Somalia adrift and operating in the void. This began with losing a war (1977–78) with Ethiopia that killed many, and maimed many more. The effects of that war produced severe social strains and 'twisted the economy out of shape'. The military débâcle, for one, sapped the *ésprit de corps* of the armed forces, and produced what military setbacks often do in a military government – a counter-coup by disgruntled army officers, in July of 1978. The counter-coup failed, but in the resulting shootout the officer corps was decimated. More ominously the Majeerteen clan-family, a largely urbanized clan that dominated the previous civilian regimes, and many of whose members were implicated in the abortive coup, was subjected to fearful reprisals by the increasingly insecure General Muhammad Siyaad Barre. From 1979 to 1984, he hounded the Majeerteen* with an intensity of enthusiastic vendetta that can only be matched by Robert Mugabe's despoliations of the Ndebele in Zimbabwe to punish the big-man Ndebele rival, Joshua Nkomo.

In response to these dreadful retributions, a number of Majeerteen officers, led by Colonel Abdullaahi Yuusuf, went over to Ethiopia in 1979 to form the armed opposition party that came to be known as the Somali Salvation Democratic Front (SSDF). (Colonel Abdullaahi Yuusuf, having lost out in subsequent factional infighting within his own party, has since been handed over to strongman Mengistu Haile Mariam for safe-keeping. He now languishes in an Ethiopian prison – a development that would not altogether displease General Barre.)

Then came the disaffection among the Isaaq people, the dominant clan-family in northern Somalia, which resulted in the cataclysmic upheaval of 1988 that saw some 500,000 Somali refugees fleeing into Ethiopia. On the heels of the Isaaq disturbances followed the insurrection in southwestern Somalia (the Lower and Upper Juba Provinces) of the Ogaadeen clan-family, a branch of the same clan-family who gave their name to the disputed Ogaadeen region of Ethiopia.

General Muhammad Siyaad Barre, increasingly besieged by real and perceived enemies from within and without, and grimly determined to hold on to power, unleashed the dreaded National Security Service (NSS) on the civilian population, so that gangs of NSS began to roam among the civilian population, terrorizing and executing summarily, sometimes on a massive scale. The rank and file of the NSS are largely composed of 15-year-olds from General Barre's 'Mareehaan clan', and are armed with AK-47 assault rifles. Understandably, these beleaguered lads don't ask questions at the first hint of public disturbances; they shoot first, and may enquire afterwards. Tribalism, nepotism, and mismanagement, as well as wholesale plunder of the Treasury, have returned with a vengeance. The plunder began under cover of the war conditions in 1977 and continues apace, except that there is not much left to plunder in Somalia these days. The country has become bankrupt and 'flat on its stomach' once again.

Near-anarchy Widespread

Thus as the decade of the '90s wears in, the Somali state has virtually ground to a halt, and the Somali nation has practically broken down into competing segmentary lineages that have turned into an orgy of internecine

*[Not to be confused with his own Mareehaan clan, *see* below. Eds.]

bloodletting. It can be said, I think with accuracy, that the entire Somali peninsula is today in a state of civil anarchy, with a near-collapse of civil polity. As a result, a veritable 'crisis of habitability' (Ali Mazrui's apt phrase for the African condition of 1980), has descended, overnight as it were, upon the Somalis, to the grave derogation of both Man's and Nature's joint environment.

One overriding lesson, I think, can be learned from the Somali experience generally. The Somali situation demonstrates, convincingly in my view, that even the poorest of African states can, with sound fiscal management and a reasonable standard of justice for, and decency among, its citizenry, put its house in order with considerable success and against all expectation. Somalia was there to prove it. This could, in turn, seem to imply that the current economic and social plight of African nations may well be due not so much to scarcity of resources, as the pundits have been telling us for a generation now, or to the excesses of some colonial regimes or their legacies, as to chronic corruption, mismanagement, and general misrule by African leaders. Africans have, in my view (and I am one), sadly misgoverned Africa. If resource-scarce Somalia can show a budgetary surplus and a growing economy in five years of efficient administration, one can only imagine what a similar application of efficient management could do for resource-rich Ethiopia, or Kenya – or Nigeria, for that matter.

Taking this line of argument of course creates some delicate problems. For one thing, it shifts the debate over the cause of the present African predicament from blaming external exploitation and the bogeyman of multinational conspiracies, to an embarrassing look at how Africans have shown themselves to be rather poor stewards of their own patrimony. This may in turn force us to switch the focus of analysis from Professor Walter Rodney's manifesto, 'How Europe Underdeveloped Africa' (1982), relevant in its time, to a new manifesto: 'How Africans Have Underdeveloped Africa'.

References

Ali, Khaliif Galaydh (1983). *Public Sector in Somalia.* Paper presented at the annual Conference of the African Studies Association, December 1983, Boston, Massachusetts, USA.

Andrzejewski, B. W. & Lewis, I. M. (1964). *Somali Poetry: An Introduction.* Oxford University Press, Oxford, England, UK: viii + 167 pp., illustr.

Befekadu, Degefe (1990). Based on an interview with him, South Orange, New Jersey, USA, 29 March.

Burton, Sir Richard (1894). *First Footsteps in East Africa or, an exploration of Harrar.* (Ed. I. Burton), vol. I. Tylston & Edwards, London, England, UK: xxxiv + 209 pp., illustr.

Laitin, David, & Samatar, Said S. (1987). *Somalia: Nation in Search of a State.* Westview Press, Boulder, Colorado, USA: xvii + 198 pp., illustr.

Lawrence, Margaret (1954). *A Tree for Poverty: Somali Poetry and Prose.* The Eagle Press, Kampanala–Nairobi–Dar es Salaam: 146 pp.

Lewis, I. M. – *see* Andrzejewski, B. W. & Lewis, I. M. (*q.v.*).

McNeil, Robert (1990). Public Television's Channel 13 Program, New York, NY, USA: 2 March.

Mazrui, Ali (1980). *The African Condition*. Heineman, London, England, UK: xvii + 142 pp., illustr.
Nyerere, Julius K. (1974). Organization of African Unity Conference in Mogadishu, Somalia, 25 April.
Rodney, Walter (1982). *How Europe Underdeveloped Africa*. Howard University Press, Washington, DC, USA: xiv + 312 pp., illustr.
Samatar, Said S. (1982). *Oral Poetry and Somali Nationalism: the Case of Sayyid Muhammad 'Abdille Hasan*. Cambridge University Press, Cambridge, England, UK: xii + 232 pp., illustr.
Samatar, Said S. (1986). The Somali Dilemma: Nation in Search of a State. Pp. 155–93 in Asiwaju, Anthony (Ed.) *Partitioned Africans: Africa's Ethnic Relations Across International Boundaries, 1884–1984*. C. Hurst, London, England & St Martin's Press, New York, NY, USA: xii + 275 pp., illustr.
Samatar, Said S. (1987). *See also* Laitin, David, & Samatar, Said S. (*q.v.*).
Sheikh Jaama (1977). *Umar 'Iise*. Based on an Interview with him, Mogadishu, Somalia, 24 April.

Annexe 5.3: Environmental Security for the Horn of Africa: An Overview

ARTHUR H. WESTING

International Peace Research Institute Oslo, Fuglehauggata 11, N-0260 Oslo 2, Norway; Adjunct Professor of Ecology, Hampshire College, Amherst, Massachusetts, USA

INTRODUCTION

Neither the present status nor the prognosis for environmental security in the Horn of Africa is encouraging. Provided here is an overview of the status of this northeastern extension of Africa as well as a number of recommendations for ameliorating that status. The present examination is being carried out within a previously established conceptual framework (Westing, 1989*a*) and elaborates both upon a general analysis of environmental security for arid habitats (Westing, 1980, pp. 104–13) and upon a summary analysis that dwells upon the Horn of Africa (Westing, 1989*b*).

STATUS

The arid ecogeographical region that comprises much of the Horn – consisting of greater or lesser portions primarily of Djibouti, Ethiopia, Somalia, and Sudan – appears to have no exploitable mineral or other non-renewable resources of significance on which to fall back for the financing of development. The regional inhabitants must therefore, at least in the near term, depend for their survival and well-being on the renewable resources under their stewardship – renewable resources that ultimately derive from the regional soil and rainfall.

Put in slightly different terms, the regional inhabitants must utilize their land – that is, their cropland, their rangeland, and their forestland – so as to obtain on a continuing (sustaining) basis their fundamental requirements for food and fuel as well as an excess of economic crops (both plant and livestock) for commercial purposes, both domestic and foreign.

The number of people living in the Horn now grossly exceeds the ability of the land to produce a sufficiency of food and of fuel for subsistence, and further plant and livestock crops for commerce. Moreover, the regional population is growing at an explosive rate. The corollary damage to the land is grim and, ultimately, suicidal:

- Cropland (which, if properly managed, could be maintained indefinitely) is being exploited at an intensity that annually erodes the soil, thereby annually diminishing the productivity of the regional cropland. Some cropland becomes completely useless each year. Additionally, in some irrigated areas, overly intense irrigation is leading to damaging levels of soil salinization.

- Rangeland (which, if properly managed, could be maintained indefinitely) is being exploited (*i.e.*, used for livestock production) at an intensity that annually diminishes the vegetational regrowth, thereby annually diminishing the productivity of the regional rangeland; some (where the plant cover disappears) erodes each year, so as to become completely useless (a process often referred to as 'desertification').
- Forestland (which, if properly managed, could be maintained indefinitely) is being exploited (primarily for fuel) at an intensity that annually lowers the yield of wood, thereby annually diminishing the productivity of the regional forestland; some (where the tree cover disappears) erodes each year, so as to become completely useless. Additionally, some forestland is being permanently cleared each year for conversion to agriculture or other purposes – land thus also lost to wood production.
- Wildland (*i.e.*, largely rangeland or forestland that is not being intensively exploited) is rapidly diminishing in extent – and with it the habitat for wildlife, and thus the regional wildlife itself.

Visible indicators of the desperate nature of the situation in the Horn include: (a) the exceedingly low life-expectancy of the regional inhabitants; (b) the extremely low, but nonetheless continually declining, regional *per caput* gross national products; and (c) the inability of the region to produce an adequate supply of staple foods and needed fuel, an inability that grows in magnitude year by year (*i.e.*, the Horn suffers from continuing declines in *per caput* food production and *per caput* fuel production).

Recommendations

A number of recommendations can be made vis-à-vis environmental security for the Horn of Africa:

- Further population growth in the Horn should be controlled, the single most important suggestion that can be offered. Associated efforts should include those aimed at improving the status of women in the region, and at improving institutional provisions for old-age security.
- Remaining croplands, rangelands, and forestlands in the Horn should be guarded with tenacity, so as to prevent exploitation beyond their sustained yield. Moreover, these lands should be managed so that, little by little, they are restored to their maximum (*i.e.* optimum) sustained yields.
- Warfare in the Horn, both intra-state and inter-state, should be curbed, so as to avoid environmental (and, of course, other) abuse, both direct and indirect, from that quarter. Among various other favourable outcomes, success in curbing warfare in the region would greatly reduce the currently huge numbers of cross-border refugees, with the associated social and environmental calamities they generate. Moreover, it must be

recognized that the various seemingly intractable wars in the region enjoy their longevity and continuing vigour largely through the support of various foreign powers: such outside support should be terminated, at least beyond the modest needs suggested below.

Many of the wars in the Horn, both intra-state and inter-state, have been attributed to the present locations of the national boundaries in the region (many of them a legacy of colonial times) in so far as such boundaries do not coincide with indigenous ethnic, religious, or linguistic affinities. In the case of an *existing* inter-state border that is contested, the problem could be mitigated by making that border more porous; indeed, border formalities should be relaxed throughout the region (here perhaps relying for a model on the level of Scandinavian border constraints). In the case of a *desired* inter-state border, the problem could be mitigated by establishing a less porous intra-state border, that is, by establishing semi-autonomous intra-state regions.

- The military sector in each of the nations of the Horn should be curtailed to a level that suffices for purely defensive contingencies; and the financial, material, and human resources made available thereby should be shifted to the environmental and social sectors. Indeed, the state of militarization in Africa (as elsewhere in the world) is unconscionably high. Various arguments could readily be mustered – humanitarian, economic, and environmental – in favour of arms reductions; as could, just as readily, the difficulties associated with achieving such reductions. But the simple fact of the matter, at least for the Horn, is that the military sector is the only societal sector that could be classed as frivolous and counter-productive. It is thus the *only* sector of society that could be re-allocated to the environmental sector without detriment to some other component of human security in the Horn.
- Education on the one hand, and participatory democracy on the other, should be fostered in the Horn, inasmuch as an informed and enfranchized public is presumably more likely to support and adopt policies more conducive than present ones to achieving social and environmental security (*i.e.*, to achieving comprehensive human security). In fact, the African Conservation Convention of 1968 provides for regional cooperation in environmental education (*see* its Article 13). It is important to add here, as well, that if the status of women in the Horn were improved, such advances would be most supportive of both social and environmental security.
- Regional (*i.e.*, inter-state) cooperation within the Horn should be fostered because many, if not most, of the threats to regional environmental security have cross-border dimensions. A number of important areas for regional cooperation can be suggested: (a) the regional wars, *both* inter-state and intra-state, have cross-border dimensions that require for their amelioration inter-state rapprochement; (b) the nomadic approach to rangeland utilization – so important to the sustained yield of the renew-

able rangeland resource – demands cross-border movement, and thus inter-state cooperation; (c) the rational and equitable utilization of the few significant rivers in the region (both for their water *per se* and as carriers of waste products) requires cross-border cooperation; (d) dealing with the recurring depredations of locusts would be greatly facilitated by joint regional action; (e) dealing with the conservation of cross-border migratory wildlife necessitates regionally shared action; (f) pooling of the currently modest regional scientific and educational resources would be a boon to the Horn; and (g) dealing both with the benefits and the problems of tourism (much of it potentially associated with the regional wildlife and its habitats) would benefit from cooperative regional efforts. The urgently required regional cooperation should be at both formal and informal levels, each category being supportive of the other. Moreover, in order for the formal cooperation to be given the chance to function properly, *all* of the nations in the Horn should become parties to the relevant regional and other multinational treaties.

Conclusion

In closing, the several nations in the Horn of Africa – especially Djibouti, Ethiopia, Somalia, and Sudan – simply must recognize that their current course is leading towards environmental (and thus also social) disaster. As inexorably linked partners in a shared ecological system, it is past time for them to recognize that they have reached the point where it becomes necessary to deal with shared problems via cooperative action. To that end, the time has come for these nations – building upon their rich cultural heritages – to relax some of their feelings of national sovereignty; to overlook some of their past animosities; to provide the regional inhabitants with an opportunity for greater control over their own governance; and to turn to forms of non-violent inter-state dispute resolution.

References

Westing, A. H. (1980). *Warfare in a Fragile World: Military Impact on the Human Environment.* Taylor & Francis, London, England, UK: xiv + 249 pp.

Westing, A. H. (1989*a*). Comprehensive human security and ecological realities. *Environmental Conservation,* Geneva, 16(4), p. 295.

Westing, A. H. (1989*b*). Environmental component of comprehensive security. *Bulletin of Peace Proposals,* Oslo, 20, pp. 129–34.

Commentary on Annexe 5

JOINT CHAIRMEN: Professor David Wasawo & Dr Arthur H. Westing

PARTICIPANTS AND OTHER CONTRIBUTORS:

Wasawo, A. Cloudsley, Juel-Jensen, Westing, Holdgate, Juel-Jensen, Cloudsley-Thompson, Barton Worthington, Kuenen, Burnett, Samatar, Trompf, Fearnside, Westing, Juel-Jensen, Imru, Samatar, Demeter, Oza, McCusker, Holdgate, Westing, Wasawo

Wasawo remarked that he had been wondering why the Horn of Africa had been chosen for consideration but he could see several good reasons. First, it was an important strategic area for human communications and commerce: for Europe; the Arab world; the African continent; the Indian, Pakistan and Bangladesh subcontinent, and the Far East and Australia. Second, it was an arena where we had witnessed some rivalry between two great powers: great in terms of material and military might. Third, it was in general an area of semi-arid or arid terrain whose populations over the centuries had adapted to those conditions to become a hardy, resilient and resourceful people, but whose contact with the rest of the world had necessitated changes in their habits and life-styles which had lead to new pressures on the environment and themselves.

For over twenty years, the Horn of Africa had in one way or another not known peace. The time they had in hand had been frittered away in war, so they were now left with little or no time to build a better environment for the future. For the human environment was a dynamic one – its tendency was to deteriorate unless it was actively looked after. That looking after required time to think about the environment; time to do something about it, and sometimes even time to judge whether it should be left alone!

We had witnessed in this area, and continued to witness, the utter wretchedness and helplessness of the refugees. The international mass media had depicted vividly untold miseries of famine and death, due not only to the vagaries of wars but also to the vagaries of climatic change. Deforestation and the loss of valuable topsoils in parts of the region were also well known and had added to the environmental problems

A. Cloudsley spoke on behalf of a representative of the Sudan who had not been able to be present. She was no expert on the environmental security of the Horn of Africa but, from her travels in the Sudan, Ethiopia and other nearby countries, she could speak of traditional ways, the loss of which coupled with rapid social change had been a matter of concern – for many of them involved a positive approach to the sustainable use of natural resources.

In both towns and rural areas, where water had always been scarce, *delkha,* prepared by mixing millet flour with oil of sandal or sesame oil, had been an item of toiletry for many centuries, dating back to the ancient Egyptians. The fine particles of flour acted as an abrasive to clean the skin, and only minute amounts of water were used for washing. The smoke bath or *dhukan,* likewise needed no water and cleaned the body like a Turkish bath. Even the traditional washing for prayer, the *wado,* was very economical and only two or three pints of water were needed to purify the whole body.

Although many urban households lived in European style, many more throughout the Sudan ate traditional dishes – *mulakhs*. These used indigenous materials which grew on poor soil. Vegetable *mulakh* was made from *Bamia* or *Okra* (*Hibiscus esculentus*), called *weika* when dried. Other vegetables used were Purslane (*Portulaca* spp.) and *warag*, which literally meant paper and was Cowpea (*Vigna sinensis)* leaves which, indeed, looked like scraps of paper. Leaves of Jew's Mallow (*Corchorus olitorius*) and Red Bean (*Vigna catiang*) were prepared in the same way; these were used for thickening stews, making it easier to eat with wafer-thin *kisra* – millet (*Sorghum* sp.) bread.

Lastly, many obtained their entertainment and psychosomatic relief from *zar* ceremonies, which were often prophylactic. Sickness and disease were often alleviated without the use of expensive imported medicines, and guests and onlookers were provided with social entertainment which did not require the use of extraneous energy!

Juel-Jensen said he had spent many years as a doctor with poor Ethiopians, often as the only foreigner. Poor as they were, they looked after their relatives and the old, for the extended family was one of the great miracles of modern Ethiopia which he would be sad to see gone. He also thought that women did have a reasonable status; they inherited much the same as did men, and he had certainly met many who ran the household! Ethiopia had suffered too much from outside interference and he thought it difficult for anyone present to suggest that they could teach its people when the West was not doing a very good job in its own countries.

Westing said that this was a region with a growth-rate in excess of 3%, which meant that every 20 years, or thereabouts, the population doubled. He believed that improved family planning was the most useful suggestion he could make and that, if the status of women and old people were protected more, institutionally, then it might be more conducive to reducing the genuine population explosion which was occurring.

Holdgate felt that this point should not be dropped. If we were going to try to persuade people to change their style of life so fundamentally that, in one generation, they should drop from large families under circumstances where large families were welcome both as part of the economic necessity of life and to give security to the extended family, then we were asking for a fundamental change in attitudes which was extremely difficult to bring about. That dilemma was something we had not yet resolved either in the Horn of Africa, or in many other parts of the world.

Juel-Jensen accepted that his profession had much to answer for but he pointed out that in Tigray a family might have 2 or 14 children but probably only 2 or 3 would survive, just as up to the 1890s on St Kilda (the outermost of the Hebridean islands) it was essential to have at least 12 children because the odds were that only 1 or 2 would survive. One reason for this was that, among other things, bleeding from the umbilical cord was stopped by putting cow-dung on it and so brought about neonatal tetanus. That practice had ceased in 1895 on St Kilda but, he believed, continued today in Tigray.

Cloudsley-Thompson said that much had been held against the West through our ignorance of history. Just as the Greeks were ignorant of civilizations as great as their own until Alexander the Great won the Battle of the Indus, so many in the West had known little about Ethiopian civilization until recently. He could think of nothing more important than getting people to know and understand the people of other countries: that was the best way to promote international cooperation, as many of his personal experiences in both the Sudan and Ethiopia had taught him.

Barton Worthington said that recently he had been advising on the ecological and environmental impacts of three major development projects in Ethiopia. He agreed that the population problem was by far the greatest ecological problem facing those peoples for the future, but it was one on which everyone had to go very, very gently. However, if it was not faced, disastrous famines would continue. He believed that part of the problem was that no one in charge of Ethiopia had ever recognized the problem. If that continued, the sum total of human misery would continue to grow with the population. While other countries with a similar problem had at least recognized the population problem, he continued to be amazed that, despite an increased awareness by everyone, including even the man in the street, the overriding importance of this issue had not been grasped. If people had come forward on the population crisis as effectively as they had done on climatic change, the world now would be in a better shape for the morrow.

Kuenen remarked that the subject was taboo to much of the world's population who were under the leadership of Rome but, nevertheless, it must be discussed.

Burnett said that there were probably very few other areas where developed nations would like to help more than in the Horn of Africa. As had been said, there was immense concern, fed largely by the media. Many would like to assist by introducing acceptable technical innovations, or by providing the kinds of information not available to the indigenous peoples, such as survey and monitoring data supplied by Landsat, or similar systems. Many had also tried, on both the large and small scale, to assist in the agricultural development of the area but they had met the 'perception barrier' which **Holdgate** had mentioned earlier. Because of the continuing history of Great Powers' intervention for their own political or economic ends, there was a terrible disbelief that the offer of any kind of technology, even that which would be genuinely useful, was not made without some ulterior motive. How was that barrier to be overcome? There was so much of value that could be provided which the indigenous people could not do for themselves but which it was for them to decide to use and develop. Solutions could not be imposed from without.

Samatar was grateful for those views but sensed some timidity about the business of non-interference. The world was in a dynamic, interacting state of flux in which not only the technology but the views, insights and recommendations of outsiders were valuable, especially when those inside could not see for themselves. Intervention was vital and necessary. After all, if the British, French and Italians had not imposed their will in 1924, there would still be slavery in those countries! That was a most timely outside pressure. Developed countries could help by being bold and forthcoming with their opinions, especially when the indigenous population was screwing things up!

Trompf asked why was there a conspiracy of silence about Eritrea, which had not been mentioned?

Fearnside also wanted to get back to the suggestion made by **Westing** for establishing internal borders with semi-autonomous regions in some of the countries. It seemed to him that an obvious one would be between the peoples of the north in Ethiopia and those with a different language further south.

Westing agreed. His comment about establishing less porous internal borders was based primarily on the idea of recreating some sort of autonomy for Eritrea. There was much to be said for the autonomy of Eritrea but, being no sort of political scientist, he could also see much for maintaining the integrity of Ethiopia. He had been aiming at some sort of a compromise which would give Eritrea some hope of governing its own affairs.

Juel-Jensen intervened to say that Eritrea was an invention of the Italians when they had settled there.

Imru was surprised by the suggestion that there was a conspiracy of silence about Eritrea; it was the most talked-about region of Ethiopia! Even the central government had gone forward to negotiate some sort of a settlement with the Eritreans under the auspices of the former President of the USA, Carter. He thought that there was great ignorance of the history of Ethiopia, especially amongst the media with ill-informed talk of frontiers, break-ups and fragmentation. Ethiopia had been constituted over the centuries, not yesterday. In Eritrea, for instance, there was no difference between Eritrea and Tigray in language or religions, and even differences within Ethiopia as a whole could be traced back to a common language and civilization. Solutions to the problems of the Horn of Africa would only come with enhanced regional consciousness, regional understanding and regional cooperation, not by breaking up an historically constituted region. Eventually some sort of devolution would be agreed; for centralization, which was associated with Emperor Haile Selassie, had only occurred within the last 70 years.

Samatar had also studied the history of Ethiopia, in its in some ways admirable, majestic, almost epic proportions. He also knew that the Italians had invented Eritrea and that the highland Eritreans had been more closely integrated with the people of Gondar and Gojam in the heartland; but to go from there to imply that the problem of Eritrea was a journalistic invention seemed to be skirting around the problem! There was a real internal problem – a vast revolutionary army had taken up arms for 27 years – and both Ethiopia and Eritrea had legitimate concerns that should be resolved in a legitimate and fair way without trampling on the rights of one or the other.

Demeter said that, on the question of borders, he thought it essential to have permeable borders in Africa for ecological reasons. Ian Sinclair, working some years before on the wildlife migratory system of the Serengeti, had compared the cyclicity of the original wildlife systems with that of the nomadic pastoralists in the Sahara/Sahel and had found striking similarities between the two. He had concluded that part of the problem of the Sahel was that waterholes, bored with western aid, had diverted the nomads into areas at times of the year when they should not, ecologically, have been there. This example showed how, in areas where there was high seasonality, ecological problems were very deep-rooted. When human behaviour was superimposed upon them, human ecological problems arose, and from human ecology onwards political intervention was only the next step. That was very relevant to the Horn region.

Oza intervened to say that it would have been more sensible to have invested aid-money towards planting more trees than on armaments, so shattering the economy of the region.

McCusker wished to share a lesson she had learned recently on aid. One of the objectives of the IBPG programme was to strengthen national programmes of plant genetic resources and to assist them to develop their own work. She had decided that IBPG's plan should be based on consultation. Over the last 4 weeks she had held consultations with groups from the Indian subcontinent, east and south-east Asia, north Africa and the Middle East. All these groups had said that they did not want training, technology, or money, all of which they had been offered. What they did want was for IBPG to analyse their individual situations with them, help to set priorities, and develop their strategies. Thereafter it would be possible to see where training, technology and money might be needed. She thought that such an approach was relevant to many of the situations which had been discussed on the larger scale of the Horn of Africa.

Holdgate said that only 6 weeks earlier IUCN had received an invitation to assist in producing a national conservation strategy for Ethiopia. Papers would come entirely from within the country and IUCN's contribution would be to discuss the needs, priorities and approaches in a similar manner to that described by **McCusker.** He thought it an excellent way forward.

Westing thought this was good news but suggested that part of the national Ethiopian conservation strategy should be to recognize that their problems spilled over their borders; no long-term planning could neglect that fact which the Ethiopians should realize. He also suggested that EGAD should be involved, at least with observer status, because it was the logical organization to deal with such problems on a regional, interstate basis.

Wasawo concluded by remarking how it always seemed to him that human predicaments were the same the world over. What could usefully be learned in one part could be used elsewhere, to the extent that there could be helpful exchanges of information, experience and ideas.

He had also noted that it had been suggested that Africa's problems derived from incompetence and mismanagement by African leaders. They did indeed have some rotten leaders in Africa but also some excellent ones – so did Europe, North America, Latin America, Asia, the Far East and Australasia, as everyone knew from their own experience. Many of Africa's problems were not of Africa's making, as had been demonstrated again and again at international fora, but some were due to internal incompetence and mismanagement: it was important to strike the balance. The world was a very complex place and it was only too easy to come to conclusions about the origins of problems; it was a mite more difficult to analyse them coolly and dispassionately and to realize that what happens, for example, in America today, affects Africa tomorrow, and *vice versa.* He recalled that when the Soviets were ruling Somalia he had found that the UNDP representative was a Russian and the Soviet ambassador the key man. Two years later the Soviets were out and the Americans were in. Why was that? Why did that happen in that part of the world? And what were the origins of the Ogaadeen war thereafter? Was it only to do with Somalia and Ethiopia, or was it much more complex than we thought? As scientists it was most important that we impinged on the human condition in the right way, objectively and analytically, and that we should be careful where, or where not, to put the blame. Perhaps the salvation of the Horn would come from environmental cooperation.

16. Ecological Economics and the Environmental Future

CLEMENT A. TISDELL

Professor and Head, Department of Economics, University of Queensland, St Lucia, Queensland 4067, Australia

INTRODUCTION

Feuds between natural scientists and economists (and within their ranks) about the scope for future economic growth, not to mention its desirability, have been common in the last 25 years or so and reached their maximum intensity in the 1970s (Perrings, 1987 p. 47). The ferocity and emotionalism involved in many of the interactions were intensified by misunderstanding and distortion of the views of opponents, and by the relative lack of (experimental and other) evidence in relation to the environmental hypotheses under debate. Some economists also felt that their 'traditional' academic discipline-based territory and established preferential position as advisers to governments and to the public on social issues were under challenge.

In the course of such debate, many economists have felt it their duty to warn the public and policymakers about the dangers of accepting the social policy-prescriptions of natural scientists – on the grounds that, although natural scientists were experts in their own fields, they were naïve and untrained in social issues. At the beginning of this debate, there was a feeling amongst most (but not all) economists that those trained in a particular discipline should keep to their own areas of expertise. Such a reductionist, partial, compartmentalized approach to knowledge rather than a holistic view accordingly tended to prevail.

However, since 1970 the small nucleus of economists involved in environmental and ecological issues has expanded, particularly in North America, university courses on environmental economics have become widely established, journals such as *Environmental Economics* and *Ecological Economics* have come into existence, and more and more economists are co-operating with natural scientists to support a holistic approach to policy formulation. Journals such as, particularly, *Environmental Conservation*, and organizations such as the Foundation for Environmental Conservation, have encouraged natural and social scientists, and also economists, to interact constructively in this process.

Here I want to outline the scope of ecological economics and briefly consider its modern origins. Recent global policy reports such as *Our Common Future* (World Commission on Environment and Development, 1987) emphasize the importance of combining economics and ecology in framing economic development policies. The current success of political parties favouring ecological conservation programmes, gives additional emphasis to the importance of combining economics and ecology in policy proposals.

Today it seems that we have to deal increasingly with environmental phenomena of a more pervasive type than in the past. Global uncertainty, and irreversible environmental effects from economic growth, are no longer rare. If the type of economic growth which has been common in the past persists, it will continue to damage The Biosphere irreversibly and will increasingly threaten the future economic welfare of Mankind. In these circumstances, alert account needs to be taken of *ecological economics*.

The Scope of Ecological Economics

Robert Costanza (1989 p. 1) sees ecological economics as addressing 'the relationship between ecosystems and economic systems in the broadest sense. These relationships are the basis of many of the most pressing current problems (*i.e.* sustainability, acid rain, global warming, species extinction, wealth distribution) but they are not well covered by any existing discipline'. Environmental and natural resource economics, as currently practised, tend to concentrate on neoclassical and conventional economic approaches to the analysis of environmental and resource issues. In the past, writers on these subjects have given relatively more attention to the non-living environment than to ecobiomes, ecocomplexes, or their component ecosystems, but one would expect ecological economics to have the opposite emphasis. Costanza (*idem*) claims that, while some ecologists deal with human impacts on ecosystems, most keep to natural systems. However, for many policy purposes, ecology and economics need to be linked, and this requires co-operation between ecologists and economists and a more open approach than hitherto on the part of the economists to the often all-pervasive environmental issues.

Methodological and conceptual pluralism is needed in studying ecological economics. As Costanza (1989 p. 2) suggests, 'there is probably not one *right* approach or paradigm, because like the blind man and the elephant, the subject is too big and complex to touch it all with one limited set of perceptual tools'. Norgaard (1989 p. 37) also so argues and develops the view that 'all aspects of complex systems can only be understood through multiple methodologies. Since ecological economics seeks to understand a larger system than either economics or ecology seeks to understand, a diversity of methodologies is appropriate and [so] pressures to eliminate methodologies for the sake of conformity, should be avoided.'

Ecological Economics: Historical Aspects of Its Emergence

The emergence of ecological economics as a recognized area of study is very recent. Indeed, it might be dated from as late as 1989 with the formation of the International Society for Ecological Economics and the publication of its journal, *Ecological Economics*, even though many earlier contributions fall into this field (*see*, for example, Martinez-Alier, 1987; Proops, 1989).

One of the first Authors to propose seriously the term ecological economics was Kenneth E. Boulding (1950), who suggested, in his ecological introduction to *A Reconstruction of Economics*, that social and economic systems be analysed with the use of ecosystem modelling, thus following an idea advanced by Alfred Marshall (1920). But at that time Boulding did not consider the interdependence of economic activity and the state of the environment; nor did he seem to be aware of scarcity problems arising from natural limits to economic growth. However, in 1966 Boulding introduced these issues to economists with his celebrated 'Spaceship Earth' article, thereby foreshadowing the neo-Malthusian debate of the 1970s on limits to economic growth.

This debate was stimulated by the Club of Rome Report and by the publication of Meadows *et al.* (1972), *The Limits to Growth*, which followed up the 'doomsday' modelling of Forrester (1971). While some economists, such as Pigou (1932), had been concerned about the external environmental effects (externalities) of economic activity and their policy implications, so that renewed interest in and development of those theories occurred in the 1950s, *e.g.* by Meade (1952) and Scitovsky (1954), mainstream economic theory in the early 1970s did not link economic development with major environmental and ecological changes, even though there were some exceptions – such as Mishan (1967) and the British economist David Pearce.

While classical economists such as the Englishmen Thomas Robert Malthus and David Ricardo were well aware of natural limits to economic growth, those limitations played no role in neoclassical and mainstream economic theories which dominated economic thought at the beginning of the 1970s (Perrings, 1987; Tisdell, 1990*a*, 1990*b*). But classical economists had not foreseen major modern environmental problems, *e.g.* global pollution arising from economic growth. Yet the importance, indeed the necessity, of interrelating economic development and environmental change was becoming increasingly clear by the end of the 1960s. Boulding (1966) helped to start the process of integration.

Important contributions towards the integration of economics and environmental thought, taking into account the difficulties of attaining a sustainable society, were made by Daly (1968, 1973) and by Georgescu-Roegen (1971). In addition, in the 1970s, a number of economic models appeared which, apart from including production and consumption as a part of economic activity, included wastes as a by-product (Ayres & Kneese, 1969; D'Arge, 1972).

Prior to the above, wastes and their possible environmental consequences had been largely ignored by economists, who had viewed them as generally irrelevant from an economic viewpoint. Accordingly during the 1970s, academics and economists who were interested in global environmental issues, mainly concentrated on the depletion of non-living resources, such as fossil fuels, and on pollution stemming from their use. Concern for *living* resources, and for the global economic and environmental crisis that seemed likely to be caused by their disappearance or by the loss of their life-support systems, though present, was not emphasized directly. But nor was it by any means absent.

The first major economic work to take account of biological extinction problems was possibly that of Ciriacy-Wantrup (1968). It introduced the concept of safe minimum standard (SMS) as a guide to species preservation. But it was not until the late 1970s that species extinction became part of a mainstream debate by economists. Bishop (1978, 1979), arguing along the lines of Ciriacy-Wantrup, criticized the approach to environmental conservation of economists from Resources for the Future (Smith & Krutilla, 1979). The latter group favoured a rather traditional (neoclassical) approach to economic analysis, including the formulation of policies for deciding on the preservation of species and of wilderness. Using the SMS approach, Bishop argued that more conservation was justified than was supported by traditional economic approaches.

Nevertheless, using the traditional approach, economists came in the 1970s to the view that, taking into account the existence of environmental uncertainty and of irreversibility, it is rational to err on the side of conservation in order to keep options open, *i.e.* to retain flexibility (Krutilla, 1967; Arrow & Fisher, 1974; *see* also Chisholm, 1988). Thus neoclassical economists came to favour conservation more than had previously been the case – even though not to the extent of proponents of the 'safe minimum standard'. Acordingly, advances in economic thought relating to the environment were also being made but using traditional paradigms.

Thus it has come about that economic thought is responding gradually to the new challenges posed by environmental changes, albeit with a lag and sometimes not as radically as may be appropriate. At the same time, however, there are still a large number of economists who see environmental issues as peripheral to their concerns. Nevertheless, policy documents which appeared in the 1980s – such as the World Conservation Strategy document (IUCN, 1980) and the report of the World Commission on Environment and Development (1987), widely referred to as the 'Brundtland Report', have provided a stimulus to the development of ecological economics, and one expects that the 1990s will see significant development of this subject.

Perceived Policy Significance of Ecological Economics

The two international policy documents just mentioned strongly favour combining economics and ecology in the forming of policies affecting economic development. In exploring strategies for sustainable development, the World Conservation Strategy (WCS) document (IUCN, 1980 Sec. 1.12) states 'there is a close relationship between failure to achieve the objectives of conservation and failure to achieve the social and economic objectives of development – or, having achieved them, to sustain the achievement'. For example, it states that in developing countries conservation must be combined with measures to meet short-term economic needs.

Framers of the World Conservation Strategy (WCS) supported the view that improved international economic conditions are needed in order to achieve conservation goals. They recommended that the measures suggested for the International Development Strategy (of the 1980s) be adopted. These measures included trade liberalization, transfer of more finance and aid to less-developed countries, reform of the international monetary system, adoption of an international code of conduct for transnational companies, and more rapid progress on disarmament. In addition, proposers of the WCS recommend that 'economic and social growth be accelerated, especially in the poorest countries, ensuring that economic and social goods are mutually supporting, and emphasizing better health, better housing, and higher educational levels and skills' (IUCN, 1980 Sec. 20.3).

Proponents of WCS claim that equitable, sustainable development requires adoption of the above-mentioned 'economic' measures as well as those recommended in the WCS document. However, there is no discussion of the extent to which these measures are compatible with, or synergistic with, one another. There will certainly be cases where increasing aid to less-developed countries threatens conservation goals, and one cannot be sure that free-trade is always supportive of conservation goals or is always economically ideal! While economic and social systems undoubtedly impact on conservation, and *vice versa*, the interrelationship is often more complex than appears to be the case at first sight. It seems likely that neither the extreme of a completely free-market economic system nor that of a centralized command system of economic management is most conducive to the conservation goals espoused in the World Conservation Strategy.

The World Commission on Environment and Development (1987) places even more stress on the importance of connecting economics, development, and ecology. The Report sees the possibility for 'a new era of economic growth, one that must be based on policies that sustain and expand the environmental resource base. And we believe such growth to be absolutely essential to relieve the great poverty that is deepening in much of the developing world' (*idem* p. 1). It points out that it is 'impossible to separate *economic* development issues from environment issues', and that it is 'futile to attempt to deal with environmental problems without a broader perspective that encompasses the factors underlying world poverty and international inequality' (p. 5). It also emphasizes that the world is involved in interlocking 'crises' which cannot be compartmentalized by regions, sectors, or broad areas of concern (environmental, economic, or social). There is not a separate environmental crisis, a separate energy crisis, or a separate development crisis: they are all one.

The interdependence of economy and ecology are summed up in the Brundtland Report as follows:

> 'These related changes have locked the global economy and global ecology together in new ways. We have in the past been concerned about the impacts of economic growth upon the environment. We are now forced to concern ourselves with the impacts of ecological stress ... upon our economic prospects. We have in the more recent past been forced to face up to a sharp increase in economic interdependence among nations. We are now forced to accustom ourselves to accelerating ecological interdependence among nations. Ecology and economy are becoming even more interwoven – locally, regionally, nationally, and globally – into a seamless net of causes and effects' (World Commission on Environment and Development, 1987 p. 5).

As for specific economic and social measures to be taken to help promote sustainable development, the Brundtland Report is more guarded about international arrangements than is the World Conservation Strategy document, but provides more details. It seems possible that it is less free-trade-orientated than the WCS document. But this is not the place to discuss the socio-economic recommendations (international and national) of the Brundtland Report. Enough has been said, nevertheless, to make it clear that the Brundtland Report recommends that ecological economics play a central role in policy formulations. Thus it provides encouragement for development of this field.

New Paradigms for Our 'New' Economic and Ecological Situation

The basic mathematical structure of theories may well reflect prevailing views about the nature of the cosmos, in the sense of the ordered nature of relationships in the universe as a whole. Commencing in the nineteenth century, economic theories began to be cast principally in terms of the differential calculus – to stress the widespread existence of stable equilibrium in economic systems and, by making use of comparative statistics for policy purposes, assumed that it was rational to act on the basis that economic systems are in practice rarely far from equilibrium. Marginalism was the order of the day and Marshall (1920) was led to proclaim that 'Nature does not proceed by leaps and bounds'. A world of certainty was assumed in which, if evolution of systems should take place, it would lead to an improved situation.

In economic circles, those views did not take a 'battering' until the Great Depression of the 1930s. In his *General Theory*, Keynes (1936) pointed to rigidities in the economic system and to uncertainties and 'animal spirits' of investors, which could cause disequilibrium or lead to an undesirable equilibrium at a low level of economic activity. The conception of an economic world of certainty, *i.e.* in harmonious equilibrium, was beginning to disintegrate. But the problems were seen to be of managing demand – insufficient aggregate demand was seen as the root cause of the Great Depression. Supply-side problems, such as might come from environmental constraints, did not appear to economists in this period to be real constraints to economic activity.

Despite the views expressed by J. M. Keynes, and the contributions of economists such as Knight (1921) and Shackle (1952), economic theory was, at the end of the 1960s, 'dominated' by models which assumed certainty, reversibility of processes, continuity of change, and change by small gradations. Supply-side constraints, arising from environmental limits, were absent from mainstream theories of economic growth which were developed – but assuming a world of certainty, technological optimism, and a high degree of substitutability for all exhaustible resources (Perrings, 1987).

More appropriate economic growth and development models for our current situation, incorporating ecological relationships especially at the global level, may be those which allow for uncertainty, irreversibilities, and 'jumps' in relationships – such as those suggested by catastrophe theory (Zeeman, 1976), and for phenomena of a type suggested by non-linearities and by chaos theory (Gleick, 1987). At least where natural relationships are of this nature, they will need to be incorporated in ecological–economic models which rely on natural relationships as building-blocks. Such models of natural systems are quite different from those which prevailed at the beginning of this century, and which exerted so much influence on the development of economic theory.

The Environmental Future, Sustainability, and Ecological Economics

Our economic and environmental futures are widely intertwined, but some optimists still do not foresee environmental limits to our economic future or serious risk from continuingly rapid economic growth. One of the more optimistic examples is Julian Simon (1977), who suggested that rising population poses no risk to the maintenance (improvement) of human living standards because the larger number of people will have more ideas, and this will make technological progress more rapid. However, there is no guarantee that this will be the case, and that technological progress will be sufficient to overcome tightening environmental constraints. Growing evidence suggests that continuing economic growth of the past type is likely to result in environmental and economic collapse, given inevitable ultimate irreversibilities. Thus to adopt the development policies proposed by Julian Simon would, it seems, be at best a foolhardy gamble.

We cannot simply extrapolate past economic results into the future. Furthermore, the larger the human population is, and the higher its level of economic activity, the greater will be the penalty of failure to counteract environmental constraints *via* technological progress and other means. In addition, most of the increasing risk and uncertainty is faced *collectively*, namely in common. It is a characteristic of the modern, interdependent world that we face greater *common* risks and uncertainties than in the past, from both the ecological and the economic points of view. Hence, by following the population growth and economic growth policies suggested by Simon, we would be faced not merely with a personal gamble but with a gamble for all.

Many of our modern ecological/economic problems involve *collective* (overall) risks and uncertainties – in contrast to what appeared to be the case in earlier times. This makes policy making, including the establishment of economic policies for development, extremely difficult. In many cases, we face great uncertainty about the consequences of our actions, and can rarely be sure that we have considered all the possibilities. To this must be added the problem that individuals often have conflicting values, *e.g.* differing ethical precepts. But, in addition, there is a new element to consider, namely the different attitudes of individuals to the bearing of risk and uncertainty.

Thus we may distinguish three possible elements of social conflict involved in major economic policies impacting upon ecological systems and *vice versa*: (1) Conflict about the accuracy of specification of the possibilities; (2) Conflict about the objectives which it is valuable to pursue; and (3) Conflict about the degree of risk that it is appropriate to take, or about how to act under uncertainty. This conflict may be intense, given irreversibilities in the environmental system and the possibility that environmental 'jumps' may occur with little or no forewarning and be difficult or impossible to predict accurately. When 'leading indicators' suggest that a 'jump' – a sudden and irreversible decline in environmental quality – is imminent, it may already be too late to prevent the catastrophe. Thus myopic trial-and-error procedures in decision-making, in which reactions are based on feedback, can no longer be fully trusted in the present era. This is of course alarming, but important to bear in mind.

Game theory and decision theory help us to perceive elements of our situation but do not capture it fully (Tisdell, 1968; Perrings, 1987 Ch. 7). Even game theory assumes much more knowledge than we actually have. We have no *ready* answer for deciding an optimal social choice where views about what is possible differ, views about what is desirable differ, and attitudes to collective bearing of risk and uncertainty are in conflict – even though we can work through the logical implications of different actions based on different points of view.

This of course is not to say that we should not have our own position and pursue it – over a period of time one may change the values of others, obtain greater consensus about what is possible, and alter attitudes to the bearing of risk and uncertainty, so that greater agreement can be reached. There are indications now that those favouring ecologically sustainable development are obtaining greater political support than formerly. However, O'Riordan (1988 p. 30) warns that this may pose dangers and suggests that 'once the notions that underlie sustainability are politicised, the concept is effectively devalued'.

One of our greatest problems today, both ecologically and from the point of view of sustaining economic production in the long-term, is our continuing loss of genetic diversity, both in the wild and of primitive cultivars and landrace varieties. Even with significant advances in genetic manipulation, most of these losses (possibly all) are irreversible. As the genetic stock falls, the growing precariousness of production from cultivated crops and domesticated animals may not become obvious until their wild relatives and primitive cultivars and landrace varieties are virtually all irreversibly eliminated.

Today we obtain high yields from modern varieties because we increasingly isolate them from the natural environment, creating relatively uniform suitable artificial micro-environments, *e.g.* in the case of plants by the use of pesticides, irrigation, and artificial fertilizers. In so doing, we depend on the use to a considerable extent of non-renewable resources, and eliminate varieties which may be able to cope better with natural environments and *co-evolve* to adapt to their changes. Consequently in the long-term, when we are no longer able to protect crops as completely against the natural 'environments' as now because of the loss of resources – especially non-renewable resources – we may have eliminated varieties and species that would have been optimal to the new set of conditions (*cf.* Oldfield, 1989; Tisdell, 1990*a*). The economic system does not ensure the preservation of these environmentally well-adapted varieties even though it may be rational from a collective point of view to conserve them (Tisdell & Alauddin, 1989).

Apart from the above 'back-drop' value of genetic diversity, there are other ways in which it is likely to be of value. For example, new uses may be discovered for a particular species, or uses may be found that were not previously imagined for it (Oldfield, 1989). Economists, using the approaches of safe minimum standard (SMS) and benefit-cost analysis (BCA), have addressed some of the issues involved, and Randall (1986) suggests that a combination of these approaches is needed to determine what species or varieties to save. While these approaches may be too mechanical and structured to employ, they do contribute to the debate, albeit from a limited perspective (*cf.* Tisdell,1990*b*).

Randall (1986) states that some extreme 'ecologists' believe living species to be so interdependent that the removal of any one species will lead to the extinction of all in the long run. If this were so it would be a matter of choosing the optimal path to complete extinction of particular species or groups of species. While there are chains of biological interdependencies, most living things are not as critically and as pervasively linked as this. Nevertheless, interdependencies of species can be wide and take unexpected forms. Indeed, in *some* ecological situations a 'butterfly effect' could operate (Gleick, 1987 Ch. 1), *e.g.* the local extinction of species or a small reduction in its population may give rise to a chain of extinctions of other species. This adds to the uncertainty of the decision-making problem, and neither the market system nor the political system may take account of such possibilities (*cf.* Lecomber, 1979 Ch. 5; Perrings, 1987 Ch. 9).

Clark (1976), using standard economic analysis, points out that when the rate of growth in the value (productivity) of a species is less than the *rate of interest* (the going rate of return on capital as for example indicated by the prevailing rate of interest payable on Government bonds), it may be most economical to eliminate the species. To do so would maximize the net present value (profit) from the available biological resource, *i.e.* the stock or population of the species. But this ignores any spillover or external benefits from saving the species. Furthermore, some economists would argue that the market rate of interest is too high to be used for social discounting. A lower rate of interest – even a zero rate of discounting of the future – should be used, it is suggested, so as to provide greater benefit to future generations (for further discussion *see* Perrings, 1987 Ch. 8). What, then, of a species which gives a negative return on its existence, namely is a pest? Some might argue that it could become valuable in the future. But what if that is not the case? Then the arguments for preservation would not be economic ones but would involve different types of ethics, such as the 'land ethic' expounded by Leopold (1933, 1966; *cf.* Tisdell, 1989).

To a large extent the economic future of Mankind appears to be intertwined with the conservation of other species and the preservation of their life-support systems. As non-renewable resources are depleted, and as entropy of matter is speeded up by human economic activity, the economic welfare of Mankind will become more dependent upon the sustainable use of living renewable resources. Therefore, those who feel a special obligation to provide for distant generations, have a particular reason to want to bequeath a large genetic resource-base, ensuring considerable genetic diversity, to future generations. Without a large genetic resource-base, the available resource-base for distant generations may be meagre. While it is unnecessary to think so far ahead to make out a case for greater conservation effort on economic grounds, to do so adds further support to the importance of conserving living resources.

Concluding Comments and Recapitulation

It goes almost without saying that the economic welfare of Mankind is dependent on the *equable* survival of The Biosphere. In addition, our scope for sustaining or improving our standard of living is linked to our actions in conserving and utilizing The Biosphere, the resources of which are increasingly shared in common. Human economic activity seems to be the greatest single cause of change in the state of The Biosphere, and the major beneficiary from the use of its resources – as inputs to economic production processes, as receptors for the wastes from economic and other human activity, and for direct consumption and enjoyment.

There is a real danger that excessive and profligate human activity will irretrievably destroy the resources of The Biosphere on which human survival and economic welfare depend, unless suitable strategies are adopted globally. While living resources may not be totally destroyed, they may be destroyed and damaged to a much greater extent than is desirable, even given widely-differing values and ethics. Economic welfare will suffer in the near future, and especially so in the distant future, as non-renewable resources are dissipated. For this reason economics and ecology are inseparable in considering the current predicament of Mankind as a component of The Biosphere. The problem is a pressing one, as present economic, social, and political, systems appear at best to be coping with it quite imperfectly. But at least and at last *some* progress is now being made in recognizing the problem.

The challenge facing ecological economics is to understand the problems which we face in managing and interacting with living resources and their life-support systems, to make these problems far more widely known than at present, and to suggest socio-economic and political mechanisms which will enable us to respond to these issues effectively in the future. In doing this, less partial and less insular discipline-based approaches are needed, and new paradigms, in the sense of patterns of activity or scientific assumptions, must be explored for solutions.

This is not to say that no role exists for more traditional economic fields linking economics and living resource conservation and utilization – such as bioeconomics, natural resource economics, environmental economics, agricultural economics, marine resource economics, fisheries economics, forest economics, and so on. But it is to be hoped that ecological economics will provide a wider perspective for all of these fields of enquiry, and will also make a substantial contribution to development economics. While this remains to be seen, there can be little doubt that such an approach is overdue.

Ecological economics has only recently begun to develop as a distinct area of enquiry, and this development has been given impetus by the formation of the International Society for Ecological Economics which has commenced publishing the journal *Ecological Economics*. The development of this area has also received encouragement from the Foundation for Environmental Conservation and its journal *Environmental Conservation*. Mainstream economics has been slow to recognize the major connections between economic activity, industrial development, and the state of The Biosphere – and also, in turn, the consequences of changing ecological conditions for future economic activity and welfare – but there has been considerable expansion in economic thought on this subject since the end of the 1960s.

During the 1980s, the World Conservation Strategy and the report of the World Commission on Environment and Development, *Our Common Future*, pointed out the many ways in which global economy and global ecology are intertwined, and the consequent imperative need for both to be kept in mind when formulating development policies. These global policy documents have given extra stimulus to the development of ecological economics.

It is clear that ecological economics will need to incorporate 'new paradigms' – *e.g.* irreversibilities, discontinuities, 'jumps', and chaos – in order to model and analyse our economic and ecological prospects effectively. It can no longer be accepted that there are no limits to our global economic future, but the future is uncertain because of ecological constraints on Mankind's numbers, greed, and profligacy. We face very difficult social choices because of conflicts in opinions about what is environmentally and otherwise possible, about what our objectives should be, and about how we should react to uncertainty.

The relevant decisions cannot be made in isolation, because they affect all of us; moreover, because of discontinuities and irreversibilities, many decisions cannot be made safely any longer by using trial-and-error procedures. The problems are highlighted by the loss of genetic diversity, which must be regarded as a major threat to the maintenance of economic welfare. Mankind's survival and economic welfare are inextricably linked to the conservation of The Biosphere. It is a challenge for ecological economics to explore these interdependencies further, make them far more widely known than at present and suggest improved socio-economic and political responses to them.

Acknowledgements

I would like to thank the International Steering Committee of the 4th ICEF: Surviving With The Biosphere, and particularly Professor Nicholas Polunin, for inviting me to prepare this paper and helping me to travel to Budapest to deliver it, and Dr Darrel Doessel, University of Queensland, and an anonymous referee, for useful comments on the first draft of it.

References

Arrow, K. J. & Fisher, A. C. (1974). Environmental preservation, uncertainty and irreversibility. *Quarterly Journal of Economics*, 88, pp. 313–9, illustr.

Ayres, R. V. & Kneese, A. V. (1969). Production, consumption and externalities. *American Economic Review*, 59, pp. 282–97.

Bishop, R. C. (1978). Endangered species and uncertainty: The economics of the safe minimum standard. *American Journal of Agricultural Economics*, 60, pp. 10–8.

Bishop, R. C. (1979). Endangered species, irreversibility and uncertainty: A reply. *American Journal of Agricultural Economics*, 61, pp. 377–9.

Boulding, K. E. (1950). *A Reconstruction of Economics*. John Wiley, New York, NY, USA: x + 483 pp., illustr.

Boulding, K. E. (1966). The economics of coming Spaceship Earth. Pp. 3–14 in *Environmental Quality in a Growing Economy* (Ed. H. Jarrett). Johns Hopkins Press, Baltimore, Maryland, USA: xv + 173 pp.

Chisholm, A. (1988). Sustainable resource use and development: Uncertainty, irreversibility and natural choice. Pp. 188–216 in *Technological Change, Development and the Environment: Socio-economic Perspectives* (Eds C. Tisdell & P. Maitra). Routledge, London, England, UK: xv + 351 pp.

Ciriacy-Wantrup, S. V. (1968). *Resource Conservation: Economics and Policies*, 3rd edn. Division of Agricultural Science, University of California, Berkeley, California, USA: ix + 342 pp.

Clark, C. W. (1976). *Mathematical Bioeconomics: The Optimal Management of Renewable Resources*. John Wiley & Sons, New York, NY, USA: xiv + 352 pp., illustr.

Costanza, R. (1989). What is ecological economics? *Ecological Economics*, 1, pp. 1–7.

Daly, H. E. (1968). On economics as a life science. *Journal of Political Economy*, 76, pp. 392–406.

Daly, H. E. (1973). The steady state economy: Toward a political economy of biophysical equilibrium and moral growth. Pp. 149–74 in *Toward a Steady State Economy* (Ed. H. Daly). W. H. Freeman, San Francisco, California, USA: x + 322 pp., illustr.

D'Arge, R. C. (1972). Economic growth and the natural environment. Pp. 11–34 in *Environmental Quality Analysis* (Eds, A. V. Kneese & B. T. Bower). Johns Hopkins Press, Baltimore, Maryland, USA: ix + 408 pp.

Forrester, J. W. (1971). *World Dynamics*. Wright Allen Press, Cambridge, Massachusetts, USA: xiii + 142 pp.

Georgescu-Roegen, N. (1971). *The Entropy Law and the Economic Process*. Harvard University Press, Cambridge, Massachusetts, USA: xv + 457 pp.

Gleick, J. (1987). *Chaos: Making a New Science*. Viking, New York, NY, USA: xiv + 354 pp., illustr.

IUCN (1980). *World Conservation Strategy: Living Resource Conservation for Sustainable Development*. International Union for Conservation of Nature and Natural Resources, 1196 Gland, Switzerland: pack of *c*. 50 unnumbered pp., illustr., in stiff paper folder.

Keynes, J. M. (1936). *The General Theory of Employment, Interest and Money*. Macmillan, London, England, UK: xii + 403 pp.

Knight, F. H. (1921). *Risk, Uncertainty and Profit*. Houghton Mifflin, Boston, Massachusetts, USA: xiv + 381 pp.

Krutilla, J. V. (1967). Conservation reconsidered. *American Economic Review*, 57, pp. 777–98.

Lecomber, R. (1979). *The Economics of Natural Resources*. Macmillan, London, England, UK: viii + 247 pp., illustr.

Leopold, A. (1933). *Game Management*. Scribner, New York, NY, USA: xxi + 481 pp., illustr.

Leopold, A. (1966). *A Sand County Almanac: with Other Essays from Round River*. Oxford University Press, New York, NY, USA: xv + 269 pp., illustr.

Marshall, A. (1920). *Principles of Economics*, 8th edn. Macmillan, London, England, UK: xxxii + 730 pp., illustr.

Martinez-Alier, J. (1987). *Ecological Economics: Energy, Environment and Society*. Blackwell, Oxford, UK: x + 288 pp.

Meade, J. E. (1952). External economies and diseconomies in a competitive situation. *Economic Journal*, 62, pp. 54–67.

Meadows, D. H., Meadows, D. L., Randers, J. & Behrens, W. (1972). *The Limits to Growth*. Universal Books, New York, NY, USA: ix + 207 pp.

Mishan, E. J. (1967). *The Costs of Economic Growth*. Staples Press, London, England, UK: xxi + 190 pp.

Norgaard, R. B. (1989). The case for methodological pluralism. *Ecological Economics*, 1, pp. 37–57.

Oldfield, M. L. (1989). *The Value of Conserving Genetic Resources*. Sinauer Associates, Sunderland, Massachusetts, USA: xviii + 379 pp., illustr.

O'Riordan, T. (1988). The politics of sustainability. Pp. 29–50 in *Sustainable Environmental Management* (Ed. R. K. Turner). Belhaven Press, London, England, UK: xii + 292 pp., illustr.

Perrings, C. (1987). *Economy and Environment: A Theoretical Essay on the Interdependence of Economic and Environmental Systems*. Cambridge University Press, Cambridge, England, UK: xvi + 179 pp.

Pigou, A. C. (1932). *The Economics of Welfare*, 4th edn. Macmillan, London, England, UK: xxxi + 876 pp., illustr.

Proops, J. L. R. (1989). Ecological economics: rationale and problem areas. *Ecological Economics*, 1, pp. 59–76.

Randall, A. (1986). Human preferences, economics and the preservation of species. Pp. 79–109 in *The Preservation of Species: The Value of Biological Diversity* (Ed. B. G. Norton). Princeton University Press, Princeton, NJ, USA: xii + 272 pp.

Scitovsky, T. (1954). Two concepts of external economics. *Journal of Political Economy*, 62, pp. 143–51.

Shackle, G. L. S. (1952). *Expectation in Economics*, 2nd edn. Cambridge University Press, Cambridge, England, UK: xvi + 144 pp., illustr.

Simon, J. (1977). *The Economics of Population Growth*. Princeton University Press, Princeton, NJ, USA: xxx + 555 pp., illustr.

Smith, V. K. & Krutilla, J. V. (1979). Endangered species, irreversibilities and uncertainty: a comment. *American Journal of Agricultural Economics*, 61, pp. 371–2.

Tisdell, C. A. (1968). *The Theory of Price Uncertainty, Production and Profit*. Princeton University Press, Princeton, NJ, USA: x + 197 pp., illustr.

Tisdell, C. A. (1988). Sustainable economic growth, production and development: An overview of concepts and changing views. *Indian Journal of Quantitative Economics*, 4(1), pp. 73–86.

Tisdell, C. A. (1989). Environmental conservation: economics, ecology, and ethics. *Environmental Conservation*, 16(2), pp. 107–12, 162.

Tisdell, C. A. (1990*a*). Economics and the debate about preservation of species, crop varieties and genetic diversity. *Ecological Economics*, 2, pp. 77–80.

Tisdell, C. A. (1990*b*). *Natural Resources, Growth and Development: Economics, Ecology and Resource-scarcity*. Praeger, New York, NY, USA: xvi + 187 pp.

Tisdell, C. A. & Alauddin, M. (1989). New crop varieties: impact on diversification and stability of yields. *Australian Economic Papers*, 28(52), pp. 123–40.

World Commission on Environment and Development (1987). *Our Common Future*. Oxford University Press, Oxford, England, UK: xvi + 400 pp.

Zeeman, E. C. (1976). Catastrophe theory. *Scientific American*, 234(4), pp. 65–83.

Commentary on Chapter 16

CHAIRMAN: Ambassador Francis L. Dale

PANELLISTS AND OTHER CONTRIBUTORS:

Simon, O'Riordan, Fry, Thorndike, Burnett, Berend, Batisse, Holdgate, McNeely, Stanton, Cohen, Mische, Tisdell, Simon, Fry, Persányi

Simon said **Tisdell** had given a very clear overview of a very rapidly developing subdiscipline, and had shown how *Ecological* Economics had come, in the last year or two, to be distinguished from the broader sweep of *Environmental* Economics.

He would like to associate himself with **Davis**'s remarks regarding interdisciplinary feuds. Frankly the problems we faced, and with which we were attempting to deal, were so severe and pressing that we could not afford the luxury of such disciplinary chauvinism. All disciplines had a contribution to make; the challenge was to harness and integrate those complementary perspectives.

He illustrated this by way of an analogy. Environmental Impact Assessments (EIAS) were now widely accepted and undertaken (although, alas, not always acted upon). They were best carried out by a team of specialists in the range of relevant disciplines; however, the leader had to perform an integrative role, drawing together not only the personalities involved but the substance of their work. This required a broad knowledge of the respective subject areas, without necessarily being expert in any one of them.

The problem was that we had generally not been encouraged to think or operate in this way. Firstly, institutional structures at local, regional, national and, in some cases also, international levels were not appropriate to the task. They operated with narrow perspectives determined by strict departmental or sectoral boundaries. Very few had specific area or problem foci (*e.g.* city, region, watershed, ecobiome, environment). The shortcomings of this almost universal *modus operandi* were well known. Hence, for example, over the last decade greatly increased emphasis had been placed on institution-building as part of development aid for Third World countries so as to promote integration and to create appropriate implementational monitoring and maintenance capacity. Conversely, however, the increased financial stringency of structural adjustment, or for that matter of monetarist policies in countries such as the UK and New Zealand, promoted chauvinism and interdisciplinary rivalry as every department sought to convince the paymasters of its uniqueness and value. In British higher education, for example, interdisciplinary centres suffered disproportionately through successive cutbacks during the 1980s, despite the increasing importance of their work. This trend had been reversed to some extent in the last year or two, but many of the new centres lacked resourcing adequate to their brief.

In wider terms, there were also grounds for optimism in new initiatives such as the Centre for Our Common Future (Geneva), the Centre for International Promotion of Sustainable Development (Canada), and the increasing importance of networks of nongovernmental organizations (NGOs) in countries of the South which promoted more holistic approaches to environment development issues. Examples of these were the the Asia-Pacific Peoples' Environment Network (APPEN) and the African NGOs Environment Network (ANEN).

It was clearly too early to offer an evaluation of Ecological Economics, so he wanted to complement **Tisdell**'s contribution (Chapter 16) by focusing on the limitations of conventional Environmental Economics, some of which he had mentioned in passing. He hoped that this would provoke debate and greater awareness of crucial issues for development and the environment which were ignored or underrated by the conventional wisdom. This was not to argue that Environmental Economics had little to offer; far from it. It was undeniable that much could be achieved by embedding environmental costs and benefits to society and individuals in economic decision-making. This needed to be encouraged and promoted. However, the focus must be widened and the limitations of current practice acknowledged, lest the widespread acclaim now being accorded to Environmental Economics ultimately contributed to its discrediting.

The basic problem was that conventional Environmental Economics had too narrow a perspective. Indeed, it operated under the rather quaint assumptions of Neoclassical Economics, seeking to attach monetary values to the environment and its component parts. In essence, there was much of value in this approach. Many of the suggestions and prescriptions did undoubtedly have the potential to promote more sustainable resource-use, to reduce pollution and ensure that the polluter paid, and to reduce long-term environmental threats to continued economic performance by applying more realistic accounting techniques that included costs and benefits to the environment. For example, a recent exercise applying these to Indonesia calculated that the country's net welfare was some 17% lower than had been suggested by the conventional measure of Gross Domestic Product (GDP). This was certainly an important advance.

However, he had two principal reservations about Environmental Economics as currently practised. Firstly, largely because of its Neoclassical underpinnings, the approach sought only to improve existing accounting practice under prevailing political economic relations. Clearly this was no small undertaking in its own right. But that focus omitted all reference to structural features and the organization of production under capitalism, state socialism or, for that matter, any other system. Many of the fundamental problems needed to be addressed at this deeper level, and were therefore not amenable to amelioration or change by these methods, which were well described by Barbier (1989)*. This was the essence of another new approach, namely political ecology.

No analytical framework was value-free; each rested upon important ideological assumptions. As in Environmental Economics, these were often left implicit. However, particularly when dealing with human societies, such assumptions had to be made explicit. Failure to do so undermined the analysis and policy based on it, as, for example, with the uncritical application of scientific concepts such as 'carrying capacity' to human contexts. The term originated in agronomy and ecology, where its meaning was unambiguous, namely the number or biomass of one or more given species that a unit of land in a specific environment could support without degradation. However, in using the term with respect to humans, we needed to specify at least six elements, several of which related directly to socio-economic and political systems and their underlying ideological assumptions, namely:

- the mix of arable and grazing land, and the fallow or rotation system being used;
- the respective species of crops and livestock, and hence the potential yields;
- the nature of productive technology being used or available;
- the nature and source of the energy supply, especially whether it was animate or inanimate, organic or inorganic, and generated endogenously or exogenously to the unit of production;
- the land-tenure system or coexistent subsystems; and

[*See* p. 380. Eds.]

- how production was organized. This included the nature and dynamics of local political entities, such as community or household structures. We needed to know who controlled resources (including household and non-household labour) and access to them, and on what terms (Simon, 1989).

His second problem with Environmental Economics followed from the previous point, in that the focus remained exclusively *economic*, with no mention or account of interrelated social, cultural, or political, factors. Particularly in the Third World, where the problems were most acute and the hurdles most formidable, indigenous societies commonly had very different value systems from imported 'Western' ones. The environment or its resources were not perceived in purely material terms. Crucial to understanding this difference was the distinction between extrinsic and intrinsic values. Many facets of life and the environment were valued in and for themselves (*i.e.* intrinsically) or as integral parts of the socio-cultural whole. Therefore they often could not be attributed monetary or economic value in the manner required by Environmental Economics.

Two examples would suffice. Firstly, trees were not simply a commodity valued for what their trunks would fetch when sold. Rather had they multifaceted uses, ranging from the provision of shade, fruit, green fodder and medicinal substances to building materials, fuel and even spiritual significance, quite apart from their role in binding and retaining moisture in the soil (*e.g.* Muslow *et al.*, 1988; Chambers Leach, 1989). Sustainable utilization of tree resources could fulfil these many needs in addition to providing some cash income from timber. However, to equate the economic value of trees solely with commercially marketable timber, fuel-wood or potential medicinal derivatives, served to undermine the value system on which local sustainability had in the past been based by disregarding the integral intrinsic dimensions.

His second example was that of land and land tenure. Because of the uncritical application of Neoclassical theory, it had become axiomatic to most 'Western' economists that individual or private tenure was inherently superior to communal tenure, various forms of which were formerly wellnigh universal in Third World and other societies. Consequently, it was argued that development could best be promoted and environmental degradation, *e.g.* through overgrazing of the commons, minimized by land reforms which provided individual tenure. He would strongly contest that as a generalization. It might well be true in some cases, but in many others it certainly had been shown not to be so. Once again, we needed to be sensitive to the variables which he had enumerated earlier and which related to the structure of any particular society and the way in which land (as one of the most basic resources) was used or, indeed, abused. Part of the reason for abuse was often to be found precisely in the way that indigenous systems had been supplanted by 'Western'-style economic systems and principles. This in turn had a close bearing on poverty, debt and the sheer struggle for survival (Okoth-Ogendo, 1989; Simon, 1989). So, in relation to the current restructuring in Hungary, for example, he would strongly urge that in extricating itself from the traps of one increasingly discredited system, the country should not uncritically embrace another with many of its own traps and pitfalls. Rather, each area, region, agricultural system or particular problem should be evaluated carefully, and local solutions sought which did not create new but equally intractable problems.

More generally, indigenous systems should be respected and harnessed, not destroyed by the systematic imposition of 'Western' economic principles. Local communities must participate in, and exercise a real measure of control over, decision-making regarding their own futures *e.g.* with the formulation and implementation of development programmes. Their lives and societies were often still far more intimately bound up with the environment than could be accounted for by conventional economic appraisal or evaluation techniques. Even in this age of ever-greater global economic integration and the apparent rolling back of socialist development strategies, such issues should not be overlooked. We needed to get away from the tendency to formulate simplistic packages which were prescribed, more or less indiscriminately, to widely different socio-cultural and environmental contexts. Instead, flexibility and local appropriateness were required. Moreover, the flow of insights should be bidirectional, not unidirectional and paternalistic, because we could learn much of importance to ourselves and to global survival from different societies around the world.

It was clearly important to ascertain economic values as precisely as possible. Techniques appropriate to different issues still required development and refinement. However, that alone was inadequate. He had highlighted some limitations of conceptualization and practice in Environmental Economics and pointed out how our horizons should be broadened. The challenge of incorporating these precepts into a more holistic and sensitive schema of appraisal and evaluation still remained. In a different context, one Conference participant had remarked in an earlier session that science was too important to leave to the scientists; their research should be relevant and necessary, not esoteric. He would agree in part, but overall he believed that the global environmental future was too important to leave to the politicians. By definition, they responded to public opinion and pressure. Scientists and scholars therefore had a crucial role to play in providing information and raising consciousness. That, indeed, was one of the key purposes of the Conference in bringing together people from such diverse backgrounds and professions. We should play more active roles by complementing the work of campaigners, community-based NGOs or journalists in pressing home the urgency of environmental problems, and the importance of finding appropriate solutions. Moreover, we should lead by example and utilize more sustainably the individual resources which we command. *Cited*: Barbier, E. B. (1989). *Economics, Natural Resource Scarcity and Development.* Earthscan, London: xviii + 223 pp., illustr.; Chambers, R. & Leach, M. (1989). Trees as savings and security for the rural poor. *World Development,* 17(3), pp. 329–42; Muslow, B. with Katerere, Y., Ferf, A. & O'Keefe, P. (1988). *The Fuelwood Trap.* Earthscan, London: viii + 181 pp., illustr.; Okoth-Ogendo, H.W.O. (1989). Some issues of theory in the study of tenure relations in African agriculture. *Africa,* 59(1), pp. 6–17.; Simon, D. (1989). Sustainable development: theoretical construct or attainable goal? *Environmental Conservation,* 16(1), pp. 41–48.

O'Riordan [in a written contribution] thought that **Tisdell** had not simply attempted to produce a better theory of ecological economics but rather was seeking to define environmental economics itself. The real challenge was to find communality in concepts that linked the natural and social sciences. Key ideas were *uncertainty* on a scale and to a degree that precluded even the use of realistic probability theory, *chaos* where consequences cannot be traced to initiating events, and *catastrophe*, where system states flipped into new equilibria, sometimes without warning and not always in a reversible fashion.

Political and social structures were not designed to cope with uncertainty (or indeterminacy), chaos, or catastrophe. Was, for example, the algal bloom that afflicted the tourist coastline of the Adriatic in the summer of 1989 a 'natural event' caused by a minor and all-but-undetectable initiating event, or was it the result of a steady build-up of nutrients in the relatively enclosed marine basin whose state had 'flipped', in the sun and salinity prevailing, into bloom conditions? How should politicians act? Blooms killed fish and ruined holidays. Tourist-dependent economies could suffer seriously if nothing was done. But what should have been done? It was impossible to remove all nutrient sources because not all were known and the role of eutrophication was far from proven. To have provided compensation for lost income would have been economically very burdensome and would have set a precedent. To have found ways to divert the blooms to areas of least economic damage might be laudable but was not always possible.

In short, the politician faced a dilemma. A first recourse would have been to commission more research. That would not solve the problem because the phenomenon was not conducive to the basic arts of modelling and prediction. Nevertheless, this was a comforting solution because it indicated that something was being done, although in reality no positive action was taken.

The trouble was that environmental perturbations did not strike society equally, and those who caused the problem rarely suffered the outcome in an identifiable way. All this raised the issues of equity, fairness, and justice. That was where the politician should be effective, for those were the tasks of politicians. But in truth, the distributional consequences of environmental calamities rarely fell neatly or observably, and often it was the most powerless who suffered most.

All this suggested that the new ecological economics should become part of a new environmental science, a science that linked the conventional empiricist science of modelling and prediction with more traditional ways of believing and knowing. In the latter case the aim was to act with foresight or caution, to assume the worst and to cope accordingly, to listen to sincerely-held beliefs and not only to expertise, and to be able to adapt to the unknowable by always leaving room for manoeuvre.

In theory this 'episcience', the combined science of knowing and believing, should be manageable for politicians. After all, this was a world of expertise and judgement coupled with pressure and principle. The new ecological economics was ready-made for the imaginative and informed politician who had scientific skills and an ethical sense of fairness and justice across generations. This was the science we should develop and teach. It was a science where nobody was yet an expert but nobody was entirely ignorant. It was a science that should link people to the planet.

Fry believed that he was the only representative present working for an industrial organization – the token business representative! The Conference had raised a host of very important scientific and biospherical questions. Among those which had particularly attracted his attention were the moral question of whether it was appropriate to use fear or exaggeration of environmental threats as a means of encouraging more public and political concern for the environment and The Biosphere; the related question was whether, in the developed world, our environment was getting better or worse. **Tolba** and others had indicated that the entire world was an environmental basket-case. He personally disagreed violently with that. He thought it was too broad a generalization which had led to a lot of misunderstanding and misconceptions. We did have serious environmental problems in the developed world as well as in the developing world. However, in Japan, Western Europe, and North America, the overall quality of life, and he believed in general, with a few exceptions, the overall quality of the environment, was improving. It was certainly better in 1990 than it was in 1960, and certainly better than it was in 1930. Then there was the question of whether in fact population growth could be curtailed through education and the availability of family-planning assistance. There were some who believed that there were only two successful models, for birth-control purposes. One was the traditional model in which a society went through a phase of industrialization during which the general standards of living were raised to a level where individual family units voluntarily reduced their population size. This was the model which had been followed in Japan, in Western Europe, and in North America, where populations were now relatively stable. The second model, and he hesitated to use the term, was the repressive, or China model, where a repressive government simply came down and said: we are going to use the power of the state to mandate and enforce population control. All of these questions were extremely interesting, but what had been of some concern to him was that the questions had been raised, and only touched on very lightly; the Conference had clattered over them rather quickly and then they had been dropped. We had forced a debate among knowledgeable participants who had many different experiences, much information and many data. He thought a tremendous opportunity had been missed; to have assembled all these bright people and not to have taken a few of those issues and forced them through to a more conclusive resolution, or at least to a merger of, hopefully, an approximate consensus position. He was afraid that some were going to go away with their individual prejudices, or individual opinions intact and we would not really have developed any kind of a constructive improvement of the present state of understanding. He would plead that when these Conferences were organized in the future, some sort of time should be allowed, after having identified a handful of these very crucial issues, to provide an opportunity for a real debate, a real give-and-take and, hopefully, force participants to come to some conclusions as to where they should go next. He hoped his comments would be taken in a positive spirit because he thought the meetings were extremely important.

On the question of economic incentives for the environment, coming from the business community, even though he had worked in the past for the US Environmental Protection Agency, he tended to take a more pragmatic approach. Much of what he would say really did not affect ecological economics, but rather how economics could be used as a tool to get a better performance out of existing environmental rules and regulations. This was where the literature was quite rich. There had been a tremendous amount written. Essentially, in the area of economic incentives and disincentives there were several tools which governments could use to encourage better industrial performance and better personal performance in the environmental field. They fell into five very obvious general categories. One was the so-called effluent or emission charges, a euphemism for taxes on pollution of one sort or another. Secondly, the marketable pollution rights which were, in effect, the licensing of rights to permit emission to certain levels of pollution and then the establishment of a market in which those permits could be traded. In effect the trading process allowed those people who had the greatest ability to reduce pollution at the lowest cost to sell those rights, or for those companies that had the most difficult and costly problems in removing pollution to buy a permit from a company that had an easy time *e.g.* in reducing sulphur dioxide. In the simplest case you had two facilities, *e.g.* two smoke-stacks putting out sulphur dioxide, where in one case the plant could reduce its pollution by 98% for virtually no cost and in the other it could only reduce its pollution to 60% at low cost. If the standard was a 90% reduction, the plant that could only reduce its emissions to 60% had to spend an enormous amount of money to achieve the 90% goal. The other plant, in contrast, could get the same total SO_2 reduction from the *two plants combined* for a fraction of the total cost. The environment was as well off as it was under the regulatory system but, by trading the right to emit between the two companies, tremendous economic advantages had come into place. The third category was the deposit and refund schemes for bottles, containers, and products. People were even now thinking about refund schemes for standard 'white' goods, such as refrigerators, so that when you bought a refrigerator you had to trade in your old refrigerator, which then became a recycled product. Fourth, there were various environmental subsidies, *e.g.* investment tax credits for pollution control equipment and, finally, various liability schemes which encouraged improved environmental behaviour.

The fifth category was taxation. Some in industry had been reluctant to support certain environmental taxes or charge systems. Many in the developed world, particularly in the industrial community, were already subject to extensive command and control systems. Usually the new tax systems would, most likely, be in addition to existing regulatory requirements. Compliance with existing systems required very large capital investment and operating costs and permitted or allowed certain minimal levels of continued emissions of the pollutants. In effect, this was a free level, usually small, of free pollution. A new tax could be imposed on top of this permitted discharge. Some companies were very reluctant to have to accept a new charge on something where the command and control system had, in effect, permitted or authorized this level which they had viewed as a right. New taxes would be designed to reduce pollution and, if successful in their implementation, the revenue from the environmental taxes would automatically decline. If a tax was imposed, people tried to avoid paying it; from an environmental point of view, pollution had been reduced, the tax had been successful, and nobody had been harmed. However, this is where industrial people became very concerned. They had then placed themselves completely and totally in the hands of the politicians, and in democracies, when you have a tax which has produced revenue, and that revenue was declining, the traditional response of politicians was to raise the tax in order to keep the revenue level high. No clear-cut distinction was made between environmental taxes to achieve environmental goals, and environmental taxes to generate revenue. If it turned out that what was put into place as an environmental tax for environmental purposes had become a source of revenue then people were, in effect, taxed endlessly and at an ever-increasing rate. Politicians might promise that the tax was an environmental measure, not a revenue measure, but industry had learned to have a healthy scepticism concerning political promises.

That case did not necessarily hold true for new pollutants. For example, if we were going to develop a new national or international scheme for the regulation of CO_2, a pollutant hitherto not regulated and not controlled in any sense, then some of these concerns might not be valid, or at least not so important. There would then be a *tabula rasa*, an opportunity to be truly creative. If the case for global warming was established, and if the case was made that there was a logical situation for CO_2 reduction, then a tax system might provide a very interesting opportunity. The unique problem of CO_2, or of all the 'greenhouse' gases, was that any nation which went first put its public and its industry at a competitive disadvantage in that it imposed a tax on a country because it had been a good international citizen. Yet it made virtually zero contribution to the problem unless all of the other countries went along at the same time. Increases in the use of fossil fuels in India and China could make a radical CO_2 reduction programme in all of western Europe, Japan, and North America, virtually meaningless.

In absolute terms, 'greenhouse' gases would be reduced but in relevant terms it would be meaningless. In this instance, unless the major economic blocs, Europe, Japan, North America, along with the Soviets, China, Brazil, and India, all agreed on a joint phased programme, the efforts of any one bloc imposed economic penalties on the volunteer group without any guarantee of international success. In this case, in the absence of any international agreement involving the widest possible participation in shared reductions, the programme would be doomed to failure. Some had suggested international taxes on carbon, fossil fuels, or on CO_2. He pointed out that, to the best of his knowledge, there was no existing international tax on anything, anywhere, and there was no existing international taxing authority. In his view international taxes were simply not politically feasible in this century and, he thought conservatively, they were probably not politically feasible until at least the year 2025 or perhaps 2050, or whenever there was evidence of a much more serious threat than could be demonstrated at present. That did not mean that we should not have such a tax, or that we should abandon the concept. What was needed for CO_2 control was first, an international framework, some convention which simply provided the framework for negotiations as a first step, without targets, without rollback levels, without calls for taxes, without specifying the means of achieving the goal. Second, after this international framework convention had been established, we would have a long period in which the many nation states would have to go through a detailed negotiation process and ultimately arrive at the second stage which was a level agreed upon of CO_2, or 'greenhouse' gas emissions, for each of the individual nation states. Unless one had at least all of the major blocs in agreement it would be of no use to the environment. Third, at the very end, after there were levels agreed upon of reduction for the 'greenhouse' gases, nations could develop whatever programmes were most efficient for them. He believed that in many cases the most efficient programme would, in fact, be a tax system, but that would be up to the individual nations and would not raise the very difficult political question of international taxes. He thought that to some extent the environmental community was spending a great deal of time and effort talking about international taxes which might have some appeal from philosophical, or theoretical, points of view but were not politically realistic.

Thorndike pointed out that enormous progress had been made in the developed countries in terms of coping with conventional air and water pollutants, and the kinds of things that might directly affect human health. The level of environmental improvement over the last 20 years had been unquestionable. Nevertheless, she thought that **Tolba**'s low-level assessment, on a world basis, applied equally to developed and developing countries alike when we looked at the environment and environmental resources. In the USA, for example, they had not coped with the legacy of the past – the hazardous wastes of materials which had not been so regarded 20 to 30 years ago – the loss of wetlands and wildlife habitats, soil erosion of agricultural lands, or the consequences of acid rain such as the 25% of lakes affected in the Adirondack Park, for which she had a particular responsibility.

Burnett made four points. Firstly, anything which brought economics and environmental concerns closer, however slight, was worth doing. He was therefore a little concerned that **Simon** did not see much virtue in techniques of national accounting that allowed for irreversible utilization of natural resources, for example, which had been developed and widely discussed by the World Bank. Even a change in accounting procedure drew attention to environmental issues and was well-worth doing in a fairly hard-boiled, world economic community. Secondly, putting values on the things that environmentalists were concerned with was one of the most difficult and fundamental problems that confronted us: it had been neglected. Thirdly, he thought that **Fry** had missed one step when describing how there might be an 'international tax' on carbon dioxide. At the stage when agreement on levels was being sought, it was surely necessary to set up an international monitoring capacity because the only kind of enforcement possible was a moral enforcement such as could be achieved by publishing the levels of CO_2 released, etc. Finally, he thought that the most difficult thing was to persuade industry itself to conserve and economize with energy. Several techniques had been tried in the USA but the British government, in its recent Electricity Act, had left it to the free market! What was the best technique; how did Hungary hope to deal with its situation?

Berend said that many Hungarian environmentalists were extremely concerned by the current situation during the period of political and economic transition when there was an acute shortage of indigenous capital. Consequently a great deal of environmentally destructive capital investment was being imported from the West because Hungary lacked appropriate regulations to control it. Hungary and, indeed, much of Eastern Europe felt that they were likely to become victims of this undesirable ecological colonization and exploitation. Moreover it became a double-edged sword. A distinct, closed set of institutions, each with its own set of ideologies had then been created which might well duplicate, unnecessarily, the work of existing institutions.

Batisse thought that environmental economics had to deal with rather different value-problems from traditional economics; for instance, **Simon** had wondered whether it was appropriate to consider everything in monetary terms. He agreed with that doubt. However, **Fry** had said that economic incentives should be used to improve the environment, and that was an obvious technique to reduce pollution, for example. But the same approach could not be used to protect the hills of Budapest, for example, or for birds in a particular locality: this required a very different philosophy from that when considering how to reduce so_2. Everybody agreed that so_2 should be reduced, but not everybody agreed about the value of birds, or the hills of Budapest! We had to be very careful, when employing environmental economics to dissect situations in order to see where the non-economic value-system had to be introduced, but this he had found rather confusing, and confused in the literature he had read.

Holdgate agreed very strongly with the principal that sectoralism should be cut out of the machinery, particularly of government, where the environment was concerned. Over the last 20 years a number of governments had established so-called Departments of the Environment alongside other sectoral bodies such as Transport, Industry and so on. The result had been, in fact, a fundamental encapsulation of the wrong attitudes to the environment. The environment was *not* just another sector, but the fundamental resource-base on which almost all other sectors drew. So long as a sectoral approach was allowed to prevail, governments would continue to get it wrong. A better model was actually that of a Finance Ministry (Treasury) which looked across the board at the implications of all the activities of government in relation to national wealth. Ways should be found for looking at the totality of actions of government as they affect the environmental resource-base, which was the foundation of the national future. The analogue of the budget was not a bad model. He believed that the Norwegian Minister of the Environment sat down annually and looked at the implications of departmental spending plans for the natural capital of the country and then reported on how it all added up.

Secondly, accepting that economic evaluations had limitations, the fact could not be evaded that for as long into the future as we could see, economic evaluations were going to be the basis of most national policymaking. There was a good saying that 'If you can't beat 'em, join 'em', and he therefore believed that considerable effort should be made to get environmental entities and resources, used in national, economic-planning policy, evaluated as correctly as possible. If we pretended that one just could not apply economic methodology to the environment, things would go badly adrift, so we had to help people like **Tisdell** and others to make the best evaluations they could. His colleague **McNeely** had been making quite a lot of effort in that field and would describe it later.

Thirdly, he agreed with **Batisse** that we had to define where existing economic methodologies reached their limits, where they ceased to be totally sufficient (if they were ever totally sufficient!), and where they ceased to be the mainstay of the judgemental process. Everyone, he believed, accepted that there were extremely important environmental value-judgements which were difficult to put into economic terms. We should never forget the inspiration of the environment – the fact that it inspired poets and painters – which gave all of us much of the quality which was an extremely important feature of life. It might not be possible to set it in economic terms, but if politicians forgot that it was an extremely important political factor, then everything else was going to fall apart.

McNeely said that thanks to Earth Day and a whole host of other media, educated people were now aware that current forms of development were not sustainable. Rather than conserving the rich resources of forest, wetland, and sea, we were consuming many biological resources at such a rate that they were rendered essentially non-renewable and no longer available to support development. The roots of this problem were found not in biology, where they were most often sought, but rather in the distribution of costs and benefits of both overexploitation and conservation. Solutions therefore must be sought in the fields of economics and politics.

The distribution of costs and benefits was a complex topic, but answers to just four questions would clarify the issues:

First, *who benefitted from overexploitation?* A considerable body of information was now available on how living natural resources were being overexploited (that is, harvested at rates far in excess of what could be sustained, leading to reduced productivity). Many forces were blamed for this overexploitation, but it was useful to ask who had benefitted from it. The short answer was that the lion's share of the benefits had flowed into relatively few pockets, and that most of the profits were earned by the wealthier sectors of the population. When overexploitation was stimulated by overseas markets, it was the consumers who benefitted.

Second, *who paid the costs of overexploitation?* Some conservationists would have answered this question by saying 'everybody', to the extent that everybody benefitted from biological diversity and, therefore, suffered when it was reduced. This was not a totally satisfactory answer. Nature certainly had some built-in redundancy, and some species could disappear (indeed were disappearing) without anybody missing them. But few data were available about *which* species were particularly important in the functioning of ecosystems, so we could not really specify the extent to which the general public was suffering from the loss of biological diversity.

But a much more specific answer was often available. In the case of tropical forests, the people who paid were very often the people who lived closest to the forest, and who had long earned sustainable benefits from harvesting the goods and services provided by the natural productivity of the system.

It was also informative to consider who did *not* pay the costs. Very often, the individuals who were gaining the greatest benefits from overexploitation did so because they did not pay the environmental costs. When we ate beef fed with cassava grown from once-forested lands in Thailand, the cost we paid did not compensate Thailand for its losts forests. Current economic systems enabled us to 'externalize' such environmental costs, if indeed we were even conscious of them.

Third, *who earned the benefits from conserving biological diversity?* It was clear that conserving biological diversity – that was using biological resources in a sustainable manner which did not reduce the overall diversity of the ecosystem involved – could bring very considerable benefits to those who directly harvested the biological resources. Such conservation – what many of us would have called effective management for sustainable yields – had been a common approach of rural people since time immemorial; and indeed, a precondition for the survival of societies throughout history had been the capacity to manage their resources for a sustainable yield. But in recent times, traditional systems for managing a sustainable yield had been replaced by government resource-management agencies which had tended to fragment responsibility along sectoral lines and had made insufficient investments to be very successful.

When protected areas – an invention of the Industrial Age – were managed for tourism or game production, insufficient benefits had gone to the local people, causing resentment and conflict. Conserving the Tiger (*Felis tigris*) in India was of much greater importance to comfortable urban-dwellers than to the local people who lost cattle, or family, to tigers; schoolchildren in northern countries might care far more about Orang-utans (*Simia satyrus*) – and Javan rhinos (*Rhinocerus sondaicus*) – than did their peers in Java. Farmers (and consumers) in Europe benefitted when Peru conserved wild relatives of potatoes (*Solanum tuberosum*), or when Mexico conserved wild relatives of Maize (*Zea mais*). The wealthy nations of the north, therefore, earned a considerable consumer surplus when tropical countries conserved a natural area containing a high degree of biological diversity.

Fourth, *who paid the opportunity costs of conserving biological diversity?* At both the international and local levels, the opportunity costs of conserving biological diversity were paid disproportionately by the people who lived closest to the greatest biological diversity. It was one of conservation's greatest ironies that the nations with the richest endowment of biological diversity tended to be the poorest developing countries, and within those countries, the individuals who lived amidst the greatest biological wealth tended to be the poorest of the poor.

If the maldistribution of costs and benefits of overexploitation and conservation was a major problem facing humanity today, then solutions should be sought through the better distribution of these costs and benefits. While some inequity was inevitable, much could be corrected through improved policies. He suggested three possibilities, though many more were certainly worth exploring.

Firstly, *using prices to guide decision-making.* Some had said that if current trends continued, we would progressively lose everything of value in the countryside that did not have a price, and we were, inevitably, reminded that Oscar Wilde had defined a cynic as someone who knew the price of everything but the value of nothing. Some economists contended that, once resources appeared to be limited, they tended to acquire prices which reflected their scarcity. That should not be accepted as an argument that market forces alone would ultimately yield the 'true value' of biological diversity, or of species, because extinction was irreversible and one species could not be substituted for another. We could not wait for biological diversity to generate its own market as diversity declined, if we hoped to arrest degradation before the threshold of irreversibility was crossed.

Therefore, to be able to place a financial value on at least some of the attributes of biological diversity would help to encourage appropriate investment in its conservation, and assist in the design of packages of appropriate economic incentives.

Secondly, *delivering greater benefits to local communities.* The moral justification for a system of protected areas or any other programme to conserve biological diversity was strengthened when it generated a flow of benefits to people. The more people benefitted directly from the proper conservation and use of wild resources, the greater the incentive was for them to protect the resource and the lower the cost to government of so doing. If a government decided that it was important to conserve an area and this involved opportunity costs to local communities, then it would seem only appropriate that such local communities should be compensated for their opportunity costs through a series of economic incentives. In fact, one could argue that the cost-benefit ratio of conversation must ultimately be positive for the local people if the resources were to survive.

Finally, *utilizing international willingness to pay.* To judge from the contributions made to voluntary conservation organizations such as WWF, the public in industrialized countries was willing to make substantial payments for the existence value of biological diversity in the tropics. How could developing countries capture more of this willingness to pay? One possibility was to establish a system of 'conservation concessions' which would work something like timber concessions, but which would be designed instead to conserve areas of outstanding natural value.

In general, those who established the economic policies governing the use of biological resources had three broad options: to manage for a sustainable yield and therefore maintain the future option of depleting a resource (when prices might be higher); to draw down supplies slowly, in an effort to spread earnings over many years; or to deplete existing stocks quickly, taking a quick profit which could then be reinvested elsewhere. Most conservationists would suggest that the first option was far preferable, but many governments were exercising the third option in the case of forests, fisheries, and wildlife.

To determine the full costs and benefits of each option would have a major influence on determining which was most profitable; careful economic analysis could well demonstrate that rapid depletion was only profitable if the environmental costs were externalized, a practice that distorted the free market approach. Improved economic policies also needed to consider the flow of benefits, and our future welfare would depend on developing social, political, and economic tools which would enable conservation benefits to be delivered to those who lived closest to the greatest biological diversity, and therefore would best be able to determine how much of that diversity would survive.

Stanton said that Man was not the only animal to operate within economic constraints, rather it was that Man had lost the innate control of those mechanisms which kept him in tune with the environment. In practical terms he was rather concerned by the condemnation of 'primitive' agricultural practices: that frequently failed to take account of their insurance value, and the fact that they had evolved to cope with chaos-derived environmental conditions, as if chaos was a new phenomenon. Data collection on primitive practices should be extended to include not only what crops to grow, and when, where, and how to plant them, but to the local storage practices and, especially, processing – an art intimately related to crop varieties. The Sahel coarse grains and those of sub-desert areas should be a particular concern because they had been rather despised in the 'green revolution'.

He also made a plea for the continuation of keeping herbarium specimens. Such sheets were 'facts': computer data on plants were only conveniently analysable 'fiction'.

Cohen said that he would like to hear more about two areas, namely, the transfer of money from the wealthiest nations to those which needed it most, and mechanisms whereby the interests of future generations could be represented in day-to-day decision-making in business. The former was absolutely essential to preserve the environment, and the latter was rarely considered.

Mische had not heard much reference to the international economic area, the global market-place, and the imperatives that were built into the present nation-state system. These were impelling leaders to mobilize for export rather than for domestic, small-is-beautiful, self-sustaining, kinds of enterprise. Unless we looked at that type of international economic system, how to restructure the World Bank and IMF to include ecological values, and how to involve all nations, we were missing out on a very important imperative that would continue to frustrate everybody.

Tisdell commented on the question of putting values on environmental items. Economists had devised methods to assign values which involved judgements and, of course, those value judgements had to be accepted if the final result was to be used. It was important therefore to look always at the value judgements underlying economic principles. Even so, there were always the extra things which could not quite be fitted in; many economists as well as non-economists often forgot that. Another danger was that people often gave more attention to items that could be quantified than to those which could not; that needed to be watched. He therefore shared **Burnett**'s feelings on the difficulties and dangers of value judgements but we still had to have valuation because all choice required it. For instance, human life was tremendously important but that had not prevented us from making decisions about whether or not to keep people alive, or not, or to spend infinite resources on them.

The points raised by **Cohen** were exceedingly complex and needed more attention than the Conference had been able to give them. How could you persuade developed countries to give up money, if that was the correct course, to less-well developed countries? He did not know, but he did not think the formula suggested by **Tolba** would help. The global market-place raised many problems. A British economist (Lipton) had pointed out that people who lived in cities exploited the rest of the community to their advantage. That was often encouraged through trade going via cities because it could then be readily taxed. If, therefore, foreign aid was given to less-well developed countries, it would be those in their cities who would benefit most. As with all aspects of environmental or traditional economics, things were never as simple as they seemed.

Simon did not think he was against changes in national accounting practices. He had said that changes from the sole use of GDP-related measures, as the indicator of economic activity, was one of the most fruitful avenues for environmental economics. The example he had given from Indonesia was an indicator of how the magnitudes of such measures were changed when environmental factors were evaluated, although the figure was a bit small in his view.

He certainly agreed with **Holdgate**'s suggestion that a 'Treasury model' was a better approach than the sectoral one. We were all aware of the often conflicting limitations put on so-called 'developmental projects' by sectoral ministries throughout the world. The thrust of his argument had been geared to the limitations of environmental economics in focusing on monetarizing everything in an extrinsic value sense which needed to be refined because of the conflicts to which it gave rise. It would be important for the future to consider some of the alternative approaches to economic evaluation. In the end it came back to the decision-makers. The average villager in rural India, or anywhere else, was unlikely to have the same economic values as the decision-makers sitting in the capital, or at the centre of an international agency ! What was needed was to be sensitive, flexible and focused on the people we wished to help at each level.

Fry agreed with **Thorndike** that most of the traditional pollutants were under control although perhaps not perfectly. He was concerned about the new issues that were coming up and were less well understood. Incidentally, he agreed absolutely with **Burnett** that monitoring systems were essential and thought that on the energy issue some part of industry was aware of the problem and taking action. What was necessary was to realize that, if energy was conserved, capital investments could be made that would further reduce energy-use, and so effect a real saving. Businessmen needed to be convinced of that and his organization was trying to persuade them.

It was interesting to note that, in the developed world, there were sufficient resources to combat environmental problems, both financial and technical, if there was the political will to address each and every one of them. The tragedy was that in the developing world there were not the resources of trained people, sufficient of them, or of financial resources. There was a complete spectrum of ability to cope: from those who were almost incapable, to those who could handle most of the problems.

The point which had been made about environmental concerns being a separate, parallel sector from other agencies was particularly relevant to the position of UNEP. It was perceived by the other line agencies in the UN structure as being an interloper. It was a nuisance to FAO, to WHO, or ILO: but that was exactly what it ought to be! These agencies should be integrating their programmes, not promoting them separately. It was not desirable that UNEP should develop as a multi-billion dollar-a-year operating agency but that environmental matters should become better integrated into the work of the other UN agencies, the World Bank, and so on.

On the possibility of a massive transfer of funds from the developed world to the developing countries he was sceptical; he did not see it happening. There would have to be an Operation Bootstrap: those countries that took the initiative would find plenty of investment funding but with some conditions, one of which would *not* be a reduction of the standard of living in the North to help the peoples of the South. He drew attention to the five south-east Asian 'tigers' as models. Thirty, 40, 50 years ago those countries were 'way behind'; but they had pulled themselves up by their own efforts. He was not sure whether those models were good or bad , but at least they were worth a closer look.

Persányi first addressed the question of how industry might be made more inclined to conserve energy. It was difficult to see how that could be achieved in Hungary and Eastern Europe where one unit of GDP was equated to 3 to 4 times more energy used than in western developed countries. The only way to change was to restructure both production *and* consumption and that was a difficult task. Problems like ecological colonization, mentioned by **Berend**, did not help. It was basically a matter of ethics. He had wondered sometimes if the world would end up with slaves in poor environments and some zones in the North where only rich people lived in a good environment! Could not a better way be found? The answer must lie in environmentally sound technology transfer.

Secondly, he reiterated his view that he hoped to see countries with no ecological policy, or rather, as others had said, a fully integrated ecological policy across the board. He realized, however, that that was only likely to be in the far-distant future and how it was probable that in most countries separate, autonomous environmental bodies would be the rule. Even so, it was important to have as much local autonomy as possible in order to involve local communities in environmentally sound practices, even where there was regional governmental control of environmental activities.

Annexe 6: The Environment in a Socialist Economy: The Case of Hungary

Miklós Persányi
Senior Adviser, Ministry for Environment and Water Management, PO Box 351, 1394 Budapest, Hungary

In Hungary we have not only a social and economic, but also an environmental, crisis which can no longer be seen as threatening 'merely' the next generation's life but as a threat to our own. It closes in on the prospects of social and economic progress: while harnessing the economy we destroy exactly those factors which are essential to life's quality. This is a price too high to pay for progress, and social achievements are at least questionable if paid for at this rate.

Among the reasons for this environmental crisis there are of course some which may be termed global processes such as the general prevalence of technical considerations over Nature and its needs, the general logic of profit-oriented world trade, medium-level industrial development which is known as the most polluting of all, or the political borders cutting across established economic, geographical and ecological entities. There are others, typical of the Eastern European post-Stalinist medium leading to the social and economic crisis in Hungary: voluntarism and rhetoric soaked with ideology promising a fabulous future while covering short-sighted policies concerned only with promoting uncontrolled group interests.

We have only the latter to blame for the widespread prevalence of uneconomic, wasteful, polluting branches and technologies in industry, farming, transport, and services. The same applies to our towns and villages: forced urbanization, no thought given to infrastructure, and the repeated, forced redistribution of the demands of transport, have resulted in an additional load on the environment. Established Nature-friendly, human-scale production, consumption, and life-style patterns almost disappeared in a feverish haste to mechanize, to chemicalize, to substitute plastics wherever possible, to centralize production – all inspired mainly by political ideology. Recently we have known a true frenzy of pseudo-market policies leading to a semi-market-oriented selectivity, very profitable to small groups abusing market anomalies, but very harmful to the environment and expensive to society. The most expensive loans, and the hardest to service, were those taken from Nature, and that capital has been misspent just as all other capital loans we received.

The central powers did not see the importance of Nature conservation.

Environmental considerations have not been given due priority, and professional and social efforts have, in the course of the anti-reform reshuffling, been declared to run against ideological purposes. Nevertheless, public concern gained impetus in the '80s and then the central power took concerted action to neutralize and hamper the efforts made.

These efforts are undeniable and they have slowed down the process of pollution, though our backlog is undeniable. Even if we fare better than most other eastern-European countries, it is a fact that what we did was riddled with contradictions. Half-solutions, partial results, and poor efficiency, are characteristic of our Nature conservation paraphernalia. We have no detailed overview of the problems and no full-scale programme to counteract them; the economic and legal instruments are lop-sided and full of gaps. There is no effective control or supervision, and education and information on environmental issues are sadly lagging behind.

If we want to see what could be done about the environment in Hungary, we have first to analyse the factors that caused so much pollution, and the factors that are to blame for the poor efficiency of our environmental policy and practices.

One of the basic factors was that we copied the Stalinistic economic policy which, based on Marxist values, virtually neglected Nature and Nature conservation. In this scheme, natural resources are of no value because they were not Man-made, and infrastructure is a 'non-productive sphere', so it was neglected too, which only added to environmental deterioration. Going by the main indices, our infrastructure in the early '80s was almost at the same level as it had been in the '20s and the '30s which meant that, compared with other European countries, Hungary had slid back considerably; that also affected the environment.

Since the late '70s our economic growth-rate has slowed down, accumulation ratios reduced, while most of the GNP has been used to keep the living standards from dropping behind, and to improve the payment balance. This left little for updating technical improvements and production. Company revenues were more heavily drawn-on than ever before, so there wasn't much left for innovation and increase in production. Under these circumstances pollution only got worse and due to accumulation, we have now to face quite new threats (dying woods, acid rain) in addition to the old.

The target of stopping deterioration processes to maintain the standards as they were in the early '80s, stipulated in documents, has never been achieved. We could only stop deterioration in a few respects, and by now the new economy plan provides no more than 'slowing down the rate of deterioration'.

Dropping living standards sharpened public awareness of environmental pollution, but the weakening economy can only spend less and less on Nature conservation. Awareness has increased considerably these last years: polls made in 1988 show that there is almost no one who finds Nature conservation is satisfactory on the whole in Hungary. The great majority

(79%) of society wish to see natural resources maintained for the benefit of the next generations; many are those who believe that Nature conservation has priority over production (62%). An overwhelming majority would accept lower living standards if only the health hazards could be reduced. Nature conservation is seen by 36% as a major concern, by 47% as a medium concern, and by 16% as a minor or negligible concern; some 54% are aware of pollution at their residence; 59% believe that information on Nature conservation is sufficient, though 34% see it definitely as insufficient. Hand-in-hand with growing education, the number of those who call for more and more forceful action on behalf of persons and institutions to promote Nature conservation grows.

The growing public awareness made it necessary for our government to define a central policy for Nature conservation. We do have legal regulations, institutions, standards, emission ceilings, sanctions, and subsidies – in those respects we are comparable with the highly-developed countries. The difference lies in how the system works.

Greatly simplified, we could say that the main economic instruments of Nature conservation consist of fines and subsidies. Theoretically such a system might work but in practice it doesn't. Although the legal regulations clearly define emission ceilings above which every additional emission unit should be liable for a fine, there are several other factors that can be considered. Location of a polluting industry is one such factor, which is perhaps a valid argument – but that the fine might be reduced or even pardoned if the industry is economically 'in a hole' does not seem to be proper. It shows you that production still has priority over Nature conservation.

Subsidies are even easier to trace back to industrial, productional interests, as these are granted, upon application, to whom the jury sees fit; and the jury consists of members of the central powers!

The main problem is that both the industries and Nature conservation are in the hands of the central powers.

Theoretically, industry is independent but in practice it is intricately webbed into the central powers by virtue of subsidies, and by the fact that there is no real ownership; so the central power steps in to provide indirect control. Consequently, industry is not sensitive to pollution fines.

The natural environment had no owner, it had ceased to be seen as common property and was an easy prey to spoil. The State has never exercised ownership rights over it, and the citizens were unable to protect their interests as regards a clean and healthy environment.

We think the free market system also has failed to answer the environmental problems: externalities could only be partly internalized and even this has needed interference from the outside. Nevertheless, a free market is truly essential to Nature conservation.

Today's pollution in Hungary is mainly a consequence of the production structure. So, we will be unable to manage Nature conservation until we have changed this structure to match our natural resources. This much-needed turnaround is also one of the main premises of updating our national economy. Another key issue in this respect is the clarification and regulation of ownership rights. The third key factor is the free market which we will have to bring into existence step by step.

These three factors are closely interrelated: without the reform of ownership rights no free market will develop, and a turnaround of our production structure will never happen if true market conditions are not motivating it. Thus these three factors are essential to Nature conservation although they have no direct bearing on it.

Last but not least: decision-making and redistribution need to be decentralized – with local communities and self-government having a very important part to play.

We also believe we have got to prove that there are, and we need, other approaches to natural resources than those which went before. We could never agree to copy, not even technically or economically, what started the world-wide ecological crisis in the Western hemisphere, and we can, even less, go on with the vulgarized Marxist model of the Eastern European industrial societies which 'were set to curbing Nature under their yoke'. What we need is a society which has room enough for small decentralized communities, which can accept limits to Man's demands, which respects the fair laws of Nature, and which conserves its resources to enable the next generations also to have a fair share of benefit.

The best ecological policy would be to have no ecological policy at all but a sound general policy and sound industries – the ecological problems could then all be solved by restructuring the economy, politics and other activities. A sound policy, founded on the systems approach, and aimed at a well-balanced development is a kind of prophylactic care, based on knowledge and respect for Nature's own vital cycles, and their interrelations which sustain human life. These are paramount and should never be sacrificed either to the planner's omniscience, or to a faith in growth-rates and self-healing processes of an omnipotent market.

Actions that Should be Taken in Hungary

Actions to be taken of Strategic Importance

- In social and economic decisions, equal importance should be given to environmental and ecological aspects. During the planning process, despite the complexity of the national economy and its ramifications, in order to further the transformation of its structure, it is necessary to determine those trends which offer environmentally sound, material-and energy-saving, low-waste-content production and consumption solutions. We have to put an end to the practice which considers natural resources as free endowments that are infinitely disposable and without value. We have to take account of publicity, and freedom of information. A complete opportunity for intervention should be assured for the citizens, even on environmental issues. We have to ask, and take into consideration, citizens' opinions, and in the case of environmental impact assessments we have to operate within an agreed civic legal system which enables those who suffer from environmental damages to be indemnified. We should support citizens or communities which participate in actions in support of the environment. Citizens' participation should be increased by open information and large-scale dissemination of ideas. Environmental protection should be treated in training and public education, and in the press as a national issue, as information ensuring environmental protection will only be effective in media whose attitude has been transformed in this way.

- We should tackle our environmental problems mainly by ourselves, although we have to take into consideration their international context as well. We should participate in the solution of global and regional problems worthy of a European country, and should be seen to be initiators of conventions for international environmental protection. Let us adapt and apply the limiting values of environmental quality stipulated by the European Community, their standards and monitoring system, for the state of Europe's environment. With such actions our membership could be accelerated and our competitivity in world markets increased.

Necessary Short-term Actions

- We have to stop the decline in real value of the budget and other sources available for environmental protection, and we have to increase the sharing of common property, for the use of which we should pay a fair price, and those who cause such property harm should have to re-establish it and meet the costs of reconstruction. Costs and consequences of damage should be counted where they actually occur and should appear as part of the true price; we should bring an end to distortions of the actual situation. Goods produced in an environmentally sound way should take priority over production which damages the environment.
- In all State decisions with an environmental impact, when framing legal regulations or economic instructions, in private or public investments or reconstruction, in the introduction of new production or supply processes, or even of new products, it is necessary to analyse the probable environmental and social impacts, the risks and the cost-benefit outcome on society. All these should be exposed to testing by professional independent investigators. When launching a new venture it should have, as a pre-condition, a guarantee to ensure the prevention of environmental damage.
- Environmentally damaging technologies, products, or supplies should be gradually replaced by environmentally sound ones. We should support research, technical improvement and know-how which promote these targets. We should reduce actual damage, by ensuring the neutralization of hazardous pollutants and wastes and promoting their recycling. Opportunities should be created for employment amongst the unemployed in environmental damage avoidance and reconstruction.
- A pre-condition for effective environmental protection is an open democracy based on participation and the support of the environment from the national income. The necessary increase in income should be

derived from State income levied on environmentally damaging products (*e.g.* on cars and petrol).

- Programmes of environmental protection already under way (the programmes launched for the neutralization of hazardous wastes, the improvement of air quality of heavily polluted regions, and the protection of sources of drinking-water) should continue to be implemented.
- In the actual process of codification – mainly in the new constitution – stipulations concerning the environment should be integrated in detail. Within the short term, new and modern laws should be created as well as regulations for their implementation. Even during the period of preparation, decrees should be promulgated concerning environmental impact assessments, the standardization of products from an environmental aspect, and for the training of citizens on environmental issues. Administrative penalties for environmentally damaging actions should be increased. A unified State body, independent of all production companies to deal comprehensively with natural resource and environmental elements, should be responsible for the establishment of general environmental requirements and their administration.
- We need to harmonize the economic and financial regulations with the legal constraints. We need to examine and abolish existing regulations which stimulate the destruction of natural values and which render it difficult to improve environmental protection. Either environment charges must be levied on those production processes which increase the need for environmental protection, or a special tax should be imposed on those products. In the case of intentional and serious damage, it will be necessary to apply fines similar to existing ones. The positive elements (*e.g.* investment grants, credit, price support) should be increased from these sources, and their application should consider environmental efficiency besides professional and social aspects.
- Instead of the disorganized and heterogeneous measuring and monitoring systems which currently register the state of the environment, an appropriately equipped service should be established together with a unified evaluating network.
- Firm instructions should be introduced without delay in all those fields where natural damage could be reduced with low expenditure, or by regulations and reorganization, *e.g.* an ending to 'throw-away' wrappings, an improvement in traffic organization, and improvement in information available to society, such as environmental consultancies, publication of levels of pollution, etc.

In those fields of the national economy which are of decisive importance from the environmental aspect, concepts have to be introduced concerning aspects of sustainable development, and improvements should be made at all possible points. The most urgent problem is energy policy in industry, agriculture, forestry, transport, urban development, and especially in the field of major industrial investments.

17. Needed Behavioural Change: Steps Towards Environmental Security

CRAIG B. DAVIS
School of Natural Resources, Ohio State University, 2021 Coffey Road, Columbus, Ohio 43210, USA

INTRODUCTION

In May of 1946, at the dawn of the nuclear age, Albert Einstein sent a memo to several hundred prominent Americans warning that:

> 'Our world faces a crisis as yet unperceived by those possessing the power to make great decisions for good or evil. The unleashed power of the atom has changed everything save our modes of thinking, and thus we drift toward unparalleled catastrophe ... a new type of thinking is essential if mankind is to survive' (*cf.* Einstein, 1960 p. 376).

This warning is strikingly applicable today, as we enter what many believe will come to be called the environmental age. Like the unleashed power of the atom, degradation of The Biosphere and destabilization of its ecosystems and wider ecocomplexes*, threaten not only the stability of our economic and political order, but the very life-support system of our planet. Everything is again changing – everything, that is, except our modes of thinking. We are drifting once more towards catastrophe – not from lack of information, for evidence of degradation is both widespread and compelling, but from lack of leadership and the will to act. *A new type of thinking is again needed if we are to survive.*

For more than two decades, environmental scientists, educators, and activists, armed with data, examples, and dire predictions, have been pleading for a new type of thinking – for an environmental ethos that is founded on awareness and concern, and promulgates a much-needed strict code of environmental behaviour. The expectation has been that enlightened and concerned individuals will choose to behave in environmentally sound ways, and will demand the same behaviour from their leaders and others. Our calls for action are familiar: curb the growth of human population; reduce air and water pollution; recycle everything possible; conserve energy; save the trees,

*[The term 'ecosystem' is used here in the strict scientific sense of Tansley (1935), Marshall (1986), and Polunin (1986). When referring to major ecological features such as rain-forests, lakes, or wetlands, that usually encompass several or many ecosystems, the appropriate term is 'ecocomplex' (*cf.* Polunin & Worthington, 1990). Eds.]

tigers, and turtles; protect endangered species; preserve biological diversity; ... and so on.

To a considerable extent that effort at enlightenment has been successful; people in nations around the world are more aware and concerned today about the quality of their environment than they were twenty or even ten years ago. This is evident from such developments as the growth of the Green movement in Europe, widening public support for strong clean-air and -water laws in the United States and other nations, proliferation of citizens' environmental action groups in both developed and developing countries, and creation in more than one hundred nations of agencies and ministries responsible for the maintenance of environmental quality. In the light of this broad base of awareness and concern, however, it is discouraging to have to admit that environmental conditions at local, regional, and global, levels are still widely deteriorating, in many cases at an alarming and accelerating rate. It is discouraging to have to admit that protection of the environment continues to be viewed in many circles as a luxury – to be pursued only when it does not conflict with higher concerns, *e.g.* economic growth, employment, or national security. And finally, it is discouraging to note that we are still not reaching those with the power to make great decisions for good or evil.

If we are correct in our assertions that protection of our environment is imperative, why are we not able to communicate this urgency to those with the power; why haven't we been able to make protection of the environment a higher priority in their minds? Why haven't we been able to motivate them to act? It is all too easy to rationalize that the unreached are unreachable or at least unteachable – that they are self-serving, greedy, unreasonable, irrational, ignorant, and all those awful things put together. It is certainly easy to identify individuals who are worthy of each of these labels, and there are undoubtedly leaders who are indeed unreachable. But, what about the others? What about those powerful people who are responsible, intelligent, and informed, but do not believe that environmental problems are as serious as we claim? What, for instance, about John Sununu?

Dr John H. Sununu is Chief-of-Staff for President George Bush of the United States. The President listens to John Sununu. John Sununu has influence; he has power over decisions for good and evil. John Sununu* is an intelligent and informed leader, who holds a PhD degree from the Massachusetts Institute of Technology. John Sununu understands the chemistry and biology of pollution. However, John Sununu does not believe that acidic precipitation is a serious problem – a problem requiring dramatic action – and he has not been reached on that score. But, to label him as unreachable would be to admit our failure, to abdicate our responsibilities, and to forgo an opportunity to learn and improve as scientists and educators. What must

*[A former Governor of the delightful New England State of New Hampshire (though born in Havana, Cuba, in 1939); according to the 1986–87 edition of *Who's Who in America*, John H. Sununa at that time already had 8 children and was an engineer specializing in the use of energy. Eds.]

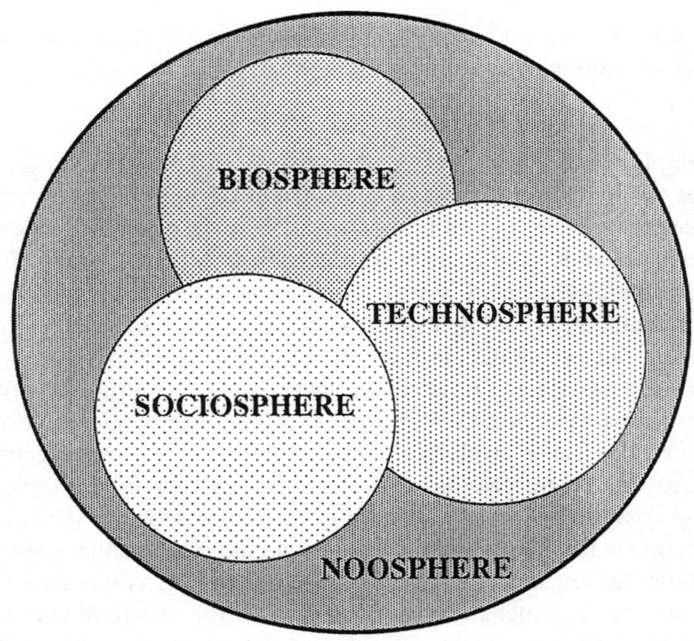

Figure 17.1. The Noosphere model (adapted from Batisse, 1973). The Biosphere is that part of the Earth's surface - crust, waters, and atmosphere - that supports life. The Sociosphere comprises cultural, social, economic, and political, systems and the institutions and infrastructure that support them. The Technosphere represents technique and technology. The Noosphere is the realm of the human mind that includes our genius and our biases; the Noosphere tempers rationality.

we do to reach John Sununu? Appeals to altruism won't work, nor will repetition of the same messages we have been sending for the past two decades, regardless of how they are enhanced with new data and stimulated by glaring examples. For John Sununu, environmental protection is not an inherent good, an end worthy of pursuit in its own right. He values environmental protection only as it relates to *his* higher concerns, which leaves us still wondering what we must do to elevate environmental protection in his hierarchy of concerns? How do we reach John Sununu?

If we want to reach and change the behaviour of those with the power to make great decisions for good or evil, we environmental educators must first change our own thinking and behaviour. First, we must become considerably more rigorous in identifying and explaining environmental imperatives, *e.g.* what *really* matters – what must be protected at all costs. Second, we must identify what motivates those with the power to act. What matters to them? What are their ultimate concerns, and how does environmental quality affect those concerns? And third, we need to develop a new conceptual framework, a framework that abandons the prevailing notion that environmental protection is inherently good – and substitutes a framework which relates environmental quality directly to higher needs, and to concerns that motivate and determine human behaviour. It must be a framework that focuses on understanding *what really matters*, and from which we can evolve a rational environmental ethos.

The Noosphere, Ends, and Means

In developing a new conceptual framework and new educational programmes to motivate people and make them change their behaviour, we should keep two things in mind. First, the belief that, given the facts, rational people will act rationally is, at best, naïve. In his seminal paper 'Environmental Problems and the Scientist', Michel Batisse (1973) integrated concepts of Biosphere, Sociosphere, and Technosphere (Figure 17.1), to reveal the complex causes and effects of environmental problems. Actions taken in any one of these spheres impacts all three, *e.g.* demanding (Sociosphere) that an industrial polluter (Technosphere) cease dumping cadmium into a local river (Biosphere) will have a beneficial effect on river ecosystems (Biosphere) but might also have various repercussions in the Sociosphere, including possible loss of jobs, higher prices for consumers, lower external costs, decline in market share for the polluting industry, higher aesthetic and recreational values, etc. Cessation of such pollution could also have an impact on the Technosphere by stimulating the development of new, environmentally benign, techniques. Given the information revealed in this modelling process, rational people acting rationally ought to be able to identify the facts, find the answers, and do the right thing to protect the environment. Unfortunately, as Batisse recognized by adding a fourth sphere, the Noosphere or sphere of the mind, rational people do not always act rationally. Each of us brings to our decision-making a complex set of concerns, values, and biases, that colour our interpretation of the facts and divert or stymie rational thinking.

A second general cause for concern is that there is in each of us a complex of concerns – a hierarchy of goals, objectives, or ends. Herman E. Daly (1977) provides us with an elegant model – the Ends-Means Continuum – that serves as a framework for understanding the range of human aspirations, the hierarchy of concerns, and the difference between absolutes and relativities. In this model (Figure 17.2), means and ends are arrayed along a continuum ranging from ultimate means to ultimate ends, both of which are absolutes. Ultimate means are those that 'cannot be created by human beings', *e.g.* low-entropy resources and natural genepools. An ultimate end is an end that is 'intrinsically good in and of itself', *i.e.* it is not a means to some higher end. Religious beliefs and values fall into this category, as does survival of our species. Ecologists and environmental scientists deal with ultimate means, and environmental educators and environmentalists attempt to link ultimate means to our higher values – to ultimate ends.

The centre of the continuum is the realm of relativities in which 'each intermediate category ... is an end with respect to lower catagories and a means with respect to higher catagories'. Politics, economics, business, and

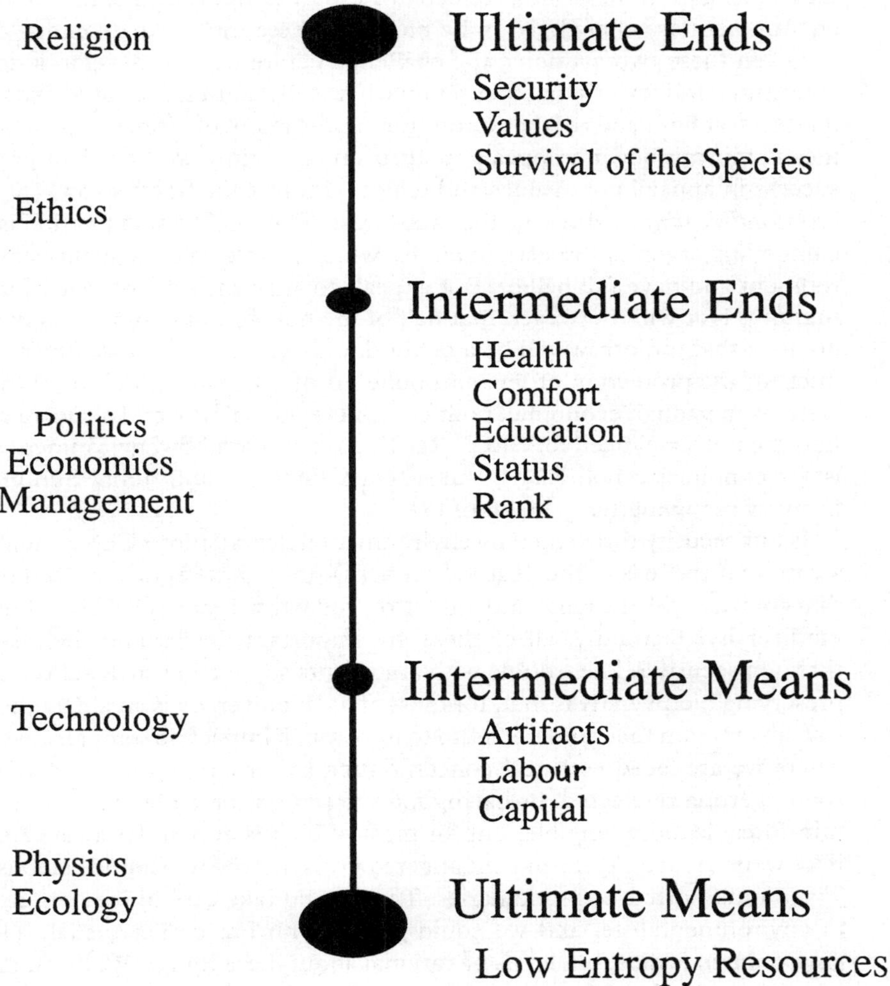

Figure 17.2. The Ends-Means continuum (adapted from Daly, 1977). Means and ends are arrayed on the right. Disciplines that address these factors are arrayed on the left, topped by religion (in the sense of belief dictating a needed code of behaviour and action).

industry all operate within this centre of the continuum, in this region of relativities and substitutability. Politicians, their economic advisers, and the 'captains of industry', *i.e.* those with the power, ignore the ultimates. They ignore the fact that our resource-base places absolute limits on our activities.

They ignore the *absolute* nature of such ecological factors as carrying capacity, recuperative capability, and sustainable yield, preferring to consider these as relativities that can be expanded by technology and distribution. At the other end of the continuum, they ignore ultimate ends, preferring instead to focus on achieving intermediate or short-term ends such as maximizing profit or getting reelected. Resources and the amenities of the environment are considered to be means to these intermediate ends.

Given these two cautions, the challenge before us is to develop a new conceptual framework that focuses on what really matters – what *ultimately* matters – at both ends of the continuum. In the realm of ultimate ends, this means moving beyond appeals to altruism. Certainly we have had great success in appealing to values and ethics – protect the Spotted-owl (*Strix occidentalis caurina*) and save the rain-forest. The finality of extinction is a compelling argument in certain circles where people value a quality environment and have full bellies. But appeals to altruism will not reach John Sununu. Nor will they reach that half of the human population who have no hope that tomorrow will be a better day! Beyond altruism we find self-interest; the protection of 'me and mine' or of 'my way of life', or of 'my system – personal, economic, political, and religious'. Beyond altruism we find the universal need for SECURITY, an end shared by environmentalists, economists, politicians, subsistence farmers, and John Sununu. Security really matters – to all of us!

Is our security threatened by environmental degradation? Can we make a case that the loss of the Tiger (*Felis tigris*), the Spotted-owl, or even the rain-forests, will decrease individual or collective security? What about biodiversity? Certainly, all of these are important; but would their loss threaten security? The trouble with arguments supporting such actions as preserving biodiversity is that, for most of us, biodiversity is an abstraction – an abstraction that does not seem to have much impact on our daily lives, where we are faced with real concerns such as garnering a larger market share, getting re-elected, or having food to put on the table. The tropical rain-forest is more tangible, but for most of us it is ever so far away. And if we want to save Tigers and endangered Owls, maybe we should establish Tiger and Spotted-owl sanctuaries. That would take care of the concerns of environmentalists, and we could get on with business as usual. The environment is nice, but let us be rational about these things. What's more important, the Spotted-owl or the livelihood of loggers and the economic security of the region or even the nation? Such attitudes are the products of our focus on *microconservation*, which engenders a 'preserve' mentality. Along such lines of reasoning, 'microconservation' deflects attention and concern from what really matters, what really *must be preserved*.

The fundamental lesson is not that we should preserve old-growth forests because they are the habitat of the Spotted-owl. The fundamental lesson is that ecological systems provide services that are essential to the survival of life as we know it on this planet. More selfishly, they provide services that are essential to human societies. Stable ecosystems are essential to our security; they really matter. Their degradation poses a real and demonstrable threat to individual, community, and even national, security.

Security: The Motivator

Security seems to be a basic human need. Indeed, the need for security may be the most powerful force driving individuals and governments to act. It is the quest for security that makes the rich man save, invest, purchase insurance, demand guarantees and warranties, and seek status, position, and power. It is also the quest for security that drives the poor man onto marginal lands, up the steep slopes, or into the marshes. It is the quest for security that drives the consumer society in the North, and it is the quest for security that drives the population explosion in the South. Ironically, it is this very quest for individual security among rich and poor alike – this expanding demand for resources – which is degrading the environmental systems that provide those resources. The devastating impacts of such degradation on individuals and communities are clearly visible throughout the world, in rich countries as well as poor ones. The impacts on the security of nations are less obvious. To clarify this relationship for ourselves and for John Sununu, we must first examine our notions of national security.

Concern for national security undoubtedly dates from the establishment of the first nation states as units of societal identity and governance. In recent times, especially since the end of the Second World War, the notion of national security has become the focus of considerable research and discussion as leaders and scholars have tried to define our national security needs and how best to meet them. What is this security after which we seek? Harold Brown, former US Secretary of Defence, has defined US national security as 'the ability of the people of the United States to determine their own future without being influenced by what happens outside their own borders' (Brown, 1983). Berkowitz & Bock (1968) defined national security as 'the ability of a nation to protect its internal values from external threat'. A related view holds that national security is the ability to protect national interests and boundaries while preserving existing political ideologies and governments.

These notions of national security share an emphasis on an external threat: the threat is out there somewhere and we must defend ourselves against it; therefore national security means national defence. Further, this traditional notion of national security has strong undertones of conflict, that national security is a win/lose game in which security rests in strength or power. But some scholars challenge this traditional emphasis on conflict and power in international relations.

Berkowitz & Bock (1968) distinguish two conflicting orientations towards national security. The Conflict Orientation views national security as a 'zero-sum game' based on national power; there is only so much power to go around. Advocates of this orientation believe that a nation can insure its own security only if it 'increases its own power at the expense of another nation or other nations'. The Cooperation Orientation, by contrast, views national security as a 'non-zero-sum game' in which national power should be minimized. The Cooperation Orientation assumes that the security of one nation can be increased if the security of all nations is increased. Attempts by one nation to increase its security by decreasing the security of others, leads to destabilization of international order.

Our notions of national security have been evolving over the last half-century. In the wake of the Second World War, national security thinking was dominated by the traditional political–military notions of conflict and power. The prevailing paradigms included Balance of Power, Mutual Deterrence, and Military Preparedness. This conflict-orientation-guided policy formulation prevailed throughout the Cold War and is still a major factor in strategic decision-making today. But, this narrow view of national security has been broadened over the past twenty-five years to include economic, social, religious, and environmental, threats that are more amenable to cooperative than to conflict modes of action.

By the 1960s, scholars and some decision-makers added economic issues to their notions of national security, recognizing for the first time that threats to our security can come from within, from poor decision-making, from poor budgetary and fiscal management, from lack of attention to maintenance of productive capacity, and from over-consumption; a deteriorating economy is a threat to national security. Paul Kennedy (1987) placed this concern into historical perspective in his excellent book, *The Rise and Fall of the Great Powers*, documenting the recurrent trend that 'uneven economic growth changes the relative military standing and strategic positions of nation states.' Another complicating factor is that modern economic systems have become international. Markets and the flow of capital are little affected any more by national frontiers or by notions of national sovereignty. Mismanagement in any country, especially in the larger economies, can create ripple-effects throughout the world. Maintenance of national and international economic stability demands cooperative modes of interaction. In this arena, the conflict orientation is counter-productive.

In more recent years, we have seen growing social and religious unrest in various countries throughout the world – all posing internal threats to national security, all contributing to at least potential destabilization of international order, and all surely more amenable to solution by negotiation than by force. To this list we must add growing concern about the stability of our productive resource-base and the quality of our environment (*see* Westing, 1989). As with economic problems, environmental problems are largely the result of mismanagement and overexploitation within nations. As with economic problems, environmental problems know no international frontiers; mistakes in one country can have devastating effects in neighbouring countries or over the entire globe. Environmental problems often require international solutions: they require coordination, cooperation, and collaboration among nations; they require a rethinking of our prevailing notions of national sovereignty; and they require a rethinking of our notions of national security. Environmental problems are not national problems; they are common problems, shared by all, and they call for common action – action that, like the problems themselves, ignores national frontiers and sovereignty. *Common problems* call for *common action* to insure *common security.*

Participants in a recent conference on environmental security, held in the Soviet Union (PRIO/UNEP, 1989), suggested that 'all security considerations centre on the notions of predictability and control (*i.e.* the ability to take corrective action)'. We want to control our lives and the future of our society. We want to know that the services which we get from our culture and society will be there tomorrow, the day after, and into the future. The English word security is derived from the Latin *securus,* meaning free from care – the same root from which we get the words sure and surety. We want to be sure of tomorrow and, to the extent possible, of what lies beyond. It is easy to see how the conflict orientation towards national security came into being. Keep things as they are, predictable – by force if necessary. Control the seas, occupy the land, acquire needed resources. The goal was stasis. But, fortunately or unfortunately, the world is not static. Security also demands that we must be able to respond to the unexpected – we must have options. Therefore, considerations of security must also include the notion of flexibility, of having options. With predictability, flexibility, and control, we can feel secure.

Stability: The Provider

The importance of stability in economic, social, and political/military, matters has become axiomatic. Unfortunately, the critical importance of environmental stability to all human activity is poorly understood by those possessing the power to make great decisions for good or evil. This lack of comprehension that stable ecosystems are the underpinnings of our economic and political systems, depreciates the value of these systems in the minds of decision-makers. New thinking is needed which recognizes the fact that unstable ecosystems are unpredictable, that unstable ecosystems have reduced flexibility and therefore offer fewer options than stable ones, and that unstable ecosystems are difficult or impossible to control. Our security relies on the services that stable ecosystems provide – services that we are rapidly losing through environmental abuse and neglect.

Our exploitation of ecological systems has exceeded their natural rates of regeneration and renewal. This is resulting in depletion of valuable resources and the deterioration and destruction of ecosystems. Our economic and social systems have evolved in, and are adapted to, climatic and ecological cycles that are to some extent predictable. Thus, we can forecast expected events and develop production systems that are adapted to the expected range of environmental conditions. We also understand that the environment sometimes surprises us, and we develop contingency plans to enable us to cope with these surprises. Accordingly, in a stable environment, however harsh it may be, human societies can predict the cycles of Nature, identify options for coping with adverse conditions, and thereby do much towards controlling their destiny. In a predictable world full of options and opportunities, control is in our hands and we are at least relatively secure. Unfortunately, for many, belief in this type of world no longer exists.

Approximately half the more than five thousand million people in the world today, and a far higher proportion of those in what we optimistically call 'developing countries', go to bed hungry every night and have no hope that their lives will improve tomorrow, next month, or even next year. They are right. They are caught in a downward spiral caused by overpopulation that exceeds the carrying capacity of their land, and by the inability of their governments, burdened by international debts, to provide the expertise and funds to remedy the situation. Environmental degradation is often an insidious process, gradually and quietly destroying the productive capacity of pastures, soils, forests, and fisheries. As resources shrink, options are reduced or eliminated, choices become restricted, and opportunities, once plentiful, disappear. As ecosystems become polluted and lose genetic diversity, they become increasingly unstable, inflexible, and unpredictable. Valuable services such as water storage and purification, protection and renewal of soils or shorelines, and the purification of the air we breath, formerly provided free by healthy ecosystems or wider ecocomplexes, are lost.

Yet, for much of this process of degradation, the downward ecological spiral is still reversible, *i.e.*controllable. Ecosystems tend to be surprisingly resilient – a result *inter alia* of their genetic diversity and flexibility. Properly treated, most damaged ecosystems will recover; the Lake Erie complex of ecosystems is a classical example of such a recovery. Unfortunately, there are limits to this resiliency: when once deterioration has passed some unknown threshold, the decline in productivity and stability becomes irreversible, at least in terms of human lifetimes. At this point, we become defenceless and lose control.

Environmental degradation is rarely confined within the borders of single countries (*see* Westing, 1988). The people of Bangladesh suffer major floods on a recurring basis because forests are being overcut in Nepal, India, Bhutan, and China. Fully 214 of the world's major rivers are shared by two or more nations and an estimated 40% of the world's people depend for drinking water, irrigation, and hydropower, on these shared waterways (Renner, 1989). As populations continue to grow, water will become increasingly scarce, and competition for water-supplies will become fierce. Soil degradation and desertification are becoming severe problems in much of the developing world. Environmental refugees, fleeing from unproductive fields and pastures, are a growing burden on developing countries, and where they migrate across international frontiers, conflict with indigenous cultures often ensues. Acid deposition, generated in the industrial areas of the United States and deposited in Canada, has led to harsh words between these two, otherwise friendly and cooperative nations. Likewise, the heightened competition for Atlantic Cod (*Gadus morhua*), caused by over-fishing and consequent destruction of fisheries in the North Atlantic, has brought Great Britain and Iceland to the brink of hostilities. The destabilization of ecological systems thus leads to the destabilization of economic and political systems, driving people from the land and turning neighbour against neighbour. As a result, we lose predictability, we lose flexibility, we lose control, and we lose security.

Protect What Matters

On the occasion of the launching of IUCN's *World Conservation Strategy*, a leading Minister in attendance stated that, 'In any individual decision, the starting-point will be to conserve what matters ...' (as reported in Tolba, 1982 p.11). This statement, typical of pleas heard over the past twenty years, begs the operational question' – *what matters and to whom?*

I offer to all the thought that what really matters are *security* and *stability* – individual, national, and common, security and the stability of our economic and political systems and of the ecological systems on which they depend. *Security is* an ultimate *end* shared by all, and *ecosystem stability is* an equally essential, ultimate *mean*. Security matters to all. It is our job as environ-mental educators to demonstrate that ecosystem stability also matters to all. You cannot have lasting security unless ecosystem stability is maintained.

Ecosystems are composed of living and inert things and processes. The *living things* are the animals, plants and microbiota, that live in the ecosystem, and the *inert things* are the physical basis of the soil, the water, the air, etc., involved. Each species occupies a niche in the ecosystem, *i.e.* it has a job or several jobs to do, though the importance of different component species varies greatly and sometimes seemingly approaches zero. The processes are the activities that take place in the ecosystem, the jobs that get done, the interactions among the species living in the ecosystems – between these species and the abiotic components of the ecosystem, and between the ecosystem and surrounding ecosystems of the aggregate ecocomplex.

Our attention as environmental educators has focused primarily on protecting or preserving the *living things,* the animals and plants – especially the animals and plants that are large, showy, pretty, soft, or valuable, and most especially if they have all of those attributes. In doing so, we have often been able to preserve bits of ecosystems that are the habitat for these *charismatic* or *flagship* species, and we should certainly continue this effort. But, do these species really matter? Environmentalists will almost always say *yes.* Ecologists will say *sometimes, but not always.* As a matter of fact, most ecosystems have complex back-up systems that can adjust quickly to the loss of this or that species. Ecosystems are indeed widely adaptable and resilient. Species regularly come and go, but the ecosystem persists and *the jobs get done.*

By focusing our efforts on species and lower taxa, we set ourselves up for some justifiable challenges from advocates of development. It is very difficult to justify, especially to the denizens of the realm of relativities and substitutabilities, that a particular species matters more than human jobs or economic development, and that species are not substitutable – when often they are! If we insist that species are not substitutable, we are labelled as irrational, idealistic, naïve, or at least impractical, and we lose respect and credibility. But, more important, by focusing our attention primarily on things, we deflect our attention from what really does matter in ecosystems – the *processes*. To a considerable extent, the 'faces' can change in an ecosystem as long as all of the essential jobs get done. The processes maintain the ecosystem, providing services that are often important and sometimes essential to human populations.

Stable ecosystems provide us with tangible products and services such as food and fibres, water storage and purification, climate amelioration, recreation, and aesthetic amenities. Stable ecosystems are predictable, flexible, and commonly self-maintaining (controlled) when left alone. Stable ecosystems are important to our security and sometimes essential to our survival. Our best bet for protecting the quality of our environment, including endangered species, is to demonstrate and inculcate globally that stable ecosystems, natural or cultivated, really matter. Saving the Spotted-owl or the Tiger in isolated islands of habitat will be totally inadequate if, in the surrounding landscape, ecosystems continue to be stressed and degraded by human activities. It is these *systems* that must be protected and maintained, so posing our leading challenge as environmental educators and broad-minded conservationists.

Ecosystem and Ecocomplex Protection: The Case of Wetlands

Few natural systems have been as maligned over the centuries as have the wetlands of our world – the marshes, swamps, fens, bogs, mires, and peatlands. Wetlands have been seen as wastelands to be drained or filled; as sources of disease, pestilence, and poisonous gases; and as dangerous places frequented by vicious beasts or capable of swallowing unwary travellers. Wetlands have been things to get rid of. In the United States, more than 90% of our inland wetlands have been drained during the past century, mostly for agricultural purposes. The Ohio State University even has a Drainage Hall of Fame, honouring those who have contributed most to the elimination of wetland ecosystems!

Today, wetlands have acquired a new stature; they are now treated as ecosystems and ecocomplexes of importance, to be preserved, protected, or even reestablished. A whole new branch of ecology has grown up around the study of these systems. Schoolchildren and college students are learning about wetlands – how they function, and why they are important to our society and to the stability of The Biosphere. The US Congress has passed laws protecting wetlands, and the President of the United States has declared a 'no net loss' policy to control destruction of wetlands by developers. All of this has happened in the last twenty years. In two decades, wetlands have thus been transformed from rogues to resources – they have thus become important, and consequently they matter. Why and how did this metamorphosis come about?

This new status reflects our growing understanding of how wetlands function. Our new concern is not with the preservation of this or that specific marsh or swamp because it harbours a threatened or endangered species, although such microconservation is laudable. Our concern is also not with the preservation of particular wetlands because of their uniqueness as ecosystems or, usually, ecocomplexes within the regional landscape, although this too would be laudable. Our new concern is far more self-serving than either of these two noble quests. We have learned in the past twenty years that wetland soils are like great sponges, holding rain-water from storms or showers and releasing it gradually through the year. We have learned that, through complex biological and chemical transformations (processes), wetlands can purify the water running through them - water draining from agricultural lands, from mine-spoils, or from urban and industrial areas. We have learned that wetlands protect river, lake, and ocean, shorelines from wave action and storms. We have learned that wetlands provide essential habitat for innumerable species of economic importance, *e.g.* ducks and geese, shrimps (including prawns), and various fish species of commercial and sports value. We have learned that we need these ecosystems and ecocomplexes and the services that they provide. We have undergone a pragmatic, although largely unconscious, awakening to the fact that we are a part of these and other ecosystems, and that their destruction would be a great loss to our economy and even to our security.

To understand why we now protect an ecosystem that twenty years ago was considered useless or worse, let us look at one type of wetland in more detail. The marshes of the prairie pothole region of North America stretch from the State of Iowa far north into Central Canada – an area that was covered by ice during the Wisconsin glaciation. The retreat of the ice some 10,000 or more years ago left a thick layer of glacial till with a poorly-drained surface. The prairie marshland that came to occupy this till surface was once so extensive that migrants moving west had to follow wagon roads along streams that had cut their way into the till, for the upland marshland was so soggy that wagons would sink in it up to their axles. All that now remains of this vast marshland are scattered marshes occupying depressions owned by governments or proving too deep to drain and plough. More than 95% of the prairie marshes have now been drained and the land planted to corn (maize), soybeans, and other crops.

A close look at these marshes reveals them to be extraordinarily complex systems. What we see as we stand on the shore is but the 'tip of the iceberg': what a marsh is and does is under the water. Marsh ecosystems comprise emergent vegetation (plants such as cattails [*Typha* spp.] that stick up out of the water); submerged and floating vegetation; fish, amphibians, reptiles, birds, and mammals, that live in or near the marsh and depend upon it for food, shelter, or breeding sites; large and small invertebrates that live on the plants and in the substrate and water; Bacteria and other microbiota of various sorts; the substrate soil; dead and decaying plant and animal tissues; and the water itself (Valk & Davis, 1978). These are the components - the things in the marsh. If we want to preserve the marsh, what do we protect? What matters? To answer this we must look even closer, and we must ask ourselves what precisely we want to preserve?

Do we care about water-holding capacity? We probably don't even think about it, feeling that most of the water-holding capacity has already been lost to drainage and there is nothing more to think about. But, is this true? Where does the water go that used to be held in the marshes and released gradually? It goes into ditches, streams, and rivers, flowing together as it rises towards flood-levels – especially in the spring, when heavy rains and snow-melt coincide. To stem this erosive and destructive flow, we have built a massive system of flood-control structures, such as dams, dikes, and levees, downstream from these drained wetlands. Indeed, the water-holding capacity of existing dams along the Mississippi and Missouri Rivers and their tributaries is almost exactly equal to the water-holding capacity lost to wetland drainage. We will continue to pay for this drainage for ever – building, maintaining, and replacing, structures to provide services that were formerly provided free by wetland ecosystems and ecocomplexes. Was this trade-off worth it? Probably, as this drained area is one of the world's most productive agricultural regions. But it was not free – and it still isn't. To preserve the water-holding capacity of the wetlands that remain, we must protect the complex set of processes that lead to the buildup of the organic substrate of the wetland, for this is the sponge. It matters not which species carry out this process, as long as the job gets done.

What about ducks? Ducks need water, food, shelter, and breeding-sites. Ducks like marshes that have a mixture of vegetation and open water in a ratio of about one to one, and they like the vegetation and water areas to be interspersed, increasing the total amount of 'edge'. This provides them with more places where they can move from their nests hidden in the vegetation to the water, to feed and return back into the vegetation – quickly if necessary to evade predators. So we need to preserve the vegetation and the vegetation–water mix. But, ducks also have to eat, and different ducks eat different things. Some eat plant tissues and some eat microinvertebrates. So we need to preserve these things too. But, these things also have requirements, so we have to preserve whatever provides for their needs. This goes on and on. What, then, do we preserve if we want to protect ducks? *We preserve the system.*

If we are interested in protecting the water-purification function of these marshes, we must preserve the processes that remove the pollutants, and those might be different for different pollutants. For nitrogen, which enters these marshes in runoff from agricultural fields and can be toxic to infants if it gets into drinking-water, it is important to preserve a complex set of processes that allow for bacterial activity on the marsh-bottom. So, why do we need a marsh? Wouldn't a pond provide us with the water and bottom needed for purification? What does the marsh contribute? Marshes contribute biomass. Strange as it may seem, in terms of nitrogen removal, the role of all of those species (cattails, rushes, sedges, lilies, etc.) that to the layman characterize a marsh, is simply to photosynthesize, produce biomass, die, fall, and sink to the marsh's bottom.

This dead litter of biomass serves as a substrate for the growth of massive populations of Bacteria. The Bacteria decompose the dead tissues, utilizing them as a source of energy. It so happens, though, that these dead tissues are usually low in nitrogen content, which means that the Bacteria cannot get enough nitrogen from the dead cattails and sedges to enable them to decompose these tissues completely. To make up the deficiency, the Bacteria extract nitrogen from the water passing over and through the marsh's bottom. Through a set of biochemical transformations, some of this nitrogen is converted into forms that can be used by other organisms, some is converted into gaseous form and passes into the atmosphere, and some becomes part of the bacterial protoplasm. Because the process of decomposition may take two or more years to complete, the Bacteria occupying a specific piece of litter become progressively buried, as new increments of dead plant tissue and other material are added to the bottom each year. The result of these processes is that at least some of the nitrogen that was passing through the marsh is lost either by passing from the marsh as a gas or by becoming buried in the deep sediments of the bottom. The water is thus purified. A comparable story could be told of the wetland processes that remove other pollutants (Davis & Valk, 1978, 1983).

Similar stories could also be told for other wetland ecosystems. In the United States, wetlands are protected nowadays because of what they do – their functions. Thus the protection is broad and general; all wetlands are protected. To alter a wetland, one must obtain a permit from the US Army Corps of Engineers, and the granting of these permits often carries with it the requirement of undertaking mitigation; for every hectare of wetland destroyed, the developer must replace it with one-and-a-half hectares of new or restored wetland of comparable quality. The burden is on the developer. With US wetlands we are following the aforementioned Minister's admonition that 'In any individual decision, the starting-point will be to conserve what matters'. In wetlands, as in other ecosystems and ecocomplexes, it is the system, and the processes which define the system, that really matter.

Connectedness: The Unifier

We live in an increasingly complex, interactive, and interdependent, world of accelerating change; a world where we must often decide and act in the face of uncertainty; a world in which mistakes can be costly or even deadly;

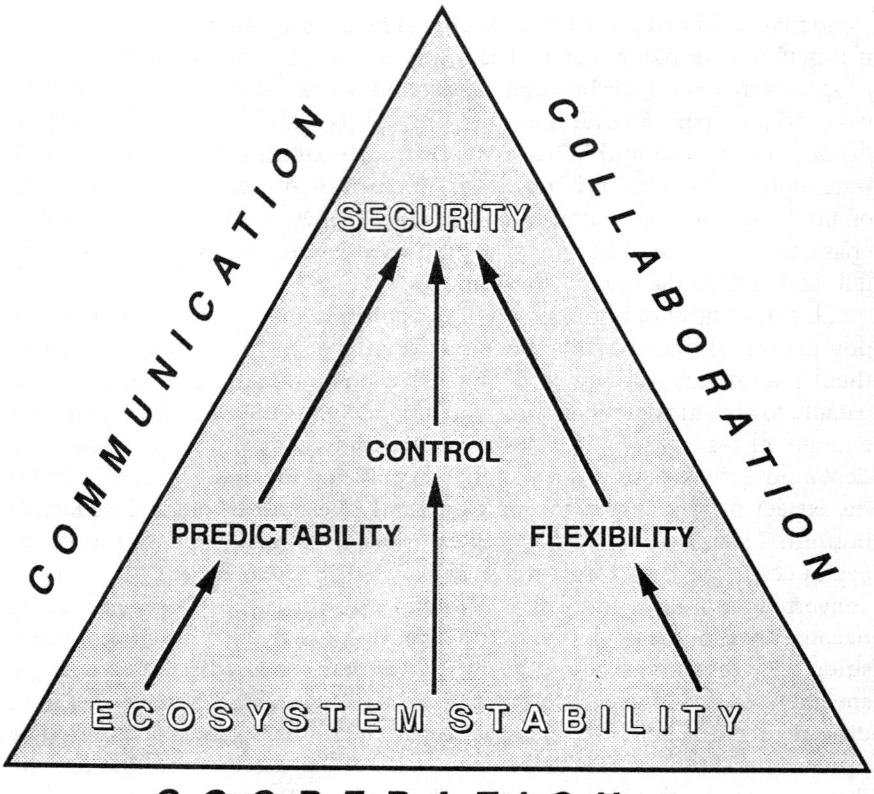

Figure 17.3. Ingredients of a rational environmental ethos. A rational environmental ethos must focus on needs, the ultimate of which is *SECURITY*; it must foster cooperation and unite people and nations; it must enhance *predictability*, *flexibility*, and *control*; and it must preserve what really matters – *STABLE ECOSYSTEMS* and *ECOCOMPLEXES*.

a world in which opportunities and threats are not constrained by national boundaries; a world where mutual dependence and cooperation are more productive modes of action than are the prevailing notions of mutual deterrence and the balance of power; a world in which traditional notions of national sovereignty are increasingly non-functional or even irrelevant. We desperately need new types of thinking that recognize our interdependence, value our diversity of experience and tradition, and focus on our common needs and responsibilities. We need new types of thinking that eschew simple answers to complex questions. We need a *rational environmental ethos.*

Towards a Rational Environmental Ethos

Rational behaviour should strive to enhance the long-term interests of the individual and society, paramount among which is our security (Milburn & Davis, MS). Further, it follows that rational behaviour should also strive to ensure the survival of our species and the stability of the ecosystems that serve as the resource-base for our economic and political systems and for all life on the planet. Behaviour that runs counter to these interests must be considered to be irrational, to say the least.

A rational environmental ethos must address people's needs – what makes them decide, and what makes them act. A rational environmental ethos does not depend on altruism as a motivator. It appeals directly to enlightened self-interest. This can be ultimate self-interest such as security, but it can also be intermediate self-interest – differential profit, getting re-elected, or being healthy and happy. A rational environmental ethos must be a unifier; it must bring people together; it must link nations. A rational environmental ethos must address needs rather than positions – people and nations can address needs and seek accord, but position must be defended. A rational environmental ethos must safeguard the preservation of predictability, flexibility, and control, as important components of security. To do all of these things, a rational environmental ethos must focus on preserving what matters – *the stability of ecosystems and ecocomplexes* (Figure 17.3).

In pursuit of a rational environmental ethos, certain behavioural changes among environmental educators and their followers are needed. We must:

- *Move beyond altruism to pragmatism and prudence.* Despite all the environmental altruism that has been generated over the past twenty years, we are still destroying major facets of The Biosphere. Behavioural change of the scale now needed will require appeals to self-interest.
- *Recognize Security as a universal motivator.* Security ranks high in everyone's hierarchy of values and concerns. Individual, national, and common, security are all enhanced by the maintenance of ecological stability. People will act to preserve their security.
- *Focus on preserving and maintaining stable ecosystems.* Stable ecosystems provide services that are important to human society and self-interests. These include food, fibre, clean air and water, and others familiar to us all. They also include predictability, flexibility, and control – the ingredients of security. Stable ecosystems really matter. Most of our effort henceforth should be diverted to demonstrating this fact.
- *Recognize that microconservation creates a 'preserve' mentality.* Our great emphasis on *microconservation* – on, for example, the preservation of endangered species and relic ecosystems – is dangerous in that it fosters the notion that we can preserve what matters in our environment by setting aside patches of the environment as preserves, *e.g.* Tiger preserves, Elephant preserves, Redwood preserves, etc. This leaves us free to alter the surrounding landscape as we see fit. We need to replace this strategy with one of *macroconservation,* that recognizes the value of stable ecosystems and strives to maximize sustainability by developing a functional mosaic of built, cultivated, and natural landscape units.
- *Focus on ecosystem processes rather than on things.* Our traditional emphasis on the preservation of species, however appealing and successful,

has led us to neglect what really matters, namely ecosystems and ecocomplexes and the processes that maintain these systems. In most ecosystems, many species can come and go without causing significant alteration in system functioning and usefulness – as long as all the essential jobs in the system get done. Cattails can be replaced by sedges and you still have a marsh.

- *Establish, at all levels of human society, an understanding that actions which destabilize ecological systems are unacceptable.* Legal actions such as fines, taxes, and incarceration, are one set of mechanisms, but new pricing structures that reflect the true costs and values of resources and commodities would be more effective and less offensive. We need better environmental accounting, with due recognition of real values.
- *Recognize the fact that Conservation can be Good Business.* Entrepreneurial ingenuity is a powerful force for innovation. Businesses spurred by governmental regulations, escalating resource costs, and customer pressure, are already finding profit in conservation.
- *Promote the adoption and strengthening of multilateral approaches to managing regional ecosystems and resources.* Individual nations, however powerful, can neither control the international transfer of environmental degradation nor protect themselves from the abuse caused by their neighbours. In today's world, notions of national sovereignty are largely irrelevant.
- *Recognize that we must replace our prevailing Conflict orientation towards security with a Cooperation orientation.* We must teach that security is not a zero-sum commodity, and that any nation that attempts to increase its own security by decreasing the security of other nations almost inevitably reduces the security of all, including itself. Common Security must be our goal.
- *Recognize that common problems can bring people and nations – even former enemies – together.* Environmental degradation is a common threat to all. It challenges us to replace old or even ancient enmities with communication, cooperation, and collaboration, on the protection and management of common resources. It is far better for all concerned to cooperate along a river than to fight over it. A common threat can make friends of former enemies. We have seen it many times in the political–military arena. Why not also in the environmental arena?
- *Promote the development of alternative modes of dispute resolution that focus on 'needs' rather than on 'positions'.* In disputes focused on satisfying needs, communication is open, information is freely exchanged, and trust is generated. When once positions become established, however, communication is formalized, information is guarded and often used as a weapon, and trust is lost (Carpenter & Kennedy, 1988).
- And finally, *we must set aside our notions that environmental protection is inherently good, that we know what matters, and all we have to do is teach what we know.* Despite all of our successes as environmental educators, and they have been many, the condition of The Biosphere is worse today than it was twenty years ago. We need to change our behaviour if we are to stem

the tide of *biospheral* deterioration. We must become much more rigorous in our own understanding of what really matters, and we must learn how to link what really matters in the environment to what really matters to us as human beings. *We must learn how to reach John Sununu!*

Repeating part of the quotation with which we began, assuredly, '*Our world faces a crisis as yet unperceived by those possessing the power to make great decisions for good or evil. ... a new type of thinking is essential if mankind is to survive.*'

References

Batisse, M. (1973). Environmental problems and the scientist. *Bulletin of the Atomic Scientists,* September, pp. 15–21, illustr.

Berkowitz, M. & Bock, P. G. (1968). National security. Pp. 40–5 in *International Encyclopedia of the Social Sciences,* Vol. 11 (Ed. D. L. Sills). Macmillan/Free Press, New York, NY, USA: 614 pp.

Brown, H. (1983). *Thinking About National Security: Defence and Foreign Policy in a Dangerous World.* Westview Press, New York, NY, USA: xvi + 288 pp.

Carpenter, S. L. & Kennedy, W. J. D. (1988). *Managing Public Disputes: A Practical Guide to Handling Conflict and Reaching Agreements.* Jossey Bass, Publishers, San Francisco, California, USA: ix + 293 pp., illustr.

Daly, H. E. (1977). *Steady State Economics: The Economics of Biophysical Equilibrium and Moral Growth.* W. H. Freeman & Company, San Francisco, California, USA: ix + 185 pp., illustr.

Davis, C. B. & Valk, A. G. van der (1978). Litter decomposition in prairie glacial marshes. Pp. 99–113 in *Freshwater Wetlands: Ecological Processes and Management Potential* (Eds R. E. Good, D. F. Whigham & R. L. Simpson). Academic Press, New York, NY, USA: v + 378 pp., illustr.

Davis, C. B. & Valk, A. G. van der (1983). Uptake and release of nutrients by living and decomposing *Typha glauca* Godr. tissues at Eagle Lake, Iowa. *Aquatic Botany,* 16, pp. 75–89, illustr.

Einstein, A. (1960). *Einstein on Peace* (Eds O. Nathan & H. Norden). Simon & Schuster, New York, NY, USA: iii + 704 pp.

International Peace Research Institute, Oslo/United Nations Environment Programme (cited as PRIO/UNEP) (1989). *Environmental Security: A Report Contributing to the Concept of Comprehensive International Security.* Oslo Peace Research Institute, Oslo, Norway: 23 pp.

Kennedy, P. (1987). *The Rise and Fall of the Great Powers.* Random House, New York, NY, USA: vii + 677 pp., illustr.

Marshall, N. B. (1986). Marine ecosystems. Pp. 172–87 in Polunin, *q.v.*

Milburn, T. W. & Davis, C. B. (MS). *Ecology, Rationality, and Time.* School of Natural Resources, Ohio State University, Columbus, Ohio 43210, USA.

Polunin, N. (Ed.) (1986). *Ecosystem Theory and Application.* (Environmental Monographs & Symposia). John Wiley & Sons, Baffins Lane, Chichester, Sussex, England, UK: xv + 445 pp., illustr.

Polunin, N. & Worthington, E. B. (1990). On the use and misuse of the term ecosystem. *Environmental Conservation,* 17(3), p. 274.

PRIO/UNEP – *see* International Peace Research Institute, Oslo/United Nations Environment Programme.

Renner, M. (1989). *National Security: The Economic and Environmental Dimensions.* (Worldwatch Paper No. 89.) Worldwatch Institute, Washington, DC, USA: 78 pp., illustr.

Tansley, A. G. (1935). The use and abuse of vegetation concepts and terms. *Ecology,* 16, pp. 284-307.

Tolba, M. K. (1982). Pp. 9–14 in *Selected Statements Made at the Session of a*

Special Character (UNEP Governing Council, 10–18 May 1982, Nairobi, Kenya). *Mazingira,* 6 (Special Issue), 60 pp., illustr.

Valk, A.G. van der & Davis, C.B. (1978). Primary production in prairie glacial marshes. Pp. 21–37 in *Freshwater Wetlands: Ecological Processes and Management Potential* (Eds R .E. Good, D. F. Whigham & R. L. Simpson). Academic Press, New York, NY, USA: v + 378 pp., illustr.

Westing, A. H. (1988). Ecology, international understanding, and peace. Pp. 85–8 in *Sustainable Development – Learning for Tomorrow's World: International Conference on Environmental Education* (Ed. I. Fjørtoft). International Union for Conservation of Nature and Natural Resources (lUCN), Gland, Switzerland, and Telemark Distriktshøgskole, Telemark, Norway: 120 pp.

Westing, A. H. (1989). The environmental component of comprehensive security. *Bulletin of Peace Proposals* (Oslo), 20(2), pp.129–34.

Commentary on Chapter 17

Chairman: Dr Arthur H. Purcell
Panellists and Other Contributors:

Holdgate, Ford, Trompf, J. Vásárhelyi, Burnett, C. Batisse, Polunin, Ramakrishnan, Shaw, Davis, Trompf

Holdgate said that throughout the Conference we had repeatedly asserted that the only hope for the future was for people to behave differently, in a manner more respectful of the Earth, more caring for our fellow humans and for our children from whom we have borrowed the Earth. The need was for stewardship of our planet, for the sake of the future.

He had been reminded several times of a model from Antarctica, which he would describe as a kind of parable. His colleagues used to argue very often (and at times tediously) over the best way to drive a team of husky dogs pulling a sledge. As a sedentary biologist rather than a geologist or a surveyor, he had not had to drive dog teams, but the argument gave him his model.

There were two systems: centre trace and fan trace (Figure 17.4, p. 418).

Centre trace (Figure 17.4a) was the more regimented. The dogs were harnessed in pairs, behind a single leader (who knew where he was going). The 'boss dog', top of the team hierarchy, came behind and was likely to give the leader a nip if he faltered. The pairs were matched in size and strength (and obviously had to work in harmony), and the overall system worked well enough even if the lace in it is imposed on the participants.

Fan trace (Figure 17.4b) – used, he believed, by the Inuit – was more flexible and democratic. Each dog pulled on its own lead. Each pulled independently. The advantage was flexibility. The hierarchy – the 'pecking order' among the dogs – had to be established first, of course, so that each knew his or her place. It worked if there was a common goal, cooperation, understanding of mutual roles and responsibilities and some incentive. It took some time for the team to settle down and work in harmony: it needed patience and training.

Oh – and what about the driver? He (or she, in today's more open Antarctica) was there to steer and to provide encouragement and persuasion, all of which demanded real understanding of the team.

That was quite a model for us, in seeking to harness individual or group efforts for environmental advance.

He feared that a commoner model for today's environmental system was what might be called a 'tangled fan trace', as in Figure 17.4c. A form of anarchic chaos prevailed. One or two dogs would pull valiantly with eyes on the distant goal. Others had other goals and tried to heave the enterprise another way. Yet others appeared keener to sort out personal differences than to move the venture forward; others tried to ride on the sledge. The driver might try to pull ahead of the dogs – a fatal flaw as Scott's *Discovery* Expedition found – or to ride on the sledge himself. One got nowhere that way. Meanwhile the ice cracked up behind, and the probability was that the team would be overtaken by disaster!

Put in simple terms, what did this mean? As **Davis** had said, the environmental

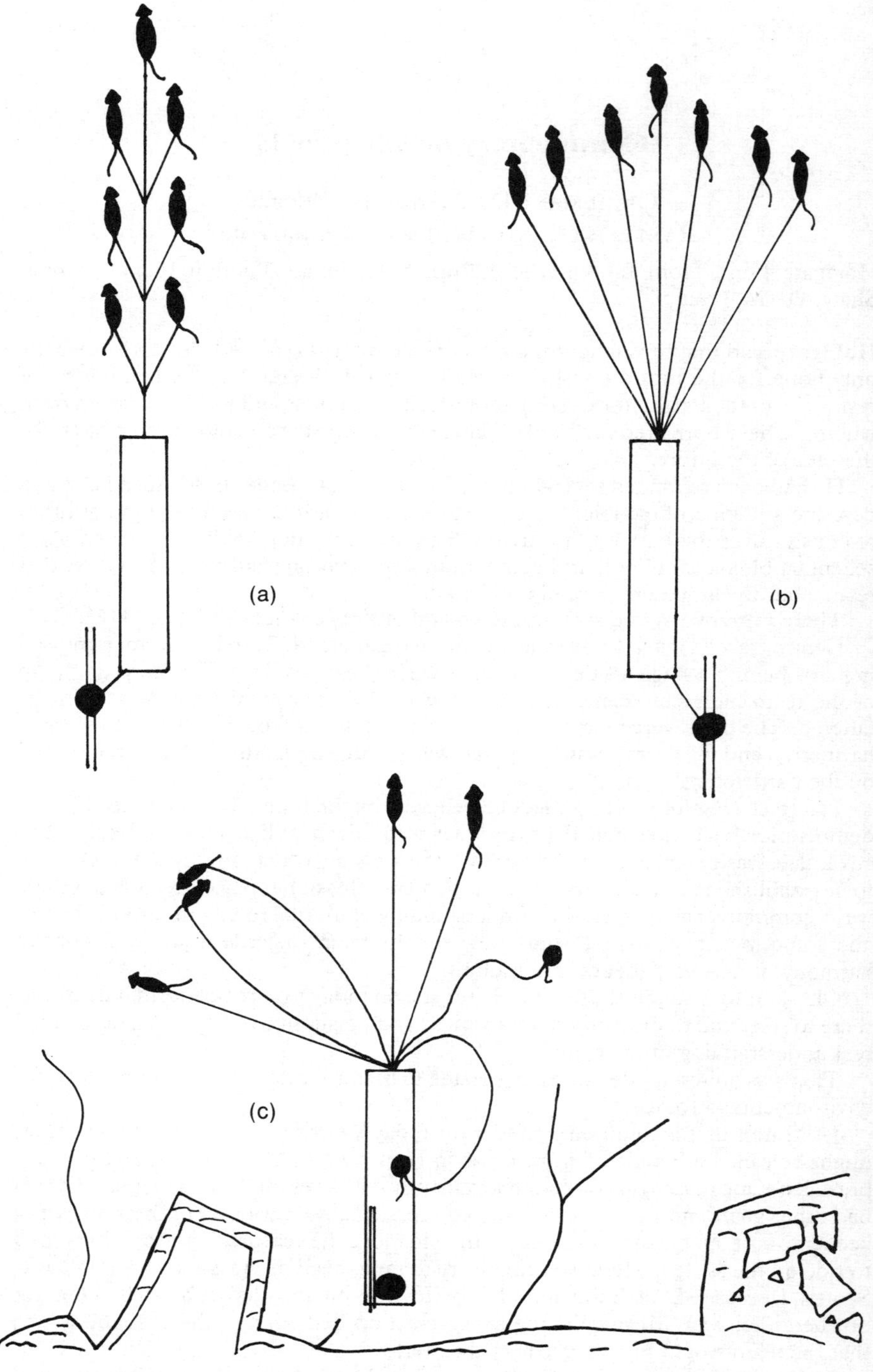

Figure 17.4. Antarctic dog-teams: (a) centre trace, (b) fan trace, (c) tangled fan trace.

movement had been cursed by finger-pointing. Statements of environmental groups commonly came down to something like:

'I (we) should be protected from the threat to my life and welfare arising from THEIR damage to MY environment;'

'THEY (usually the government) should stop THEM (usually industry) from doing things I consider inimical to MY/OUR environment.'

We had come, in this Conference, as indeed the environmental movement had done, to recognize that we could not solve the world's environmental problems by leaving the action to third parties. We all needed to modify our behaviour because in the end the environmental problems were caused by the addition and integration of innumerable actions taken by us as individuals, or as members of groups such as industrial firms. The issue was: how could we solve these problems. He suggested that there were four needs.

First, if we were to move people forward towards a better world we had to secure a much wider, if not universal, acceptance of *a Common Goal*. This came for people, as for huskies, through observation, understanding and interaction. The team leader, steered in the right course by the driver, observed a mark to head towards, and understood how to move across the intervening terrain. We did it by scientific observation, understanding the systems we were dealing with, and interaction which sought a consensus about the direction in which we should collectively move. He agreed strongly with **Davis** that to do this we depended on good information, which all individuals could see and evaluate for themselves, just as all members of a dog-team could see the rocks head.

But – as a second point – there was an underlying assumption in all this. Sledge dog-teams worked well together because they were both conditioned by instinct and moulded by training *to Pull Together*. In human society this meant that, if we were to pull together, we needed to widen our consensus on *how* to move as well as on our goal. The ingredients were diverse – a common ethic, shared values, accepted duties and agreed methodologies. At this Conference, several people had suggested that fear – what Rudyard Kipling called I believe 'the ties of common funk' – could provide a powerful unifying force, very much as people in a leaking lifeboat would fall to baling without worrying too much about their relative status in external society! Equally clearly, to give people a common belief that action was needed for their mutual security could be a powerful influence. Shared concern and a shared sense of security in this sense constituted 'shared values', while 'agreed methodologies' were those which were agreed to lead to a reasonable prospect of 'truth'. Acceptance of the honesty of other partners was certainly important, as a part of the shared value system.

In forming a common ethic, the prophet, the poet, and the inspirational leader all had a crucial role to play. Beliefs about what constituted 'right conduct' were the deepest and most durable influences on human behaviour, as the persistence of Biblical or Islamic codes of conduct over the millennia testified. And it was another characteristic of our species that it worked in groups. He had heard it said that each one of us, in our decisions, relied predominantly on no more than 20 key advisers – our 'team' (hence the concern over who occupied the central role in the White House!). To build groups, inspired by shared beliefs, values and methodologies was therefore a key to the adaptation of behaviour.

Third, *Training* was needed – as each individual needed to understand his or her role in the system. Here the educator was clearly crucial. Education in its widest sense would obviously be a key to sound behaviour in future. Education and communication also provided a way of building new groups, in place of old and no longer appropriate ones – and one important role of the communications system was to allow members of one particular group to become aware of what was going on elsewhere, so that people could find, and relate to, like minds outside their traditional background. In this sense a liberal capacity to communicate was a key to success, while barriers to communication constrained adaptive behaviour.

Lastly, *Leadership* was crucial. Here his simple polar model was relevant again. A good driver of a dog-team led from the rear. He did not interfere with the team more than was essential. He provided occasional cheerful encouragement – or cracked the whip to deter damaging aberration! He provided rewards – especially food – and did not let the team overtax its strength. These attributes were important also in human leadership, and as the Chinese proverb said, when a good leader has done his work, the people said 'we did it for ourselves'.

We had repeatedly said during the Conference that we needed to break the mould of human behaviour. His model called for people who understood their situation, knew their goal (or accepted leadership towards it), respected one another's positions, pulled together – cooperatively – on the basis of shared instincts and beliefs, adjusted their behaviour through communication and mutual awareness, and were led by those who cared, understood, recognized that they could only achieve their aims by getting the best out of their teams, and were content to share the glory of a good performance with those who had done the work.

Could we adjust the behaviour of the many groups in our complex societies into such a model before too much of the ice broke up?

Ford explained that he was not a scientist and offered his comments as a generalist. **Davis** had eloquently and graphically pointed out some of the behavioural changes that urgently needed to be brought about in attitudes towards the environment.

The Church, politicians and psychiatrists had at least one thing in common. Their problem was how best to influence and condition the thinking and thus the attitudes of people. Using the right techniques, the message they sought to impart was more readily received and acted upon. He referred here only to the means of imparting the message, not its content which might be for good or for evil. We had examples in the Catholic Church, the Wesley movement and, more latterly, the sinister process called 'brainwashing'. The whole process of thought-conditioning had been ably analysed by, among others, William Sargant in his admirable study *Battle for the Mind.* (1957; Heinemann, London, England, UK. 284 pp.)

He submitted that we needed to apply these same techniques to condition people to the needs of The Biosphere. Perhaps as scientists the information techniques now available should be examined and used more widely in the cause. Much had, of course, already been done; the growth of the Green Movement was evidence of this. However, as we all knew, there were many misconceptions among the public at large and some downright errors in the messages which were propagated. We needed not only to promote and encourage zeal and enthusiasm amongst ordinary people to save The Biosphere but also to educate and stimulate our leaders at all levels to share our concerns. We believed we had a mission and there was no reason why we should not embark on a crusade to save our polluted world.

Fosberg had said that the long list of human self-destructive activities and, in particular, the growing threat of P. R. Ehrlich's 'Population Bomb', should frighten us all. He could assure the Conference that after what he had heard that week he was frightened! But as we all knew, it was not enough to limit ourselves to the technique of replacing old concepts with fear. Fear must in turn be replaced by a knowledge of what positively could be done and by a determination to get on with it.

With respect, scientists could not do this in closed meetings. **Simon** had referred to the lack of exchange between the different disciplines within the scientific community (p. 377). Many heads had nodded in agreement. How much larger and potentially much more damaging was the information gap that existed between the scientific community and the public at large. Scientists needed to go out and carry their message to the man-in-the-street and, above all, to his leaders in a language that was authoritative and could be comprehended.

He suggested it was about time scientists frightened the politicians, the industrialists, the bankers and the media – the opinion-formers and the decision-makers. In his own limited contacts he had been appalled at the ignorance and, indeed, sometimes the cynicism displayed towards environmental matters by otherwise well-informed people: but these same people were the ones who were making decisions that affected our very existence. They needed to be made aware of science's concerns and challenged to say what they proposed to do about them. An informed and continuing public dialogue needed to be established.

Conference members were all busy people and occasions such as this 4th International Conference on Environmental Future, with its eminent and distinguished participants, occurred only irregularly and all too infrequently. He wondered therefore whether it should not be possible at future conferences for the opinion-formers and the decision-makers he had referred to earlier to participate in such deliberations. In practical terms he would like to propose that a half-day, or even a whole day, towards the end of the next conference might be set aside when prominent politicians, industrialists, financiers, etc., would be invited to hear the concerns, and to join in the search for acceptable and realizable solutions to them. Participants, in turn, could only benefit from hearing at first-hand of the problems and constraints, political and economic, that confronted our leaders. At the very least, all would be participating in that essential dialogue which was so urgently needed.

Trompf presented ten propositions representing in a shorthand way the problems and prospects of the world's religions in connection with the environmental future. He recognized five problems:

Problem 1. The single most important factor in world politics today was *residual tribalism*, which had as its base tribal religions and their expectations of intense group loyalty. Around 7,000 of these religions still persisted in some form today, and as a prevenient factor tribalist residues affected all nationalisms (*see e.g.* Gellner, 1983; King, 1986; Kasmanie & Trompf, 1989; Trompf, in press)*. Pacificist tendencies, especially from greater traditions, such as Christianity and even Islam, could not wish away the conflict-engendering tendencies often unleashed by this residual tribalism – extremisms, ethnic or racial prejudice, logomachy and real war, sexual inequality, etc. (*e.g.* Isaacs, 1975; Hadden & Shupe, 1986).

Problem 2. Partly because of residual tribalism, the most powerful force for motivating change in human populations was *used* (or what for didactic purposes we could call *bastardized*) religion, with detrimental effects on the environment (*cf.* King, 1968 pp. 373–5). Small-scale traditions, with their secrets of nurturing the environment, had steadily fallen vulnerable to newfangled techniques, and larger traditions had lost a grip on old asceticisms or simplicity of life (e.g. Geertz, 1963; Shinn, 1965 pp. 25–50; Worsley, 1984). At the same time, some apparently inhuman *traditional* or long-inured customs had persisted, and some very environmentally unfriendly ones (the Shinto cult of cleanliness, for instance, reinforcing the daily abandonment of chopsticks in Japan).

Problem 3. The more gloom and doom were 'spread abroad in the land' about The Biosphere's uncertain future, the more rampant would be the growth of popular *apocalyptic* (or millenarian) *expectations* about the end of the known order at the hands of the divine (Trompf, 1979 pp. 83–4; 1990 pp. 4–10 *et passim*). Anticipations of a literal eschatological Event, as it was called, were integrally related to *fundamentalisms*. The latter constituted the fastest-growing set of attitudinal configurations on the planet, and one which might breed passivity or neglect towards Nature, as well as bringing environmentally concerned groups under fire as too

*[*See* p. 423. Eds.]

humanistic, too 'flexible', and not trusting God's certain ways of 'sewing up history' (*cf.* Hadden & Shupe, 1989; Marty, 1990 for background).

Problem 4. The modern world had been overtaken by what we could term a macro-cult of Cargo (or of money by which to procure cargo – cargo consisting of internationally-marketed commodities that now, dare he say it, littered the Earth). This *cargoism,* in regional contexts, set levels of expectations, usually very high, as to how many new goods every family should enjoy to have a sense of well-being in the universe, and the demands on industry and waste management, and thus on the environment, were rising exponentially, especially in disadvantaged nations. In the West, cargoism had fast-growing secular fronts of ideologically justified 'brash materialism', 'techno-cultism', even 'hedonism', which all exacerbated these strains and abandoned the traditional religious values that could be useful for alleviating them (Trompf, 1987 pp. 104–8; 1990 pp. 10–14, 58–75, *cf.* also Leiss, 1976; Springborg, 1981 pp. 221–44; Romaña, 1989).

Problem 5. Residual tribalism and the prior tradition of religion's usability in society had produced formidable *civil religions* or *surrogate religions of the state,* especially in the world's two superpowers. Despite extraordinary religious plurality, religion in the United States was to an extraordinary degree alternative patriotism, and typically failed to be prophetically critical of prevailing social trends; while communisms transformed ecclesiastical structures and theological dogmas into statist ones (Talmon *et al.*, 1960; Berger, 1976; Bellah, 1980; Hughey, 1983). These special orientations tended to be imperialistic and not responsible towards The Biosphere, and had fostered scientism (on this latter, *e.g.* Whyte, 1963 pp. 26–35; Roszak, 1969 pp. 1–41).

To balance the five problems he recognized five prospects:

Prospect 1. There was an *ever-growing awareness* that religions and related groupings, which touched all cultures, and in the cases of such traditions as Christianity and Islam, straddled thousands of them, must address the environmental future. But the task had so far only been initiated by the 'foresightful few' – the *sadvipras,* if he could use the Sanskrit – whose effects remained limited in the face of other concerns of the masses (Hargrove, 1986; Engel & Engel, 1990).

Prospect 2. There was a new atmosphere of *multi-cultural* and *inter-religious dialogue* abroad, with environmental issues on the agenda. On the one hand, there was a current international concern to conserve endangered indigenous cultures and thus precious environmental insights kept by them; on the other hand, high-level dialogue between representatives of the larger traditions had been steady through the '80s, and the noteworthy 1986 Assisi conference, with a related network on Conservation and Religion organized in connection with the World Wildlife Fund anniversary, had had the environment as their focus (*Alyanak,* 1986).

Prospect 3. A mass of intellectual and *educative work* was now being carried out, first, to reconnect religious values to the salvation of Planet Earth, and second, to counter the destructive results of pressures inimical to that salvation. In terms of higher-level theological re-evaluation, for instance, one could range from Fr Sean McDonagh's pioneering book *To Care for the Earth* (1986) across to the World Council of Churches' 1980 statement on Faith and Science in an Unjust World (Shinn, 1980) to take Christian examples alone. The journal *Environmental Awareness* was an important Hindu-oriented Indian production of relevance; and there had been impressive Muslim contributions to discussions on environmental ethics (*e.g.* Husaini [n.d.], Zaidi, 1981). The American journal *Breakthrough,* specifically devoted to 'global education' in the Americas, had not neglected inter-religious communication (*e.g.* 1989 pp. 30–42).

Prospect 4. On an historical basis, one could argue that intense *apocalypticism eventually lost steam* (Desroche, 1973 pp. 26–41) (although its intensity was likely to be extremely heightened between 1994 and 2001 AD), fundamentalism would shift towards less abrasive styles for want of commonsense ideologically and *Realpolitik* in the political arena. At the same time, economically and environmentally imposed restraints on cargoism might well be heralded or supplemented by religious reactions against the cult of money and the undermining of religion by materialisms, including communisms (as recent events in Eastern Europe suggest).

Prospect 5. Historically speaking, political control over human populations had been unexpectedly superstructural, with no assurances – for many reasons – that governmental policies would be realized (*cf.* Mews, 1989). On environmental issues, however, *religions* had the potential for reaching much deeper existentially, producing a profound love of the Earth as Mother or God's handiwork – or at least a sense of obligation towards it (Dubos, 1972 p. 158, *cf.* Santmire, 1970; Lovelock, 1979). All movements concerned for biospherical conservations, including ours, would do well not to neglect an appeal to humanity's spiritual yearnings and sensibilities (and thus help to realize some of the ideals set out by **Davis** (Chapter 17).

Cited: Bellah, R. N. (1980). *Varieties of Civil Religion*. Harper & Row, San Francisco, USA: xv + 208 pp.; Berger, P. (1976). *Pyramids of Sacrifice: Political Ethics and Social Change*. Allen Lane, London, England, UK: xiv + 242 pp.; Desroche, H. (1973). *The Sociology of Hope* (trans. C. Martin-Sperry). Routledge & Kegan Paul, London, England, UK: vii + 209 pp.; Dubos, R. (1972). *A God Within*. Charles Scribners, New York, USA: x + 326 pp.; Engel, R. E. & Engel, J. G. (Eds) (1990). *Ethics of Environment and Development: Global Challenge, International Response*. University of Arizona, Tucson, USA: 264 pp.; Geertz, C. (Ed.) (1963). *Old Societies and New States: The Quest for Modernity in Asia and Africa*. Free Press of Glencoe, New York, USA: 220 pp.; Gellner, E. (1983). *Nations and Nationalism*. Cornell University Press, Ithaca, NY, USA: viii + 150 pp.; Hadden, J. K. & Shupe, A. (1986). *Religion and the Political Order*, vol. 1: *Prophetic Religions and Politics*. Paragon House, New York, USA: xxix + 458 pp.; Hadden, J. K. & Shupe, A. (1989). *Religion and the Political Order*, vol. 3: *Secularization and Fundamentalism Reconsidered*. Paragon House, New York, USA: 320 pp.; Hargrove, E. (Ed.) (1986). *Religion and the Environmental Crisis*. University of Georgia Press, Atlanta, USA: xxi + 222 pp.; Hughey, M. W. (1983). *Civil Religion and Moral Order* (*Contributions in Sociology 43*). Greenwood, Westport, Connecticut, USA: xvi + 223 pp.; Husaini, S. W. A. (n.d.) *Islamic Environmental Thought: An Overview on Recent Developments*. International Institute for Environment and Society Paper, IIES, Berlin, Germany: 35 pp.; Isaacs, H. R. (1975). *Idols of the Tribe: Group Identity and Political Change*. Harper & Row, New York, USA: xiv + 242 pp.; Kasmanie, A. & Trompf, G. W. (1989). *The Crippling of Lebanon (Research Institute for Asia and the Pacific Occasional Paper 7*). University of Sydney, Sydney, Australia: i + 48 pp.; King, P. (1968). *An Ideological Fallacy. Politics and Experience: Essays Presented to Michael Oakshott on the Occasion of his Retirement* (Eds King, P. & Parekh, B. C.). Cambridge University Press, Cambridge, England, UK: vi + 424 pp.; King, P. (1986). *The African Winter*. Penguin, Harmondsworth, England, UK: 249 pp.; Leiss, W. (1976). *The Limits of Satisfaction: An Essay on the Problem of Needs and Commodities*. University of Toronto Press, Toronto, Canada: 240 pp.; Lovelock, J. E. (1979). *Gaia*. Oxford University Press, London, England, UK: xi + 157 pp., illustr.; Marty, M. (1990). *Fundamentalisms Compared (Charles Strong Memorial Lecture 1989*). Strong Trust, Adelaide, Australia: 11 pp.; McDonagh, S. (1986). *To Care for the Earth: A Call to a New Theology*. Claretian Publications, Quezon City, Philippines: xii + 224 pp.; Mews, S. (Ed.) (1989). *Religion in Politics: A World Guide*. Longman, Harlow, England, UK: x + 332 pp.; Romaña, A. L. de (1989). Post-crisis Equilibrium: From Growth to Harmony: Part 2, Impoverishing Economic Growth. *Inter-Culture: International Journal of Intercultural and Transdisciplinary Research*, 22, pp. 80–170; Roszak, T. (1969). *The Making of a Counter-Culture*. Doubleday Anchor, Garden City, NY., USA: xv + 303 pp.; Santmire, P. (1970). *Brother Earth*. Thomas Nelson,

New York, USA: 236 pp.; Shinn, R. L. (1965). *Tangled World.* Charles Scribners, New York, USA: 158 pp.; Shinn, R. L. (Ed.) (1980). *Faith and Science in an Unjust World: Report of the World Council of Churches' Conference on Faith, Science and the Future,* vol. 1: *Plenary Presentations.* WCC, Geneva: xiv + 392 pp.; Springborg, P. (1981). *The Problem of Human Needs and the Critique of Civilization.* George Allen & Unwin, London, England, UK: vi + 293 pp.; Talmon, J. L., Voegelin, E. Aron, R. & Sperber, D. (1960). Abdankung des Messianismus? *Die industrielle Gesellschaft und die drei Welten* (Das Seminar von Rheinfelden) (Trans., M. R. Jung). EVZ-Verlag, Zürich, Switzerland: 301 pp.; Trompf, G. W. (1979). The Future of Macro-historical Ideas. *Soundings: An Interdisciplinary Journal,* 52, pp. 70–89.; Trompf, G.W. (1987). The Ethics of Development: An Overview. Pp. 102–29 in *The Ethics of Development: the Pacific in the 21st Century (Papers from the 17th Waigani Seminar)* (Eds. Stratigos, S. & Hughes, P. J.). University of Papua New Guinea Press, Port Moresby, Papua New Guinea.; Trompf, G. W. (Ed.) (1990). *Cargo Cults and Millenarian Movements: Transoceanic Comparisons of New Religious Movements (Religion and Society 29).* Mouton De Gruyter, Berlin, Germany & New York, USA: xvii + 456 pp.; Trompf, G. W. (Ed.) (in press). *Islands and Enclaves: Nationalisms and Separatisms in Island and Littoral Contexts.* Sterling, London, England, UK & New Delhi, India.; Whyte, W. H. (1963). *The Organization Man.* Penguin, Harmondsworth, England, UK: 394 pp.; Worsley, P. (1984). *The Three Worlds: Culture and World Development.* Weidenfeld & Nicolson, London, England, UK: xiv + 409 pp.; Zaidi, I. H. (1981). On The Ethics of Man's Interaction with the Environment: An Islamic Approach. *Environmental Ethics,* 3, pp. 36–47. Journals referred as wholes include *Breakthrough* (co-edited Mische, P. M. & Merkling, M., Suite 456, 475 Riverside Drive, New York, NY 10115 USA; *Environmental Awareness* (edited Oza, G. M., Oza Building, Salatwada, Baroda 390 001, India); and *The New Road* (edited Alyanak, L. c/o WWF-International, CH-1196 Gland, Switzerland).

J. Vásárhelyi said that she represented the civil environmentalists, not the professionals! One further concept upon which needed behavioural change should be built ought to be added to the excellent list given by **Davis**, namely, an essential identification with material Nature, the physical land itself, and the environment.

That was not necessarily the only way environmental change could come about. The Hungarian environmental movement had begun with a dozen people in 1984. After 6 years of apparently fruitless fighting for the Danube and against the proposed dam, 40,000 people had demonstrated against the latter in front of the Parliament building; but *not* because of environmental consciousness. It was more a protest against the Communist way of decision-making!

Ignorance was also a great problem as the Hungarian experience of the Chernobyl accident illustrated. No information was given at all. Someone speculated on television that the radiation would decrease, but the fact that it had risen greatly was never revealed. Some of the population was convinced that they were in deadly danger and that they would die, but there was no panic because of the general scepticism the regime had engendered. Apart from one privileged kindergarten, children were not allowed to remain indoors in an attempt not to create panic. Scepticism, unfortunately, was bad also for the environmental cause because it resulted in disbelief of any information.

The first real evidence for optimism had come when Earth Day had been recognized on the previous Sunday in Hungary in a manner beyond her expectations and she now believed that Hungarians were at last beginning to identify with their environment.

Burnett commented that it was difficult to change people's goals. He had been privileged to be involved annually in tripartite discussions on strategic issues and general relationships between Russians, Americans and British, involving high-ranking military, diplomatic, and academic, individuals. Discussion was almost entirely concentrated on war and strategic issues. Within two years, despite outside perceptions, it had become clear to him that none of the military participants wanted nuclear war but were, indeed, terrified of its consequences. Nevertheless, that did not prevent a further 8 years of discussions, right up to *perestroika*, but still largely on (non-nuclear) war and related strategic issues. The lesson was clear: change could only be effected if one could change goals, to change means was not enough.

C. Batisse announced, for the especial benefit of Hungarians present, that there was now an association, '*Les Amis de la Terre*' (Friends of the Earth International), which was particularly conscious of the need for behavioural change.

It tried to provide reliable information for the attention of decision-makers on subjects such as air, water and soil management. For example, the President of the association, Professor Pierre Samuel, had just prepared a report on air quality in France which had pointed to infringements of existing regulations and called for reinforcement of standards. That report had been considered by UNEP when the association had been elected to the Global 500 Roll of Honour.

The association considered that an in-depth change of certain personal habits was of the utmost importance to the well-being of the physical environment and had also urged industry to behave in an environmentally-sound way. On the occasion of Earth Day, the association had invited everyone to commit themselves to leaving a clean Earth for their children, and to choose at least one from ten practical commitments on sheets (*see* Figure 17.5, p. 426) which were distributed throughout Paris supermarkets. Fifty thousand sheets were returned, collected by children and displayed on a symbolic 'Tree of Promises' erected on April 22nd in front of the Science and Industry Museum, Villette.

Polunin said that surely we should recall the slogan 'Every Day an Earth Day' as a guide to our personal behaviour.

Ramakrishnan said that his remarks would be orientated to developing countries. He thought that to talk of behavioural change in societies reflected a degree of scientific arrogance. What was needed more, perhaps, was a behavioural change amongst the scientific community and behavioural evolution or adaptation in the rest of society. Behavioural change was not a static concept, and in developing societies one was dealing with something that was changing rapidly and adaptively all the time. Behaviour, inevitably, had to be adapted to meet change but that was rather different from changing to a value system that was considered very important by environmentalists but not necessarily so in some societies. Moreover, amongst traditional societies there were many environmentally desirable values which had become submerged as a consequence of other changes but which needed to be identified, brought out, refined and developed further. An evolutionary approach to behavioural modification was most likely to succeed in promoting most successfully the best kind of environmental management.

Shaw, having considered **Trompf**'s comments on fundamentalism and residual tribalism which led to both in-group and out-group enmity, and **Davis**'s emphasis on the widespread preoccupation with national sovereignty, wondered whether, over evolutionary time, our psychological make-up had developed some kind of bounded rationality whereby ethnocentrism and non-common values were fostered and reinforced. How much might such an innate attitude result in resistance to well-meaning educational programmes, or to rhetorical notions that we should all have, or pursue, common values? It might sound rational to advocate to individuals that they re-order their priorities in terms of a common, perceived interest in the future of Planet Earth, but how effective would that be if there was a deep psychological propensity resisting the notion? Could we really break the mould of human nature as **Holdgate** had advocated?

Davis replied that he had no answer to those questions. He thought that we

22 AVRIL 1990... JOURNEE DE LA TERRE

COUP DE CŒUR POUR UNE TERRE PROPRE !

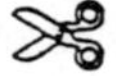

JE SUIS UN AMI DE LA TERRE

Je promets de changer mes comportements
pour laisser la terre propre à mes enfants
Je choisis l'une de ces promesses :

❑ **J'évite de prendre la voiture**
Je choisis de préférence de prendre le train, le métro, le tramway, le vélo.

❑ **Je pollue moins en voiture**
Je roule moins vite, je choisis de l'essence sans plomb.

❑ **J'utilise uniquement les bombes aérosols**
qui portent le label "protège la couche d'ozone".

❑ **Je respecte les forêts**
Pas de feux, pas de déchets, je débroussaille, je choisis du papier recyclé.

❑ **Je n'utilise plus de sacs plastiques**
Je choisis les produits qui ont un emballage carton, je choisis "des produits propres".

❑ **Je récupère mes déchets**
Je dépose systématiquement mes bouteilles en verre dans un conteneur, je récupère les petites piles, je récupère le papier ...

❑ **J'économise l'eau**
(voiture, pelouse, machine à laver...)

❑ **J'économise l'énergie**
Je ne surchauffe pas la maison, j'éteins en quittant les pièces, j'isole, j'encourage l'énergie solaire.

❑ **Je protège les animaux sauvages**
(Ivoire, fourrures, sauvagine) et je surveille la propreté des animaux de compagnie.

❑ **Je me remue**
Pollution, décharges sauvages, paysages dénaturés : j'organise une pétition, je constitue un groupe, j'écris, j'agis.

TAPEZ : 36 15 AMIS DE LA TERRE

ou cochez la promesse de votre choix et retournez ce coupon à :
Les Amis de la Terre B.P. 1970 75533 Cedex 11
Vous recevrez gratuitement des informations pratiques et utiles.

Nom .... Prénom..........................
Adresse ....
Code Postal ....
Tél. ....

Figure 17.5. Sheets distributed throughout Paris supermarkets.

should fall back on diversity. Problems were not going to be solved in any single way, by any single society, or in the same way. If we were dealing with some kind of bounded rationality then perhaps we should be thinking more in terms of evolving behaviour in different societies as **Ramakrishnan** had suggested took place.

Trompf commented that when the term 'bounded rationality' was used in inter-cultural discussions it often reflected the views of European or American intellectuals who had examined other people's epistemological systems and decided that, by the criteria used to assess them, they were not quite rational enough. Dialogue, therefore, broke down. This point was pertinent because we all had to face up to a world which was completely different from that rationalized by scientists. Rationalities, from culture to culture, were relative; we could not expect perfection, nor was there any Utopian solution to our problems. All that could be done was to weigh up problems against prospects and struggle on! If he might be forgiven for saying so, that, theologically, was what the vicissitudes of life were all about. What he regarded as most vital for progress was the ongoing work of discovering what would be common bases for discourse about the environment. That was going to take an incredible amount of routine study but should prove eminently worth-while.

18. Needed Socio-political Change*

General Olusegun Obasanjo†
PO Box 2286, Abeokuta, Ogun State, Nigeria

Introduction

Societal dynamics need to be rearranged in a way that will enable African countries to leapfrog out of the current morass that seems to have encapsulated them. As the 21st century looms larger on the horizon, the tail-end of the 20th century seems, almost suddenly, to have become a minefield of dictatorial, authoritarian, undemocratic, governments and leaders. Conversely, democracy in all its ramifications appears to be gaining an increasing degree of acceptability and popularity even among erstwhile cast-iron ideologies. Indeed the current wave represents a consistent march towards the institutionalization of democratic reforms at the social, political, and economic, levels in many societies.

Undeniably, democratization of political and general development processes is a major key to sustainable development. Access to institutions and structures of governance, and government itself, must be improved. Such a process should involve a systematic, irreversible, and progressive, release of energies of the people for development. The aim should revolve around the reduction and limitation of the disadvantage of the poor in development schemes and programmes and in participatory politics. The momentum and impact of such changes must be capable of propelling us in Africa towards a peaceful co-existence and interaction with our natural environment.

Whatever we do in the political arena must be informed and guided by certain interconnected strategic imperatives, if democracy is to be established and nurtured. Those imperative steps must be aimed at achieving the following:

A. Creation of Trust and Building of Confidence;

B. Decentralization of Political Power and Authority;

*[Largely as presented, in the absence of the Author, by Dr Martin W. Holdgate, Director-General of IUCN. Eds.]

†[Head of Federal Government of Nigeria and Commander-in-Chief of Nigerian Armed Forces, 1976–79, whereafter he stepped down to become a farmer – Director of Obasanjo Farms near Abeokuta from 1979. Eds.]

C. Pluralism [of Authority] and Decentralization of the Economy;
D. Political Communication;
E. Education and Political Enlightenment;
F. Promotion and Defence of Human Rights;
G. Renewal of Mandate and Succession Programme; and
H. Creation of Appropriate Political Machinery.

Trust Creation and Confidence Building

The consensus of opinion among discerning observers of African government and politics is that there is a suffocating air of mutual distrust, and often cynicism, between government and the governed. Simply and plainly stated, the people do not trust their rulers and the rulers in turn do not trust their subjects. Political trust is an essential ingredient in any strategy that would enable us truly to achieve responsive governance; but, most unfortunately, by many Africans government is still seen largely as an alien institution. Perhaps, at least in some circles, this is explicable as one of those sadly enduring legacies of Africa's widely colonial past. But then, we must admit that successive African leaders have not really attempted seriously to change this conception and perception of government. They have in most cases reigned and ruled but have largely failed to govern.

The distrust apparently stems in part from the colonial experience of government being distant, anonymous, and faceless, and of alleged atrocities being committed in the name of government. The political history of African societies since the dawn of colonialism shows that Africans have never really been able to believe fully in their leaders. Subsequent political independence, rather than change that perception, complicated it. While government is seen largely as an alien institution, politics is little more than the allocation of valued objects or rights carried out before greedy eyes and grasping hands. It therefore becomes inconceivable that those in government would create room for the subjects to participate effectively in the political process. Government leaders in turn have often tended to perceive the people's reaction to their policies as mischievous or even diabolical, believing, as it were, that they must be based on misinformation and ignorance. Because of this air of distrust, constituent units cannot effectively partake in policy formulation, execution, and other aspects of the political process. The overall effect has been that citizens became subjects of governance rather than objects of governance.

This is reinforced and accentuated very widely in Africa by what Hyden (1980) calls the 'economy of affection'. The immediate loyalty and concern of most government officials is first to their village and *locales* rather than to more broadly-based and enduring goals of national development. Those who do not have relatives or close associates in government, naturally feel neglected, ignored, excluded, and 'short-changed'. This, therefore, breeds an air of resentment of, and detachment from, government policies, and concomitant disinclination to support them. The popular conception among constituent units is that the only way to pursue your clan's goal is to get into power, thereby creating a 'we/they' syndrome.

Reversing and eliminating this air of distrust imposes on African leaders and governments the dire need to convince the people that government can be responsive to their collective needs, and that they, the people, can and should be involved in the process of policy planning, formulation, and execution. Trust and confidence must be built up through popular participation by the citizens in matters that affect their lives and through a two-ways communication between government and the governed. But in place of these abstract musings we need action in concrete terms, and how to get it raises the question of the structure of governments in Africa, where the political topography is dotted by varying forms and degrees of over-centralization of power and authority.

There is a widely overarching need for reorganization of most African societies on a different basis of legitimacy from the present. African governmental structures and leaders need to be more open, more accessible, and to have their legitimacy rooted firmly in the approving participation of their people. The people's participation must find expression in the political process, and involve a convincing democratization of access to the institutions of power. The first step in this direction is the need to de-bureaucratize the political process, by the most effective means of embracing head-on the concept of decentralization and devolution of political power and authority. This, after restoration of trust, is the second strategic imperative required for the institutionalization of responsive governance as a practical reality.

Decentralization

The excessive centralization of political power and authority in Africa has literally turned most nations into over-governed and over-ruled but grossly mismanaged entities. Most societies are almost always at the whims and caprices of a few technocrats and upper-echelon bureaucrats who, in conjunction with the political leaders, play 'god'. One of the major spin-offs of this unfortunate situation has been the phenomenon of institutional fragility, such that the 'slightest of tremors' have almost always resulted in the collapse of our political penthouse 'like a pack of cards'.

It is indeed an irony of history that, for a continent that is predominantly rural, we have continuously refused to establish viable political structures at the rural level – structures that are capable of aggregating and articulating the peasant's views and suggestions on the political process. It is now imperative for us to establish tiers of government, rising from the grassroots, that will allow for effective and meaningful participation by the people at the social base (and representing much of the social basis) of society. A government that is abstract and shrouded from the largest proportion of the country cannot be said to be responsive in its governance. It may reign or rule, but it will not, strictly speaking, govern.

Let me say here, that the mere establishment of local governments is not enough: government must be made to function effectively at that level. The practice in most African countries wherein ministries of local government are established at the central level – to control the local governments – is negating and contradictory, and will only stultify and hamper the efficiency of this tier of government. Local governments must be autonomous and independent, and must be made to function as such. Accessibility to this kind of governance would serve to imprint, on the minds of the people, the perception of self as an active participant within the political process. This would in turn make local governments serve as the basic units of governance, and function as the hubs of the development process.

The added benefit of decentralization is the related question of democratization of access to political institutions. Development, properly conceived, must be seen as the ability of a people to induce and manage change within a given area and within any period of time. Having local governments that are autonomous within their various spheres of influence, would naturally increase the access of the citizens to democratic structures. It would enable them to participate more actively than otherwise within the political process, and it would in similar vein enable them to monitor closely the activities of government officials and any changes. The 'alien institution' conception and perception would thus be whittled away.

Decentralization has the multiplier effect of significantly marginalizing public corruption, which all-too-often swallows up public funds earmarked for national development. Many development- and mass-oriented projects have been shipwrecked on the quicksands and rocks of corruption. (Corruption is also one of the causes of government unresponsiveness, and one of the propelling factors behind governments' resort to coercive instruments of the state – either to stifle freedom of expression or to gag articulate and critical consistuent units.) The logic of this reasoning is that, if administration is decentralized, funds would also be decentralized, and this would make whatever fund was at the disposal of government officials small enough to become less attractive and conducive to corrupt practices. At the level of local government, the leader is a member of the community and he or she does not have the advantage of distance and anonymity which people enjoy at the central level. There are many positive spin-offs of a decentralized political structure that is executed with due honesty of purpose and direction. But decentralization by itself is not enough.

Pluralism and Decentralization of the Economy

There is the need not only for good government but for less government. In a pluralistic society, there must be guarantees of non-oppression of minorities in addition to individual freedoms and protection of human rights and property. There must also be pluralism and decentralization of the economy.

The State should not have monopoly of means of production and distribution, while social services and welfare must be left with the best unit within the polity and society that can perform to advantage. Economic exclusion, in the manner of political exclusion, breeds resentment to authority by those excluded. Total release and use of the initiative and energy of the citizenry on the economic front, is a *sine qua non* for economic development and progress.

Political Communication

Communication is a major instrument of development. For the establishment of responsive governance, there should be a two-ways flow of communication between the people and the government. Communication should not be a monologue but a full and active dialogue.

More often than not, people are not properly briefed about government's intention as it relates to policies that affect their lives. Political communication here should be seen as different from government propaganda. Propaganda's effectiveness is ephemeral. Propaganda without concrete objects with which the people can relate would create a stark, irreconcilable contradiction between their objective reality and the subjective preferences that the propaganda would have foisted upon the people. Good political communication enables the government to sensitize its short-, medium-, and long-term, goals to the sensibilities of the people.

For political communication to be effective in bridging the gap between the government and the governed, it should be done in the languages with which the people can identify and which they understand. It must be reinforced by positive symbolization; that is, the leadership must lead by example. Such an action would be perceived as honesty on the part of the leadership and would help them to carry the people along with them as they carefully skirt and traverse the journey which national development represents. It would remove the unnecessary air of esotericism and facelessness of the government. Indeed, carrying the people with the leadership is important, for without good political communication one cannot rightly hope to consolidate whatever gains are made in the establishment of a responsive governance as a model.

Communication is the roof that protects the political penthouse from deterioration and decimation by the inclement weather of an hostile political environment. It is the propelling force behind the persistence of instituted responsive governance.

Education and Political Enlightenment

Political education and communication are two faces of the same coin. Political education denotes the deliberate and conscious effort on the part of the government to inform the citizens about their rights, duties, and obligations to the state. It sensitizes the sensibilities of the governed to the political process. (Any government that intends to be responsive to the needs of its people must take this as part of its responsibility.) This enables the citizens to be politically responsible and politically conscious, and accelerates mass mobilization while aiding the process of political acculturation. General education of the citizenry facilitates and enhances political education and communication. An educated mass is undoubtedly easier to govern than those that are illiterate, ill-informed, and ignorant about the political process. An educated populace tends to be more amenable, more pliable, and more easily sensitized, to the exigencies of governance. Education generally improves responsiveness to governance. It also makes the citizen ask the reason why such and such an action needs to be taken, and what will be the most likely result of taking it.

Political education equally contributes towards the evolution of a political culture that is more democratic than otherwise, as each side of the political divide is conscious of its powers and limitations. The abuse of power and fundamental human rights violations would thereby be checked, though this raises yet another of the strategic imperatives.

Promotion and Defence of Human Rights

Any society which is characterized by a frequent occurrence and recurrence of an abuse of human rights cannot be said to have a 'political soil' conducive to the germination and sustained growth of responsive governance. Respect for fundamental human rights of the citizens by the government is the linchpin and the kernel of a proper conduct of governance.

A stifled, caged, and gagged, populace can only be marginally propensible to positive political interaction. In a similar vein, political restriction, or, worse, repression, imbues in the citizens a feeling of depression, deprivation, and a sense of loss. These are basically non-conformable with democratic attitudes and principles. By extension, such a repressed citizenry cannot be expected to contribute their innovative best to the national development. Thus a society that denies an integral stratum of its constituent units the exercise of its fundamental human rights, is nothing but a moribund if not a dead society. Human rights must not only be respected, they must be safeguarded through policies, actions, directives, and, most importantly, through constitutional provisions. Let me insert here a few words about constitutional provisions as they relate to the question of fundamental human rights.

Giovanni Satori (1971), in his typology of constitutions, identified three main types strictly on the basis of fundamental human-rights provisions. These are: Nominal Constitutions, Facade Constitutions, and Garantist Constitutions.

Nominal Constitutions pay scant attention to fundamental human rights. In fact in some instances, the question of fundamental human rights is subjected to the pleasure of the executive head of government. A typical example was the Ghanaian Constitution under Nkrumah, where the question of fundamental human rights was subjected to a presidential declaration. In this case the President was expected to declare, upon assumption of office, that all Ghanaians enjoy fundamental human rights. This of course permits the government to abuse fundamental human rights without fear of consequent challenge.

The Facade Constitution glosses over the provision of fundamental human rights. These are actually embodied in the constitution, but there is a provision that such rights are what the government deems them to be. A typical example is the Soviet Constitution prior to the new era of *glasnost*.

The Garantist Constitutions, in contrast, not only have fundamental human rights enshrined, but also guarantee them by spelling out the modes of redress and by asserting the independence of the judiciary as the last hope of the common man or woman against all forms of governmental tyranny and administrative abuse and oppression.

Obviously the garantist constitution is the only form that is conducive to, and compatible with, the institutionalization of responsive governance as an ideal vehicle for transportation towards a high degree of political development.

Renewal of Mandate and Succession Arrangement

Responsiveness of governance and popularity of a government must have a 'litmus' test – renewal or rejection of the mandate it has been given to govern. This demands that those in government must never for a moment lose sight of the transient nature of political power and authority. They must inculcate it in themselves as one of the most basic and enduring canons of democracy that political power belongs, in the final analysis, to the people; it is to be held and exercised in trust for the people. Possession and consequent exercise of political power and authority should not be seen as the exclusive preserve of anybody or any group. Fully understanding and practising this allows for an orderly succession and peaceful transition of political power from one set to another.

The mere thought of having to leave office as and when due, naturally imposes on the leadership due responsiveness and responsibility. On the other hand, the erroneous belief that one can stay in power *ad infinitum*, tends to make a government careless, unresponsive, and dictatorial, with an imbued sense of self-glorification. Moreover, those who make peaceful change impossible encourage violent change. Non-peaceful successional arrangements in the political system breed resentment, detachment, and apathy to the system. The key is the conviction that it is preferable for the other side to win than for the society to be torn apart.

It is imperative that we learn these basics if we are to institute responsive governance. Orderly and peaceful succession is possible where there is a basic agreement on the fundamentals of statehood and nationhood, and above all, in a situation wherein there is an obviously deep attachment and commitment to the system.

This should be seen as the nexus, the pivotal basis, the kernel, and the crux, of our political machine.

An Indigenous Political Machine

Whatever structures and institutions are created must be created by ourselves for ourselves. The political machinery must take cognizance of the people's political temperament, attitudinal disposition towards authority, and behavioural norms. Any contrary situation will create the antinomies of structural organization, or at best give birth – as it has done in the past – to what has been called 'the prismatic society'.

As was observed, some 50 years ago, discussing politics without reference to human beings does the deepest error to our political thinking. We must each identify the specific peculiarities of the political animal in us as a people. Aristotle, in his *Politics* (*cf.* Baker, 1946), remarked on the necessity of fitting the constitution of a city state to the character of the people. Thomas Hobbes (*cf.* Warrender, 1962) dealt with the issue of national character and the personality of the nation. John Stuart Mills [1806–73] also observed that 'political machinery does not act in itself as it is first made, so it has to be worked by men, and even by ordinary men. It needs their active participation and must be adjusted to the capacities and qualities of the just men available'.

Due cognizance and attention must be paid to the African political past, allowing the wisdom of our past to harmonize our present and structure the evolution of our future. For too long in Africa, we have seen the world and our immediate political environment through the 'binoculars' provided for us by others. But now we may be on the threshold of history. The 'topography' of our 'political landscape' is in sight, and from our generation of experience we see that western political models may not work. The precarious balancing-game which has been foisted on us by the western models has shown us that we need to rearrange our political 'chessboard'. The present situation in which there is a surfeit of Kings and Queens on the board has turned every move into a dangerous adventure.

Whatever political institution we operate, it must be essentially of our own design – flavoured and spiced by our history, our political thoughts, and the societal dynamics that made precolonial situations and usage exist and persist. Although modern societies need structures in governance that are essentially different from some needs of the past, yet we must not move in the manner of a people without a past. Ours is a rich political past. The nineteen-nineties must serve as the period of laying foundations for our political and economic survival and sustenance in the 21st century. That survival must be anchored on responsive governance.

Whatever we do, responsibility for the environment, like governance, must be democratized. Our interaction with our environment must involve a bottom-up arrangement, rather than the extant situation where the flow of authority is from top to bottom. Consequently our first suggestion should be of the need for a greater involvement of the individual in the protection of his or her environment.

Environmental Issues and the African Scene

There are world-wide worries about the emission of carbon dioxide from the burning of fossil fuels, with a broad scientific consensus regarding the 'greenhouse' effect to be attributable to conventional energy consumption, the depletion of the ozone layer in the stratosphere, and acidic precipitation, as all largely human-engendered. Yet Humankind needs energy for development and indeed for survival. It is incumbent on societies all over the world to link energy sources and other contributions to these environmental problems in a manner that allows the curtailment or phase-out of energy sources which do great harm to the environment. Of the sad depletion of our world's forests, a sizeable proportion is due to the African use of trees as firewood, thereby robbing us of much-needed protection of The Biosphere.

In addition to the threats just mentioned, there are other serious environmental dangers connected with toxic wastes, deforestation, pesticides, chlorofluorocarbons, the drying up of underground aquifers, and changes of climate – at the mercy of which Mankind has always been, though not to the extent that has latterly been envisaged. For now it seems that the numbers and profligacy of our species on Earth are affecting the global climate.

It is regrettable that natural resources are rarely dealt with as economic assets. Their depletion is often seen as income generation, yet income properly conceptualized must be 'the maximum amount that can be consumed in the current period without reducing potential future consumption'. One could visualize an endless chain of interconnecting scenarios from this ill-founded accounting system to which should be added the accompanying consequences of such actions. The required correction would be the need, first, to make the economic accounting system reflect environmental realities. As a corollary, there must be a bold move towards integrating environmental and economic objectives.

Global Overpopulation

The danger of overpopulation especially in the South should not be based on relativity in figures but on the quality of life that can be afforded. Although it has been suggested in some quarters that the countries of the North should slow down their development to reduce their consumption – not only in the overall interest of the world, but also to allow the South to make some progress in development without further degradation of the environment – this suggestion seems impracticable and unrealistic, being akin to the last man in a long-distance race asking the first man to wait for him to catch up. In normal competition, it does not happen.

In the past, the forest, the hills, and the mountains, all related to the total being and existence of an African. His economic and social life were to some extent inextricably interwoven with these concrete features of his physical environment. For instance, trees were believed to serve as *Chi* (a personal god) to each individual.

Unfortunately, the current rate of deforestation is such that the obvious picture is one of progressive delinking of Man from his natural environment. This is why efforts at combating deforestation, the prevention of desertification, and general environmental degradation, should be made a fundamental part of daily life. Because, by delinking a man from his woods and forests, he is delinked from the spirit of his ancestors; socially and psychologically he is torn asunder.

We have to reconcile the individual and societal self-interest without stifling economic progress in developing countries. To me, dampening energy demand through taxation and technology in industrialized countries is in the world's collective long-term and enlightened self-interest. We must not fail to regulate our thirst for energy if we are to reduce environmental stresses.

In the past in Africa, proverbs and folk-lore widely took cognizance of the problems of environmental protection. A basic change in policy requires the reintroduction of this into the mainstream of policy inputs in Africa. Such a reintroduction should, as a matter of necessity, be done through a bottom-up approach, for which it should be embedded at local and community levels.

Furthermore each family, if not individual, should have environmental protection responsibilities, not merely in terms of hygiene and sanitation, but also in terms of tree-planting, afforestation, avoidance of toxic waste, erosion, and desertification. In other words, let individuals and families take care at least of their own environment.

Renewable energy resources from the Sun, the wind, and water, should form the basis of our socio-economic activities at the family, community, local, and national, levels. Reverence for the environment, as part of the spirit of our ancestors and our being, must be encouraged without detriment to political, social, and economic improvement (Obasanjo, 1989).

References

Baker, E. (1946). *The Politics of Aristotle*. Oxford University Press, Oxford, England, UK.

Hyden, Govan (1980). *Beyond Ujaama in Tanzania: Underdevelopment and Uncaptured Peasantry*. Heinemann, London, England, UK: p. 8.

Obasanjo, Olusegun (1989). *Constitution for National Integration and Development*. Friends Foundation, Lagos, Nigeria.

Satori, G. (1971). *Studies in Opposition* (Ed. Rodney Baker). Macmillan, London, England, UK.

Warrender, H. (1962). *The Political Philosophy of Hobbes: His Theories and Obligations*. Oxford University Press, London, England, UK.

Commentary on Chapter 18

CHAIRMAN: Dr Michel Batisse
PANELLISTS AND OTHER CONTRIBUTORS:

McMichael, Thorndike, Samatar, Mische, Cloudsley-Thompson, Westing, Burnett, Furedy, Holdgate, McNeely, Thorndike, McMichael, Holdgate

McMichael remarked that **Obasanjo**'s contribution (Chapter 18) had dealt fully and frankly with the political changes required in Africa today, as he saw them. It would be impertinent for one coming from a 'North' country somewhere south of Asia (or a 'Western' country located in the 'Far east'!) to comment on that situation.

What he could do was to make some observations on the socio-political situation in that part of the world in which he lived, namely, the Western Pacific Rim and South-East Asia, and on some aspects of biospheral survival which would involve societal change by which he meant changes in social behaviour and attitudes.

The Asia-Pacific Region was said to contain the fastest-growing economies in the world and had the potential to become the great economic powerhouse of the next century. The region now held between one-third and one-half of the world's population, including the world's most populous country, China. An important but often overlooked element was the group of small, independent island States in the Pacific – surely amongst the smallest nations on Earth in terms of both land area and population, some of which were likely to be severely affected by global climatic change.

All of these countries were in vastly different stages of development and operated under widely differing political systems. Consequently there was no simple or single answer to the question of what socio-political changes might be required in the years ahead, even if we were able to agree on what was likely to be best for The Biosphere and for ourselves as part of it. What we could say for certain was that the relative impacts on The Biosphere of the people living in the different countries of the two regions varied enormously.

Some years ago the distinguished American ecologist Ray F. Dasmann, then working for IUCN, had drawn a distinction between two kinds of societies – which he had called 'Ecosystem people' and 'Biosphere people'. Referring particularly to the South Pacific Islands, Dasmann had written:

> 'Ecosystem people live within a single ecosystem. They are dependent upon that ecosystem for their survival. If they persistently violate its ecological rules, they must necessarily perish. Thus a hunting people who continually kill more wild game than can be produced by the normal reproduction of wild animal populations, must run out of food and starve. A fishing people who persist in over-fishing will destroy their base of support. Those who practice subsistence agriculture must develop some means for keeping the soil in place and restoring its fertility. Island people have lived under particularly strong restraints and could not tolerate any great increase in their own numbers since the resources of islands are not only limited but tend to make their limitations obvious. Only continental people can develop myths of unlimited resources.
>
> 'Biosphere people' draw their support not from the resources of any one ecosystem, but from the entire biosphere. Any large modern city is the focus

for a network of transportation and communication that reaches throughout the globe – drawing perhaps beef from Argentina, lamb from New Zealand, wheat from Canada, tea from Ceylon, coffee from Brazil, herring from the North Atlantic and so on. Local catastrophes that would wipe out people dependent on a single ecosystem may create only minor perturbation among the Biosphere people, since they can simply draw more heavily on a different ecosystem. Consequently Biosphere people can exert incredible pressure upon an ecosystem that they wish to exploit and create great devastation – something that would be impossible or unthinkable for people who were dependent upon that particular ecosystem' (Dasmann, 1974 p. 92).*

He supposed that, if our only goal was the maintenance of The Biosphere, we would all revert to being 'Ecosystem people'. In fact there were still some Ecosystem people living in parts of the Asia-Pacific Region – in some Pacific Islands, in parts of Papua-New Guinea, and among the Australian Aborigines. To an extent it might also be said that many Chinese were still essentially 'Ecosystem people', though as a nation China was increasingly dependent on biospheral resources. By contrast, Japanese, Australians, New Zealanders, Canadians and American were clearly 'Biosphere people' and there was no real prospect of their being anything else in the foreseeable future. Despite the attempts of a few people in Western countries to revert to an ecosystem-based existence, it would clearly not be possible for everyone living in those countries now to do so without socio-political changes of such magnitude as to lead to chaos.

On the other hand it would be absurd to suggest that we could all go on the way we had been without concern. The developed nations simply could not continue to consume a greater and greater amount of the world's resources and the developing nations could not, indeed would not, be denied access to those resources that they needed to improve their living standards. Countries such as Korea, Taiwan, and Singapore had already largely made the transition from 'developing' to 'developed', while Thailand and Malaysia were well on the way to becoming 'Biosphere' nations. So what was to be done? What socio-political changes could we propose, or anticipate, that might lead, in due course, to a more equitable and possibly sustainable situation?

He thought that there were several possible directions in which the world might move over the coming decades, that could help towards biospheral survival. Perhaps the most important was attitudinal change, on the part of developed and developing nations alike, to the exploitation of the remaining biospheral resources. We surely had to move away from regarding tropical rain-forests, for example, as resources to be used, and had to see them instead as survival resources, a substantial part of which should be protected for all our sakes. We had to learn to resist the economic arguments of individual entrepreneurs or enterprises, whose primary interest was to exploit resources for profit, unless and until they could prove that their proposals were consistent with *biospheral* interests. There were some welcome signs that such attitudinal shifts were happening in several of the countries of the Asia-Pacific Region: for example, the decision by Thailand to ban any more commercial logging of its forests; the Australian Federal Government's insistence on the protection of large areas of tropical, subtropical and temperate forest, despite the wishes of several State Governments to keep them open for exploitation; and moves by Pacific countries generally to put an end to the appalling drift-net fishery. Australia and France had taken a brave stand in favour of an unexploited Antarctica, with their advocacy of setting aside the whole continent as a wilderness reserve or 'world park'. All these were hopeful signs that possibly marked the beginning of a new period of ecological awareness on the part of the peoples and Governments of the Region.

In addition to changes in the attitude of Governments, there was a need to

*[*See* p. 441. Eds.]

encourage changes in the attitudes of individuals to the consumption of goods and services. It seemed to be characteristic of human beings that they strove to acquire more and more goods and services, even when their genuine needs were already well satisfied. It was, by and large, this desire for goods that determined election outcomes in most countries – witness the way people voted (Australians for example) in favour of those political parties that promised less taxes and thus greater spending power to the individual. Not only was this the case in the Western democracies; few would doubt that the great political changes now under way in the USSR stemmed, in large part, from a need to find better ways of meeting the demand for consumer goods. Was it then possible that these attitudes could be changed?

The rise of consumer consciousness in the developed nations had been remarkable in recent years as community awareness of the environmental crisis had increased. Regrettably it was still at an essentially superficial level; and while one could not decry the concern for recycled and environmentally benign products, the benefits they yielded were more than outweighed by increasing *per caput* energy consumption and growth in the wasteful use of other essential Biosphere resources, especially forest products.

By contrast with the need to reduce the excessive levels of consumption in developed countries, few would deny the right of the peoples of the developing world to gain access to more goods and services. We really could not expect people living in developing countries to accept the present situation as permanent; consequently we must expect that the global demand for goods of all kinds would rise dramatically as the economies of those developing countries which are able to develop, strengthened. Probably not all of them would. Some would choose not to – possibly including some of the smaller South Pacific nations where the value systems might be more like those of 'ecosystem people'. But most countries would continue to strive for economic growth and some would succeed, even if they had little hope of attaining the present levels of consumption of Western nations.

He could not let the occasion pass without drawing attention to a situation currently confronting China in relation to motor-cars that was reported recently in the international press (Anon., 1990). The Volkswagen company had a motor-car manufacturing plant in Shanghai as a joint venture with the Shanghai Tractor & Automobile Corporation. The plant had the capacity to produce 60,000 cars a year but in fact produced fewer than 16,000 in 1989. Why? Because the Chinese Government had banned practically all new car purchases, domestic and imported alike, as part of its 'hard work and plain living' programme to curb runaway inflation. The Volkswagen factory ceased production for 39 days last year while a similar joint-venture plant involving Peugeot in Canton closed for two months. In the long run, the Government intervened and bought all of the factories' excess inventory, some 7,440 units which were now stored in Government warehouses. Needless to say, the car manufacturers, both foreign investors and local joint venturers, were anxious to increase production which would create jobs and lead to reduced costs per unit. No doubt as the Chinese became wealthier the demand for motor cars would increase, and provided the Government relaxed its controls, the number of cars *per caput* would rise. The impact on the global environment of car-ownership levels in China comparable with those of Western countries would be severe. Here we had the paradoxical situation that the short-term economic interest of China coincided with global environmental interests, but the interests of the Western investors (and no doubt the desires of the Chinese people) would seek to reverse this position.

A similar situation could be seen in relation to energy consumption and the production of 'greenhouse' gases. Australia, like many other developed Western nations, was seeking ways of reducing its contribution of 'greenhouse' gases to the atmosphere and was focusing on energy efficiency – at both the production and the consumption ends of the cycle. At the same time, Australia's economy was struggling to overcome significant trade deficits and foreign indebtedness – hence it sought to maximize exports. One of its large export income earners was coal, which was shipped in huge volumes to Japan and China. The Chinese expected to increase their energy production significantly in the years ahead, partly through major increases in the purchase of coal. By comparison with that of some other countries, Australian coal was generally low in sulphur and so had less damaging environmental consequences when burned. Accordingly Australia considered that it should seek to maximize its share of the Chinese coal market, both in its own economic interests and because that would be marginally better for the environment than burning of coal from some other source, even though the overall impact on the global environment would be significant.

So what should be done? Should Australia refuse to help the Chinese to increase their energy output? Should we advise them to do so by some means other than burning coal – for example by selling them uranium fuel for nuclear power-plants? Or should we take a highly-principled stance and say that Australia would not play a part in contributing to global pollution in this way and let other, less scrupulous nations sell their high-sulphur coal to China? There was no doubt that Australia would do the economically rational thing and sell as much of its coal as it could to the Chinese and so the world's burden of atmospheric CO_2 would continue to increase.

But if the developed nations were to make a serious effort to reduce energy demand and *per caput* consumption of goods and services, then maybe – just maybe – there would be room for China and other countries like her to achieve the additional levels of development to which they aspired, without pushing the world over the brink into biospheral collapse. In the developed countries in recent years there seemed to have been less concern shown by people for the well-being of fellow citizens, though the pendulum might now be swinging back towards a greater willingness to share the national wealth. Regrettably, he saw no sign of a revitalized concern for greater equity between peoples of different nations, so that any self-imposed reduction in consumption levels by the citizens of Western nations seemed unlikely in the short term. This might seem to be a rather pessimistic view and in many ways it was. Frankly, he thought we had little cause for optimism because human beings were essentially self-interested, or at best, tribe-interested. It was not in our nature to think globally, or to be as concerned for people of different nationality and race as we were for ourselves and our kin.

Nevertheless, there are some hopeful signs here and there and we could but try to build on these and spread the message through speaking out, through personal commitment and, above all, by political action. The rise of environmentally-aware political parties – both the so-called 'greens' and others with broader interests including the environment – was one of the most encouraging signs, at least in Western countries. In Australia's recent federal election, between 15% and 20% of the electorate had voted for parties with strong environmental policies. Environmentally-aware groups now held the balance of power in the National, and in two State, Parliaments. One could only hope that this trend would continue and, perhaps by the beginning of the 21st Century, we would see environmental policies as *the* most important factors in deciding who got elected, and environmental performance as the key to who should stay in office. *Cited:* Anon. (1990). Volkswagen Faces Uphill Climb in China, *International Herald Tribune*, 17 April 1990, p. 6; Dasmann, R. F. (1974). National Parks, Nature Conservation, and 'Future Primitive'. *Proc. South Pac. Conf. on Nat. Parks and Reserves.*

Thorndike spoke as one who played a dual role, as a member of a non-governmental organization – the Centre for Environmental Information dedicated to providing environmental information to all who needed it, and as a State government policy-adviser and decision-maker for the Adirondack Park – the major part of the world's fourth largest Biosphere Reserve. She had long been interested in improving the interface between environmental science and environmental policy. Finite resources and a growing global population had placed Humankind on a collision course with Nature's constraints. Given the urgent need to advocate responsible policies, how could accurate environmental information, grounded in often uncertain science, be communicated to policymakers and to a public which might influence those policymakers? On what basis could policymakers with limited or non-existent scientific training make decisions? At what point *must* they make decisions, given the dynamic nature of information, and of the world itself?

Humankind, while attempting to survive with The Biosphere, lived within the context of socio-political systems. What changes within these systems needed to occur? She emphasized that neither capitalism nor communism gave any evidence of having a monopoly on establishing an environmentally benign future. She suggested that we had at least two roads to salvation, regardless of political persuasion, and that *both* must be travelled if we were to achieve the inter-generational equity towards which **Tolba** had challenged us.

First, our government institutions, from the international to the local level, needed to elevate a holistic and a long-range perspective on a par with, and to the level of, short-term decision-making.

Second, we needed greatly to strengthen capability, at the local/regional eco-complex level, to inform and educate, in order to create an environmentally-literate citizenry. The ability to think globally and act locally was a superb goal but most people did not know how to do this.

Regarding the long-range perspective, it was essential that governing bodies recognized that 'long term' was not five years, the next election, or whichever came first; that we should think in terms of what Garrett Hardin had termed 'ecolacy' – by always asking the question 'What then?'. We needed to establish institutions or mechanisms at all levels of government where scientists, professional planners, economists and technical experts, with *relative* dispassion, could examine, analyse, synthesize and constantly update information which was both multidisciplinary and interdisciplinary. Such mechanisms could provide guidelines and lay out alternative strategies which would add to the needs of the moment, sufficient lead-time for adaptation, preservation, mitigation, or whatever the particular environmental problem required. Perhaps, most importantly, such institutions, or mechanisms, would call attention to the environmental implications of problems which had not even been identified by the governing body as being environmentally significant.

Obasanjo had called for a transfer of power on the African continent to strengthened local governments. She suggested, however, that while local government, with its greater level of accountability, might appear to be the logical vehicle for resolving environmental problems which were largely local in both their origin and effects, that can be a two-edged sword. Environmental problems did not respect political boundaries; where they did, their cumulative impacts might have adverse effects well beyond those boundaries. If the broader public interest was to be served, it was unlikely that local government, operating in a narrower context, would provide the necessary broader perspective.

Local government could act effectively within the context of the kind of long-range decision-making or planning mechanism suggested previously. Her own organization was attempting to establish a Regional Resource Forum, a concept it was believed could be replicated and adopted anywhere on the planet, wherein scientists, educators, government officials and individual and group constituencies might interact effectively to address environmental issues, including so-called global ones. Climate change, for example, was the cumulative impact of countless local decisions about energy use and resource allocation.

Time did not permit description of the proposed activities or methods, but she offered the concept as an example of how the gulf between scientists, educators, government, and society, might be bridged.

Finally, she would merely emphasize the second road, universal environmental literacy, as an imperative task and priority for all societies and their governments. Two decades of laws and regulations in developed countries had slowed, but failed to stem, the tide of degradation. Fundamental changes in behaviour and attitudes, and in environmentally benign technology transfer, were essential. This was the province of education and we could no longer afford to support it merely as an adjunct to environmental research and law enforcement. It simply had to be at the forefront when government, and the society which supported that government and determined policy, allocated funds. Concurrently, industries must recognize that improved environmental literacy was in their own self-interest.

In summary, socio-political reforms must now be viewed from the perspective of the North–South differential. The old East–West dichotomy was no longer the ecological frame of reference. In that respect, she concluded with a message from the First Annual Report of the Council on Environmental Quality to the US Congress in 1970:

> 'In dealing with the environment we must learn not how to master nature but how to master ourselves, our institutions and our technology. We must achieve a new awareness of our dependence on our surroundings and on the natural systems which support all life, but awareness must be coupled with a full realization of our enormous capability to alter these surroundings.'

That statement described the interface between science and policy, for which education must become the bridge if future problems were to be avoided. To convince governments and the private sector of the urgency of doing so remained the challenge.

Samatar thought the main thrust of **Obasanjo** was that he wished to see a political culture develop in Africa which did not exist there at present. How did one set about creating that political culture? He saw how a wave of democracy had swept across eastern Europe, but were those countries going to embrace the American political culture, the heart of which seemed in many ways to be the rapacious exploitation of the environment? By contrast, in Africa the problem was to tame Nature. Was it possible to create a political culture that would strike a happy medium, and would, perhaps, eastern Europe provide some of the answers?

Mische wished to enter a caveat to the proposition that had already been made twice, namely, that we should not expect global taxation. Change had occurred in a lot of other things very quickly, and judgements of what people would do on the international scene could not be based on either the past or the present. There had been rapid growth in understanding the environment which had brought about many changes on national scenes. Change on the international scene, with growing global awareness, might happen far more quickly than anyone expected. That was especially the case if there was going to be a focus on the need to re-conceptualize security and sovereignty, which seemed probable

Cloudsley-Thompson remarked that we were all liars and hypocrites: if not, we would not be civilized people! We told white lies in order to be courteous to people. Even at the Conference, participants trimmed their words so as not to offend people. We could not therefore expect politicians to tell the absolute truth; but we could expect white lies rather than black ones. That was not always the case.

The other problem was the big companies or multinationals. They were not bad in themselves but, because responsibility was shared amongst many people in such a company, no one individual felt responsible, and no shareholder felt any particular responsibility. In fact, if a company did not exploit its environmental opportunity, its shares would be reduced in value. A shareholder could always then sell his or her shares but, of course, someone else would always buy them; the environment did not benefit. There were many such in-built institutional factors in our society which made us, individually, less able to to take care of those environmental questions that should be our concern. He did not see how these situations could be resolved, but that was no reason for not being aware of them.

Westing considered **Obasanjo**'s views to be forthright, and perceptive in the way they described both the severity and dismal nature of the current situation in Africa; but the suggestion that governance in Africa might be localized made him recall earlier remarks about the dangers of tribalism.

More important as a matter for concern in Africa than even the local, or national, region, was the ecocomplex or biogeographic region. In some large countries there were ecosystems that could be managed successfully within national boundaries, but in many parts of the world biographical regions extended beyond local areas and overlapped national boundaries.

McMichael had referred to Dasmann's notion of 'Ecosystem people' and 'Biosphere people', and this notion was also clearly relevant to the African situation. He hoped that 'Ecosystem people' would continue to be viable, perhaps with biospheral back-up. 'Biosphere people', on the other hand, should begin to think much more about the importance of the ecosystems on which they had to lean for insurance and enrichment, but the basic unit of concern ought to be the ecosystem. That would pose difficult problems, politically as well as environmentally.

He thought that, in suggesting that instability was the norm today, **McNeely** had been right, although he did not agree that there was more instability than before. Judged by the incidence of armed violence over recent decades, if not centuries, political instability, at least, did not seem to have changed much.*

He had also been thinking about earlier comments on behavioural change. He recalled that a very high percentage of Hungarians, when polled, had appeared to be very much attuned to the need for conservation and were already thinking environmentally. He believed such attitudes were now pervasive throughout the 'North', so that behavioural change had in fact already occurred without too much effort to bring it about. That was an important behavioural lesson with implications for politics as well. What was needed was to institutionalize the change so as to lock it into place and capitalize on it: that too was, in essence, a political issue.

Burnett pointed out that what was really under discussion was power – the oldest problem in the world – how a group achieved power and exercised it. The burden of **Obasanjo**'s view had been that more power should accrue to smaller groups within a wider organizational or institutional framework. Quotation from the classics was not very helpful: the Greek city states were very small units indeed, hardly comparable with even the small countries of the modern world. The only way that power had been changed over the centuries was through increased and more widespread education; frankly, it did not matter much what particular political system was in power. In modern Britain local regions probably had less power than they had had for the last two centuries – under a democratically elected government. Opposition was only possible through articulate, informed activity: that required knowledge, *i.e.* education. One of the problems for 'Biosphere people' was that their whole development had resulted in the opportunity for change to be reduced because of the increased complexity of their lives. It was not clear always what their targets should be, or how they should be attacked.

Furedy doubted whether opinion polls were a good guide to assessing how 'green' people had become. Perhaps they indicated that people were willing to do such things as cooperate in recycling schemes, conserve energy, or purchase goods which they believed to be environmentally friendly. But if it came to a loss of jobs and livelihood, then a real sticking point would have been reached. It was when environmentally desirable actions impinged on particular industries, or the economy, so as

*[This referred to Spring 1990 since when the situation seems to have deteriorated markedly. Eds.]

to reduce employment, that there was a real test of behavioural change. Political analogue had shown that!

Holdgate thought that **Obasanjo** was convinced that Africa must create its own political system but with due regard to the historical and cultural diversity within that great continent. He had set out eight universal guiding principles but the processes that they might give rise to would have to be debated further within the African context. The principles had a wider application, such as to eastern Europe. No doubt those countries would wish to develop by taking cognizance of what was happening in other parts of the world, and regarding it as wise economics to avoid some of the overkill aspects of the US system.

The need to widen the educational base was another of **Obasanjo**'s principles. In line with what **Burnett** had said, education and understanding were an essential basis for a more articulate population which could then support or check government more effectively.

The question of decentralization, another of the eight principles, would need more discussion in the context of the system it was desired to establish. An administrative district might not be the best unit to start with from the environmental point of view, but it did take cognizance of today's political reality. As the necessity for more rational planning, or for an environmental policy emerged, as in the Horn of Africa (p. 327 *et seq.*), so the importance of the environmental unit could be recognized and accommodated.

On the matter of international taxation, he drew attention to the coming together, increasingly, of what had been separate national systems, the European Community being the obvious example. The EC did, of course, levy taxes on a pluri-national basis and another example was the move towards a levy on goods, such as timber, that could then be fed back from the consumer to conserve the producer's resource base, *e.g.* through the International Timber Trade Agreement. These examples suggested that, although there was not yet an international taxation system, steps towards one, however faltering, were being taken.

He agreed with **Cloudsley-Thompson** that the quest for absolute truth was a philosophically entertaining matter but he thought, although he was no philosopher, that all truth included a relativistic component. Truth was affected by the location of the person who sought to establish it, the skills and capacity for appraisal of that person, the historical context, and many other things. Points that might by today's yardstick ring true might not ring true by tomorrow's. What one could expect, however, was honesty: which was what he thought was in people's minds when talking about the need for truth – the expectation that one sector should not deceive another sector of the community for gain, for example. We were entitled to demand honesty and transparency and could do a lot better with more of both. We might then take different action, but that was another matter.

Lastly, he agreed that behavioural change was happening, or at least attitudinal change was happening. Opinion polls gave more positive results than when a government decided to take action on the results – everyone applauded a better environment but when asked to pay more in taxes, say 10%, to achieve it, then only 50%, or less, were in favour. He agreed with **Furedy** that action was the acid test. That led him to suggest that one of the greatest tests would be to see if the quality of life, which was taken for granted in western Europe and in the USA, could be sustained on 30%, say, of the present levels of energy consumption, resource consumption, and relative wealth consumption, without forfeiting employment. That was perhaps one way in which developed countries really could lead the world.

McNeely thought that there was no doubt that all the Asian forests, perhaps all the world's forests, had been occupied by people in pre-colonial times. The forests were only exploited when colonial governments came up with the new idea of central control of forest exploitation, and a very thorough job they had made of it. In Nepal, the government had seen the light and returned the forest to the Sherpas who now controlled it in Sagamafa National Park.

He held rather different views from some on international taxation. Firstly, how was the money going to be spent? There would be a great deal of it which would, to judge by previous experience, involve a tremendous centralized bureaucracy. Consequentially, some might be spent on inappropriate central government authorities that would otherwise not have survived. Indeed, international taxation might even prevent some of the kinds of changes which it had been established to achieve, simply because there was too much money! Therefore, he thought that we should go cautiously in advocating such taxation.

He agreed with others that attitudes had changed but not behaviour, although the latter might change if forced. Once it became clear that everyone would have to make some sacrifice then it would be psychologically easier to behave differently.

He had thought about the consequences of the proposition that the US should reduce its consumption. Much would depend on the proportion imported. He believed that much came from developing countries. A cut in consumption would, therefore, reduce or destroy the primary market on which the developing countries depended for their own development. It was an example of how apparently obvious, environmentally desirable actions turned out to be more difficult than one might suppose!

Thorndike remarked that, on the matter of white lies and black lies, her own experience of political decision-making was that decisions were mostly made on a basis of public perception. In its turn, that might be based on correct or incorrect information, but it was inherent in the nature of politicians to follow such a course. Whether they did it because they liked it, by force, or by elective power, politicians always wanted to remain in office. Public pressure, opinion and reaction were certainly going to continue to play an important role in a politician's decisions: it was not a matter of truth as the politician saw it. Another aspect was that any politician or decision-maker who did not acknowledge that value judgements were an inherent part of decision-making had not been truthful with themselves. In the last analysis, value judgements were an important factor in determining, in the best of situations, what a political figure decided. Political honesty was important.

Turning to international taxation she confided that, as one who lived in one of the most affluent States in the most affluent part of the world, the proposals made by the environmental community, even by economists, for a nation-wide US carbon tax had simply not filtered through to the kinds of politicians who could make a difference. Even raising the tax on petrol as a means of limiting consumption had not, so far, been possible. She hoped, however, that things might change and that the US might lead by example in a year or so.

Lastly she commented on her own especial concern for environmental education and information. She believed that there was no doubt that there had been an enormous, renewed, and widespread awareness; it was global. But awareness, as had been rightly said, was not the same as action, and she feared that we were a long way away from filling that gap.

McMichael agreed with **Westing** that those who had chosen to live as 'Ecosystem people' should be encouraged so to do: there were many possibilities around the world. The fact was that most people were not going to make that choice. There was clear evidence that people wanted to move up the scale to become 'Biosphere people'. The problem was that, if the bulk of the world's population wanted to move up, then those at the top would have to move down to meet them somewhere. That was an important socio-political ingredient of the environmental problem.

On behavioural change, he agreed with others that people had become more aware of their environment and might change their behaviour in relation to their own tribal environment. Within a nation or a group of people, all sorts of sacrifices would be made if it was perceived to be in the collective national or tribal interest, but they would not do that for another group elsewhere who were not their own kith and kin. That was a very ancient, ingrained reaction and was the big problem to be overcome if we were to develop a sense of international responsibility as distinct from narrow tribal or national responsibility.

Holdgate commented that there was one example of how the tribal group could be enlarged, namely through the shared experience via the media of the immediacy of peoples' condition in remote parts of the world when struck by disasters. When the victims of disasters, famines, floods, or tornadoes, from all parts of the world, appeared 'live' on living-room television screens it brought about an immense wave of communal sympathy – a sense of common humanity. That could and should be built upon.

Annexe 7: New Trends in Soviet Environmental Policy*

MARAT R. KHABIBULLOV, JOAN T. DEBARDELEBEN &
ARTHUR B. SACKS

Institute for Environmental Studies, University of Wisconsin-Madison,
1225 West Dayton Street, Madison, Wisconsin 53706, USA

CHANGES IN THE CONTROL SYSTEM: BUREAUCRATIC RESTRUCTURING

In January 1988 the Central Committee of the Communist Party of the Soviet Union and the USSR Council of Ministers adopted a resolution, 'On the Fundamental Restructuring of Environmental Protection in the Country'[1] which provided a 'green light' for a series of changes to consolidate a highly fragmented and proliferated system of environmental control.

The resolution involved broad-ranging criticisms of 'shortcomings' in past policy. Specific targets included the excessive fragmentation of environmental functions among various ministries and departments, but also the absence of effective economic levers and stimuli, insufficient application of scientific achievements, a short-range approach in adopting economic decisions, and the operative policy of using only 'left-over' investment funds for Nature protection.

The fragmentation of functions in the environmental area reflects the general structure of the economy as a whole, which is based on planned production organized according to economic activities, such as metallurgy, agriculture, forestry and so forth. Until recently, about 130 ministries, committees and agencies existed at the All-Union level alone, with corresponding entities at the Union Republic level in many cases. These bodies sought to implement policies in pursuit of their own organizational interests which were, at times, highly parochial in nature. In effect, many of these agencies were free-standing entities, operating under the control neither of a cohesive central government nor of the public. In this way the phenomenon, widely referred to as 'departmentalism', developed. It is now generally recognized that the resulting difficulties in managing this huge bureaucratic super-organism have contributed significantly to the deep inefficiencies of the Soviet economy as a whole and to a decrease in the standard of living, which has become alarmingly visible, especially over the past few years. The

*[This contribution is mainly of historical importance now. How far the numerous and encouraging environmental initiatives will persist after the more recent upheavals (1991) can only be a matter of conjecture, but the signs now are less promising than they were. Eds.]

bureaucrat, always considered a self-interested and hampering figure, in the current circumstances has become a threat to perestroika in the USSR: it is the bureaucrats who hold the strings of state power. This unexpected contradiction between a highly centralized political system and the fragmented process of economic decision-making is a paradoxical feature of Soviet politics.

Bureaucratic Reorganizations

The most important measure of the January 1988 resolution on the environment was the creation of the new State Committee for Environmental Protection (*Goskompriroda*). Its formation climaxed a long-standing debate in the USSR about the necessity of forming a single governmental organization for environmental management, an organization with real power, independent of other ministries. Designating the new body as a State committee rather than as a ministry has an important meaning in the Soviet hierarchical system. A State committee generally has authority over ministries within a specific area. *Goskompriroda* is now the central agency of state management in the field of environmental protection and the rational use of natural resources, and it, along with the Council of Ministers, bears all responsibility for protecting the environment.[1]

The functions of *Goskompriroda* incorporate those previously carried out by relevant subdivisions of the USSR State Agro-Industrial Committee; the USSR Ministries of Water Resources and Land Reclamation (both disbanded in the spring of 1989); the USSR State Committee for Hydrometeorology and Environmental Control; the USSR State Forestry Committee; the USSR Ministry of Fish Industry; the USSR Ministry of Geology, and others. The following control functions were reassigned to *Goskompriroda* from these agencies:

- land-use and protection, from the State Agro-Industrial Committee;
- surface-water use and protection, from the Ministry of Water Resources;
- ground-water protection, from the Ministry of Geology;
- regulation of atmospheric contamination, and coordination of environmental control in the country, from the State Committee for Hydrometeorology;
- wildlife use and protection, including hunting and game management, conducting wildlife inventory and the Red Data Book compilation, from the State Forestry Committee; and
- protection of fish, water animals and water plants, from the Ministry of Fish Industries.

All of these agencies were responsible not only for proper management of natural resources within their purview, but also for economic production involving those same resources. Production has almost always had priority over protection, creating 'the confrontation of hands': the left hand, pretending to protect the natural resource, has been paralysed by the right hand, as it sought to exploit the very same resource for the sake of 'the Plan'.

Consequently, these bodies did little more than gather data about environmental degradation in the country – data that were kept hidden from public view, from other ministries, and even from the highest levels of government. This system was doomed by its own ineffectiveness.

Unlike the agencies previously responsible for protection *and* production, *Goskompriroda* has not been given any routine economic functions, and therefore, in theory, it serves as an organization free from direct economic interests and pressures. To produce a unified environmental protection system, regional committees and agencies are now being created at all levels – from the Union Republic level down to the local areas.

Despite this consolidation of functions, several governmental agencies still retain environmental regulatory roles. For example, the Ministry of Public Health carries out oversight of atmospheric and environmental pollution in settlements and enterprises; *Gosgortekhnadzor* (a monitoring organ, subordinate to the Ministry of Geology) continues to oversee the use and conservation of common minerals; and the State Automobile Inspection (GAI) Agency monitors vehicle emissions. These bodies never had economic production functions and therefore retain their role in environmental management.

To accomplish its overall mandate, *Goskompriroda*, has been designated to fulfill the following specific tasks:

- to exercise comprehensive control over Nature protection activities in the country, and to coordinate the relevant ministerial and departmental activities;
- to oversee utilization and conservation of land, terrestrial and subterrestrial waters, air, vegetation (including forests), wildlife (including fish), sea-life and contiguous seas;
- to prepare recommendations concerning Nature conservation and rational use of natural resources, and to monitor the fulfilment of relevant assignments envisaged in the State plans for economic development;
- to prepare recommendations on the improvement of the economic mechanism of Nature utilization;
- to introduce ecological rules, norms, and standard regulations, laying down procedures for natural resource utilization and protection of the environment against pollution and other harmful effects;
- to carry out ecological surveys of the development and distribution of the national productive forces and economic sectors, to monitor the compliance with ecological norms in the design of new equipment, technology and materials, and in construction and renewal projects;
- to issue licenses for wildlife use, for air consumption for industrial purposes, for the disposal of industrial, municipal and other types of waste, and for the discharge of harmful components into the water and the atmosphere; to issue permission for geological explorations; and to exercise control over the allotment of lands for all kinds of economic projects;

- to run the country's Nature reserves and to exercise State authority over the game-hunting facilities, and also to maintain the State Wildlife Inventory and Red Data Book;
- to organize the dissemination of ecological information among broad sections of the population, and to promote public education in the environmental field; and
- to plan and carry out cooperation in the field of environmental protection with foreign countries and international agencies.[1]

According to the January 1988 resolution which created it, *Goskompriroda* possesses comprehensive power in enforcing environmental legislation; its decisions are binding on all ministries, departments, associations, enterprises and organizations. Thus, by law, it has the right to impose bans on construction, reconstruction, or expansion of industrial or other enterprises, if these activities involve violation of environmental protection legislation. In such cases, it may also halt production activity and file suit against responsible parties.

Despite the broad-ranging powers laid out in the resolution, *Goskompriroda* has not yet been able to realize fully its mandate. Just prior to the establishment of *Goskompriroda*, Soviet author Leonid Leonov wrote: 'It is necessary to give this organ all rights, even veto power'. But in the two years since its formation, *Goskompriroda*'s ability to exercise these rights and responsibilities has been unclear. While it has been able to halt production when there have been environmental violations, in general its status *vis-à-vis* other State organizations remains ambiguous. For example, the head of the former Commission on Environmental Protection of the Presidium of the Council of Ministers was a First Deputy Chairman of the Council of Ministers itself. In contrast, the head of *Goskompriroda* only has rights as a 'row' minister, suggesting a lower status.

Soviet writer Valentin Rasputin has noted caustically that, in the hierarchy of the Council of Ministers, the head of *Goskompriroda* is seated right after the Minister of Water Resources, recognized broadly as a force for environmental degradation.[2] Soviet ecologist Vladimir Sokolov has proposed elevating *Goskompriroda*'s status by making it directly subordinate to the highest agency of state authority – the Presidium of the Supreme Soviet.[3] The first chairperson of *Goskompriroda*, Feodor Morgun, former local party leader in the Ukraine, had a reputation as a good administrator, but he was generally viewed as being uninformed on environmental matters. To many, his appointment was considered a compromise in a bureaucratic struggle for power. Following Morgun's resignation from the post during the summer of 1989, Prime Minister Ryzhkov apparently had difficulty finding an acceptable and willing successor. In August 1989 a scientist with strong environmental credentials, Dr Nikolai Vorontsov, was appointed and approved by the new Supreme Soviet. Vorontsov is not a Party member, suggesting that his appointment may reflect an effort to select someone on the basis of his acknowledged expertise, rather than on his politics. This appointment also suggests that *Goskompriroda* may now be able to take an independent stance in articulating its mission and asserting its authority. Whether its status and power will be adequate to assure success remains to be tested, however.

One of the most difficult problems facing *Goskompriroda* is the strong resistance to implementing the very governmental decision which created it. As one might expect, this opposition comes from those ministries and committees which have lost their monopoly control over natural resources with the creation of *Goskompriroda*. *Goskompriroda* was to be formed through the transfer of facilities and duties of the existing ministries and committees on All-Union or local levels. However, these ministries and state committees – powerful 'states within the state' – have hampered this transfer by attempting to maintain control over subordinate organizations which impact upon overall environmental protection.

For example, the Ministry of Water Resources, frequently censured for its destructive effect on Nature, preserved its control over water distribution and use. This created the paradoxical situation in which fresh water was regulated by the Ministry of Water Resources, but effluent discharge was controlled by *Goskompriroda*. Even though the Ministry of Water Resources has since been disbanded, the situation still remains unclear. The new Ministry of Water Resources Construction retains many of the same personnel. Will it simply carry on the same policy under a different name? One hopeful note is the Supreme Soviet's rejection of P. Polad-Zade as head of the new Ministry, presumably due to his former position in the Ministry of Water Resources.

Another example of ineffective bureaucratic 'restructuring' involves the sharing of functions between *Goskompriroda* and the Ministry of Fisheries. The latter has succeeded in retaining 90% of its control functions. Similarly, the State Forestry Committee has retained its control functions for forest management and conservation, but these same functions also reside in *Goskompriroda*. This division of functions seems to suggest that a forest exists outside of the rest of Nature, and should and can be regulated by the Forestry Committee, while 'Nature' somehow excludes forest ecosystems, and is rightfully the purview of *Goskompriroda*. It should be pointed out, however, that such inconsistencies are not the province of the Soviet Union alone; one only has to look at the current US environmental protection structure to see the same phenomenon!

The State Committee for Hydrometeorology has offered perhaps the strongest resistance. In addition to obstructing the transfer of equipment and funds, it provided *Goskompriroda* with its own staff. As a result it is difficult to determine where one committee stops and the other begins. *Goskompriroda*'s first deputy head previously had the responsibility for overseeing the censorship of Hydrometeorology's environmental publications. Another deputy is a former party administrator who has never been recognized as a nature lover. Yet another recently shocked the environmental community with obviously incorrect statements about ecological training in the country.[4] The former Chairman of the Irkutsk Executive Committee, under whose auspices environmental decline in the area greatly accelerated, has become the chairman of *Goskompriroda* for the Russian Republic.[5]

Thus far the party-dominated '*nomenklatura*' system of personnel appointment continues to negate much of *Goskompriroda*'s purpose. It is generally held that the new State Committee incorporated too many people who possess administrative and bureaucratic experience alone, and is insufficiently staffed by those with knowledge and experience in environmental protection and sound environmental management practices. This phenomenon of moving bureaucrats from one agency to another, with consequent mis-matches, is an unfortunate but understandable result of such major societal restructuring. Deviation from the official line was long considered a crime. Now these bureaucrats must adapt to a new official line but it is not clear how deep their commitment is to the goals of the new State Committee.

As a result of such manoeuvres, the very essence of the January 1988 governmental decision has been significantly changed and positive impacts limited.[6] Whether justifiably or not, committed environmentalists fear that the old-style fragmented, ineffectual bureaucratic machine ultimately will be replaced by a unified bureaucracy but one with little real knowledge or sympathy for environmental protection. The result, they worry, will be harder, not better times for the Soviet environment. The process of system reconstruction is a part of the reforms under way across the entirety of Soviet society. Few disagree with the premise that the revamping of administrative structure is essential. The concern, however, is that 'restructuring' may turn out to be little more than renaming. The recent appointment of Dr Vorontsov as head of *Goskompriroda* does, however, suggest the possibility of a new, more vigorous effort in environmental protection.

Another part of the State effort to improve the Nature protection system is the 'Long-term State Program of Environmental Protection and Rational Use of Natural Resources for the 13th Five-Year Plan and for the Period Until 2005'. This was prepared in accordance with a decision of the Communist Party of the Soviet Union's (CPSU) Central Committee and the USSR Council of Ministers of 1 August 1987. Unfortunately, this important project was developed using traditional methods of 'departmental formalism': the ministries and departments responsible for protection of particular natural resources each established 'control figures' outlining tasks for Nature protection for all other ministries. These ministries, on their own, are then expected to work out concrete environmental protection measures to fulfill these goals.

Although the goal is a comprehensive State programme, this departmental approach, however, works inherently against an integrated solution to environmental problems. Soviet environmentalists are concerned that the result will almost certainly be a severely fragmented and thus ineffectual collection of conservation efforts, carried out by agencies whose primary interest historically has been the enhancement of resource consumption. This 'Long-term State Programme', created out of the old fragmented system, has since been turned over to *Goskompriroda* for implementation.[7]

The programme itself sounds very promising: clean water and clean air, pesticide reduction, and so forth, all by the year 2000. Its chief goals are: the transition to rational natural resource-use, the exclusion of negative impacts of development on Nature, and the overall preservation of The Biosphere, including the human environment. To implement these goals, the programme mandates the formation of a unified system of environmental management, increased technological safety, natural-resource-conservation policies, independent environmental impact assessments, improved environmental science education, and increased public participation in environmental protection.

While the adoption of the State Programme is an important step for environmental protection, its success depends upon its implementation. Hopefully, this programme will not share the fate of previous governmental programmes which have existed only on paper. The 'long road from intention to implementation', a phenomenon plaguing past reported efforts in the USSR is a key concern among Soviet environmentalists. The environmental programme faces a particular obstacle, as a part of the task laid out in the programme itself involves the development of mechanisms for self-implementation.

The two-volume' complex and comprehensive 'Long-term Programme' provides detailed coverage of all projected measures. These are described for territories, sectors of the economy and individual natural resources. But two features are striking. First, the magnitude of the USSR's environmental problems is so great and the rate of environmental deterioration so swift, that there is real concern that the time-frame for implementation will be so gradual that environmental problems will outpace the proposed solutions. Second, the Programme has been kept largely secret; only authorized persons can see it!

This last item points us to the next aspect of environmental restructuring in the USSR: the transition from the old air of secrecy into the new policy of openness.

ECOLOGICAL '*GLASNOST*'

A new era in Soviet history, wherever it may ultimately lead, is enriching the world's political experience and lexicon. The word *glasnost* can be translated only roughly into English as 'openness'; its meaning is rooted in the very specific circumstances of the Soviet Union, with its long history of secrecy. 'Ecological openness', one component of glasnost, may well serve as a visible indicator of the successes and failures of current changes in Soviet environmentalism.

The ideology of the 'closed society, surrounded by enemies', reigned in all aspects of Soviet society for 70 years, producing generations of mute and frightened people. This orientation, manifested in policies of secrecy and suspicion, strongly influenced the system for gathering ecological data: those considered to be strategically valuable being hidden. Each agency that was responsible for some aspect of environmental protection recorded the facts of environmental degradation for which they were responsible, then hid them from the public and sometimes even from other agencies of the Government itself. The Ministries willingly promoted this suppression of information; secrecy provided opportunities to expand the bureaucratic channels of the Ministries, augmenting their control of information as they increased their perceived importance and their work-forces. When society is closed and voiceless, then bureaucratic agencies are accountable only to themselves, and the wisdom of plans and projects is not subject to scrutiny or debate. In fact such realities resulted in a mountain of problems, hidden under propaganda in support of the virtues of socialism and 'the system'. Now, just over the past three or four years, these problems have surfaced, a necessary but painful outcome of *perestroika*. The Soviet joke, 'In capitalism, a man exploits men; in socialism it's just the reverse', has become a reflection of a new public attitude towards the revelations and realities of Soviet life.

For a long time it was considered an eternal truth that socialism is free from serious contradictions and assures the resolution of environmental problems. The sieve of ecological censorship selected only those data which were congruent with the official policy. This information-gap was an essential part of economic and social stagnation. The transition to a new way of thinking is proceeding slowly and with difficulty. Ministries still quarrel about how much and what sort of information to release, referring to segments of the population who are not yet ready for the truth.

Ecological openness, however, is now considered an urgent necessity for society, in order to facilitate the taking of decisions consistent with sound environmental principles and with the furtherance of human health. The worsening ecological situation underlies the inability of society to predict the general consequences of partial actions on the environment. Without the steady flow of up-to-date information about the state of the environment, it is impossible to raise ecological awareness and consciousness among the general public and decision-makers, and to take the steps necessary for problem amelioration or resolution.[8]

The problem of the secrecy of information has been demonstrated most clearly in connection with the consequences of the Chernobyl disaster. The Soviet population did not officially learn of the disaster until four days after it became known to the international community. The powerful Ministry of Atomic Energy still retains its right to decide what information may be released. Its officials continue to claim that the operation of nuclear power-plants is safe and ecologically irreproachable. At a recent plenum of the USSR Union of Writers, one participant suggested moving this Ministry from Moscow into the new town of Slavutitch near Chernobyl 'for more concrete management of nuclear power-stations'.[9] Popular attitudes toward nuclear power are referred to as an 'atomic allergy' which is not a passing illness.

The main concerns surrounding the Chernobyl accident involve the impacts on human health. Following an optimistic report in *Pravda* about the effects of the catastrophe,[10] the editors still continued to receive angry letters with new questions. No convincing answers were provided. The editors of other central newspapers[11] requested a response from the author of the the *Pravda* article, the Prime Minister of Belorussia. They received a fascinating response from the Council of Ministers: 'The republican government constantly pays attention to the situation and is very interested in its elucidation in the mass media'. Several pages of recommendations followed. Then the unexpected conclusion: 'Information about all these (medical) questions has reached the population of the contaminated zone in a volume, *necessary* for assuring human health'.[11] Determining what subset of information is necessary 'for them' represents the old thinking that has dominated the Soviet bureaucracy.

Similarly, the documentary film 'The Threshold', which provided an unbiased look at the catastrophe, elicited a strong reaction from the Ministry of Middle Machinery: '... it is necessary to exclude from the film all data about the levels of contamination exceeding the maximum permissible, ... about the worsening of health of people working in Chernobyl .... The showing of the film is directed toward forming opinions not only against the development of nuclear power production ... but against the *state system of the country*'.[11]

This 'half-*glasnost*' approach has given birth to rumours, reinforcing anxiety and distrust. Information about all levels of radiation above permissible norms were labelled secret. In 1986 the medical service of the Ministry of Nuclear Energy hid the results of treatment of people harmed by the disaster in Chernobyl. At Chernobyl, even the head of the city's sanitary service did not have the right to measure the radiation dose near the station, although he has this obligation, according to law. The facts were hidden not only from the public, but from the Government. The Minister of Energy and Electrification of the USSR issued an order forbidding publication of any negative ecological effects of power stations on the environment and people.[12]

It is no longer possible to hide all the information that some want hidden. More or less precise maps of radioactive contamination were revealed in February 1989 and published in March 1989.[13] At a press-conference in Minsk on 2 February 1989, facts were released for the first time indicating contamination levels, stating that in a single Republic 520,000 people were exposed to radiation, and noting that nineteen institutes of the Academy of Sciences were investigating the consequences of the accident. Now new questions have emerged: where are the Soviet books about the accident? Why has the production of personal radiometers not been arranged?[14] Nonetheless, a first step has been taken. The reports in newspapers about the current situation near Chernobyl appear regularly, but it is still difficult to know how complete they are.

Regular coverage of the environment is now a feature of most newspapers and magazines as well as television and radio. 'Ecological bulletins' describing environmental pollution in cities are published in many local newspapers. Newspapers regularly request that officials give complete responses to public concerns about ecological issues. Public debates on television deal with major projects which may impact on the natural environment. If only officially promoted projects could previously be discussed in the press, now almost every article in favour of a project is accompanied by an article against it or by a discussion of other alternatives. Fewer issues are closed to public scrutiny. Areas only recently considered top state secret (such as nuclear tests and target-range ecology) have become a matter for public discussion.[15]

Boasting about the successes and advantages of socialism has been replaced by a frank and sober analysis of mistakes and shortcomings in the Soviet economy and society. This change has been well received by the Soviet people who believe that the extension of *glasnost* is helping to clarify the causes and solutions of environmental problems. An important effect has been the expansion of environmental awareness and citizen involvement. The new information about the dramatic and dangerous environmental situation in the country has awakened society and a vigorous 'green movement' has arisen. In turn, the people are demanding more and more information.

Social Movements

The tremendous increase in social activity in the country since M. Gorbachev came to power has diminished the role of official social environmental organizations at the same time as it has brought an explosive growth in the number and size of independent, informal groups and 'grassroots' environmental movements. Just born, these groups already have become a valuable and effective source of positive change and a powerful lever for expressing the public will and affecting the decision-making process. The number and variety of new groups is impressive, even in comparison with the American and West European experience. The numerous Western groups developed over a much longer period of time; the Soviet ones have appeared all at once, during virtually one *Earth Day*.

The new movements, however, are tied to developments in the mid-1960s when Soviet writers and Nature lovers drew attention to the threats facing Lake Baikal, due to the decision to construct a pulp- and paper-mill plant on its shores. The campaign for Baikal's protection has served as an important symbol for Soviet environmentalists – a symbol of the transition from a period of governmental ignorance and indifference to a period of greater governmental and citizen awareness, and of more effective pressure from the environmental community.

It was not by chance that this movement was initiated and headed by writers such as Valentin Rasputin and Sergei Zalygin. In the pre-Soviet period, writers also served to awaken society to new issues and challenges. Stalin annihilated social consciousness, exiling, imprisoning or killing writers and intellectuals. During the Brezhnev period, a new generation of writers also suffered much oppression. Nonetheless, a few were enough to stimulate public awareness of Lake Baikal's problems.

The powerful and very effective technique of affecting public opinion through literature is an almost unique feature of the Russian tradition. Sergei Zalygin, who heads a 'green' organization, recently wrote: 'At one time, Man's thought, especially scientific thought, divorced him from Nature by designating the air as meteorology, the earth as geology and soil science, and water as hydrology. But now our thought, which has been infinitely enriched with knowledge, should once again unite us with Nature and its processes, so that ecology becomes not just a science, but our urgent imperative. This is our task in literature, too'.[16] Zalygin's negative attitude towards Soviet science has a certain foundation. Over the past 40 years, academic research institutes have been in the 'pockets' of ministries; these ministries justified proposed projects as 'scientifically proven', based on work at 'their' 'pocket' institutes. In this way, science lost much independence and in many cases became the servant of particular economic and bureaucratic interests. A prominent example is the Institute of Water Problems, supervised by the Ministry of Water Resources. This institute became a sophisticated defender of the most destructive massive experiments with Nature proposed by the Ministry of Water Resources.

At the same time, a more authentic environmental science exists in the USSR. In the terms of pure investigation it is fairly advanced and has received international recognition. A major problem, however, lies with the 'long road from intention to implementation'. Even if Soviet scientists know *what* should be done, they have not been able to rest assured that the results of their researches will be put into practice. The ineffective economy is a primary reason for this failure.

The subordination of science to the administrative system has not been uncommon in Soviet history. A prominent example was the furtherance of '*Lysenkovshina*' in the field of genetics under Stalin. Scientists were, in some cases, executed or incarcerated in psychiatric hospitals for rejecting the State's position on environmental determinism. This legacy left deep wounds that are still felt today. Even now one can count all the major Soviet ecologists on one hand.[17] Therefore, writers have frequently filled the empty 'ecological niche', helping to initiate public environmental movements. And it is not coincidental that the first large environmental association in the country, 'Ecology and Peace', founded in 1987, is headed by a writer, Zalygin, and not by a scientist.

This group, also called 'Green Peace', was formed within the framework of the Soviet Peace Committee. It took on the responsibilities and the functions of both a traditional scientific ecological society – a society still unformed in the Soviet Union – and the moribund Nature protection societies of the union republics. Its major goal seems to be the joining of scientific activity with social action. Specifically, its scientific tasks include determination of methods of sustainable development at the global, regional, and local levels; gathering information on the state of the environment in the USSR and abroad; the identification of problem areas within the environment; and the promotion of solutions to these problems. With regard to the social dimension, 'Green Peace' supports the public examination of large-scale development projects; public actions to further Nature conservation; and peace propaganda to help create an 'environment for life'. This last notion lies hidden in the Russian name of the organization; the Russian word *mir* means both 'peace' and 'world'. The name of the association, 'Green Peace' (*Zelenyi mir*) is therefore especially significant.

'Green Peace' links concerned citizens, social groups and governmental organizations, on the basis of an informal structure and democratic principles. This contrasts with the formality of official 'social' organizations. 'Green Peace', although young, has already taken important steps to further its goals. In cooperation with the Academy of Sciences it conducted an ecological assessment of a large-scale water management project in the southern Ukraine.[17] 'Cruises for ecology and peace' have become fairly regular along the Volga River. The group has also sponsored public meetings, demonstrations and discussions. Future historians of social movements in the USSR might well, without exaggeration, call December 1988 the Soviet Union's 'ecological month'. During that month, organizational conferences of the Ecological Union of the USSR and of the Social-Ecological Union were held in Moscow. In Tartu, Estonia, the 'green' movement formed its association. The 'Ecological Fund of the USSR' was created as well. There were many other initiatives at the regional and local levels.

Despite their efforts to develop positive action programmes, the tone of ecological actions is often, of necessity, pessimistic. As V. Rasputin has noted, the optimists never attended to ecology. Based on over 25 years of study, one leading environmentalist, Professor N. Reimers, believes that 'even the most obscure comparisons are in fact much brighter than the real situation'.[17] Awareness of the negative side is necessary if the population is to grasp the scope and urgency of the problem. Dissemination of 'real' information, a kind of 'shock therapy', is considered to be the most important function of the new-born movements.

Numerous environmental groups emphasize the hazards of nuclear power. The only positive aspect of the Chernobyl accident is that, by coincidence, it occurred just as *glasnost* was developing, creating an opening for ecological activists. In the wake of the Chernobyl accident, their views found resonance from broad sectors of the general public. Widespread anti-nuclear sentiment, expressed in public protest actions, has put the future of the State Nuclear Programme under serious threat. After Chernobyl, plans to construct five new stations were abandoned; after the disastrous Armenian earthquake of 1988, the station near Yerevan was closed; another, under construction in the Crimean resort area, is now stated to be a training centre. Compromises by the government in promoting its nuclear programme reflect the newly-found influence of public activity on the decision-making process.

Also successful were demonstrations to stop the construction of biochemical enterprises near Kazan. Scholars at Kazan University buttressed the protests with their scientific analyses regarding the possible effects of the plant on the surrounding resort area. The Kazan movements suggest the potential strength of environmental coalitions which involve diverse forces and methods of pressure: actions by 'green groups,' individual citizens writing letters to political organs and the media, and mobilization of scientific expertise. Activists in the Kazan area are now continuing their efforts, this time in opposition to the construction of the 'Tatarskaia' atomic power-station located about 200 miles east of the city. Using Chernobyl-type (RBMK) reactors, this station is located near the junction of two rivers which flow into the Volga, the Kama and the Viatka. It thus seriously threatens the whole river basin. In this case, university scientists offered their support to conservation activists by providing an environmental impact assessment, which should have been done by state agencies before the confirmation of the project.

Cooperation with the international environmental movement is implicit in the Soviet notion of 'Green Peace'. At the outset, the Soviet Green Peace Organization only coincidentally had the same name as the international movement; now, however, the Soviet association is making contact with its counterparts all over the world. The attitude of the government towards such international links has recently changed from ignorance and opposition to official approval and support. A regional office of the 'Greenpeace' international was opened near Moscow early in 1989. To celebrate this, the rock music album 'Breakthrough' was produced and distributed first in the Soviet Union, in order to carry the ideas of the movement to young Soviets who might never before have heard about it. Joint meetings, marches and demonstrations of both Soviet and Western environmentalists are not yet common, but they are beginning to occur in the USSR too. 'Next step – Soviet', was the name of a meeting-dialogue, organized by Danish youth organizations which took place recently in Moscow. The purpose was to discuss issues affecting the survival of Humankind. More than 40 Soviet 'green' groups became acquainted with one another for the first time at this meeting, and for the first time they met foreign counterparts seeking an alliance.[18] The name given to the event is symbolic. The Western participants communicated that the next step is to be made by the Soviets.

In almost all republics and regions, 'green' groups have become schools of free expression. While they are not yet as organized or extensive as 'green' parties in Western Europe, they represent a first step in the search for an active and effective role by the USSR on the political stage.

Environmentalism in Policy

It is difficult to determine precisely the impact of the new environmental movement on political life in the USSR. The first major sign of its effect, however, appeared in the protests by Baltic activists against nuclear power-plants and hazardous factories built on their territory under the direction of the central government. As about 70% of all industry in the Baltic Republics belongs to the All-Union ministries and most natural resources are managed by them as well, the question of control is vitally important. Because the quality of the environment in Estonia, Latvia and Lithuania (as well as Armenia and other parts of the USSR) has deteriorated so greatly under Soviet rule, questions of control over natural resources and the environmental impact of economic development serve regional nationalistic movements better than most other issues. Environmental sentiments have penetrated the platforms of virtually all forces seeking political power in 'national' regions in the entire country. Significant sectors of the public, now more educated and concerned about environmental quality, take attitudes towards Nature preservation as a measure of the degree to which institutions serve the public interest.

Since the first real parliamentary elections in the spring of 1989, the composition of government itself is, for the first time, almost entirely dependent upon public opinion. In these elections, to win a seat a candidate usually had to be, or at least pretend to be, an environmentalist. Ecological items were an essential part of the election platforms of the majority of candidates for the Congress of Peoples' Deputies, regardless of the constituency. At times occupying up to one-third of the candidate's whole programme, these items varied according to their emphasis on the depth and extent of the problems. Some pointed to the global scale of environmental degradation, while others focused on local issues such as the need to shut down a local polluting factory. To some extent, the differing emphasis reflected the diverse educational backgrounds and occupations of the candidates.[19]

Many candidates who were not previously known for their love of Nature became environmentalists overnight! This phenomenon raises doubts about the prospects for realizing the environmental components in the candidates' programmes, particularly if environmental items were only third, fourth, or fifth in priority, after political, social, economic or other more immediate issues. As economic conditions have worsened over the past year, and as political disorder has increased in the Republics, the environment has become a lesser priority. Meetings of the Congress of Peoples' Deputies (broadcast on television) show the inability of the newly-formed state body to address the most important problems of the country without confusion and disorder. Effective public policy has not emerged from this process thus far. As the Soviet people experience democracy for the first time, they are already frustrated by its inability to improve living and working conditions quickly. As a first attempt at real parliamentary government, the Congress could not possibly achieve viable solutions to difficult problems in its initial sessions. Nonetheless, the potential exists that a reformed parliamentary group for Nature protection may help to promote the healing of a sick environmental management system. People need time to get used to the new structures and process; unfortunately, Nature cannot wait.

The 'greening' of the top Soviet authority is occurring too. In his last speech to the United Nations' General Assembly, in December 1988, President Gorbachev declared that 'international economic security is inconceivable unless related not only to disarmament but also to the elimination of the threat to the world's environment.'[20] Gorbachev pushed the idea of an international ecological service to fight frightening environmental disasters in the world. The USSR stopped commercial whaling in 1988, ratified the Montreal protocol on protecting the ozone layer, announced its intent to reduce hydrocarbon production by 1993, signed a nitrogen oxides treaty in November 1988, reaffirmed its commitment to the USA-USSR Bilateral Agreement on Environmental Protection signed by Brezhnev and Nixon in 1972, and initiated other bilateral agreements with foreign countries. The Soviet Union is also a member of the '30 per cent Club', which consists of countries committed to the reduction of sulphur dioxide emissions by 30% by 1993 (compared with 1980 levels).

The 'new thinking' underlying Gorbachev's reforms is reflected in the government's environmental policy too, on both the international and the domestic fronts. In Foreign Minister Shevardnadze's articulation, 'The Biosphere recognizes no division into blocks, alliances or systems.'[20] He expresses the USSR's policy of penetrating ideological barriers to realize broader cooperation with the Western world in the field of environmental protection. In the domestic sphere, Gorbachev's government has turned its face to ecological problems largely for economic reasons. For example, long-standing controversial plans to divert Siberian rivers to the south were cancelled after Gorbachev came to the Kremlin. Though touted as a victory of the progressive environmental social movement, the plan, which would have cost around a hundred thousand million roubles, was abandoned almost entirely for economic reasons. His priority was the recovery of a paralysed economy. But the proposed restructuring of the economy is largely consistent with the needs of the environment.

Economic Mechanisms for Environmental Protection

'How can pollution be made unprofitable?' This is the key question for Soviet economists who are trying to develop economic mechanisms to further environmental protection. The intention, also affirmed in policy documents, is to replace methods of environmental management by more self-regulating economic incentives. The old administrative approach was ineffective because enterprise managers, driven by planned production quotas, often judged that adherence to environmental requirements would make plan fulfilment more difficult. Indeed, to enforce environmental legislation, it would have been necessary to close numerous industrial and agricultural enterprises. This, of course, would be unrealistic in the country already plagued by major deficiencies in most products.

In a manner more consistent with Marxist assumptions, policy makers and economists are attempting to realize the notion that economic forces should form the foundation for political and ideological institutions. For the first time since V. I. Lenin's New Economic Policy, economic laws are considered to be fundamental.

A necessary feature of economic reform in the USSR involves a shift in emphasis away from extensive extraction of raw materials and the production of machines, dams, factories and other goods which damage the environment and yet do little to improve the availability of consumer goods, towards a more efficient utilization of limited raw materials and the production of consumer goods.

A second important feature of the reform is the creation of economic incentives to protect the environment. Here price-reform is very important. For example, existent prices for energy and raw materials are generally less than half of world market prices.[21] These unrealistic price-levels inhibit conservation efforts and lead to miscalculations of opportunity costs. Price reform is not only essential for *perestroika*, but it should also further conservation efforts. However, its introduction has been delayed, for fear of sparking popular anger by increased prices of consumer goods.

Until now, use of most natural resources has been granted free of charge. This encouraged wasteful practices and hindered introduction of low-waste technologies. In recent years, industrial enterprises have had to pay for water use; water consumption beyond the planned norm brought a five-fold increase in the price. However, these charges have generally been too low to be effective. Within the next year or two, an improved pricing system is supposed to be extended to other kinds of resources, including land.

Changes in the land tenure system are also seen to be a key step in recovering Soviet agriculture from decades of forced collectivization and dictatorship by centralized planning power. Collective farms have not been able to determine their own production plans and the State was an assured buyer of low-quality output. Under this system, incentives to produce quality goods and to take care of the land were destroyed. Recognizing this problem, the leadership now endorses land-leasing, which may in fact facilitate a revival of family farming, a *de facto* form of private agriculture. If a farmer is allowed to lease a parcel of land for a long period of time, he may develop an incentive to care for the land and safeguard its productive quality. This transformation in property forms in the agricultural sector is being hampered by the bureaucracy, but nonetheless is proceeding step by step, especially in the Baltic republics.

The land tenure system has other problems. The price for land, expropriated from agriculture for other uses, was set at four to five thousand roubles per hectare, regardless of the quality of the land. Thus, even where this policy was actually enforced, industrial enterprises could occupy the best lands for the same cost as for the worst. This price should be calculated, taking into account the cost of products never received, and expenses for developing new lands to replace the lost ones. Under such a system, the ultimate cost could reach 84 thousand roubles per hectare.[22] A second problem is that monies paid by the enterprises appropriating the land have gone into the State budget, not to the the previous agricultural user.

Even if enterprises are required to pay for resource use, and for waste dumping, these charges will be effective in stimulating resource conservation only if enterprises have independent budgets. This has not been the case in the past; therefore, costs were ultimately passed on to the State. For example, fines for pollution simply took money out of one governmental pocket and put it into another. Furthermore, these sums were automatically included in the operational expenses of the violating enterprise, causing a growth in production costs. The result was further subsidization of inefficient and wasteful enterprises. True enterprise, both self-financing and self-accounting, and major components of Gorbachev's economic reform package, would address this problem.

Over two years have passed since the Party and government adopted the resolution to restructure Nature protection in the country. One task was to work out norms and sums of payments for resources and standardized pollution payments, as well as fines for exceeding those norms. This work was to be done in six months, but nothing has been implemented yet. This delay depends in large part on the failure of the reform, overall, to realize its goal of creating economic units which are truly self-sustaining and self-accounting. Nearly three years ago, provisions in the 'Law on the State Enterprise' were intended to help create an economic mechanism of Nature protection; also they have been blocked by the same factors. Privatization of the economy, termed 'leasing' by Gorbachev, and the setting up of a market-oriented system of management are necessary if the proposed regulations are to be enforced. Furthermore, these measures may help to rehabilitate a stagnating economy and to create work-incentives. Here, goals of effective economic reform overlap with effective environmental policy.

Another dimension of economic and political reform, the formation of cooperatives, may also facilitate new forms of conservation activity. Ecological cooperatives, a new kind of private enterprise which, in fact, replaces clumsy and ineffective state enterprise, may produce technology and devices needed for environmental protection. Their appearance was somewhat unexpected, as they appeared among the thousands of cooperatives formed after passage of the 1987 Law on Cooperatives. The cooperatives allow speed, flexibility and independence in their response to demand; most valuable, the cooperatives are connected with other enterprises and customers by the rules of free market, which means that sales depend on real societal needs.

Ecological cooperatives have filled a 'niche' between official state factories and organizations on the one hand, and individuals attempting to help the environment on the other. Sometimes they can take on tasks never fulfilled by large enterprises due to the small volume of work, or where service workshops do not have adequate capacity. The 'Ecology' cooperative in the city of Cherepovets repairs treatment facilities in a metallurgical combine.[23] The 'Ecological Fund', a scientific cooperative organized at Kazan University, has a different focus. It provides analyses of contamination by different substances, environmental impact assessments of projects, and helps industrial enterprises to create new technologies and ecological devices. The money raised helps to finance scientific research, provides for scholarships, the purchase of equipment, etc. Workers are attracted to cooperatives not only by the fairly high salaries but also by the chance to work productively in an environment unhampered by bureaucratic obstacles.

Another new economic form, joint ventures, involving cooperation between Soviet and foreign enterprises, serves to join foreign technology with Soviet labour and ideas. Just recently the first such venture appeared in the field of environmental protection. Called 'Prima,' this Soviet-Italian enterprise is dedicated to developing measures to improve environmental conditions in cities.[24] This form of international cooperation is considered to be especially necessary for the country, in view of the USSR's lack of advanced technology and equipment, and the decline of a work ethic.

Soviet environmental problems have received considerable international recognition. More importantly, in the light of realities of global interdependence, the international community has begun to commit its energy and expertise to work with the Soviet Union to resolve its environmental problems which impact upon The Biosphere as a whole.

Cooperation proceeds slowly and with difficulty because of the tremendous problems associated with creating interactive possibilities between Western and Soviet economic approaches. But current changes in Soviet economic and social life suggest that both sides are committed to addressing the global environmental crisis.

References*

1. CPSU Central Committee and the Council of Ministers of the USSR: 'About Restructuring Nature Protection in the Country', *Pravda,* 17 January 1988.
2. V. Rasputin, 'Our Fate is in the Fate of Nature', *Sovetskaia Rossiia,* 20 June 1988.
3. V. Sokolov, 'Ecology and Glasnost', *Pravda,* 28 March 1989.
4. E. Minaev, 'Think about our welfare and about the health of our children', *Semiia,* No. 1, 1989. In this interview, Minaev also said that only one institute in the USSR in oil chemistry is training ecologists; meanwhile more than 60 departments conducting environmental education exist in the country.

*[This contribution was received too late to change the method of citation or fully complete references, for which we apologize. Eds.]

5. G. Dolgov, 'Ecology and Bureaucracy', *Literaturnaia Rossiia*, no. 4, 27 June 1988.
6. B. Safarov, 'Control without Authority', *Izvestiia*, 19 June 1989.
7. T. Korsakova, 'Nature for Tomorrow', *Komsomol'skaia Pravda*, 11 March 1989.
8. V. L. Sokolov, *Komsomol'skaia Pravda*, 11 March 1989.
9. G. L. Dolgov, *Komsomol'skaia Pravda*, 11 March 1989.
10. A. Simurov, 'Must Never Be Forgotten', *Pravda*, 11 February 1989.
11. A. Illesh, 'The letter is timely, but do not publish it', *Izvestiia*, 2 April 1989.
12. P. Polozhivets, 'We are the Fledglings from the Same Nest', *Komsomol'skaia Pravda*, 28 March 1989.
13. Iu. Izrael, 'Chernobyl: The Past and Forecast for the Future'. *Pravda*, 20 March 1989.
14. O. Egorova, 'Open Up your Maps!' *Komsomol'skaia Pravda*, 20 February 1989.
15. 'Ecology of the Test Site', *Komsomol'skaia Pravda*, 2 April 1989.
16. 'On the Plenum of the Union of Writers of the USSR', *Literaturnaya Gazeta*, 2 July 1986.
17. Iu. Kantsielson, 'This Ecological Month', *Vechernaia Kazan*, 7 February 1989.
18. T. Tumanova, 'Nature Doesn't Recognize Boundaries', *Komsomolets Tatarii*, 22, 1988.
19. 'The Candidates for the Public Deputies of the USSR: Electoral Platforms', *Vechernaia Kazan*, 11 March 1989; and *Vechernaia Kazan*, 29 April 1989.
20. Quoted by H. French, 'The Greening of the Soviet Union', *World Watch*, May–June 1989, pp. 21–8.
21. 'This is a Talk about Human Survival', *Sotsialisticheskaia Industriia*, 30 December 1988.
22. Iu. Khanukov, 'How much is a Hectare Now?', *Pravda*, 9 March 1989.
23. V. Filippov, 'The "Ecology" Cooperative', *Izvestiia*, 20 April 1989.
24. V. Zaikin, 'Will "Prima" help Moscow?', *Izvestiia*, 17 January 1988.

19. Law, Morals, and The Biosphere

CHRISTOPHER D. STONE

The Law Center, University of Southern California, University Park, Los Angeles, California 90089-0071, USA

INTRODUCTION

Why do planet-wide environmental problems arise, and why do they persist? Some commentators stress the surge of human population, with its consequent pressures on fuel and other resources. Others point to the two-edged 'advances' of the technological revolution, for whose benefits we pay a price in toxic and non-biodegradable wastes. A non-materialist critique pins our predicament on defects of the human spirit, such as greed, short-sightedness, or a pathological urge to dominate Nature. And there are yet others who deny that resources world-wide are inadequate to provide for anticipated increases in population; for them, our troubles stem from inept and fractious local governments that have been slow to adopt available technologies and techniques. The truth is almost certainly some composite of all these claims and yet others. But granted the manifold origins of our earthly plight, the most important focus should be on the cure. Given the growing world-wide concern that we are veering towards an environmental crisis, why do our social institutions have such a hard time in trying to correct our course?

With this dimension of the problem in mind, I have been asked to contribute some thoughts from the perspective of the law, with special attention to the question of how law and morals relate to the coordination of world-wide efforts to control Biospheral degradation. The aim is to clarify, for a predominantly non-lawyer audience, the problems that confront those efforts, and to point the way to some solutions.

PART I: ENVIRONMENTAL DEGRADATION FROM A LEGAL PERSPECTIVE

The starting-point is a simple reminder that the scientist and lawyer examine the planet from different vistas. The scientist – a geophysicist or geochemist, for example – enjoys the grand panorama of the astronaut, a view dominated by the marvellous wholeness of the Earth and the interconnectedness of the globe-spanning phenomena that sustain its tenants: the envelope of atmospheric gases, the great body of ocean, and the broad belts of weather and vegetation.

Lawyers and diplomats look out over the globe from a more cramped and mundane vista. Ours is an inherited world in which all that primordial unity and harmony have been disrupted into political territories – pencilled, as it were, through a history of human caprice and adventure, into two sorts of regions. First, there are the territories of nation-states; second, there are the 'global' commons – those areas that lie above and beyond the legitimate reach of any nation-state, and include the atmosphere, the high seas with their sea-beds, and outer space.

Given the rules and customs of international law, the legal and diplomatic response to Biospheral hazard depends to a considerable extent upon the location of the injurious actions, and of the threatened injury, with reference to this two-fold division into national territories and commons areas. Specifically, there are five possible combinations of (1) where the harm-threatening activity occurs, and (2) where the harm is concentrated. These five possibilities can be easily illustrated by a simple table:

Site of Act	*Site of Harm*	*Example*
(1) Nation *A*	Nation *A*	*A* annihilates an endemic species.
(2) Nation *A*	Nation *B*	*A's* radioactive debris blows across the boundary into *B*.
(3) Nation *A*	Commons	*A*'s sewage invades the high seas.
(4) Commons	Commons	*A* over-fishes the high seas.
(5) Commons	Nation *B*	*A's* tankers discharge wastes on the high seas, which invade *B's* beaches.

A matrix based on these five variations of locale is not the only *entré* into a legal-institutional analysis. Conflicts can be grouped by reference to other variables, such as the number of participants whose cooperation is required to bring the problem under control, and the generally-perceived immediacy and intensity of the threat. But a quick overview of these five locale-based situations provides a useful introduction to the complications of international environmental law and diplomacy, and enables us to identify, at least in contour, some appropriate policy responses.

Case (1) 'Internal Affairs': Nation A *Alone*

The paradigm for Case (1) is the nation that extinguishes (or allows the extinction of) its 'own' living resources – its elephants or tigers or trees. Outsiders object, but, legally, as long as there is no transboundary trespass, the outside world has no recognized basis for relief in customary international law. Obviously, there are some borderline cases which test where the abuse of one's 'own' resources ends and extraterritorial effects begin. If State *A*'s massive deforestation increases the risk of flooding in a neighbouring nation, or has measurable atmospheric implications, *A*'s conduct shades over by degree into cases of types (2) and (3), and gives outsiders a stronger platform for objection than they would otherwise have. But for the purposes of this discussion, we assume that there are no, or only *de minimis,* adverse physical effects beyond *A*'s borders – only adverse public reaction.

What can be done to affect these internal policies and practices? Many local factors, such as population control, cultural attitudes, and styles of government, certainly have in the aggregate considerable implications for *A*'s influence on The Biosphere, but, as a practical matter, lie outside the reach of direct international influence.

Indeed, Stockholm Principle 21, in declaring that 'States have ... the sovereign right to exploit their own resources pursuant to their own environmental policies,' records the resistance to correcting the case (1) situations with intrusive legal principles (UN Doc. a/Cont. 48/14/Rev. 1, UN Pub. E. 73, II A, 14 [1973]). On the other hand, the Stockholm Conference took place nearly twenty years ago, and the changes of attitude in that time-span have been considerable. It would thus be unwise to dismiss too readily the prospect that, in the long term, international legal principles may evolve as a check against a nation's despoliation of its 'purely internal' environment. We have as an inspiring model the developing body of International Human Rights law. Certain aspects of the environment, such as rare species, may come to be considered as part of the common heritage of Mankind, wherever found, and their destruction, denominated 'ecocidal' acts, could be likened to genocide in the international community.

Short of such a momentous shift in the law, there are, broadly-speaking, the following five present options by which the outside world might affect these 'internal' situations.

(i) Assistance in Planning

One approach is to help nations to recognize and implement conservation measures that are in their own best interests. For example, consultation with outside experts, often under the aegis of United Nations agencies, has proved useful in assisting local governments to appreciate the economic benefits of wildlife preservation and environmentally-benign agricultural techniques. In such circumstances, if *A* institutes the changes, the apparent conflict between Nation *A* and the outside world vanishes.

But in many cases, a nation that has agreed to undertake an enlightening evaluation of its policies, perhaps even an environmental impact review, will emerge unpersuaded that its own interests and those of the outside world converge. In these circumstances, what can be done?

(ii) Non-degradation Transfer Payments

One alternative is for the 'outsiders' to pay the country to adopt the desired course of action, either through direct funding or by debt relief. The idea may have sounded rhetorical in 1961, when Kenya's Jomo Kenyatta is reported to have suggested that, if African wildlife was indeed a world possession, 'the world could pay for it' (Hassett, 1972 p. 38). But the concept is being taken quite seriously today, with authorities such as Joan

Martin Brown of UNEP even floating the concept of 'renting' trees (Stammer, 1989).

The point, of course, is that if Nation *A* has a biologically rich wetland which it wants to develop into a port or seaside resort, and which outsiders want kept 'as is,' then the best solution – indeed, the only solution – may be to arrange for the outsiders to compensate *A* to take the steps which are, after all, for the benefit of the outsiders, though not of *A*.

Domestic governments make comparable purchases as a matter of course. If the United States Government should require my homesite for a public park, it can condemn the property – provided it pays me fair value. Of course, the difference between the domestic and international situations is that no international body has the power to force an involuntary condemnation. All compensation agreements would have to be acceptable to the nation whose cooperation is sought. Therefore, unlike the domestic homeowner, the nation may either refuse altogether, or may try to hold out for a price that reflects what the outside world is willing to pay – rather than its true opportunity costs, *i.e.* the value to the nation of developing the lands as compared with the value of the land as left undeveloped.

My guess is that this risk of hold-out has to be regarded as more theoretical than real, considering the large number of potential 'sellers' of wetlands or woodlands in the context of the limited funds that are currently available for such purchases. There is not enough demand to enable the suppliers to get away with 'strategic' behaviour. Thus, at present, the far-more-valid concern is providing an adequate international environmental fund to finance the purchases – a project that I will discuss in Part III (p. 486 *et seq.*).

(iii) Technological Assistance

Many countries will regard non-degradation payments as objectionable – as sounding too much like a bribe to remain less developed, particularly if it leaves the nation's citizens with no more prospect, and no more self-esteem, than that of caretakers. As is well known, assistance is required not only to protect The Biosphere but also to lift less-developed countries to a level of sustainable development. One compromise measure that advances both goals is for the industrialized world to undertake some of its wealth-sharing in the form of free provision of technology specifically aimed at reducing environmental hazards. For example, in providing aid to a less-developed country, assisting nations might give systematic preferences to projects such as construction of waste-treatment plants.

(iv) Conditions on Aid and Trade

In most instances, the nation whose cooperation is sought will have valued relations with the outside world that are independent of Biosphere maintenance or repair. This interdependence provides the opportunity for creating environmental linkages to the transactions that are independently motivated. Specifically, environmentally protective conditions can be attached to the supply of anything from capital (in the form of loan conditions) to deliveries of material and technology – for example, in the form of conditioning the building of a plant on the installation of appropriate effluent safeguards.

(v) Trading Sanctions

Finally, nations that object to another nation's 'internal' policies can resort to a medley of more confrontational rejoinders, such as refusals to trade. Some of these, such as those under the Convention on International Trade in Endangered Species (CITES), involve refusals to trade in the specific plant or animal of concern. But, theoretically, the reprisal could extend to a more general trade sanction against the offending nation – not to buy or sell or otherwise trade in any product with a nation that was an environmental outlaw. In fact, no nation has yet gone so far as that. Under United States law, if the Secretary of Commerce certifies that nationals of a foreign country are diminishing the effectiveness of an international fishery conservation programme, the Government may ban the importation of that nation's *fish products*. But it is significant that US officials have been reluctant to invoke this restricted trade power even against nations that appear to have flaunted their whaling obligations on the high seas. It therefore seems unreasonable to expect more sweeping product-sanctions in other circumstances, such as against nations that are environmentally destructive on their own turf.

Case (2): Transboundary Pollution: Nation A *Affecting Nation* B

The second class of cases is epitomized by Nation *A* being engaged in, or planning, an activity that will be carried out within its boundaries, but that has some probability of projecting pollutants or debris across the border into a neighbouring Nation *B*. In these cases, *A*, the transgressor, can less convincingly dismiss the outside world's criticisms as meddling in 'strictly internal affairs'. While external effects are always a matter of degree (the law usually ignores activities that do not amount to physical trespassory invasions), it is obvious that the greater may be the external effect, or its likelihood, the stronger will be the sentiment that *A* ought to take *B*'s interests into account, and the stronger *B*'s diplomatic and legal position.

In the face of such a threat by *A*, the options available to *B*, the 'victim' state, include the following:

(i) A Domestic Accounting by A *of the Impact of its Activities on* B*'s Interests*

A, in the course of planning a project, could be expected to assess, and, while under no legal compulsion, account in a 'good-neighbourly way', for its effects on *B*. At present, there is no recognized obligation to do so under international law. United States law, however, adopts this technique – to a degree. Thus both the National Environmental Policy Act (NEPA) and the

Endangered Species Act (ESA) provide that actions by US agencies which will have an effect on another nation's environment have to undergo an Environmental Impact Assessment – just as though they constituted major projects in the US's own interior.

(ii) Notification

In some circumstances, *A* may be encouraged or obliged to notify *B* of its plans in advance – plans, for example, to change the flow of a watercourse. A separate body of developing principles is aimed at notification of imminent perils. Prompted by the Chernobyl (USSR) experience, the nations party to the International Atomic Energy Agency (IAEA) have mutually agreed to notify the outside world immediately of any radioactive leak that is likely to affect other nations. Still, it is unclear whether there is currently any general obligation to notify of an imminent peril – for example, should a toxic cloud of the sort released at Bhopal (India) venture on a transboundary course. At the 1972 UN Conference on the Human Environment (Stockholm), a Draft Principle 20 was under consideration which would have encouraged notification in just such circumstances, but which suffered an unfortunate dilution before final adoption (Goldie, 1975). It is certainly time now to press for formal adoption of the language that was rejected in 1972.

(iii) Consultation

In some circumstances, norms of good international behaviour may go beyond merely informing one's neighbour; a nation may be expected to sit down and discuss with its neighbours any project that may pose considerable risks of transboundary harm. To call such a principle, as has been done, 'an evolving rule' of general international law is, however, almost certainly more hopeful than accurate. But consultation requirements can be formally required and indicated as part of treaty structures, as they have been in some shared river-basin agreements.

(iv) Veto Power in B

The strongest form of accounting would be to require the nation that is proposing hazardous action to secure the prior approval of the hazarded states. Nothing of the sort is claimed to exist under general principles. A few treaties have, however, adopted what are effectively veto requirements; but such provisions, abrading the free-hand prerogatives of sovereignty, are understandably rare (Springer, 1983 pp. 150–2). On the other hand, perhaps in the modified form of providing for majority or supermajority approval prior to permitting major Biosphere-threatening actions, vetoing may emerge as an important device in some future environmental conventions.

(v) Litigation

The most dramatic alternative, if diplomacy fails, is litigation. International environmental lawyers are disposed to point to a scattering of cases from which one can imply the existence of some 'customary' international liability for transboundary pollution – that is, liability even in the absence of any special treaty among the states involved. Invariably there is reference to the *Trail Smelter* arbitration, in which the United States won relief against Canada for damage from transboundary, wind-borne lead smelter fumes in the 1920s and '30s (Trail Smelter Arbitration, 1938, 1941). But as a barrier against Biosphere degradation, litigation based on customary law principles is of limited practical value. There are several reasons to be doubtful about the value of litigation in such circumstances, namely:

First, an environmental lawsuit requires that the plaintiff show some physical, trespassory invasion of its territory. As long as such a requirement stands, *A*'s slash-and-burn policy, for example, however badly it *affected the uptake of carbon dioxide* from rain-clouds to *B*, would not amount to an actionable legal wrong.

Second, even if the trespassory requirements are met, the power of the aggrieved state to find a forum in which to try the wrongdoer is filled with problems. The polluter's own courts are not likely to be favourably disposed, and there is no ultimately effective compulsory jurisdiction before the International Court of Justice.

Third, even if a polluting nation shows up in court, it is not clear by what standards of conduct its actions are to be judged. This exemplifies the yet-unresolved question of *state responsibility*. Is *A* to be charged for all pollution that wafts into *B*, however innocently and diligently *A* has behaved? Or is *A* to be held only for damages that result from its failure to live up to some international standard of reasonable care?

Fourth, the questions of proof are often too daunting to overcome. In the *Trail Smelter* arbitration, the US had caught Canada with a smoking smokestack. But more typically, a plaintiff nation will be put to a controversial proof: was *B*'s territory invaded by some agent, call it chemical *z*, and if so, *which* nation's *z* is to blame? Is *z* by itself benign, and mischievous only in combination with another agent *y*? Even should the courts adopt a standard of strict liability (liability without proof of the acting state's negligence), the plaintiff would still have to prove the causal connection between the activity complained of and the injury received.

Fifth, should the plaintiff successfully prove violation of some standard, there is almost no precedent for an international court or arbitration *to enjoin* the wrongdoing nation (Gray, 1987 pp. 69–74) – such prohibition by instruction (*i.e.* injunction) being the most effective remedy that nations deploy against polluters in their own courts. Ordinarily, the injured nation would have to look forward to an award of damages, at best. Damages, to begin with, are merely *ex post facto* remedies; their aim is to compensate after the harm has been done, the mere prospect of paying future damages being calculated to daunt polluters into changing their present practices. But polluters know that international fora are not as inclined as the US courts to impose on a polluter the full costs of the harm it is causing. At one point the

Trail Smelter arbitrators said that, even if the US claim for compensation for damages that had been done to urban businesses and property were theoretically provable, they were 'too remote and indirect to become the basis ... for an award.' The implication is that, even if a polluting nation loses a suit, the damage award against it may be too paltry to force it to bear the full costs of the damages it is causing in the neighbouring country (in economic parlance, to internalize the social costs ascribable to its conduct).

Sixth, litigation is notoriously slow. The environment may suffer serious and even irreversible damage long before the courts can catch the culprits. Indeed, even should the courts finally hand down a damage judgment, there is no assurance that it will be enforced. Recently, after 14 years of litigation, a Netherlands court held in favour of Dutch plaintiffs against French interests for salt pollution of the Rhine (Bureau of National Affairs, 1988*a*). When the victorious plaintiffs tried to collect on the judgment in France, the French court refused to recognize the judgment on the grounds that the plaintiffs had not made out a causal connection – an issue which the defendants had presumably raised, and lost, before the courts of The Netherlands (Bureau of National Affairs, 1988*b*).

Settlement is always possible. But there is no guarantee that emissions which are mutually agreeable to two disputing nations, *A* and *B* (who, after all, have many independent items on their bargaining agenda, ranging from trade to military assistance), will also be the ideal level of pollution from the perspective of the rest of the global community.

None of this should be allowed to discourage the continuous fostering, through litigation, of the body of customary international environmental law that is aimed at transboundary pollution. It is not unrealistic to hope that the law will gradually achieve compensation for victims of abrupt international incidents that rise from ultrahazardous activities – such as sudden, clearly wrongful chemical spills that can be traced unambiguously to a definite point of origin. But unfortunately, most Biosphere degradation occurs too insidiously – too gradually, too 'innocently', and too ubiquitously – to be staunched by the principles that international customary law is likely to make available to plaintiffs in the at-all-near future.

Indeed, even should specially-tailored bilateral and multilateral treaties implement major improvements in transboundary pollution law ('improvements' from the plaintiffs' point of view), the rights of neighbouring states to sue would not ensure a dependable line of defence for The Biosphere's major maladies. This is true: first, because whatever their legal rights, nations remain generally disinclined to jeopardize relations by litigating environmental conflicts; second, the most serious global problems commonly originate from activities that affect the commons area without causing immediate, legally demonstrable injury to any sovereign nation; and finally, under present law, there is no clear plaintiff to initiate a lawsuit.

Case (3): Nation A*'s 'Internal' Activities Affecting the Commons*

Much of the most serious damage to The Biosphere stems from activities which take place within sovereign territory but the influence of which extends upwards and outward into global commons areas – particularly the atmosphere and oceans. While the high seas suffer from waste that has been accidentally or deliberately discharged directly into them – for example, by tankers (*cf.* Case [4], below), a far greater load of ocean wastes originates within national boundaries, and reaches the sea through river runoff or *via* deposition of airborne waste.

From a legal perspective, the most important characteristic of these cases is that the degraded area lies outside any conventionally recognized jurisdiction; it is, therefore, doubtful whether any nation is empowered to protest about any of the damage formally. Indeed, under classic principles of international law – which incline to conceive 'damage' in terms of affronts to sovereignty – environmental losses of common heritage, even where they have economic value, are not actionable wrongs. The climate and deep-sea dolphins have no police to monitor their abusers, and no standing, as nations have, before the World Court.

To deal with these situations, I have elsewhere proposed a system in which some public or non-governmental organization is designated to act as guardian for each sensitive portion of the commons areas (Stone, 1987*a* p. 119).

Consider, for example, the fact that coastal nations annually dump into their adjacent waters millions of tons of sewage sludge, industrial wastes, and dredged materials (World Resources Institute, 1988 p. 330). What mechanisms are there to make these nations have second thoughts? Of course, to the extent that the effects are restricted to the dumping nation's own waters, the damage is internalized, and the dumping state has some consequent motivation to minimize ecological hazard. Still, the dynamics of ocean currents suggests that some of the effects will extend beyond the dumping state's EEZ (Exclusive Economic Zone). Perhaps the effects on the deep ocean (where life is sparsest) will be minimal, but who is there to monitor these global effects and say for certain?

Under a guardianship regime, an Ocean Guardian (perhaps the Joint Group of Experts on Scientific Aspects of Marine Polluting (GESAMP), with legal staffing) would be authorized: (i) to monitor ocean conditions; (ii) to appear before the legislatures and administrative agencies of states considering ocean-impacting actions, to counsel moderation on behalf of their 'client'; (iii) to appear as a special intervener-counsel for the unrepresented 'victim' in a variety of bilateral and multilateral disputes that may arise over ocean dumping, and perhaps most important; (iv) international treaties might endow the Guardian with standing to initiate legal and diplomatic action on the commons' behalf in appropriate situations – to sue to enjoin commons-damaging activities at least in those cases where, if the damage were to a sovereign state, the law would afford that state some prospect of relief. The Guardian would speak for the otherwise voiceless portion of Nature, in the way in which guardians are commonly designated in law to speak for persons who cannot represent themselves – such as infants and the insane (Stone, 1972).

A system of commons' Guardians would be a step forward, but no panacea, for Biosphere degradation. Placed under guardianships, the commons 'areas' would be elevated to a legal and diplomatic standing on a par with a sovereign. But we have already seen that that status, under present law, is of only limited advantage. Hence, protection of commons areas depends not only upon guardianships, but also upon our bringing about significant changes in the existing substantive law which would be available for the Guardian to invoke. The oceans not only need their own independent voice; they need more protective law than currently exists.

Case (4): Nation A*'s Commons Activities Affecting the Commons*

This fourth group of cases includes activities which range from weapons-testing in the atmosphere to tanker spills and over-fishing that take place on the high seas – to the extent that the injuries of interest occur also to the atmosphere in general (before its effects on any particular state warrant that nation to receive legal relief), or to the living resources beyond EEZS. Viewed legally, the problems are in many ways similar to those examined above under case (3). In these situations, too, the establishment of guardianships would be an improvement. Abusive use of drift-nets should be controlled by convention, and part of those conventional arrangements could profitably include provision for guardians.

There is, however, a significant distinction between this group of situations and the others – one that makes this group perhaps the easiest to regulate. Nations feel that they can submit to restrictions on their activities in space or on the high seas without conceding the legitimacy of checks on internal sovereignty. This is not to say that maritime nations are indifferent to the monitoring of civilian vessels on the high seas; but it is less threatening as a precedent than to accept international monitoring of an on-shore facility.

There are other reasons why it may be relatively easy to secure control over these case (4) problems. As regards some of the more remote commons areas, such as the Moon and planets, no nation has yet developed a vested reliance on the activities that might have to be regulated. Ground-rules on good behaviour in space, laid down far in advance, will not threaten any economies and work-forces, as might requirements for retrofitting chemical plants and utilities to bring them up to satisfactory levels of emissions performance. Moreover, so few nations have the technical competence to affect outer space, that the number of parties required to make an agreement effective is, practically speaking, small. On both counts, one would expect it to be much easier to effectuate conventions on areas that are geographically and technologically remote than on those closer by.

History confirms that theoretical expectation. Pollution of the ocean from ships on the high seas was an early subject of international accord; control of ocean pollution from land-based sources remains difficult to negotiate (Soni, 1985).

Much more needs be done in this class of cases, perhaps with most immediate attention to the effects on the high seas of drift-netting and waste disposal, and to protecting the imperilled Antarctic. All nations whose activities affect the commons (whether originating on the commons or elsewhere) should be urged to prepare environmental impact statements, and to consult with appropriate guardians. A suggestive harbinger of consultation mechanisms is provided by the International Council of Scientific Union's (ICSU's) Committee on Space Research's Consultative Group on Potentially Harmful Effects of Space Experiments. Although ICSU lacks enforcement powers, nations planning space experiments are urged to bring them before a committee of respected experts to investigate and judge any potentially harmful effects.

Case (5): Nation A's *Common Activities Affecting Nation* B

Finally, there are activities that take place on the commons which result in injury to national territories: beach-invading pollution from discharges on the high seas, or debris that tumbles from an Earth-orbiting satellite. As a matter of international law, the fact that the responsible party launched the harm from the global commons, rather than from its own territory, is no reason to relieve it of liability. On the contrary, the prevailing international attitude appears to regard these cases as more legitimately subject to regulation than their mirror image, the Case (2) situations, in which activities within national territories spill out on to the commons. For example, the 1971 International Convention for the Establishment of an International Fund for Compensation for Oil Pollution Damage contemplates compensation for such damage that may be suffered on the territory, or within the territorial sea, of a nation; however, damage to the seas or sea life from the same spill goes uncompensated.

In all events, litigation aimed at compensating for damage that has already occurred is not as satisfactory as averting the harm in the first place. One preventative strategy is represented by Canada's response to the threat of pollution of its fragile Arctic frontier. Canada in 1970 unilaterally declared a unique anti-pollution zone 100 miles (*c.* 161 km) seawards from its coasts, in which it claims special limited jurisdiction to regulate all potentially polluting activities – even to the point of claiming power to seize ships that operate in the anti-pollution zone in violation of Canada's regulations for the area. The Canadian action remains controversial (and not acceded to by the United States), perhaps from concern that it could lend itself to uneven-handed application by the Canadians. But in the absence of effective international control of activities on the commons that endanger states, some sort of self-help by the coastal states cannot be regarded as unreasonable. Indeed, one may wonder why measures by the coastal states that are aimed at protecting their territories by regulating contiguous areas, should be more controversial than their arrogating to themselves, through EEZs, the entire wealth of the same ocean space.

Some Limitations on Treaty-making

Thus far, I have emphasized the significance of *locale* in designing global strategies, and suggested the unlikelihood that the slowly-growing body of principles of customary international law will evolve into giving significant protection for The Biosphere. One remedy is to strengthen the hand of international agencies. The United Nations Food and Agricultural Organization (FAO), International Marine Organization (IMO), International Register of Potentially Toxic Chemicals (IRPTC), and many other agencies and services, can influence conduct by recommending standards and processing information, even without multilateral agreement.

Of course, in many ways the negotiation of special treaties represents the most effective strategy – even allowing for the fact that, strictly speaking, conventions are not universally applicable against all nations, but bind only those that choose to sign. The principal virtue of special-subject international treaties, such as the generally successful ban on atmospheric nuclear tests, and, more recently, the (strataspheric) Ozone Convention and its protocols, is that they establish cooperative obligations in detail. Legally-compelled compliance need not be central; but where it is desired as part of a treaty package, liability clauses can be drafted that do not put a plaintiff to the time-consuming and exacting elements of legal proof, such as proof of damages, of causality, of responsibility, and so on. Jurisdiction can be firmly established: treaties commonly include provisions that the signatories consent to the authority of the ICJ (or some other forum) to resolve controversies that arise under it. Most important, treaties commonly restrict national action preventively – in advance of provable harm – while ordinary customary principles incline to hold the law powerless until harm has occurred or is imminent.

Considering the advantages of treaties, why are there not more of them? The problem is that the negotiation of international environmental treaties is hampered by at least four potential grounds of conflict: over effects, over causes, over evaluations, and over strategies.

(1) Conflicts over Effects

When nations come together, there is often disagreement as to whether some alleged phenomenon is actually taking place – for example, whether we are actually witnessing a statistically-significant increase in world temperature, or a decline in certain fish-populations. The uncertainty is amplified when future effects are projected: when, if ever, will the level of atmospheric CO_2 double, and what consequences will such a doubling have, for, say, soil moisture and sea-level? When so many scientists are in disagreement, it is hard to expect diplomats to put their fullest weight behind ameliorative programmes that compete for resources with other legitimate, and in most ways more immediate, demands – such as the elimination of AIDS, poverty, drugs, and so on.

(2) Conflicts over Causes

Even if there is consensus that a significant effect is taking place, there is often disagreement as to its cause. For example, there is little disagreement that algal 'blooms' can signal trouble for other sea life. But these blooms are events that were occurring long before significant human tamperings with the environment. Even if human activities, as seems likely, are exacerbating them (*Ocean*, 1988), there is considerable room for uncertainty as to what chemical agents, and which nations' activities, are causally responsible. Nation *A* may point the finger at nitrogen emanating from Nation *B's* agriculture; Nation *B* blames phosphorus and trace-metals from Nation *A*'s industrial waste – and there may well be a grain of truth to each position.

(3) Conflicts over Evaluations

These prior conflicts may be satisfactorily resolved – that is, a consensus may be reached both that some significant effect is taking place and that some identifiable factor is responsible. Yet, even then, efforts at negotiation may be obstructed by major disagreement over evaluation. To agree that a species of whale is being over-hunted is not to agree that whales are worth saving. There may be consensus that we are indeed experiencing a 'greenhouse' warming, and even agreement on the major causal factors, without nations being able to reach agreement on how much sacrifice in current wealth a particular level of reduction warrants. The problem is exacerbated by questions of intergenerational justice, and by the notorious fact that the impact of major global transformations, such as changes in sea-level and shifts in temperature and rainfall patterns, is destined to fall unevenly across nations, as likewise are the burdens of the ameliorative efforts that appear to be required.

(4) Conflicts over Strategies

Finally, we have to consider situations in which agreement will be reached on all the above disagreements: that a certain apparent effect is genuinely occurring, and that eliminating it warrants a certain joint global effort, even if the costs rise to such and such an amount. Negotiations can still break down over agreement on strategy – on how to tackle the problem.

One sort of strategic question involves choice of remedy. Some international accords prohibit certain conduct directly – for example, those that forbid trade in endangered species, or nuclear weapons-testing in the atmosphere, or mistreatment of prisoners of war. But note that outright prohibition is feasible in those instances because the conduct proscribed is almost universally regarded as unredeemably bad. By contrast, much of the global degradation results from activities that, while hazarding the environment, have associated benefits. Consider transporting oil by tanker and burning fossil fuels. The underlying activities cannot be eliminated outright without significant sacrifice in immediate welfare. In these 'grey' areas, negotiators may find themselves debating the merits of various devices such as user chargers, marketable pollution rights, and so on.

Perhaps an even more fundamental arguing-point involves preferences for prevention or adaptation. Those who incline towards prevention generally emphasize a cost advantage. It is said to cost less to prevent an oil-spill than to clean it up. Those who incline towards reactive strategies can point to evidence that we sometimes exercise over-caution – namely, that in our zeal to prevent some hazard, we wind up promulgating a maze of bureaucratic restrictions the costs of which exceed the benefits. Particularly when the timing and magnitude of an effect are unclear (for instance, when and with what effects will there be a doubling in atmospheric CO_2?), and when no one really knows precisely where and in what degree any resultant harm will appear (such as sea-level rises or croplands turned to desert), the arguments for waiting for the problem to define itself better, and then responding with well-tailored adaptations, is not entirely frivolous.

The strategic posture favoured – whether to prevent or adapt – will of course reflect competing national priorities for current expenditures. But strategic agreement is plagued also by the existence of diverse cultural 'tastes' towards various specific strategies.

Some nations, such as the United States, are accustomed to settle differences (and internalize social costs) through civil torts suits. Other countries are more inclined to mediation. Some nations may favour technical, preventive solutions, such as mandatory smokestack scrubbers; others may regard such requirements as too specific and intrusive, and incline towards contingent *ex post* liability should damages occur.

Of course, these various grounds for conflict only make it difficult to negotiate treaties, not impossible. Where the disagreement is over effects or causes, the most effective international response may be a first-stage agreement to cooperate and underwrite advances in modelling and data-gathering. Where the conflicts are over evaluation and strategy, the marginal value of additional *fact-gathering* is minimized; what is required, at the least, is a continuing frank dialogue on international values and the world community – a subject to which I now turn.

Part II: The Moral Dimensions

Thus far, I have been stressing Biospheral degradation as it appears from the viewpoint of the law. But a comprehensive perspective on control efforts requires a synoptic vision that adds the viewpoint of *morals* to that of law. Indeed law, where it is to be effective, always has to draw on morals. But this interdependency of law and morals is especially crucial in the area of international cooperation.

In part, the burden which morality must carry owes much to the absence of a strong central world government with powers, ultimately, of coercion. Treaties can and do raise the spectre of sanctions. But we cannot expect even the most muscular treaty-made law to be backed by the familiar threats that domestic law deploys against polluters – such as criminal fines, punitive damages, or even imprisonment of individuals. In fact, it is likely that the more effective and threatening the drafters of a proposed convention make its legal sanctions, the dimmer will be its prospects of widespread ratification.

All this makes cooperation more dependent on a feeling of rightness than on force. For example, an elephant-producing nation that does not participate in an ivory trade ban, while others do, may capture the benefits of an increased price for its ivory. There is no world police force to compel the noncomplying nation to cooperate, only a fear of informal retaliation (or of being thought a bad world citizen); that and, hopefully, a shared sense that the world community's action in banning the trade was morally justified.

But in protecting The Biosphere, on what moral principles can we rely? There are, in fact, two sets of moral questions relevant to Biosphere degradation, each of which raises some of the most fundamental philosophical issues that Man has ever dealt with. We can call the first question that of finding *a shared international-morality-in-respect-of-the-environment.* Under this heading come such issues as: if the nations of the world are to cooperate in the reduction of globe-hazarding substances, how are the burdens of those reductions to be apportioned?

The second set of questions is even more challenging: Putting aside the differences that divide nations, what are *the obligations that Humankind, as a whole, owes to the rest of the natural world?* Have we duties to whales?

The debate over the future of the Antarctic can serve as illustration of the second type of question. Rather than divide the wealth locked in the Antarctic among the peoples of the world, some argue that we should leave the Antarctic untouched, as a *World Park*. Those who favour the park concept, to the extent that they rely on an obligation to the Earth, are raising a fundamental issue of philosophy: As we weigh the costs and benefits of various options, are we to adopt a homocentric perspective, from which changes to the environment are to count only derivatively – that is, through the costs and benefits *to humans* of the contemplated environmental change? Or are we to try to construct an independent, non-homocentric moral perspective, from which changes to the environment can be evaluated as 'good' or 'bad', aside from (perhaps even in spite of) their effects on Humankind?

Obviously, in this space we cannot treat either question with justice. But just as scientists want an opportunity to draw the attention of the scientific community to pressing areas of scientific research, I want to focus the attention of lawyers and ethicists – those who deal with normative rules – on certain ethical aspects of the global environmental movement, if only as a kind of outline for future collaborative efforts to fill in with detail.

A. International Morality in Respect of The Environment

Aristotle's distinction between *corrective justice* and *distributive justice is* the starting-point for an analysis of morality among people in respect of the environment. Questions of corrective justice are precipitated by blameworthy acts. For example, *A* has wronged *B* to *A*'s benefit or to *B's* injury (or both); *Corrective Justice* speaks of what *A* must do to set the situation aright, such as give up his ill-gotten gain or compensate *B* for *B's* injuries. *Distributive Justice* involves obligations that arise from situational disparities: the discrepancy between *A*'s position and *B's* position (in wealth or power or some other 'good') exceeds defensible bounds, and redistribution is in order independently of either side's blame.

The global environmental movement has provoked both kinds of justice claim – on a grand scale. Indeed, both can be illustrated within the single context of the Ozone negotiations.

Take *corrective justice* first. India is reported to have rested her demand that the developed nations compensate her $2 thousand millions as a precondition of India signing the Montreal Protocol on the grounds 'that since it is the Western nations that caused the ozone depletion, it is their moral responsibility to transfer technology for CFC substitution' (Bureau of National Affairs, 1989). The sentiment is of course intuitively understandable, but the application of corrective justice principles in this context is complicated.

To begin with, the advanced nations' development of air-conditioners and halon-using fire-extinguishers is not quite comparable with fostering piracy or waging aggressive war, each of which activities has presented problems enough for ethicists. How do principles of corrective justice apply to the unintended consequences of nonculpable acts?

There is an even more general, but no less confounding, consideration underneath: corrective justice principles are vague and controversial in their application to ordinary mortals, and it is even less clear how they apply among nations. If well-motivated (but ultimately harmful) activities, such as the manufacture of CFC-containing air-conditioning units, do give rise to *prima facie* corrective justice claims, are those claims so attached to nation-states that they carry across generations, requiring Nation *A's* contemporary and future citizens to make up for the non-culpable mistakes of their forbears? In this global moral accounting, can Nation *A*, the ozone depleter, offset the claims it may have against *B*, the forest burner? How about claims against *B* for *biogenic* (rather than anthropogenic) CO_2 and methane that happens to escape from its territory naturally? On a corrective justice accounting, are the biogenic emissions *B*'s responsibility?

The *distributive justice* claims are appearing even more ubiquitously than those for corrective justice. That is because most of the controversy is not over recriminations for alleged past wrongs, but over preventive measures to reduce degradation in the future. These efforts to reach agreement on forward-looking steps are almost inevitably hampered by distributive conflicts.

Imagine an international negotiation aimed at reducing carbon dioxide emissions. Suppose the parties agree to make the accord 'fair' in the sense of demanding *equality of effort* to reduce pollution. Does this imply that each party must employ *equal technology* – for example, the 'best available technology' (BAT) for CO_2 abatement? Or does it imply that each party must achieve *equal unit reductions,* on which interpretation the nations with the highest historical baselines will continue to outpollute the less-developed nations – and maintain, therefore, their economic edge?

On the other hand, the equality could be understood as an equality not of effort but of *outcome.* A convention oriented towards just outcomes would presumably aim for an equality of wealth or of opportunity, or at least of some fundamental baseline index such as assured adequate food, fuel, clothing, etc. But there are obviously enormous problems assessing the 'justness' of the various conceivable outcome measurements.

The difficulties of making these ethical judgments give some impetus for market solutions, in which the 'unseen hand' of market forces supposedly rescues the human mind from the need to make some hard choices. In the carbon dioxide – 'greenhouse' area, a market 'solution' might take the following form. We might first estimate the capacity of the Earth to withdraw carbon *via* natural processes. For example, under the present carbon budget, approximately 7 thousand million tons of carbon are being pumped into the atmosphere annually, of which perhaps 5 thousand million tons are being withdrawn, leaving a surplus of 2 thousand million tons. We would next determine how much surplus is tolerable. (Presumably a 'balanced budget' would be ideal, but, quite likely, unachievable.) If, say, a surplus of 500 million tons is tolerable, then global carbon output must be reduced to 5.5 thousand million tons.

Whatever the level agreed to, the world authority (or, more modestly, the covenanting parties) would then allocate 'rights' to emit carbon, the sum of the rights aggregating the total allowable. A trading would ensue in which the countries that had the highest need to emit more than their allotted share would have to buy the rights to emit from the nations that could most easily make do with less than their allotment. The claimed virtues of such schemes (and their limitations) are well examined in the literature.

So far, theoretically, so good. But note that the discussion above skips over the initial question: how shall the original entitlements, the basis for the trading, be assigned? That is not a question of economics but of ethics. And one quickly finds that there is no consensus in international morals (such as they may be) even on the most foundational issue which a just distribution of entitlements requires. Arthur H. Westing favours a market solution to the carbon emissions, but would assign the original entitlements to nations on the basis of their geographic *area:* that is to say, a nation which comprises 1 million hectares would start out with the right to emit ten times as much as its neighbour with 100,000 hectares (Westing, 1989).

But why, one may ask, is area more significant than *population?* Indeed, it is not clear that geographic area has any moral salience at all. On the other hand, assigning entitlements on a *per caput* basis can be criticized as sending the wrong message on population: it favours a nation whose population is least under control! Instead of a *per caput* allocation, one could allocate *per nation.* But are we to say that the tiniest nation has the right to pollute as much as the largest? One is tempted to employ a corrector for standard of living, for example *per caput wealth.* Or perhaps there should be a corrector for *benefit,* as some peoples and nations stand to gain more from controlling global warming than do others.

It is instructive that, thus far, none of the major international environmental accords have attempted quite such fine-tuned 'solutions' to the distributive justice tensions. Treaty-makers are not academics. Their job is entirely practical, and when distributive claims are aired they are apt to forge compromises in rough and ready, and probably intuitive, approximations – particularly if there is no clear and persuasive argument that a practically achievable agreement is unjust.

That would appear to characterize the 'solution' to the distributional tensions arrived at by the Montreal negotiators, who assented to a general reduction in controlled substances, but deferred the compliance schedule for less-developed nations – defined for these purposes as countries whose annual level of consumption of controlled substances is less than 0.3 kg *per caput.* There is also a proviso allowing an increase of up to 10% for less-developed countries which need that amount to satisfy 'basic domestic needs', as well as provisos for some 'trading' of rights for nations at the lower end of the development scale.

All in all, if a convention is workable, there is no need to complain. But considering that the question of compliance is still unresolved, and that the future will see more and more international efforts at global environmental treaties, surely the way should be paved by according more and more attention to the questions of global justice in the context of environmental control measures.

B. Obligations Towards Nature

Throughout the world, there is a growing movement to protect Nature. Sometimes the rights of 'higher' living creatures, such as whales, dolphins, and elephants, is championed. Sometimes our supposed obligations are turned to threatened species of a 'lower' sort, such as exotic plants, and sometimes to places (and their ecosystems) such, most grandly, as the Antarctic. Of course, sometimes the preservationist argument seeks a homocentric foundation, *e.g.* that we may find in some humble plant a medicine which will operate to human benefit. But ever-increasingly, preservationists are favouring Nature for its own sake. This is the position embodied in the 1982 UN World Charter for Nature:

> 'Every form of life is unique, warranting respect regardless of its worth to man, and, to accord other organisms such recognition, man must be guided by a moral code of action ... Nature shall be respected and its essential processes shall not be disrupted.'

The question is, considering the apparently irreducible competition between *Homo sapiens* and the balance of the planet, how can such a grandiloquent sentiment be translated into a comprehensible set of constraints? Indeed, what philosophical sense can be made of the underlying endeavour?

The development of a non-homocentric morality – one which defines Mankind's obligations towards Nature – is no less important a venture, and, unfortunately, no less complex, than that of fashioning a morality that relates peoples and states among themselves. First, what is needed is not just a moral viewpoint which accounts for Nature in principle and overall. We need a moral viewpoint that is rich enough to advance us through the ontological conundrums: by reference to what principles ought the moral and legal world to be carved up into those 'things' that will count and those that will not? If whales are to be a legitimate unit of moral concern, do our obligations attach to each individual whale, or do they extend, less restrictively, only to the whole pod of whales, or to particular whale habitats or species? The implied restrictions on human activity vary with each answer. Moreover, one cannot run away from the dilemma by adopting a holistic or Gaian viewpoint that champions the goodness of the planetary whole; if the everything – the whole – is good, what can be bad? How, faced with any human choice, would a holist judge that one outcome is better (less bad) than any other?

Second, suppose that we can do the carving-up correctly, that is, identify those objects towards which some *prima facie* moral regard is justified, *e.g.* perhaps a certain ecosystem, or the Antarctic. There will often remain the question – even if moral obligations to a piece of Nature are conceded to exist in principle – of how they can be discharged: how does one 'do right by' a mountain? Shielding it from all change is one answer – but nothing seems quite so unnatural as putting a freeze on Nature. Some more flexible standard seems to be called for. But when we try to construct a non-human, and in particular an 'inanimate'-regarding, moral viewpoint, the familiar guideposts of conventional moral theories – pain, preferences, dignity, reciprocal advantage – are of little use, and new ethical systems seem to be called for.

Third, there are the persistent distributional dilemmas which, when Nature is added as an interested party, only become more complex. It is not enough to carve up the world, establishing what is to be of moral significance; nor is it enough to agree how that regard translates into *prima facie good* and bad acts. What are we to do in the case of conflicting indications? For example, suppose that, working with reference to one or another of the various Nature-valuing viewpoints, a *prima facie* case can be made for preserving each of two species of animal, but we cannot preserve both. Do we favour the rare species of lower animal over the less exotic but 'higher' one? It is easy to imagine a moral framework whose basic principle is 'more life is better than less'. One can imagine, too, support for the preservation of a singular, unsullied desert. But what do we do in the face of an irrigation project that offers to transform the desert into a habitat teeming with life? The general problem is one of a moral judgment's strength. In other words, even if the continued existence of a species, or the state of a river, is granted to be 'a good', how do we make adjustments for that moral fact in the face of conflict, or of other, competing 'goods'?

All I can do in this space is to outline these problems, and to plead that I have dealt with them as best I can in my book, *Earth and Other Ethics* (Stone, 1987*b*). I do not imagine that anything I have to say there will satisfy everyone that a non-homocentric ethic is defensible. But I hope that work will inspire others to join in the examination of environmental ethics; the relevance to the current debates on The Biosphere is fundamental and profound.

Part III: Towards a Global Environmental Trust-fund

I have noted how many factual uncertainties there are to resolve if The Biosphere is to be advanced to the top of national agendas. I have stressed, too, the range of what might be called moral uncertainties. But there is no need to suspend all cooperative activity until these conundrums are solved to general satisfaction. On the contrary, there is much that can and should go forward. In this vein, I would like to close with a proposal for my own version of a *global environmental trust-fund* – one based on the wealth of the global commons (*see also*, Stone, 1992).

The idea is simple. We have seen that much of the environmental crisis derives from the fact that, because the global commons areas of the world are in practical effect 'unowned,' they can be, and are being, used and abused with relative impunity.

If we were to rectify this practice, and charge even a fraction of the fair worth for the various uses to which nation-states put the global commons, we would advance two goals at once. The charges would dampen the intensity of abuse, and at the same time underwrite the costs of providing public order, such as marine resources management and the repair of environmental damages.

Let me just play with some moderately current estimates to convey the potential magnitude of what would be involved:

Start with the oceans. Some 80 million metric tons of fish are harvested annually from the world's salt waters. An ocean-use tax of only one-tenth of one per cent of their value would raise at least $50 millions. The same token rate on offshore oil and gas would yield $75 millions.

There is another, dirtier use of the oceans: as a gigantic sewer. It is often much cheaper (and, in some circumstances, relatively less harmful) to dispose of toxic refuse in the sea, rather than on land – particularly because there are no neighbours to complain. The officially reported ocean dumpings – almost certainly very much understated – amount to over 200 million metric tons of sewage sludge, industrial wastes, and dredged materials, yearly. Much of this takes place within traditionally territorial waters, rather than on the 'high seas' commons as such. But the repercussions are inevitably ocean-wide. A tax of only $1 a ton would raise another $200 millions.

Nations use the atmosphere as they use the oceans – as a cost-free sewer for pollutants that will wind up as someone else's problem. By burning fossil fuels and living forests, Mankind pumps some 7 thousand million metric tons of carbon into the atmosphere annually. A carbon tax of only 10 cents a metric ton would raise $700 millions – more than ten times the current budget of the entire United Nations Environment Programme. The same dime-a-ton tax on sulphur dioxide and nitrogen oxide emissions would produce another $30 millions or probably far more.

The total visualized thus far is more than 1,000 million dollars annually. And that is before adding a tax on chlorofluorocarbons, or halons, or the incineration of toxics at sea – or on the millions of millions of litres of liquid wastes that are discharged into the oceans as run-off. Nor does it include many potential levies for several non-polluting uses of common-heritage assets, akin to fishing and oil. Consider royalties for the minerals that will someday be taken from the sea-bed, and fees for the uses of outer space! Why should a needy global community give away to the first grabber, rather than auction such commodities as those sea-bed minerals, limited resources such as positions for geosynchronous and earth-orbiting satellites, and frequencies on the radio spectrum?

Many persons will object to a pollution charge component, contending that it is outrageous to allow anyone to pollute, even if paying for such a right. Yet, some pollution is inevitable, and even more of an outrage is what polluters get away with, as they currently do, scot free!

Indeed, if there is a real objection to the above proposals, it is that the initial rates I have suggested are too paltry. Viewed as a way to combat pollution, they would probably fail to confront the polluting nations with anything near the full costs of the harm they are causing, and therefore would presumably fail to elicit a high-enough investment in conservation and pollution control. Viewed from the reverse side, as a strategy for maximizing revenues for the environmental infrastructure, they fall short of extracting the full value of what users would pay (for their ocean-dumping, for example) if they were required to bid for the rights at auctions. And, considering the seriousness of our plight, the final bills appear far too modest. The United States' share, which, unsurprisingly, is the largest on my calculations, would amount to only $230 millions a year – less than half the price of one Stealth bomber!

Nonetheless, in total, the funds would finance a host of measures which most observers currently regard as critical: improvement in global monitoring and modelling; fostering of adaptive technologies (such as fast-growing trees); enforcement of safer methods of commons-hazarding waste disposal; gathering and storage of genetic material; institution of a global environmental patrol-force with the capability to respond quickly to major oil-spills and nuclear accidents; improved enforcement of treaties already in force (such as the Convention on International Trade in Endangered Species); purchase and maintenance of examples of rare or species-prolific ecosystems or wider ecocomplexes, and so on.

The infrastructure to do such jobs already exists. It includes, along with UNEP, such organizations as the World Conservation Union (IUCN), the World Meteorological Organization (WMO), the UN Food and Agriculture Organization (FAO), and the World Wide Fund for Nature (WWF). Many of these UN, governmental, and nongovernmental, organizations are capable of doing first-rate, dedicated work. What they need now most desperately is proper financing.

It is true that similar-sounding proposals did not find many backers in the relatively recent ozone discussions in Montreal, or, earlier, in the Law of the Sea negotiations. But those plans would generally have involved taxation of industrialized countries at a fraction of their GNPS, or on the basis of their use of particularly noxious materials, or else would have gathered off-shore mineral royalties with the intention of redistributing the yield to the less-developed countries to foster general economic development.

A Global Commons Environmental Trust fund would be different and, I hope, more practicable and saleable. It would be consistent with continuing aid to developing nations, although that would not be its primary aim. The idea is to assure some continuous vital oversight, particularly of those regions and qualities of our beleaguered planet that have otherwise no 'guardian', with the clear mandate and resources to look after them. This proposed fund is designed to fill that major gap. It would seek from users of the commons, a financial charge for their use, so as to apply the proceeds largely back to the commons, for the maintenance and mending of their *major portion of our planet.*

Is there any near-term remedy for so much of The Biosphere that could be more sensible or timely?

References

Bureau of National Affairs (1988*a*). Court in the Hague upholds decision on damage awards to greenhouse owners. *International Environmental Law Reporter* 11, (Current Reports, No. 10, October 12, 1988), pp. 530–1.

Bureau of National Affairs (1988*b*). French court rejects tribunal's award of $340,000 to two Dutch water companies. *International Environmental Law Reporter,* 11, (Current Reports, No. 12, December 14, 1988), p. 652.

Bureau of National Affairs (1989). India wants $2 billion from others to sign ozone depletion Montreal Protocol, *International Environmental Law Reporter,* 12, (Current Reports, No. 8, August 9, 1989), p. 389.

Goldie, L. F. E. (1975). International maritime environmental law today – an appraisal. Pp. 102–6 in *Who Protects the Ocean?* (Ed. J. L. Hargrove). West Publishing Company, St Paul, Minnesota, USA: xiv + 250 pp.

Gray, C. D. (1987). *Judicial Remedies in International Law.* Clarendon Press, Oxford, England, UK: xix + 247 pp.

Hassett, C. (1972). Air pollution: possible international legal and organizational responses. *New York University Journal of International Law and Politics,* 5, pp. 1–52.

OCEAN (1988). Algae blooms signal trouble. *Ocean,* 21, p. 69.

Soni, R. (1985). *Control of Marine Pollution in International Law.* Juta & Co., Capetown, South Africa: xiii + 301 pp.

Springer, A. L. (1983). *The International Law of Pollution.* Quorum Books, Westport, Connecticut, USA: iv + 218 pp.

Stammer, L. B. (1989). Saving the earth: who sacrifices? *Los Angeles Times,* (March 13, 1989), p. 1, col. 6; p. 16, cols. 2–3.

Stone, C. D. (1972). *Should Trees Have Standing?* William Kaufmann, Los Altos, California, USA: xvii + 102 pp.

Stone, C. D. (1987*a*). *Umwelt vor Gericht: Die Eigenrechte der Natur.* Trickster Verlag, München., Germany: 164 pp.

Stone, C. D. (1987*b*). *Earth and Other Ethics.* Harper & Row, New York, NY, USA: viii + 280 pp.

Stone, C. D. (1992). Guest editorial. *Environmental Conservation,* 19(1), pp. 3–5.

Trail Smelter Arbitration (1938, 1941). Reported in *American Journal of International Law,* 33, pp. 182–212 (1938) and *ibid.* 35, pp. 684–736 (1941).

Westing, A. H. (1989). Law of the air. *Environment,* 3(3), pp. 3–4.

World Resources Institute [& The International Institute for Environment and Development] (1988). *World Resources 1988–89.* Basic Books, New York, NY, USA: xii + 372, illustr.

Commentary on Chapter 19

CHAIRMAN: Professor Donald J. Kuenen
PANELLISTS AND OTHER CONTRIBUTORS:

Westing, Miller, Dale, McMichael, Stone, Oza, Stone, Vallentyne, Goldberg, Stone, Samatar, Stone, Westing, Stone, Harmathy, Medawar, Stone, Burnett, Stone, Burnett, Stone

Westing observed that, for Humankind to be able to survive with The Biosphere, it would be necessary to recognize that the environmental problems which threatened such survival did not respect national boundaries. The corollary to that often awkward situation was that the nations of the world would have to relax their still strongly-held notions of national sovereignty, at least to the extent that they could act in concert in dealing with the human environment, both with respect to its protection and with respect to its utilization.

Cooperative environmental action by sovereign nations – whether within a particular biogeographical region or world-wide, and whether within domains subject to national jursidiction, or within domains beyond any such jurisdiction – required formal international agreements via multilateral treaties. Moreover, for such treaties to be enacted in the first place, and respected in the second, they had to have broad support.

Stone had provided (Chapter 18) a magnificent synthesis that covered environmental harm from both legal and moral perspectives, distinguishing among all permutations of the possible sources and sites of such harm with respect to national territories and extra-national territories. It was particularly useful in its elaboration of environmental litigation and of the strengths and weaknesses of environmental treaties, and also in pointing to the lack of consensus on who could (or should) speak for Nature. **Stone** had made it clear, *inter alia*, that a treaty could not stand without moral or ethical underpinning. In other words, the lesson for the Conference was that legal norms, even if achieved, were empty without supportive cultural norms. That, in turn, pointed to the over-arching necessity for appropriate education and information that lead to environmentally-supportive, attitudinal changes and, thereby, to the necessary cultural norms. Moreover, the cultural and legal norms in question would be mutually supportive of each other.

The exposition of Aristotle's distinction between 'corrective justice' and 'distributive justice' was especially enlightening. The notion of corrective justice did not seem to him to present a major conceptual hurdle in terms of public understanding and acceptance. On the other hand, the notion of distributive justice would require substantial public education. This would seem to be the case especially when it came to environmental domains beyond any national jurisdiction – domains that, first of all, must come to be recognized as constituting the *common heritage of Humankind*; and, from this recognition, that distributive justice must be accepted as having two components within such common-heritage areas, one encompassing *sustainable use* and the other encompassing *equitable use*.

He would use the atmosphere as an example of a common heritage of Humankind that required both sustainable and equitable management (Westing, 1989). **Stone** had recognized the need for equitability in this regard, but did not share his view that, in order for such equitability to enjoy moral salience, an approach to the issue must be based on national area – and, contrariwise, certainly *not* be based on such an ecologically counter-productive, or ethically indefensible, criterion as national population, national wealth, or past national arrogation. *Cited:* Westing, A. H. (1989). Law of the air. *Environment,* 31(3), pp. 3–4.

Stone had provided a most informative legal examination of those instances in which an action occurring within one nation caused environmental harm within another. However, it was useful to add that the concept of a nation having the responsibility not to export environmental harm beyond its national borders had been clearly enunciated in the Stockholm Declaration of 1972 on the Human Environment (Principle 21); and had been established in a binding form with respect to one class of air pollutant, namely, radioactive débris, via the widely adopted Partial Test Ban Treaty of 1963 (Article I.1). Perhaps of greatest interest with respect to environmental transfrontier transgressions – of interest because it provided such a beautiful model for other nations to emulate – was the Nordic Environmental Protection Convention of 1974. This treaty had, in effect, abolished national borders among Denmark, Finland, Norway, and Sweden, in so far as these might obstruct the attainment of legal redress from environmentally harmful activities.

The need for a relaxation of national sovereignty in reference to environmental harm that occurred within the limits of one nation had been suggested by **Stone**. The Convention of 1975 on the Prevention and Punishment of the crime of Genocide was cited, *inter alia,* as an already accepted instance of such relaxation. He added that the Convention of 1972 Concerning the Protection of the World Cultural and Natural Heritage provided a further, albeit modest, instance. Moreover, the time might well be ripe for the nations of the world to embrace an 'ecocide' convention (see, *e.g.*, Falk, 1984) that would in some senses parallel the Genocide Convention.

Stone had correctly stressed the widely recognized difficulty of challenging the notion of national sovereignty in any attempt to deal at the legal level with multinational environmental problems. However, he had made the important additional observation that the development of legal régimes for domains beyond any national jurisdiction (i.e., for common-heritage areas) did not face that particular obstacle. Moreover, the time had arrived, in his opinion, for instituting international environmental taxation for the purpose of supporting a global environmental trust fund.

We had been provided with a splendid analysis of the interplay between law and morals as those issues applied to The Biosphere. He had thus added a dimension without which we would have no hope in the attempt to cope sensibly with our environmental future. *Cited*: Falk, R. A. (1984). A proposed Convention on the Crime of Ecocide. Pp. 45-50 in: Westing, A. H. (Ed.). *Environmental Warfare: a technical, legal and policy appraisal.* Taylor & Francis, London, England UK: xiii + 107 pp.

Miller remarked that the aptness of **Stone**'s analyses of law and ethical considerations of the future of the environment was striking in its applicability to the questions and deliberations of the Conference. Two areas of World Commons – the world ocean and Antarctica – were prime subjects for both legal and ethical analyses. The nations involved would soon have to consider several of the measures outlined to sustain these resources as an enduring heritage and reconcile their differences over the use and abuse of the natural wealth of these two resources.

Such accords as the 'Agreed Measures' of the Antarctic Treaty Parties, the Convention on Seal Conservation, and the Convention on Antarctic Marine Living Resources were worthy examples of measures agreed upon to protect the wealth of world commons and illustrated how the world community could assume the role of 'plaintiff'. Additional accords in respect to the future of Antarctica and its unique ocean could be anticipated, most especially in a plan for environmental impact assessments and research before the fact. The Global Environmental Trust Fund, nourished by an 'ocean-use tax' and with an 'authority' based upon the plea of the Law of the Sea, could become the rationale and means for managing fair access of all nations to Antarctica, as well as to the wealth of the World Ocean. **Stone** had helped us to anticipate what threats these commons might yet endure and had guided us in finding agreement on relief and ultimate sustained protection.

The excellent series of Conferences on Environmental Future had reviewed the history leading to these events and those held in Iceland, Finland and Scotland were timely conventions that had pin-pointed world environmental issues and served as focal points towards looking for solutions. Budapest, too, would become such a landmark. Through the generous support of the Conference, the nation of Hungary, the Hungarian Academy of Science, its academic fraternity, environmentalists, and futurists had assumed the leadership in extending consideration of these global problems to include the Eastern-bloc nations.

The exciting changes which were occurring today in Europe would require new laws and regulations, some of which would certainly pertain to benefiting the human environment and protecting natural resources. As governments addressed the issues, they might benefit from the experiences of people elsewhere who had established some guidelines and found partial solutions to funding conservation efforts. For instance, the compilation of environmental law, world-wide, produced by the Legal Commission of the World Conservation Union, could provide examples of legal wording that had been upheld and really worked. Another example was tax laws that allowed exemptions for contributions to environmental and conservation groups made by the individual taxpayer, and exemption from taxes for the non-profit organizations themselves. Such organizations could thus direct most of their income towards education, lobbying against proposed legislation that threatened damage, or seeking relief from laws or regulations already in place that had adversely affected the environment and natural resources. People need organizations that would express the ideals and the moral values on which conservation legislation was based. Tax exemption laws benefited not only the larger, or well-recognized organizations such as the Foundation for Environment Conservation, but also newer ones that espoused conservation causes which had not yet reached the broader, national level of understanding and acceptance.

As nations re-examined and modified their laws and regulations which pertained to their environment, natural resources and conservation in general, the opportunity would be presented to ordinary citizens to become involved and to express their individual concerns about local and national needs. Concerned individuals could become a powerful force by bonding together to make their collective voice heard, for as Margaret Mead had said:

> 'Never doubt that a small group of thoughtful, committed citizens can change the world. Indeed, it's the only thing that ever has.'

Mead, a distinguished anthropologist and activist, had spoken from her experience as an outspoken critic on behalf of the global quality of life. In that capacity, she had actively participated in the 'Spirit of Stockholm Association' and served as a nongovernmental organization observer at the United Nations Conference on the Human Environment in 1972.

From the small beginnings to which Mead referred have developed such strong non-governmental organizations as the World Conservation Union, until recently known as the International Union for Conservation of Nature and Natural Resources (IUCN). He personally recalled its early struggles and growth in difficult times. From its ethic, its planning and its advocacy had come the measures and strategies, legal and moral, which served as models for environmental legislation at all levels of government. The format for conservation law that originated in one or another commission or working group of IUCN, its Survival Service (now Species Survival Commission), its Education Commission, its Commission on National Parks and Reserves, and others, had been adopted in various parts of the world from national to village level at the urging of local conservation groups. Never doubt that '... citizens can change the world'.

The same ethic, and similar rationale, had had an effect on the continent of Antarctica and on its ocean, where the only citizens were the non-human biota. There, without benefit of autochthonous environmental advocates, the protection of the natural wealth had been in the hands of a few involved nations under the Treaty of 1961. Regarded as a world commons by most member states of the United Nations (yet lacking official designation as such), Antarctica remained vulnerable to exploitation and degradation. Under the Antarctic Treaty, some thirty nations now had before them a convention for a mineral regime, still to be ratified, that would allow for an orderly process of mineral ores and oil extraction from the continental shores and waters of Antarctica. Three or more states refused to sign the convention, however, expressing, in part, the concern of their citizens that the environs of Antarctica should be spared the degrading scenes characteristic of industrial sites and the accidents which accompanied extraction and shipping. Meanwhile, that magnificent, untrampled white continent was being seen repeatedly by millions of television viewers and by hundreds of tourists, suggesting, perhaps, an alternative use based upon scenic resources. The final designations and fate for the living and scenic wealth of Antarctica would surely be guided in large measure by the influence exerted by groups of thoughtful, committed citizens.

The highest ethic – a note the Conference would most likely emphasize – was that of Man's unselfish responsibility towards the globe as steward of its resources. It was most important to understand the ethic of sharing with all creation the wholeness of the living planet, and the collective good that can thereby benefit Mankind. To save plants and animals because of their possible material value was a noble purpose, but was it not a bit nobler, or far wiser, to save the indigenous biota of the oceans and the poles, of the mountains, the deserts and the rain-forests for their own sakes than it was to maintain an attitude of proprietary interest to enrich the few? Did we not have the responsibility of emphasizing the constraints needed on the behaviour of Man: in the workplace, at the boundaries between neighbours, in the world commons, and certainly in our local village and family? Family planning and the containment of human populations was the urgent humanist mission we must recognize, accept, and teach, if we were to attain any solutions adequate to the environment's problems and diminish the Earth's ills. Humanity must find a level of agreement on a common ethic, acceptable at home and ensured by individual understanding, purpose, and commitment. We must build a universal consciousness of our common future and a collective responsibility towards all global resources, including our environment that would assure us a future.

Dale said he was privileged, as a layman, to participate in the discussions: the genius of the ICEF Conferences was the fact that multidisciplinary viewpoints were encouraged and forcefully expressed.

He was not qualified to critique **Stone**'s contribution. However, as a lawyer, he envied the clarity and enthusiasm with which he had expressed his views. He wished to speak about the media. The Conferences had raised the level of knowledge not just for the world's scientific community, but for the non-scientific public as well, and had recognized the need to sound the alarm to governments and human beings – ordinary citizens – around the world. While flying to Budapest on 'Earth Day' he had stopped in Chicago and bought the New York Times in which coverage was deep and good: the nations were responding to alarms sounded by scientists. Two weeks ago, he had attended the Eleventh World Media Conference which met in Moscow for five days. Eighteen former heads of state were there, and it had been refreshing to see and hear how often environmental concerns were raised. For example, Mr Takeo Fukuda, Former Prime Minister of Japan, had emphasized the environment right alongside world peace and the world economy. He read from Fukuda's speech to show his understanding of the issues:

> 'In comparison, our global resources, energy, food and other conditions maintaining Man, are finite. The problem of famine is already a grave one. Air and water pollution, acid rain, destruction of forests, the spreading of deserts, destruction of the ozone layer by freon gas are examples of the appalling extent of the devastation of Nature.'

and had concluded by saying:

> 'If nothing were to be done about these things, what would happen to Mankind and the Earth in the 21st century? A wise coordination of population growth and environmental conservation is an urgent and vital issue. Each of these is a global task involving all of Mankind, and must be dealt with, transcending race or nationality, differences in religion or ideology.'

He then referred to another delegate to the World Media Conference, Dr. Nikolai Vorontsov, a distinguished biologist and Chairman of the USSR State Committee for Nature Conservancy. He had spoken at length on 'New Thinking as a New Approach to Nature.' He quoted only a small portion:

> 'We all agree by now, I hope, that The Biosphere is an "empire", one and indivisible, so that state borders should have nothing to do with ecology. It is not worth even talking about preserving The Biosphere without international cooperation. One can, of course, preserve separate, "museum" pieces of it, but it is a "whole organism", which means that no part of it can be healthy unless the whole is healthy.'

He had concluded by suggesting that, as to rain-forests:

> 'Maybe, it would be no more than fair to set up a world fund made up of contributions by oxygen-consuming states so that the oxygen-producing states would find that they were making a profit, instead of a loss, on its production. It would then be just as profitable to 'sell' oxygen as to sell timber. Otherwise the owners of the tropics will simply find themselves in a set of conditions in which they will have to destroy this invaluable world resource.'...
>
> 'What then is our order of priorities. Our point of departure is the following: Mankind is a part of The Biosphere; The Biosphere knows no borders; we must abandon the technocratic thinking and recognize the priority of ecology over the economy; we must introduce the concept of a territory's ecological capacity as a condition for ecological security. This concept would, of course, be elaborated in real terms, all the way to the relevant formulas, depending on the territories and their ecosystems.
>
> 'Our ecological and political conception is based on the assumption that ecological problems in the modern world call for political and economic solutions.'

Finally, he referred to a dramatic, heart-rending plea to the World Media Conference from a newsman from Laos. His country was among the 42 poorest and least developed nations in the world. He had recited in agonizing and almost tearful detail the plight of his nation of 4 million people living on subsistence from mountains, forests and streams. Forest destruction and erosion of soils was rampant in his country:

> 'It is known that the forest and environment protection activities in general, and specifically the establishment of permanent occupations and fixed-locations for the upland minorities is a very complex problem, but it has an important role to play as to the stability and development of the country and also has an impact on the ecological environment of other countries within the area, in the continent and in the world. Due to the lack of funds, expertise and

experience, the efforts and ambitions of the Lao Government and the Lao people to stop the destruction of the existing forest, to protect the environment, and to implement the different programmes for shifting cultivation, forest development and Nature conservation will not be successfully achieved and implemented.'

Then he had pleaded:

'I hereby request to the international press and senior politicians who are here present, and who are well aware about the dangers which are threatening the environment and particularly the tropical forests and their fauna and flora, to help us in sensibilizing the opinion and expertise, and in collecting funds to assist Laos which is facing tremendous problems of this kind, in order to safeguard the tropical forest which is not only the property of the Lao people but also belongs to the international community. Without international cooperation in the field of Nature conservation, tropical forest rehabilitation and environment protection from the world community, Laos alone will never be in the position to match this problem which is far beyond its own capacity.'

His was a frightening story. With these pleas in mind, the Conference and each participant was urged to continue the broad-scaled efforts to guide Humankind, but also to make the scientific data understandable to laymen. In that connection the world was closing in on us, becoming smaller to Mankind.

He suggested that between the ICEF Conferences – held every few years – the WCB should organize fact-finding tours. Some participants should visit some of the 'hot spots' or 'danger spots' of the Earth and, at each such place, after assessing it, make a public statement about the conditions found, what the prospects were if nothing was done to change the situation, then prescribe what could be done to stop the environmental destruction, and finally predict the great benefit to Mankind if the course prescribed was followed. By personalizing the situation in this way, the Conference's influence would be multiplied many times more than if the impact was limited merely to the scientific community alone. As a layman, he pleaded for activism on this issue. We lived in a world of inequality – where people and nations differed in knowledge, resources and abilities. While we had to deal with nations and peoples in relation to their understanding, we should never discriminate against them. We must protect equal access and opportunity for all nations and peoples in connection with issues relating to The Biosphere. He had sounded the alarm! Concern was breaking out all over the world like a fiery red rash, and the world was closing in on us:

'I looked in the distance
  and I saw an animal.
I came closer
  and I saw a man.
I came still closer
  and I saw that it was my brother!'

McMichael was particularly interested in the suggestions for guardians of the global commons because it remained a continuing problem of how to view an individual's standing in regard to the legal enforcement of issues of a common nature, in Antarctica for example. He did not share **Miller**'s pessimism about Antarctica for several reasons. First, because of the position adopted by France and Australia plus the gathering momentum of nations who were having second thoughts, there seemed no likelihood that the mineral convention would come into being. Second, he did not believe that those nations which had already established a foothold, or research stations, as part of existing treaty arrangements had done so either with some sort of territorial ambition, or with an intention to move in, if and when the treaty ceased to operate. Indeed, there was no reason to suppose that the treaty would not go on indefinitely; the 1991 review, as he understood it, was only required if the partners decided that they desired one. Nevertheless, a movement was afoot for positive action in relation to Antarctica as part of the global commons, and there was a suggestion that it should become some sort of an international wilderness park under an environmental protection panel. He thought it difficult to envisage the United Nations in its entirety as an appropriate body to exercise jurisdiction, but he could not see countries which had an interest in the matter leaving it only to the nations of the Antarctic Treaty in its present form. Committees such as SCORE or SCAR were of no use: they had not been set up for that purpose. How then could Antarctica be managed as an international wilderness, even if existing countries surrendered their claims to it in the light of this new concept?

Stone said that if such an assumption was made, then, right away, there were very large moral problems for those advocating a wilderness status. If it were supposed that there were minerals in Antarctica for which prospecting – which caused most of the damage – was necessary, moral problems arose that could not be dealt with by global platitudes. For example, Malaysia had said that if there was wealth there, why should it be locked up? Ways would have to be found to divide that wealth. Even before questions of guardianship arose one had to decide on what basis could one deny that supply of wealth to the poor people of the world. That scenario put in sharp relief the conflict between hungry, energy-starved people, and the rest of the world, over a pristine part of the Earth. That differed profoundly from the normal trade-off between one person and another, which was what ethics dealt with normally.

Guardianship posed a problem. Normally, a guardian would be expected to have expertise in all relevant matters, but it would be necessary to conscript a range of different experts for the Antarctic as a whole. In such a case he did not think a single park-type guardian appropriate. The matter would be made more difficult because there were already potential prior claimants, but he still found the problem of distributing potential wealth more vexing. **Westing** had, quite rightly, reminded us that the Law of the Sea touched upon equitable distribution of wealth, and his recollection was that the mining nations had been prepared to pay a royalty on production from the sea-bed, so long as it was of the order of 8 or 10%, *i.e.* comparable with prices paid for land mining, but leaving it open how then to divide it up; that was an unsolved problem. He admitted that his proposals for the repair of the global commons skirted that most difficult question. If one had taxes which were to be divided amongst nations, how did you rank claimants? Did one do it *per caput*, by population size, by national wealth, or what? He had no solution.

Oza reported that, at a national seminar in New Delhi in 1985, on law and science for the protection of the environment (organized by the Ministry of Environment and Forests [India] and the Jawaharlal University), he commented that even the judges held wooden concepts on environmental issues and he had pleaded that environmental benches should be established in the High Court to deal with such cases. This had now been accepted by the Government of India, including the Supreme Court. He believed that the time was now ripe, in view of the many global environmental problems, to establish international environmental courts to consider situations when one nation inflicted environmental damage on another, or on the welfare of humanity as a whole.

Stone did not think such special courts were necessary. There were a number of suitable fora but we lacked the laws. Nations could go to arbitration: the USA had successfully sued Canada, for example, and the International Court of Justice was empowered to hear complaints by one nation against another on treaty matters, but we lacked environmental treaties.

Vallentyne said he had heard **Westing** say that only three environmental treaties were necessary and he assumed that these were for air, seas, and space, respectively. If such treaties were made they could only relate to the control of human intervention into those areas, but that did not solve all the problems. 'Environment' was not a holistic word: it meant nothing unless it was connected to some kind of living thing. 'Earth' was far more holistic and that was the term we should use. Even better, instead of a 'global environment trust fund', as had been suggested, would be a 'biospheral trust fund', which embraced the concepts of environment, life and the Earth.

Goldberg had had memories revived of a discussion, now some 20–30 years old, about how the coastal environment would be affected if a nuclear reactor suddenly blew up. That question had been set aside but recent events had revived his interest. We could now predict that nuclear accidents due to human error would occur, on average, about every thirty years. As most reactors were located in the coastal zone, an accident would impact not only upon the nation owning the reactor but also on adjacent neighbours and the ocean commons. Reactors were a legal liability to the nations that owned them but, in the light of the probable recurrence of nuclear accidents, was it worth while for scientists with knowledge of the consequences of marine radioactive pollution to work with the legal community to establish the international liabilities of such reactors?

Stone responded that at present, if such an explosion injured the commons, there was first a problem that, if fish were killed, they did not belong to anyone, and secondly, who could bring a suit and on what grounds? He supposed that if someone ate contaminated fish there might be some sort of causal linkage which would enable a case of liability to be brought. What he should emphasize was that the nation operating a reactor ought to have legal liability for it. At present nations only had partial liability, far less than they ought to have. He therefore agreed that an assessment of the likely harm that might be caused, with an eye to changing the law, was desirable.

Samatar thought that what was emerging was that science and technology had unleashed problems on such a scale that they had reduced the world to a village, but the logic of that was that some sort of world order was the only way to deal with such problems. It therefore followed, so long as there were sovereign states, that global problems would not be solved. It seemed to place humanity between a rock and a hard place, as one of his country's sayings went.

Stone accepted the argument in part but pointed out that a village had more homogeneity and a more unified set of interests; the world did not display a similar unity. He did not think that a single government, with its full panoply of powers, was necessary to achieve environmental protection. Arrangements could be made which would order those operations of governments which would affect the environment without compromising, or wholly eroding, sovereignty in other areas. Local criminal, or religious, laws, and local political systems consistent with the further merging through internationalization of environmental laws, could still be maintained. That was the direction in which he would like to go.

Westing acknowledged the fact that 'biospheral' might be a more meaningful and desirable term than 'environmental' and could be used in treaties, conventions and the like.

He was more concerned with two huge domains that were already being abused, and would be seriously threatened in the foreseeable future, namely, the sea and the air. These had something special about them; they were both beyond national jurisdiction. We needed both a Law of the Sea and a Law of the Air very badly, as well as an 'ecocide' convention that reached into all the intricacies of The Biosphere in quite a different way. Incidentally, he disagreed violently with the notion of distributive justice in terms of distributing the largesse of the atmosphere. He did not think it ethical to divide up the atmospheric commons.

Finally, in establishing such laws and conventions, he preferred to recognize three kinds of situations; areas under national jurisdiction, shared areas under national jurisdictions, and areas beyond any national jurisdiction. Rather than trying to deal with the world for environmental purposes as a single global village, as **Samatar** had suggested, he thought it better to deal with it as a series of regional villages – what he would call biogeographical regions.

Stone said he had not adopted the idea of an 'ecocide' treaty along the lines of a genocide treaty. Genocide was the deliberate killing of a people whereas most of the destructive environmental acts arose, like the destruction of forests, or animals, through the extension of human habitats. Societies responsible for such acts were well motivated towards providing for themselves and were not simply concerned with destruction *per se*. It was therefore rather difficult to come to a consensus on an ecocide treaty on tropical forests, for example, when it could be argued that the Brazilians were only trying to provide for themselves in a better way.

Harmathy said first, that while it was understandable that the discussion had centred on international problems, national elements were important also. **Stone** had noted that international law made use of traditional, conventional concepts of national domestic law but it needed to be asked, did international law have the backing of national law, especially of effective laws so far as environmental law was concerned? A real problem was that many issues in this area had not as yet been solved by national laws. In addition, there were important differences between environmental laws in different countries.

Secondly, the discussion, understandably, had been concerned with the future of the Earth and of its environment, but we should not forget Mankind. Many new problems had arisen from developments in science and technology and they were both ethical and moral. He would not elaborate as all participants were well aware of them , but they should not be forgotten.

Lastly, there was a further set of problems related to the first two but of a different nature, namely, problems of human rights. That was a particularly poignant problem for those in the former Eastern bloc. Broadly speaking there were two possible approaches: one was to search for common human rights and the right to privacy; the other – which could be interpreted in the opposite sense – was societal rights. In Hungary and other adjacent countries the emphasis had been for many years on the latter, but now the pendulum was swinging predominately to the other extreme: yet a balance had to be found.

Medawar said that if we were going to improve the situation in The Biosphere we would need both international law and science. She was concerned that **Stone** appeared to think that scientists could simply collect facts mindlessly. That was not true; scientists had to have ideas before they collected facts. It was important to understand that.

Stone assured **Medawar** that he was not under that misconception, he had actually trained in the philosophy of science!

Burnett drew attention to two issues. One of the difficult features of international law was that little attention appeared to be given to its enforcement. That would become a crucial issue in the case of international guardianship.

The second issue was that it seemed to him that different kinds of phenomena appeared to have been classed together, perhaps because basically they were in the commons, *i.e.* had a material base. But the ozone problem and over-fishing were quite different; whereas the latter seemed to involve deliberate intent, the former was an inadvertent effect. Moreover, they had quite different consequences upon their perpetrators.

Stone pointed out that one reason for the apparent lack of machinery for enforcement lay in the fact that the more teeth a treaty had, the fewer signatures it garnered. There had to be compromise and one could only do the best possible.

With respect to the differences between the situation in the ozone layer and problems of over-fishing, he recognized them; indeed, there were other differences such as fish regenerating, etc. But, if there were no attempts to attach criminal remedies, the distinction between intentional acts and inadvertent consequences was not important. The only important issue was how to devise some way of getting rid of the activity in the future.

Burnett said that that distinction was not entirely correct for, in connection with the Bhopal accident, India was claiming damages for the consequences of an inadvertent act.

Stone pointed out that once people knew about the accident it was no longer inadvertent but he accepted that at the time it was due to non-culpable behaviour. That was a good point but in the case of ozone there was no question of damages and one was only looking to legislate for the future.

20. Transnational Governance

Ervin Laszlo

Rector, The Vienna Academy, Porzellangasse 35, 1090 Vienna, Austria

Concern over the governance, and indeed the governability, of contemporary societies is growing. Transnational governance is increasingly perceived as desirable; however, it is not clear whether such governance is yet fully feasible, and there is no agreement on the tasks with which it should be entrusted. This address focuses on these two basic issues in turn.

The Feasibility of Transnational Governance

Questions of feasibility can be approached from various angles: legal, institutional, political, etc. In the time allotted to this address I can only outline these approaches, without entering into detail. I shall present an overview of the issue from a broad historical and psychological as well as social-psychological perspective.

In its legal and institutional form, the modern nation-state dates only from the Peace of Westphalia, which was concluded in 1648. The concept became institutionalized throughout Europe in the 17th and 18th centuries, and has spread to the far corners of the world in the great wave of decolonization following World War II. Whereas the developing countries involved objected to almost every concept which they inherited from their former colonial masters, they never contested the validity of the principle of sovereign national states. As a result, today's world community consists of more than 180 nation-states, and merely a handful of territories with non-sovereign status. Humanity has accepted the 'inter-national' system as a permanent feature of the world.

By the middle of the 20th century, national sovereignty had become sacrosanct. Yet national governments agree on few things other than their refusal to relinquish their sovereign powers – either to subnational entities such as cantons, provinces, regions, 'republics', or federated states, or to transnational entities such as regional federations, economic communities, and international or 'global' bodies. This unyielding adherence to national sovereignty is inscribed neither in the 'laws' of history nor in the laws of human nature. It is the product of an epoch that seems already to be passing.

For in the last decade of the 20th century, unilateral insistence on national sovereignty may have to yield to the exploration, at least, of more adapted or 'tailored' forms of social and political organization.

Manifold Corollaries of Globalization

Processes of globalization make changes in the organization of societies both urgent and imperative. The concentration of decision-making power in the hands of central national governments is becoming dysfunctional, both for small and poor countries where the national level is undesirably restraining, and for large and diversified states where the national level is unduly constraining. It will be useful, and indeed necessary, to create new types of socio-political units, decentralized according to specific spheres of competence. Communities that regulate the behaviour and interaction of people in civil society should not be as large as the mainstream contemporary nation-states: the centres of decision should not be removed from the people whose lives are affected by the decisions.

On the other hand, economic communities could be more extensive than many are today, in both population and geography: economies of scale are important in regard to natural resources, to labour-forces, and also to markets. Cultural regions could be larger still, corresponding to the diffusion of the culture that provides the values and the world views that enable people to perceive and to act in their natural and social environment. The units of decision-making could be even broader when it comes to questions of defence, whether in the face of aggression from another society or aggression by a society *vis-à-vis* Nature.

The aim of decentralization and diversification in the decision-making of contemporary nation-states is not Utopian *per se*; its constraints, being institutional and legal rather than historical and psychological, are transcendable and temporal and not intrinsic and permanent. There are no insuperable factors in the psychology of individuals that would limit their allegiances to monolithic national states. No individual is obliged by his or her intellectual or emotional make-up to swear exclusive allegiance to only one flag, in the conviction that it symbolizes 'my country, right or wrong'.

People can be loyal to several layers and units of society without being disloyal to any. They can be loyal to their community without giving up loyalty to their province, state, or region. They can be loyal to their region and also feel at one with an entire culture, and even with the human family as a whole. As Europeans are Englishmen or Germans, Belgians or Italians, etc., as well as Europeans, and as Americans are New Englanders or Texans, Southerners or Pacific Northwesterners, etc., as well as Americans, so people in all parts of the world can have both narrow and broad identities, even if the latter are underutilized and atrophied owing to the persistent dominance of the nation-state ideology.

Doubtful Desirability of Nation-states

In the further development of the institutional and political order of contemporary societies, the ideology of the nation-state need not be preserved. The historical process is already moving beyond the system of nation-states, even if the legal instruments are as yet underdeveloped and public recognition of the process is slow to emerge. The Europe of today is not a nation-state; it already provides a wider identity and a supranational arena of decision-making for contemporary Europeans. The new Europe is neither a source of confusion nor a ground for conflicting allegiances. Indeed if, for example, the English and the Germans, or the Belgians and the Italians would not persist in the legal device of forming sovereign nation-states, their Europeanness would unfold much more effectively than at present, and action on the community-level would be greatly facilitated.

In North America such regional areas as New England, Dixie, and the Pacific Coast, are not nation-states, but Americans identify themselves with these subnational regions in addition to identifying themselves with the US as a whole. If the union of the states of America would not persist in claiming nation-state sovereignty for itself, the regional identities of its people could also evolve and come to the fore, reducing alienation without, supposedly, weakening national identity. Regional autonomy, in the guise of the 'problem of nationalities', has become a survival problem in the Union of Soviet Socialist Republics, and is also emerging as such in the federation of the provinces of the People's Republic of China. In all of these giant federated units, a creative resolution of the problems posed by the claim of exclusive sovereignty is urgently needed.

– and of Giant Centralized Confederations

The centralized structures of giant nation-states are a hindrance to the freedom and autonomy of their people. In a society that exceeds 60–80 million souls, national governance tends to be dysfunctional. The periphery has a tendency to get detached from the capital, diversity interferes with unity, and structural disequilibria appear between rich and poor, or between city-dwellers and country-folk. Historically, communities that managed to conserve their identity through the centuries did not exceed a relatively modest population-range – for example the English, the French, the Dutch, the Finns, the Austrians, and the Hungarians, to mention only a few in Europe. Even the Chinese and the Indians, enormous as their national populations are today, have evolved as regional cultures of relatively modest size and have consolidated into nation-states and grown to giant dimensions only during this century.

Transnational governance appears feasible only if modern nation-states are decentralized. Transfers of sovereignty are required in both directions: from the federated level of large nation-states to their federated subunits, or upwards, and from the national level to the level of the 'continental' subregion or region. As long as nation-states defend their full and exclusive sovereignty, transnational governance will be a 'paper tiger'. In this regard the experience of the two multipurpose global bodies which were called into being in this century is significant. The League of Nations broke apart when a belligerent nation-state began its territorial expansion, leading to World War II; and the United Nations, though persisting in a legal and institutional form, is at the mercy of its nation-state members. Its General Assembly is not a transnational governing body but a forum for debate, with authority that does not exceed that of making recommendations.*

That some specialized agencies can, nevertheless, function in a comparatively effective manner is due to the consent of the member-states – a consent that they can withhold or withdraw at their discretion, together with whatever funds and manpower they may have contributed. In the last analysis, the very name of the organization implies its limitations: 'united nations' is, *sensu stricto*, a contradiction in terms. Nations, if sovereign, cannot be united without losing their individual status. The process of union is possible only if there is a transfer of sovereignty to the collective level of the partners in the union.

The Tasks of Transnational Governance

Assuming, if only for the sake of argument, that effective transfers of sovereignty may take place as the international community heads into the 21st century, the question arises as to what ends and purposes such transfers should serve. In the light of the history of the European Community – the body that comes closest to having achieved the status of transnational governance – the economic sphere is first to come to mind. On deeper reflection, however, the term 'governance' does not appear proper when applied to economic processes. The governance of the economy backfires even when attempted at the national level, as the experience of the East European socialist states demonstrates. On the transnational level, the problems would be further compounded. The economy can be regulated in some respects, its processes coordinated and harmonized (as the EEC has done), but it should not be governed. The result would be *dirigisme* in some form, with drawbacks that would far outweigh any possible advantages.

The concept of governance is no more felicitous when applied to the exchange of information. Although some information-flows need careful surveillance – those, for example, that carry vital technological and strategic messages – the governance of information raises the spectre of censorship which, even within the framework of a well-intentioned 'information order',

*[Nevertheless the United Nations Security Council, recently at last, has shown itself capable of engendering prompt and effective action, through persuading the coalition, led by the United States and the United Kingdom, to prevail massively over a belligerent nation-state, Iraq, in such a decisive manner that it seems unlikely that any nation-state will attempt to take over by force a weaker neighbour for a long time to come if ever in future. This emergence of a single effective and willing superpower seems to portend well for future peace. Eds.]

would end up by doing more harm than good. Science, education, and culture in general, should not be transnationally 'governed', nor would health and social services be properly subjected to transnational governance. There are, on the other hand, two areas where transnational governance is both urgent and imperative. These are the areas of *security* and *environment*. Both concern collective defence in the face of aggression: in the one case aggression aimed at a human society, and in the other aggression aimed by a society at Nature.*

Transnational Defence Governance

The belief that national security considerations would necessarily lead to a call for a powerful national defence-force is as much of a fiction as national sovereignty, and is strongly dependent on the latter. If a community does not claim unconditional sovereignty over its territory, it has good reasons to entrust the defence of its borders to joint peace-keeping forces. The logic of such a step has already penetrated in Europe: a European Defence Community, rejected by France when it was first proposed in the 1960s, is again in the making and may become reality before the end of the century. People in Scandinavia, the Benelux region, and on the rim of the Mediterranean, seem open to proposals of joint defence forces for those regions.

It may, however, be more difficult to convince the majority of citizens of certain nations to give up their national armed forces: their current national ethos is apt to include a dream of major military status. But surprises cannot be ruled out, as a referendum conducted in Switzerland in November of 1989 illustrates: although the Swiss army is a national institution that has long been held in high esteem by the population, the Swiss socialists managed to collect enough signatures to force a referendum on whether or not to maintain it. The expectation was that not more than 5–6% of the population would vote for abolishing the army, but in fact more than 30% did.

Medium and small European countries may readily espouse the position that it is pointless for them to maintain an expensive army apparatus when, with much smaller expenditures, they could assure good internal and external security – the former through a well-equipped police force or national guard, and the latter through community-level peace-keeping. But even if European states managed to evolve a joint defence system, neither the United States nor the Soviet Union seem ready to entrust their national defence to a peace-keeping force on the global level that alone would be effective in their cases. Although the United Nations peace-keeping forces proved their effectiveness in Cyprus, the Near East, and the Middle East (and were honoured with the Nobel Peace Prize in 1988), the record for global peace-keeping of the international community is not encouraging. Support for inserting such a force remains limited to those who are con-

*[This recalls to mind our recent attempts to have peace – *inter alia* in the bestowal of the Nobel Peace Prize – recognized as being with Man's and Nature's environment as well as militarily. Indeed we have strong expectations that increasingly in future, peace with the environment will prove to be the most important of all forms of peace! Eds.]

cerned about chronic trouble-spots where the superpowers themselves are stymied. The greater the military establishment, the more difficult it is to integrate it into a collective peace-keeping system.

Global Security-system Imperative

Effective regional and global peace-keeping may be merely Utopian as long as nation-states insist on their full sovereignty, but may become realistic as effective transfers occur from the nation-state level to the regional, and in some cases to the global, level. In the absence of such development, collective peace-keeping may be no more than a thinly-disguised form of potential aggression, where forces with supposedly pre-emptive first-strike capability – but also readiness for full-scale retaliation – confront each other in a precarious balance of terror.

A global system of security has become imperative, given the level of thermal and chemical pollution that would be produced by prolonged full-scale war, even if it were fought with conventional weapons. The Biosphere's delicate, already highly-stressed, balances could be irreversibly impaired. A third world war would be likely to have no winners, only losers. And the losers would include all the people of the globe, whether they took part in the fighting or not.

On the other hand, a reliable system of national, regional, and global, security would enhance chances of economic prosperity. The economies would be freed from the burden of maintaining costly military establishments and could use their human and financial resources for productive ends. The advantages, as the post-war 'economic miracles' of both Germany and Japan demonstrate, could be enormous – as would be the environmental and conservational benefits if only a fraction of the military savings could be turned to their uses.

Transnational Ecological Governance

The second area where global governance is imperative is that of environment. In this context 'environment' stands for The Biosphere as a whole – the total system surrounding our planet in which life exists naturally, and in which Humanity and Nature are integral elements and interdependent partners.

Humanity, comprising all races of *Homo sapiens*, in the manner of other living species, can only survive in an environment where basic biospheral balances are adequately maintained. As just noted, these balances have already been seriously stressed. We are on the way to a higher global heat-balance; to some extent the 'greenhouse' effect has become irreversible. We have thinned the stratospheric ozone shield particularly through our use of chlorofluorocarbons; and this, too, has passed the threshold of full reversibility. We have killed off countless species of plants, animals, and, doubtless, microbiota, and these at least can never be regenerated. We threaten a third of the planet's total land-surface with desertification, and may have already sealed the fate of several tropical rain-forests. Only time can tell the extent of the damage we have wrought, but it would be foolish to wait until time does tell. With each passing day the processes of environmental degradation become more difficult to turn around.

The preservation of the essential balances, as well as of the richness and diversity of The Biosphere, are global tasks. In no other area is 'acting locally' as much in need of 'thinking globally' as it is in that of preserving our sustenance and home, The Biosphere. Global thinking must find concrete expression in coordinated action by all societies of the globe, without exception.

Objectives of Global Environmental Governance

The basic objectives of global governance in regard to the environment are widely known and uncontroversial. They can be grouped in three basic sets, of which one set would focus on the regulation of the mining (*sensu latissimo*) and use of natural resources; a second set would be concerned with safeguarding the balances and regenerative cycles of Nature; and the third set would foster the creation and maintenance of capacities for dealing with environmental downgrading and emergencies.

The principles that would govern the use of natural resources are the stewardship principles as regards resources within national territories, and the collective heritage principle regarding territories outside of national borders. Today, sovereign nation-states consider themselves the absolute owners of the forests, wetlands, croplands, rivers, and lakes, that fall within their national territory, as well as of the metals, minerals, and fuels, which are found on or under the land and the continental shelves of their territorial seas. Consequently nation-states' use of 'their' natural resources can, and often does, do violence to Nature and global interests. Roman law specified '*jus utendi et abutendi*' – namely 'the right to use is also the right to abuse'. Nation-states could be depended upon to rectify their behaviour in these respects only if, instead of claiming the sovereign right to use and to abuse, they would subscribe to the principle that they are *stewards of the Biospheral resources that happen to fall within their borders.*

Parallel considerations apply to resources that are located outside national territories*. The great reservoirs of industrially valuable metals and minerals on the continental shelves of the sea, for example in polar regions, must not be irresponsibly depleted any more than resources situated within national territories. The collective heritage principle, applying to extra-territorial resources, would govern global accords specifying access, use, and preventing unfair exploitation, by technologically and/or financially powerful actors.

Another set of measures, aimed at safeguarding the balances and regenerative cycles of Nature, would have goals such as the setting and enforcing

*[Particularly the 'commons' and their resources that are dealt with so effectively in prospect by Professor Stone in the preceding chapter. Eds.]

of rigorous controls on the emission of chlorofluorocarbons (CFCs), as well as on the burning of fossil fuels; the setting of upper limits on the use of atmospheric trace-gases (such as carbon and nitrous oxides, hydrocarbons, and methane); the designing and implementing of major reforestation programmes; the designation of perhaps up to 10% of the planet's land surface as protected areas for the on-site preservation of genetic resources; and the carrying out of soil conservation programmes on impacted areas.

A third set of measures would need to focus on the creation and maintenance of environmental emergency capabilities. The objectives include the identification of geographic regions that are vulnerable to flooding, if and when the polar ice begins seriously to melt; the warning and, if necessary, the relocating, of coastal populations threatened by inundation; the retraining and relocating of farmers affected by changing weather patterns; and the maintenance of adequate rescue capabilities for use in the event of major ecodisasters and ecocatastrophes.

Such measures need to be globally concorded, and to have the status of enforceable law. Effective enforcement presupposes, however, that the corresponding aspects of national sovereignty are transferred to a collectively constituted, global-level authority with a mandate to safeguard the viability of The Biosphere.

Need of Transnational Action and Funding

Both such security and environmental dimensions of transnational governance are urgent and imperative. In addition to the upward transfer of national sovereignty, implementation presupposes a resolution of the question of funding. Voluntary contributions of funds would be neither adequate nor reliable; and mandated contributions, if they are not be be coercive, need to identify the source of the funds. In regard to collective security, this would not be a problem: joint peace-keeping, whether on the regional or on the global level, could be entirely financed by the savings achieved by member states thanks to the reduction of their military and defence budgets. There would even be excesses – in all likelihood very substantial excesses – that members could use to finance development, economic and technological modernization, health and social services, and other expenditures which they might deem essential.

Unless savings from the security area were to be transferred to the area of the environment (which would correspondingly reduce the savings available to member states for their own purposes), the environmental area of global governance would have to find its own funding base. A likely source appears to be environmental taxation.* Environmental taxes could not only raise funds for joint operations; they could also have beneficial effects in reorienting economic activities along environmentally sound lines. They need not be an added burden on the economies concerned: states could

*[In addition to royalties and other payments for use of commons and commons-based resources in the manner proposed in the preceding chapter. Eds.]

compensate for the additional taxes by reducing others, such as corporate and personal income taxes, and value-added taxes.

The specific mix of taxes could be left to individual states to decide on; it would be enough for them to introduce sufficient tax incentives to redirect their economic activity from polluting and wasteful, to clean and sustainable, modes of operation. As examples we may cite how taxes could be introduced on the production and use of fossil fuels, and likewise of hazardous nuclear energies: this would create incentives for the research, development, and application, of renewable, clean, and safe, energies, and also enhance energy efficiency. Taxes on polluted wastewater, and on the use of clean water above specific thresholds, would induce water recycling, internal purification, and efficiency in water-use.

Similarly, taxes on unsorted solid waste, and proportionately higher taxes on hazardous waste, would prompt improvement in product design, production processes, internal recycling, and land and water detoxication – as well as technology choices and corresponding consumer preferences. Taxes could be levied even on urban and industrial sprawl, making it more efficient to clean up unused or derelict industrial and urban areas than to pave over virgin or desirable agricultural land, far too much of which has latterly been covered more or less finally by buildings, roadways, or runways, etc.

Conclusion and Recapitulation

In this last decade of the 20th century, the processes of interaction among human societies have transcended national borders in almost every vital area. Only the principles of governance have remained what they were before the Peace of Westphalia, more than 350 years ago. The time is now ripe for *transnational governance*. Its spheres of application need to be carefully specified and delimited, however, and the creation of giant bureaucracies and rigid hierarchies must be consciously and consistently avoided.

In the last analysis, decision-making can grow to the transnational dimension if, and only if, the traditionally sacrosanct nature of national sovereignty is surrendered. A full recognition of the historical obsolescence of that principle is the single greatest precondition of recovering the governability of contemporary societies in regard to two of the most pressing issues of our times: global security and Biospheral sustainability.

Contemporary societies are becoming increasingly ungovernable. At the same time, new issues emerge that call for governance even about the national-transnational plane. A precondition of effective transnational governance is the surrender of the principle of nation-state sovereignty as an operative principle in international cooperation. This step could be facilitated by the progressive decentralization of national decision-making, shifting it from the nation-state level both downwards, toward grassroots communities, and upwards, into the sphere of regional and international organizations and communities. Such a move would be consistent with the demonstrated ability of individuals to evolve allegiances simultaneously on several levels: local, national, regional, and even global.

The two areas where transnational governance is essential are those of security and environment. Nation-states can no longer protect their citizens either from the adverse effects of aggression against Nature, or from aggression directed against them by other states. Global-level concords are needed to ensure the safeguarding of the twin concerns of peace and environmental viability. Savings accrued from the reduction of military and armaments budgets, together with revenues from special-purpose environmental taxes and levies from use of 'commons' space and resources, could provide financing for the transnationally concorded operations.

The transcendence of the now-obsolete principle of full sovereignty for individual nation-states has become the key issue for implementing the imperative modalities of transnational governance, to ensure both peace and Biospheral sustainability in the contemporary world.

References

Further details and bibliographic references are given in Ervin Laszlo:
The Age of Bifurcation. Gordon & Breach Science Publishers, New York & London, 190 pp., 1991 (or in German: *Global Denken*. Horizonte Verlag, Rosenheim, 190 pp., 1989, and Goldmann Verlag, München, 190 pp., 1991; in French: *La Grande Bifurcation*. Tacor International, Paris, 189 pp., 1990; also in Spanish, Italian, and Chinese).

Commentary on Chapter 20

CHAIRMAN: HE Ambassador Adib Daoudy
PANELLISTS AND OTHER CONTRIBUTORS:

Wasawo, Mische, Kefeli, Dmitrieva, Westing, Holdgate, Tisdell, Burnett, Mische, Ramakrishnan, Imru, Laszlo

Wasawo said that **Laszlo** had brought immense knowledge and experience to his consideration of transnational governance (Chapter 20). He could only concur with this thoughtful contribution for he, himself, was neither a politician, student of politics, nor had he governed. He would have to speak to the theme, therefore, from his heart rather than his head.

Why Transnational Governance? **Laszlo** had pointed out that there were some human imperatives which demanded that certain problems could be more effectively addressed and solved if a little national sovereignty was sacrificed for the greater international good. In East Africa, for example, we were fortunate in inheriting, at independence, the East African Common Services Organization. This Organization was transformed into the East African Community, after due negotiations by politicians from Kenya, Tanzania and Uganda. It served an extremely useful purpose in the rapid development of the three states. Its success has been such that even some neighbouring states have applied to join it. That it broke up after a period of only twelve years, was a matter that is regretted everywhere in East Africa. He suspected that even today, if a referendum were to be carried out, a large majority of East Africans would want it to be reinstated in some form or another, based on the lessons learnt through its demise.

The decision to scrap the East African Community was made by politicians without reference to the people at large. Nobody had, to date, carried out an exhaustive analysis as to what drove them to take such a backward decision. But part of it would be found in personal greed. Community institutions were so structured that it was a little more difficult to make arrangements for quick 'kick-backs', than would have been the case at individual government level. Part would be found in lack of goodwill on the part of politicians that would have allowed the implementation of programmes in accordance with the letter as well as the spirit of the agreements. Part would be the attitude: 'I know better (and therefore my country knows better) than you and your country.' Part would be a lack of vision for the future. Thus one of the requirements for effective and sustainable transnational governance was a crop of politicians, at the national level, who were men and women of integrity, goodwill, humility and of vision. How to institute an elective mechanism at the national level that threw up these sorts of people, was one of the challenges. The many groupings existing in Africa today would thrive or deteriorate according to how far these tenets were met, among others. He believed this applied to the world at large.

Within the context of the Conference, however, several references had been made to global problems that required concerted global action. There were already protocols, such as the Montreal one. There were also arrangements such as those emanating from the Law of the Sea, although some of our powerful friends were not being very helpful on that one.

The Brundtland report had inspired us by its title: 'Our Common Future'. He did, however, have a suspicion that we might not have much of a 'Common Future' unless we addressed the problems that disturbed our 'commonness' as human beings. He wanted to draw attention to some of these problems, as they were, in his view, some of the prerequisites for transnational governance. But before he did so, he observed that: 'The concept of goodwill among people is in good measure dependent on our attitudes towards one another.' The response of humanity to disasters such as those we had been having in the Sudan and Ethiopia was always a reminder of the intrinsic goodwill that existed amongst all of us. We all felt horror when, for example, there was an announcement of an air disaster in any part of the world. We felt for the bereaved. We mourned the dead. It was against this deep-seated human trait that he would now turn to two of our problems, namely, racial and colour prejudices, and religious intolerance.

The first problem was at the root of a lot of violence and Man's inhumanity towards Man – a problem which had punctuated the history of Mankind, and was still with us today. Prejudice of any sort was, he believed, an ingrained human trait; but as one of his teachers used to say, 'keep your prejudices to yourself, for the sake of the common good.' It was when prejudices were expressed or acted upon that they did damage. Nobody doubted the damage that had been perpetrated in the name of racial or colour prejudices. They had affected the physical and psychological environment of human beings. They were our concerns as environmentalists.

But colour and racial prejudices were complex manifestations. He asked participants the following: throughout his student days at Oxford he had never experienced racial prejudice, not even in 'digs'. But it was there aplenty whenever he visited London, and this was confirmed by other Africans studying there. He enjoyed his visits to Scotland, particularly to the pleasant town of Inverness. He did not know anybody when he first arrived in Inverness, but the camaraderie that developed around evening fires, over a wee dram of whisky, was spontaneous and colour-blind. Why were these manifestations and their contrasts present in the same people – the British; and could we learn something from them for our future commonness?

Having stayed in the United Kingdom for some years, he had gone back home to Kenya which was still under colonial rule, to be faced by notices at hotel entrances reading: 'Africans and Dogs not allowed'. He tried to have a cup of tea in one of these citadels and was slapped and kicked out. Eleven years later Kenya became independent, and the notices went although we still had the same ownership. One of the hoteliers had become a good friend of his and was a good Kenyan although he was white. He had never got around to asking him why his hotel organization had put up those strange notices at the hotel entrances before independence. But here was a fellow who was able to change his human colour-vision in response, he believed, to changed circumstances.

During the Second World War, the whole of the British Empire was mobilized to fight the Nazis. His people were informed that they were fighting evil – the evil going under the banner of 'herren-volk'. Many Africans died in that war. After the war, a British Prime Minister, Harold Macmillan, had articulated his country's stand with regard to Southern Africans' problems in his famous 'Wind of Change' speech which had given them all heart and hope. Years later, another British Prime Minister, Margaret Thatcher, was going it alone at Commonwealth Conferences to the extent that one of our Presidents in exasperation referred to her as a racist! And yet, when one came to think of it, what was the difference between 'herren-volk' and 'apartheid'. Why did we mobilize to fight one with all our might, and yet were lukewarm about the other? The same swastikas were now showing in South Africa.

He had started with the proposition that racial and colour prejudices were complex manifestations, and the three illustrative happenings he had described seemed to bear that out. But the very fact that they were complex argued for their being treated with the urgency worthy of a problem important for the future of humanity. Part of the action would be through education, for the mind of Man had to be addressed. Education would require relevant approaches and tools, which could only be obtained through enquiry and research. An underpinning by legislation would be necessary also.

The second problem related to religious intolerance. Man's need for spirituality was as old as his culture and probably as old as the species itself. Even so-called pagans believed in something. The flowering of spirituality had expressed itself throughout history in divers ways depending on geography, conquest, cultural infiltration, and so on. We thus had, among others, Fetishism, Confucianism, Buddhism, Christianity, and Islam. However, we bore witness to situations where many sins were perpetrated in the name of religion, including violence, murder, coups, and so on. Some of these activities had affected international relations, and might make transnational governance that much more difficult. Religious intolerance was, therefore, a problem which, although very difficult to deal with, nevertheless had to be addressed.

On the other side of the coin were the multiplicity of sects that had recently been witnessed mushrooming all over Africa and the rest of the world. It had been said, with some justification, that if you wanted to get rich quickly, you started your own religious sect! Some might be vehicles of subversion. Governments were now being faced with difficult choices. They had either to let this sort of thing run riot in the name of 'religious freedom', or judiciously, had to step in to protect the public.

He closed with a personal statement. He believed that, by the Grace of God, he was what he was, but equally, that with the help of God and some effort on his part, he could be something better. He suspected that this might be true of most humans: but each one of us should be left to decide on his own kind of God.

Mische drew attention to PROJECT GLOBAL 2000. This was an international partnership of non-governmental groups involved in re-conceptualizing security and sovereignty in the contexts of economic and ecological security. It was an effort to bring together teams from some 50 countries over the next few years, to look at these questions within regions and nations. He believed it provided a focus and forum for pursuing the points that **Laszlo** had made concerning sovereignty or, indeed, many other themes which had been raised at the ICEF Conferences.

His central thesis was that throughout history leaders had been following the concept of absolute sovereignty. The logic of such societies was that, to ensure security, they were forced to follow a 'macho'-power paradigm which, necessarily, had to exclude the use of humanistic, authentically religious, feminine nurturing – whatever terms were used to describe the values that those societies' leaders were compelled to exclude as basic criteria for public policy at the national level. To focus too strongly on the apparent irrationality of human beings, or on fallen human nature, was to disempower reality. He suggested that, until it was generally realized that there was a systematic powerlessness built into the principal of unlimited sovereignty which could only be overcome by building international systems, then discussion was going to be very limited. This realization was what had caused him and his wife to found Global Education Associates. Although both of them had backgrounds in international development, they had worked in local development and it was because they were concerned with the latter that they realized the need for international systems.

In 1973 they realized how global economic and monetary forces paid little heed to declarations of self-determination or sovereignty, and that convinced them to look for the type of international, transnational governance that they believed was needed. In 1977 they had written *Toward a Human World Order: Beyond the National Strait-jacket,* which confirmed the view, shared by **Laszlo** and others, that there was a systematic strait-jacket – the basically lawless, international marketplace – that prevented Heads of State from adopting the values that were needed. Small-is-beautiful scenarios and environmentally-responsible regional models should be at the tops of national agendas. In redefining security they had had to focus on military, monetary and economic issues and the linkages between them because these were all global in their scope. In today's interdependent world, for example, security was increasingly dependent upon competition within the lawless, international economic system, so that exporting had become, primarily, a national security sector. We had little choice in the present situation if confronted by substantial debt or substantial trade deficits, when foreign currency was essential, other than to mobilize society, training and the economy to produce for export. Food and agricultre also became the tools of national security. We were forced to make alliances largely with the 'haves', who were able to produce for export, *i.e.* business and other substantial corporate enterprises. These might not be concerned with human needs, nor concerned to produce for domestic consumption; but they exported for wealth. That was a typical example of the kind of strait-jacket in which we were enclosed. Being forced to rethink sovereignty in these terms had led them towards international systems and, once those were achieved, he believed that they could move more easily to decentralized, small-is-beautiful efforts.

There was, therefore, a double motive for looking at transnational governance. One was the systematic breakdown that **Laszlo** had described, the other was the systematic powerlessness of the present system which he had outlined. Powerlessness of leaders as well as people would continue without effective international law. His experience had been that, as people from different constituencies were helped to see the strait-jacket in the present system, they developed an additional motive, related to self-interest, namely, to commit themseves to better understanding the present system, to understanding the strait-jacket which confined them, to thinking about the kind of international system needed to loosen their bonds, and the necessary kinds of political action to achieve such a system.

Kefeli spoke of the ways in which his institute, the Institute of Soil Science and Photosynthesis (ISSP) at Pushchino, where it was associated with the Vernadsky International Centre for Biosphere Studies, contributed to international endeavours. Transnational governance had not been a real concept for his father or grandfather and only for today's children would it perhaps become a reality in the 21st century. Transnational science could help to bring that about. Investigations at ISSP involved many international programmes such as investigations of ecological systems without, or with only minimal involvement of, Man, *e.g.* the Yenisey meridian, based on the R. Yenisey which flows to Siberia from Mongolia, China and India. Russian scientists were collaborating with the Mongolians and the Chinese on this project, and with Americans in the regions bordering the Bering Straits.

Many of their other programmes, especially the agricultural ones, were for the benefit of all nations. For example, plants in the Mongolian desert produced substances which inhibited the growth of others for 4 m around. Isolation and characterization of the active principle might enable it to be used as a natural herbicide instead of the present synthetic ones. He would not elaborate on the many other researches, either international in scope or with international applications, but the international emphasis of his institute and of science itself, inevitably, promoted transgovernmental notions, albeit indirectly.

Dmitrieva spoke of the universal environmental degradation which affected Mankind and to whose causes he was a party. Ecological awareness and understanding were prevented by interactive causes as discrepant as individual or group egoism, real gaps in our knowledge, and the lack of information in a form intelligible to uninitiated people. The novelty of the problems and their complexity demanded joint action and the combination of different intellectual efforts. It had become clear that long-term goals for the preservation of The Biosphere and the survival of Mankind often contradicted the present-day activities of many governments.

Vladimir Vernadsky, a twentieth-century thinker believed that only a transformation of The Biosphere to a new state, the Noosphere, would enable Mankind to persist. The Noosphere would not appear everywhere immediately but would have various origins. These would be in conflict with the destructive tendencies of the past and the present. The development of the Noosphere presupposed the recognition of a well-balanced concept, the necessary means to achieve it, and ecological awareness by people at all levels; otherwise the choice would continue to be between a 'good economy or good environment'. These ideas had grown in popularity as they had become more topical and, at a symposium in March 1989 on 'Man and Biosphere: History and Modern Times', Victor Kovda and Nicholas Polunin had proposed the establishment of a Vernadsky International Centre. This had been achieved by December 1989. It was a non-governmental, non-profit-making, institute whose objectives were the accumulation, systematization, and dissemination of knowledge about the state of The Biosphere; the organization and coordination of investigations of The Biosphere and its components; the development of methodology and means of exploring The Biosphere at different organizational levels, having regard to Vernadsky's doctrine of The Biosphere and the Noosphere; the development of recommendations of global and regional importance to be considered both at national and at international levels; and the development of ecological thought and education for people throughout the world. The most important task of the Centre at present was the interdisciplinary exchange of information on the state of, and investigations into, The Biosphere, and the effective, rapid distribution of such information through translation and publication. Much attention was being paid to the situation in the USSR where 15% of the territory and one-sixth of the population were in zones of ecological catastrophe. However, combined and concerted efforts on an international scale should provide reasonable strategies and indicate ways to realize them. Cooperative development from disaster zones to effective models of Noosphere development was the way to improve the prospects of our world's survival.

Westing shared, with **Laszlo**, a sense of the overriding importance of undermining national sovereignty and moving to transnational governance, and he hoped that we would reach that *nirvana* in the near future.

Firstly, he supported the notions of nested communities of varying size with different conceptual foundations which were not mutually exclusive, and the idea that the human community, the economic, cultural, defence, and environmental, communities should co-exist and all be of importance. The human community already existed in the 171 national sovereign states, but he would suggest that the size proposed for such groups – 60–80 millions or less – should be reduced by an order of magnitude and that 10–12 millions would be nearer the optimum size. The economic community already existed in the form of international trade. He was not sure whether the cultural community had to be operationalized, and he did not agree with the ideas for the defence community, although he recognized that some sort of defensive structures were necessary. In his view, the most important community to be superimposed on the human community was the environmental community.

The trick to achieve this situation was to maintain the fiction of national sovereignty while eroding it. For that reason we should try to maintain ceremonial aspects so that people believed that sovereignty still existed. One of the best ways to undermine it was, as some 24 or 25 nations had already done, to submit to the compulsory and unconditional jurisdiction of the International Court of Justice. That was a simple and straightforward procedure to follow which undermined national security very effectively.

Thirdly, he did not share the view that there was no individual whose emotions would not prevent him from giving up allegiance to a flag. He thought, on the contrary, that such allegiance was tied closely to a person's emotional make-up, by things such as sex-drive, chauvinism, patriotism, or the urge to build dams across undammed rivers! Such things were difficult to overcome and it was better to try to channel them in more benign directions than to pretend that they did not exist.

Lastly, and of lesser importance, he thought it had been suggested that high military expenditure was not conducive to economic growth and that the post-war economic miracles in Germany and Japan were tied to low levels of military expenditure. Japan had, of course, kept its military expenditure to about 1 or 2% of GNP. Germany, on the other hand, was perhaps the 3rd or 4th highest military spender in the world. A few years ago a careful investigation of the 27 richest nations in the world on the basis of *per caput* GNP showed that they coincided with the developed, industrialized nations. Investigation of their military expenditure in relation to their post-war economic achievements demonstrated no correlation at all. The argument of economic benefit could not, therefore, be used to justify a reduction in military expenditure.

Holdgate wondered if he was right in thinking that what we could expect to see was what he believed could already be detected, the growth of transnational institutions partially in terms of their geographical and thematic coverage which would gradually extend in scope and influence? That kind of development – an evolutionary approach rather than the sudden appearance of some great new system which those nation-states that clung to sovereignty would perceive as a threat, was surely what we should hope to see.

He saw three areas in the environmental field where such an approach might be needed: the regulation of the use of natural resources; the protection of regenerative cycles; and, of course, coping with emergencies. He could not see any nation resisting a gradual approach, as outlined, in any of these areas. In a way we were already collaborating at international level on safeguarding parts of the regenerative cycle, *e.g.* the Montreal Protocol on the ozone layer, and there were other possibilities, *e.g.* the Law of the Sea or atmosphere, a possible climate convention, Biosphere conventions, and the like. These were instruments of a legal kind which involved governments accepting certain obligations in order to achieve shared objectives, but they surely eroded sovereignty to some extent even if the sovereign states severally enforced the obligation. To accept a regulation on the use of a natural resource seemed to him to be much more difficult. Nevertheless, progress in the transnational, rational regulation of a natural resource was being made at the margins in Europe through the EC fisheries policy, and elsewhere through regional fishery conventions, although they had not been very successful. He would not suggest the Common Agricultural Policy (CAP) of the EC as a good model as that might deter rather than encourage!

In general, was not the trick to encourage others to demonstrate that success succeeds and so to extend, progressively, what was thought necessary rather than to attempt to achieve all at once? If so, what could be done to publicize what was effective and to demonstrate it to other parts of the world? Was there a role here for some kind of evaluation of the effectiveness of transnational, or international, instruments alongside UNEP's State of the Environment Reports?

Tisdell noted that **Laszlo** thought that the environment and security were ready for transnational governance but wanted to leave economics alone for the time being. That might be true, but the environment and the economy were very closely linked, if the Brundtland Report was to be believed, and could not really be decoupled. It followed that some transnational element would have to be introduced into the economy with environmental changes. Even if such transnational governance turned out to be only partly effective, it might act as a spur to its extension in the future. We were left with the problem, however, of involving the economy in any changes to transnational governance of environmental activities , or else we would simply have to ignore the Brundtland Report.

Burnett supported the views of **Tisdell**. In the EC, through the CAP, nations were being brought together economically (although he did not personally consider it technically a very satisfactory policy!), but amongst the agricultural directives were those which were effectively environmental directives: that was inevitable. So, without it being the explicit intention of Ministers, Europe was moving to an increasing tie-up between economics and the environment. He thought that in that sort of way, in effect by-passing the real problems, progress might be made even in such sticky areas as security, *pace* Mrs Thatcher!

Mische believed that we should face up squarely to the fact that sovereignty was a fiction. The examples given in the discussion had demonstrated that, clearly, sovereignty had to be redefined rather than undermined because traditional concepts of sovereignty no longer corresponded with what went on in the real world. People were increasingly recognizing that they were powerless, that was one reason why, earlier, he had given the example of debt as a strait-jacket. He agreed most strongly with **Tisdell** on the need to link economics with the environment. So long as we did not mobilize to face-up to the export situation, the longer ecological security would stay on the back-burner. He was all for facing-up to, and talking about, how to find security together in areas where it could not be found by a nation alone.

Ramakrishnan had been impressed by the examples of international cooperation which were now emerging, but they were between countries which were more or less equal partners in terms of economic growth. The real challenge and problems arose with geographically contiguous countries with unequal economic growth, and it was there that most of the environmental problems arose. Those situations needed more consideration and thought.

Imru said that a most acceptable trend to regionalization was prevalent in many parts of the world, especially at the institutional level, but the popular aspects needed more consideration. **Wasawo** had mentioned the lack of sympathy between the peoples of the world; how could such differences be bridged?

Laszlo said that he had intended to stimulate discussion and he believed that had been achieved.

He had been toying with the notion of nested communities with overlapping competence for many years, and had even thought of writing a book on it. He had contemplated the range from 2 millions to 100 millions for the possible optimum size but was not certain what was the right figure. It was clear to him that very large communities lacked social cohesion and this lead to regional breakups. Traditionally 2 millions had been regarded as a very good size to keep people together. Historically, Hungarians, Poles and Finns and perhaps even smaller groups had managed to retain cohesion. Other, originally small, countries had grown tremendously and it was possible that, through the use of modern communications technology, cohesion could be maintained and the optimum size increased. Plato had said that the size of a community was measured by the distance that a man's voice carried: today, a man's voice could carry around the globe! The problem needed much more thought.

To subscribe to the rulings of the International Court of Justice was a very good measure of the willingness of a state to surrender some aspects of its sovereignty, but we did not have comparable institutions in the economic, environmental, security etc., spheres. **Tolba** had mentioned the possibility of establishing an authoritative environmental institution but that could only become effective if there was a transfer of sovereignty in ecological areas such that nations would abide by its rulings.

On the matter of allegiance to a flag, he had meant to imply that we were not committed solely to one allegiance; we could hold multiple allegiances. We had an EC flag with 13 stars as well as 13 national flags and, on occasion, we also flew the UN flag!

He was uncertain and surprised by the notion that there was no linkage between low military expenditure and high economic growth. Surely it had been the case that when Germany's economic 'miracle' had started, the military budget was small? However, he was quite certain, whether the correlation was true or false, that military budgets should be kept small for many other reasons.

One problem about moving gradually to transnational governance, as advocated by **Holdgate**, was that growth was now approaching thresholds of stability and of irreversibility. He would like to avert a major systematic breakdown by anticipatory, proactive actions, so that what was known technically as a bifurcation – a major transformation – could be avoided. Such transformations were occurring in eastern Europe, in the Middle East, and in the Third World and were very dangerous. It might be possible, through the gradual growth of transnational governance, to bridge these unstable conditions. It might be pioneered, perhaps, in the environmental area where global concern and public awareness were growing so quickly, triggered as they had been in 1988 by Margaret Thatcher followed by almost all other Heads of Government or State, paying lip-service at least. He was not well informed about the EC's agricultural policy but if, indeed, it had resulted in combining economics with environmental issues, even inadvertently, then that could play a pioneering role in further transnational interactions.

He accepted **Mische**'s comments on people's sense of powerlessness and understood how they felt but, did governments realize their own powerlessness? Even if they did recognize it they were not prepared to admit it publicly and would continue to behave as if they had power. So long as that fiction was maintained it was difficult to act against them. Ways had to be found, therefore, to push them at the community level into decision-making in specific areas at a level which was, in fact, just above that of the national state.

Inter-regional problems were very real. He had been trying in the UN to study regional and inter-regional cooperation, in particular how compensatory measures could operate and in what areas large and small companies could work together (six volumes of studies had been published!). **Ramakrishnan** would know that in his own region, SARC (South Asia Regional Cooperation) had brought together six tiny countries and India, whose GNP, population, and every other attribute, was larger than those of the other six put together. He believed that cooperation was beginning to work but there was much to be done in bringing in compensations, much as the EC had done when Spain, Portugal and Greece entered the Community. Every effort must be made to involve the poorest and least-developed nations in such arrangements.

Finally, he emphasized the importance of promoting the public consciousness. He believed the people of the world were well ahead of their governments and his experience in the UN was that, when once people started to talk to each other, they rapidly developed a sense of solidarity and understanding. However, if delegates developed in this way through talking together and went too far, their governments rapidly recalled them and sent in fresh people! Capitals did not really talk to each other, they maintained the fiction that they were autonomous and sovereign entities. So although there was a need for greater empathy, the most important thing was to maintain communication between people rather than between institutions.

21. The Third Baer–Huxley Memorial Lecture: Building an Environmental Institutional Framework for the Future*

MOSTAFA K. TOLBA
Executive Director, United Nations Environment Programme,
PO Box 30552, Nairobi, Kenya

It is an honour and pleasure to address this distinguished assembly with the Third Baer-Huxley Memorial Lecture on the occasion of the Fourth International Conference on Environmental Future. Permit me to begin by extending my sincere thanks to you – Professor Polunin – as well as to the Hungarian Academy of Sciences, the Foundation for Environmental Conservation, the Government of Hungary, and the other organizations responsible for convening this Fourth ICEF.

Today, we honour Jean Baer and Julian Huxley. These remarkable men recognized that our common future is not an accident of fate, but the fruits of shared responsibility. Jean Baer and Julian Huxley focused their scientific talents, intellectual integrity, and moral courage, on helping to illuminate the human condition, and on building international institutions charged with protecting our planet and our future – our shared future. Both men sought a new world order, one that transformed patterns of nationalism into the promise of internationalism. It is therefore entirely fitting, Mr Chairman, that this Memorial Lecture – as you suggested – should focus on building institutional frameworks to protect our future.

What I want to discuss are some key environmental issues that must be addressed, for example by bolstering existing institutions or creating new ones. Let me state at the outset that what I will be talking about will, as you could imagine, concentrate mainly on the United Nations and only on the field of environment. Let me also state that institution-building is a noble endeavour. At the same time, no institution – no matter how lofty the intent, how forceful the mandate, how careful the design – can in itself save our planet.

We live in an imperfect world, a world awash in economic and environmental inequities. We cannot be awash in institutions that repeat old mistakes. More than 10,000 international organizations already exist. Actions – and action-orientated institutions – are urgently needed to safeguard our future.

*[Already published, by permission, in *Environmental Conservation*, **17**(2), pp. 105–10. Eds.]

During the last two days, you have discussed the growing body of scientific data confirming that our planetary Biosphere is threatened, and more seriously than at any previous point in human history.

Before us stands the scourge of local, national, and regional, environmental decay. Yearly environmental damage in Europe alone – by means of polluting rivers, overflowing landfills, rendering lands useless, losing forest, emitting choking air-pollution, damaging buildings, and raising health-costs – is estimated to exceed Belgium's annual GNP. The Rhine and Rhône rivers often resemble open sewers. One in each five trees in this country is damaged because of acid rain and other air-pollution. Appalling as that is, Hungary is fortunate by comparison with neighbouring countries. For example, over 70% of Czechoslovakia's forests are estimated to be damaged by air pollution. For the rest of Europe, with the exception of France which has 23% of its trees damaged by acid rain, the ratio ranges between 30% and 70%. It is much the same story throughout other industrialized regions.

Widespread Ecological Destruction

Compounding national and international ecological deterioration is the advent of environmental destruction of planetary dimensions. This is an entirely new phenomenon, and we need institutions that are equal to the unprecedented task facing all humanity: to slow down, stop and, ultimately, reverse the wasting of our planet. Demographic momentum, ozone-layer depletion, the prospect of global warming, increased airborne toxics, deforestation, soil erosion, desertification, the assault on biodiversity, the proliferation of hazardous wastes, and the loss of freshwater resources, are only some of the chronic problems which, seemingly, we face ever-increasingly. We are an environmental house divided against itself – and a house divided cannot stand for long.

Synergistic, perhaps irreversible, forces are at work. No nation can quarantine itself against ecological decay. We are all in this mess together.

People everywhere are alarmed about ecological destruction. They have drawn a direct link between ecological damage and health risks, including their own personal security. People will not accept rhetorical camouflage or noble-sounding but ill-defined environmental institutions. People everywhere are demanding a new deal – a global contract geared to global partnership.

Since 1950, nearly one-fifth of the planet's entire productive topsoils have been lost through soil erosion, salinization, waterlogging, wind erosion, and urban expansion. Damage to the stratospheric ozone layer may well decrease the world's agricultural productivity, in addition to increasing incidents of skin cancer, eye cataracts, and decreasing the human immunity system.

We are on a collision course between, on one hand, human population increase and inequitable resource-use and, on the other, ecological decay. Each day – as over 180,000 additional people share the planet – 65 million pounds (*c.* 30,000,000 kg or 30,000 tonnes) of productive top-soil are lost. Since 1984, there has been a world-wide decline in *per caput* grain production.

Need of Global Monitoring and Assessment

Let me stress that such figures are merely estimates. Indeed they will remain estimates for a long time – until we improve our global monitoring and assessment capabilities. That is the *first* priority in looking at future institutional requirements. We need to coordinate and strengthen global monitoring and assessment capacities. We need to clarify natural resource endowment grids, resource extraction rates, sustainable yields, and replenishment capabilities. We need to understand pollution assimilative capacities and transboundary impacts. We need, in short, to understand, far better than we yet do, our planet's vital ecological processes.

A *first* step should be to improve coordination of existing monitoring capabilities within and outside the United Nations, so that complex and dynamic linkages among climate, ozone, forests, deserts, food yields, and demographic, environmental, and economic, indicators – to name just a few – can be more clearly understood and effectively acted on than is possible at present.

Since its founding, UNEP's Annual *State of the Environment* reports have informed governments, NGOs, and people in general, about current and emerging environmental problems of global dimensions. However, it is impossible for any existing institution alone to create a comprehensive, truly global report. Clarity about our global systems demands cooperative action. So, UNEP also collaborates with the World Resources Institute, the International Institute for Environment and Development – two major non-governmental organizations – and the United Nations Development Programme, to produce another State of the Environment Report, the *World Resources Reports.* On the 5th of June this year, UNEP and UNICEF will launch a joint report on 'Children and the Environment'. UNEP is making preparations for a further *State of the Environment* report which will endeavour to make the most authoritative and impartial assessment to date of what has happened since the 1972 Stockholm conference, and where we should be going. In this effort we are collaborating with our sister organizations in the UN system, with governments, with major NGOs, and with the scientific community.

– of Pooling Resources for Coordination

So UNEP is striving to coordinate monitoring and assessment capabilities. We are all getting closer together in the UN. But we have far to go. Environmental factors are too complex for any single organization or one group of organizations to explain. No one has sufficient human resources, scientific capabilities, and other necessary means, to meet *global* monitoring and assessment requirements. National government departments, scientific research institutions, and NGOs, have much to offer in this respect. My *second* point is that we've got to pool national and international capabilities.

The need is especially urgent with respect to developing countries. But that is not the prerogative of developing countries alone – developed countries also need, and have the capacity, to enhance their monitoring and assessment systems and to make their results available to others.

Increased monitoring and assessment is indeed a crucial starting-point, and UNEP's Earthwatch is an important step in this direction. It is a joint venture with all relevant UN organs and organizations. Its networks are composed essentially of the relevant national institutions. That whole system needs to be very much strengthened to meet the challenge.

Equally challenging is the need to translate scientific descriptions into policy prescriptions. While the gap between science and public- and private-sector policies is narrowing, it is not doing so nearly fast enough. Such gaps cause delays in action. My *second* proposal thus concerns environmental early-warning capacities.

– and of Early Warning for Action

In making the case for an international institution for early warning we should examine functioning international centres, such as UNEP's Earthwatch, United Nations Disaster Relief Office, FAO, the relevant mechanisms of the European Community, the US–Canada International Joint Commission, and several others. Monitoring and assessment capacities need to feed directly into warning systems. Early-warning systems need to indicate, whenever possible, long lead-times – to avert and prepare for ecological tragedies ranging from droughts and desertification to impacts of oil-spills, genetic resources destruction, chemical pollution of air and water, and the inappropriate use of biotechnology.

The General Assembly of the UN has given us a green light to begin preparation for an early-warning mechanism, and has asked specifically for UNEP's Earthwatch to be strengthened to carry out the job. A strengthened Earthwatch can probably – I repeat, probably – handle the global or regional threats. It could not, and should not attempt to, address early warning at the national level. That should be the responsibility of strengthened or new, specialized national governmental or nongovernmental bodies. Early warnings are, however, useless without early-response mechanisms, as is particularly evident in cases of environmental emergencies.

Towards an International Environmental Emergency Centre

The *third* area that should be addressed by revamped institutions or through creating new ones is, therefore, to define terms of reference for an international environmental emergency centre. Preparations for this have begun, and should be completed by the 1992 Conference on Environment and Development. Such a centre needs to work quickly, with authority and with adequate funding.

It could function as a 'switchboard centre', in which a series of small, concentrated networks of national experts and institutions with very high levels of expertise would perform. Such specialists and specialized institutions would work best within existing relevant UN and other agencies. For instance, in the case of nuclear power emergencies, the network could work under the auspices of the International Atomic Energy Agency.

In the case of chemical emergencies, institutions could work closely with UNEP's International Register of Potentially Toxic Chemicals.

A network to deal with oil-spill emergencies should function under the auspices of the International Maritime Organization.

In all this I am talking about creating networks *which work*. This means that the emergency centre must possess the authority to coordinate action, and that the coordinating UN bodies, or the collaborating centres themselves, must have sufficient financial resources to respond or to mobilize response.

Avoidance of Environmental Conflicts

The *fourth* area which I want to address concerns creating an international body charged with avoiding conflicts resulting from environmental problems. There is growing agreement among governments that environmental protection is a corner-stone of national security. Increased competition for shared water-resources, the devastation of acid rain, the proliferation of hazardous wastes, and depletion of natural resources, are creating potential flash-points between nations. Such an international body as the one I am suggesting would work to promote cooperation, rather than confrontation, through international programmes that ensure everyone's interest and lead to environmentally sound and sustainable development. This is most probably an important role for the Governing Council of UNEP. It could probably be linked to a function of ensuring that transboundary environmental impact assessments of major development activities are carried out.

Conflict Resolution

My *fifth* area of institutional consideration is due responsibility for conflict resolution. With all the goodwill in the world, conflicts are bound to arise between neighbouring states over one or more of the issues which I have just mentioned, or others that will doubtless emerge. Various proposals have been mooted in this respect in last year's Hague Declaration, during last year's UN General Assembly, and at various international conferences that were held practically throughout 1989. Proposals include (i) expanding the UN Security Council to address environmental conflicts, (ii) to entrust one of the UN General Assembly Committees with this role, (iii) to ask the UNEP Governing Council to carry that responsibility, or (iv) to create a new UN body charged with conflict resolution. In any case the body charged with this function should carry also a closely-related function – that of monitoring the implementation of international environmental agreements.

For conflict resolution to work, we have got to elaborate binding legal frameworks. Countries entering conflict resolution should agree beforehand on the *binding* nature of the decision of the body concerned – similar perhaps to the International Court of Justice, or probably closer to the dispute rulings of GATT (General Agreement on Tariffs and Trade). To help compliance, a conflict resolution mechanism needs to include liability and compensation provisions.

These five functions – increased global monitoring and assessment, early warning, assistance in environmental emergencies, defusing potential conflicts, and conflict resolution or settlement of disputes – would form some of the basis of institutional requirements to protect the planet. They will have to involve governments, intergovernmental bodies, and NGOs.

Some Further Propositions

There are other functions that need institutional change. Faced with the worst ecological deterioration in the history of humanity, two propositions have gained acceptance. One is that all countries – developed, undeveloped, and developing – must work together towards sustainable development, increased conservation, and greatly improved environmental protection. The other accepted proposition is that action will take time, and will also take money. We do not have the former and lack access to the latter. This is why the spotlight has squarely moved to questions of costs and structures of funding mechanisms.

That constitutes my *sixth* point, and on it we have got to move forthwith. Delay only increases costs, and there is little agreement on how to meet escalating costs. For example, increased family-planning measures moving towards a sustainable population equilibrium could require a three-fold increase in funding, to $10.5 thousand millions a year. Five to ten thousand million dollars a year is needed to combat world hunger. An extra two thousand million dollars a year (1982 dollar value) is needed to address desertification. The virtual elimination of ozone-destroying chemicals is likely to cost between 2 and 7 thousand million US dollars in this decade.

Yet that is only 'the tip of the iceberg'. Action to save our tropical forests, conserve Nature's species from extinction, save our seas, clean our air, manage our soils, and prepare for climate change, will need to be quite staggeringly costly. A recent study by the Canadian Government suggests that the costs of reducing carbon dioxide emissions by 20% in that country alone would exceed $100 thousand millions. At the same time, the report suggests that such a CO_2 reduction target could result in savings – through increased energy efficiency and other measures – of more than $190 thousand millions.

Rich countries have the luxury of considering such measures. Undeveloped and developing countries – caught in a money-squeeze and the vice-grip of poverty – simply cannot. The plight of the less-developed countries continues to worsen, because of rising debt-service payments, dropping export revenues, disappearing investment flows, and accelerating resource destruction.

ENDLESS FINANCIAL NEEDS

When we talk about funding requirements, we must address the fact that development assistance – in terms of North-South capital flows – is diminishing in real terms. Last year the developing nations transferred to the rich countries $43 thousand millions more than they received in new aid and loans. In these circumstances the South will not, and anyway should not, accept a diversion of existing aid to promote environmental projects that deal with global environmental problems with whose creation the developing countries in general had little to do. They would need additional funding to accept any such diversion. What is needed is *more* public finance from bilateral and multilateral agencies, and the creation of a new Global Funding Facility that is dedicated to dealing with global environmental problems.

At the same time, public finance from national budgets – even if it drastically increases, which is unlikely – cannot halt the carnage of resource depletion and ecological destruction. The World Bank, the IMF, and bilateral funding agencies, have limits to capital infusion. They cannot meet the real challenge – spelled out in the report of the World Commission on Environment and Development, and in UNEP's Environmental Perspective to the Year 2000 and Beyond – to prompt a new era of growth and so *accelerate* development. Accelerated growth cannot come about by the goodwill of international institutions. It demands increased investment – and increased productivity. Without more money and more production, growth cannot be achieved. And without growth, poverty will remain the worst polluter. The harsh fact remains: conservation is incompatible with absolute poverty.

We need new and additional sources of funding to deal with the serious global environmental problems which we are witnessing now, and to help to alleviate devastating poverty. An obvious first source of funding stems from the many thousands of millions that are being saved by demilitarization. Peace is breaking out all over the world, but prosperity is not. Diverting even a few days of the annual global military budget – estimated at nearly $1 million million millions – to environmental protection, would make an enormous difference to our world. The superpower nuclear arsenal of 10,000 warheads could be reduced to less than one-tenth while still ensuring deterrence.

Savings from those overdue actions could be shifted to infusing more capital for stand-by lending, structural adjustment policies, sector adjustment loans, concessional funding, grants, and subsidies. This does not seem to be forthcoming, at least in the foreseeable future. Nor do I believe that money can buy such policy reforms as are necessary for global environmental protection. Poor policies in energy, agriculture, industry, fiscal and monetary management, pricing regimes, and incentive structures, have all led to poor environmental conditions. Strengthening current institutions or creating new ones to repeat systemic errors is not the answer.

Changed Attitudes Required

Achieving environmentally sound and sustainable development needs to hinge upon changing attitudes about the resilience of the planet. Consumption patterns and an addiction to disposable goods must change. There is an urgent need to integrate environmental factors in *all* dimensions of private and public-sector decision-making, ranging from energy and agricultural policies all the way to R&D funding linked to national industrial policies. There is a need to understand the full, direct and indirect, costs borne by society in the production of goods and services. Familiar tools such as cost-benefit analysis, input-output analysis, and environmental impact assessments, coupled with fresh approaches such as environmental accounting, environmental auditing, valuation of natural resources, and reforms in calculating national income accounts, are helping us to understand the long-term implications of fast-track economic growth.

Adjusting economic attitudes to integrate environmental values is a first step towards achieving developmental sustainability. Essentials such as water and air simply can't be viewed as free goods. Economic incentives and disincentives, carbon taxes, levies on using genetic resources, user's fees for maintaining the quality of clean air, and other mechanisms, must become operational – in the short term – to meet the costs of global environmental protection.

Also needful of change are some basic economic assumptions. There is growing agreement that basic wealth calculation which looks only at fixed capital depreciation while ignoring natural-asset depletion, is foolish. Environmental auditing and environmental accounting will be likely to deflate income earnings and GDP calculations in the short-term, but will work towards ending 'ecological deficit financing' that destroys future economic prospects. It will get the prices right.

A word, however, of caution. Getting the price right isn't synonymous with getting the *policy* right. Approaches that look to the 'magic of the marketplace' can never constitute the whole answer. There are no panaceas, especially when one is looking at many non-market impediments in developing countries. More to the point: market efficiency simply isn't a goal for the world's poor, whose concern is subsistence, and whose right must be greater equity in income distribution. So this is the *seventh* area for future institutional considerations – an area where we need extremely innovative and courageous approaches to set things right. Here there could be no remaking of the Bretton Woods agreement. We need new financial institutions or mechanisms that reflect global partnership – that reflect the real cost to Society of all goods and services – in which each country contributes according to its share of global ecological destruction, and in which each country has an equal say in the use of the available resources.

Technology Transfer Crucial

Eighth, and last, we know, right now, that more attention needs to be directed to technology transfer. Unless appropriate technologies which eliminate CFCs, reduce CO_2 emissions, increase process efficiencies, and close production cycles, are transferred to the developing and other poor parts of the world, there seems no doubt that our mutual destruction will be assured.

Technologies are bought and sold like any other commodity, except that high costs of research and development are buttressed by patents, trademarks, and other rights. Industry cannot be expected to abandon patents without compensation, which will take funding from the developed world. But that is not the whole equation. The *choice* of patent acquisition must reside with the country buying the technology. To choose, you need knowledge to purchase knowledge. Increased information is needed by buyer countries to strengthen their bargaining position in technology transfer agreements.

Equally important is the need for increased training – both technical and managerial – by transnational corporations transferring technologies to other countries. The gap between North and South in trained personnel is startling. While developed countries have, on average, 285 scientists and engineers for each 10,000 people, developing countries have only some 95 per 10,000. In Africa, there are fewer than 10 scientists and engineers for every 10,000 people.

At the same time, no country – no matter how wealthy – has a surplus of trained personnel who are able to meet growing environmental challenges. For example, it has been reported that the United States National Science Foundation predicts that the US will have a shortfall of 700,000 technically-trained people by AD 2010 if present trends continue.

Governments need to increase emphasis on education and training, and accelerate incentives to promote R&D into environmentally-benign technologies. This leads me to the final policy arena that needs to be addressed through international institutions – training. There is a need to intensify training in developed and developing countries, to promote attitudinal changes in industry, financiers, governments, and development agencies, about development sustainability. The time has also arrived to establish a *World Environmental Academy,* functioning as an educational centre of excellence – a centre where national and international experts work together to exchange views, educate one another, and plan joint strategies to improve the status of our Earth.

Conclusion

To end I'll remind you of the story of the then Head of the US Patent Office, who startled a Congressional hearing at the turn of this century by asking that his office be eliminated. He told Congress that no more patents were forthcoming, because all possible human inventions that could have been invented were already invented. I would like to say the same of UNEP, IUCN (The World Conservation Union), WWF (The World Wide Fund for Nature), or any of the world's environmental institutions or even our current series of Conferences on Environmental Future. I would like to see a time when environment is so automatic and so completely a part of our decision-making – as, say, profit or loss – that environmental institutions will become redundant. But unfortunately we are very far from this. What we need to do is to think about the institutional reforms – and, yes, sacrifices – that will get us there.

We are approaching the eve of a new century, and a new millennium. The pace of events challenging our future is accelerating. It is difficult to bring forward solutions and propose institutional frameworks that would be geared to the twenty-first century. Uncertainties abound, though I am sure we will move ahead better informed about institution-building. We don't need noble-sounding bodies of bureaucratic labyrinths. We do need institutional bodies that are involved in macro- and micro-economic procedures and that respect natural resources – institutions that develop instruments for conservation and protection, that define workable standards, that ring alarm-bells, that can really assist when an emergency takes place, and that help us to avoid conflicts and resolve them when they occur.

We need institutions with teeth, that can elaborate regulations, define and administer global funding levels, and possess compliance *and* enforcement powers. And, above all, we need institutions which are founded on the notion of equity between and among nations. In all this, NGOs and citizen groups have a major role to play. The thousands of international, and the hundreds of thousands of national, NGOs have had, do have, and will continue to have, a key role in framing our world's institutional future.

The challenge for all these environmental and/or conservational organizations and institutions is to establish bridges among themselves, to ensure concerted action and open proper channels of communications with national governments to ensure that they – the NGOs – become part and parcel of the process of formulation and implementation of environmental protection policies.

Striking the balance between global priorities and national sovereignty is not easy – especially in developing countries, where resentment is growing over lecturing from the industrialized countries that are apt to be primarily responsible for environmental deterioration. Success can only be built by forging a consensus between developed and developing countries. And this can only be reached if everyone agrees to render the North–South divide obsolete. It is happening in East–West relations. It must happen in a global context. Without shared goals, any progress will be transitory. Success can scarcely be expected if countries continue to pursue only their own self-interest. But it may not prove to be so very far off if we all engage in a genuine dialogue that is translated into global partnership reflected in participatory development and joint environmental protection.

We've got to be future-looking – to address the problems of today in the light of the situation in the twenty-first century and beyond. A major challenge is that reformed or new institutions must acknowledge the fact that they have to meet both our needs today and those of the yet-to-be-born. It has been repeatedly said: We did not inherit this Earth from our parents; we are borrowing it from our children. Let us not turn this into a mere slogan. Let us respect it. Let us resolve to hand the Earth over to those who own it, but in a better shape than when we borrowed it. Let us hand them institutions that work, institutions that help them to understand the Earth and use it well. Let us move ahead with the heavy responsibility of intergenerational equity. Thank you.

Summary

Among the prerequisites to building an effective environmental institutional framework for the twenty-first century, our current thinking singles out the following eight for prior consideration:

1. Improvement of coordination – after due strengthening where necessary – of the existing global monitoring and environmental assessment facilities both within and outside the United Nations system.
2. Creation of an effective and permanently-maintained capability of early-warning of threats of environmental disasters.
3. Establishment of an international emergency centre for environmental action through revamping existing institutions and creating new ones when and where necessary.
4. Creation and maintenance of an international body charged with avoiding conflicts arising from environmental problems, so diffusing potential conflicts in an increasingly overcrowded world.
5. Means of assessing responsibility for conflicts over environmental matters and settlement of cognate disputes.
6. Means of settlement of questions of environmental costs and establishment of adequate funding sources and mechanisms for their disbursal.
7. Necessary changes in human attitudes to make way for innovative and courageous approaches to get things right.
8. Need for widespread technology transfer based on research, applicational experience, and far wider and better environmental education than currently exists.

We need to be globalists in persuasion, and holistically-minded futurists in concerted action, if any lasting success is to be achieved in addressing the environmental and allied problems of today in the light of the situation to be expected in the twenty-first century. The ultimate inhabitants of Planet Earth, for whom we hand on our institutions, will need ones which work far better than ours do nowadays; so we must do everything in our power to improve their working and our global environment wherever the capability exists, as it surely does still very widely.

Commentary on Chapter 21

CHAIRMAN: Professor Nicholas Polunin
PANELLISTS AND OTHER CONTRIBUTORS:

McCloskey, Tolba, Vallentyne, Tolba, Stone, Tolba, Thorndike, Tolba, Tisdell, Tolba, Simon, Tolba, Barton Worthington, Tolba, Westing, Tolba, Purcell, Tolba, Holdgate, Tolba, Polunin

McCloskey enquired how the new kinds of approaches suggested by **Tolba** could be applied to the marine environment.

Tolba pointed out that the marine environment had three components. An inter-agency group had reported the previous month that the open ocean was, fortunately, in quite good shape. The coastal areas were where the damage was actually happening. UNEP had been trying for 17 years to get governments to come together on a regional basis where there was a common concern, *e.g.* the Caribbean, Mediterranean, and Arabian/Persian gulf. Some 120 governments were now involved as contracting parties, planning programmes for protecting their enclosed, or partially enclosed seas. The last component was the living resources of the seas, including the marine mammals. With IUCN, which had resources that UNEP lacked, they had developed an international, global conservation plan for the marine mammals.

In all these initiatives there was always the problem that conflict had to be avoided when, for example, one country was polluting the coastal waters of another, thereby affecting tourism, bathing and living resources generally, which could lead to major friction. A second problem was environmental emergencies, and he had already referred to the network set up to deal with oil-spills. Here too, conflict resolution was important. Lastly there was the question of monitoring and assessment. The arrangements here were extremely weak. Even in the Mediterranean, where 100-120 national institutes had been involved for some 15 years, the results were very scanty. The general issues he had described earlier applied to all ecosystems, including the marine ones and wider environment.

Vallentyne asked whether UNEP had thought of changing the terminology of its new agenda to a biospheral one rather than an environmental one?

Tolba wondered whether Biosphere was really new. When he was on the Board of UNESCO 20 years before, they had started the Man and the Biosphere Programme, one of the oldest of effective environmental programmes in existence. He did not really think there was any need for UNEP to change its terminology; changing attitudes was far more important if anything was going to be achieved. If they changed to The Biosphere, then another 20 years would be lost because everyone was now becoming passionate about *the environment;* they needed to capitalize on that right now.

Stone suggested, first, that a new institutional forum in which to settle disputes was not required. There were already several, the World Court for example, but rather, there was a lack of treaties and laws to underpin possible disputes. Secondly, he agreed that basic problems underlying equitable distribution, the North-South division, and how to spread the burden of protecting the Earth, were those of energy and the environment. If that was the case, surely it cut across the idea of an environmental agency for, if the agency was successful, it would rapidly be deflected into questions of distributing wealth rather than establishing legal environmental protection or improving environmental efficiency.

Tolba said that he, himself, was against new institutions but several suggestions had come up at the time of the Hague declaration in 1989 and at the UN General Assembly. There were already about 10,000 international organizations and if, collectively, they could not meet the challenges then there was little point in creating any more! His idea was that the UN Security Council could take over responsibility for resolving conflicts and that either the 5th Committee of the UN, or the Governing Council of UNEP itself, would act as a Trusteeship Council linked with the issue of liability and compensation if arbitration was involved. He did not see the International Court of Justice taking on such issues. He was after an arrangement that assumed the responsibility for resolving a conflict before it went to armed conflict – when the Security Council would take over – not end in 20 years of litigation, which was the likely outcome if the International Court was involved. In other words, a goodwill organization was needed whose decisions would, nevertheless, be binding on parties.

So far as establishing an environmental agency was concerned, he could only say that UNEP was meeting the kinds of problems such an agency would face daily and everywhere. He was off next day to meet Ministers about the existing Ozone Protocol and for the last 8 months he had been trying to deal with the tricky issue of additional resources to meet the requirements of the poorer countries and the transfer of technology. These things were not the responsibility of any one group but of us all; if not, we would end all by destroying life.

There were two universal responsibilities: those who could afford to help should help those who could not afford to improve the environment in the present economic climate, and that insufficient technology was produced in the developing countries, including eastern Europe, so that technology had to be transferred. All the problems that had been discussed – ozone, climate change, biological diversity, hazardous waste disposal – required those two things for their resolution, namely, additional resources and technology transfer. An additional institution was not necessary.

Thorndike wanted to know whether **Tolba** was optimistic or pessimistic that his agenda could be achieved, and what was the alternative for human life on the planet?

Tolba said that he was an optimist by nature, which was probably why he was still head of UNEP after 17 years! He was optimistic because he could see change coming. Governments were beginning to recognize the situation and beginning to work together. Progress would take time. There would not be a beautiful environment or a protected Earth tomorrow. We had to go at things in earnest for the next 20 years: we had a clear goal.

Tisdell had not heard much about the responsibilities of the less-developed countries, only those of the developed ones. If the latter were to give aid to the former they might expect them to adopt certain policies such as better measures for population control, or more stable government, so that aid would not be wasted. Did **Tolba** think that reasonable?

Tolba believed that developing countries were keen to carry their share of the burden but, they were keen to do so *without* being told what to do! They had seen that the developed countries had not fared too well, that the industrialization and development of the last 50 years had spoiled the whole world, and now the countries largely responsible were going to tell them not to do the same. What the developing countries could reasonably expect was to see the North first changing its attitudes and habits and regulating its use of natural resources to a reasonable level before they were told to do something else. Developing countries were fully aware that they were continuing to destroy their forests and soils and were prepared to try to redress the destruction. There were, however, some truly global problems that could only be corrected by a global partnership, regardless of the developed or less-developed status of the countries. Everyone would have to be involved and he did not think that the developing countries were reluctant to play their part. If they did not, they knew they were finished; but they also knew that they were at present the least capable of helping without assistance.

Simon wondered whether there was any better prospect of advancing the new environmental agenda than there was for an equally important global issue, the new international economic order, which had failed?

Tolba was neither for nor against the new economic order, but governments could not operate in the abstract; they either faced issues and the need for resources or they did not. In the case of the ozone problem he was hopeful, despite everyone saying that no resources could be made available. He expected that in London on 20th June [1990, Eds] an international fund of perhaps US$250 millions would be made available for assisting the developing countries through technology transfer from the developed countries. Governments were concerned to see concrete action, to know the costs, of organization, of the mechanisms for collecting and disbursing the money, and to receive an assurance that it would not then be pocketed by the leaders of developing countries as it had been before. Once satisfied, they were ready to chip in.

Barton Worthington commented that a good many people believed that the greatest problem in facing the future was the rate of multiplication of people themselves. Where would **Tolba** put that in his list of priorities?

Tolba, Top!

Westing recalled that earlier **Tolba** had suggested that a solution for dealing with marine problems was to adopt a regional approach. He wondered how far that would be useful in solving environmental problems both in coastal areas and on land, *e.g.* for shared rivers? He recalled a possibly glib statement by Garrett Hardin to the effect that to globalize a problem was to kill it, so that perhaps more emphasis should be placed on regionalizing global problems.

Tolba said that a global problem was just a name. It was not created by something up there, nor would solutions be implemented in that way. Problems had to be solved by nations and governments, so global problems were those in which everyone had to participate in solving – each at national level. The nature of the solution depended on the nature of the problem. Some were of such a magnitude, such as climatic change or the ozone layer, that they could only be solved by cooperation of all governments. Others, such as problems of coastal areas, affected and could be solved by groups of neighbouring nations, and that could apply to a shared river such as the Zambesi, or the Nile which was a potential flash-point. Some problems, however, were purely within national borders and so should be dealt with locally. Solutions must always match problems.

Purcell asked what was the best example known to Tolba of the most effective attempt to impart environmentally-important information to the less-developed countries?

Tolba replied that the clearest example was the problem of the ozone layer. The developing countries had sat and only listened in Montreal. Now they were coming in as contracting parties, and even the two potentially largest users of ozone-depleting substances – India and China – were attending all the negotiations although they were not parties to the Montreal Protocol. That was because they wanted to be sure that they knew what they would have to do, and that they could do it; they wanted to be sure that there would be concrete results, not just talk. There was, indeed, a general feeling that, up to now, we had been talking too much about the environment and, especially the politicians, not doing enough.

Holdgate remarked that he agreed wholly with Tolba that we should work to abolish the North-South divide, and his opinion that that would be achieved only through a major transfer of resources southwards to help the developing nations to develop themselves despite the appalling problems which they faced. That would require the affluent nations to alter their perceptions of their responsibilities to the environmental problems of the world, if the transfer was going to be on an adequate scale. A question which had come up repeatedly during the Conference was, how should we set about improving communication to get that message across.

Tolba thought that **Holdgate** knew as much about that as he did and that they had probably discussed it more than any other participants! The only way was to get people to identify with the problem. Ozone, yet again, was a good example. The issue had been discussed for 12 years from 1973 to 1985 before a treaty, without teeth, was agreed to in Vienna; basically it said that we would all work together to save the world, but no real action would be taken! Two years later we had reached the Montreal Protocol. There was a big fight to get a reduction of 50% agreed , not least with the British government. Now, in less than a further two years, the pendulum had swung to 100%. Where had the pressure come from? It had not come from publication of the percentage reduction measured by the scientists but from the popular impact of the groups that had pointed out that ozone depletion meant increased ultraviolet radiation, more skin-cancer and cataracts, and a reduction in the human immune response. Everybody started wondering, why not me, or my family, or my children, who would suffer these monstrous diseases. The link with human health did the trick! It had not been possible to get such a link with climatic change or biological diversity, but the way to go about things was to get at individuals, the man-, or woman-in-the-street, and link environmental deterioration firmly in their minds with a direct impact on them or their family.

Polunin then called on the Hungarian Acting Minister for Environment and Water Management to conclude the discussion (*see* Annexe 8).

Annexe 8: An Intervention on the Environment

DR MIKLÓS VARGA

Acting Minister for Environment & Water Management, Hungary

Ladies and Gentlemen, may I say some words about the consultative meeting initiated by Hungary in order to find new ways of cooperation among the CMEA countries, which are – as you know – in a state of great transition in political, social and economic respects as well. The so-called planned economy system can't be maintained any more, while the market economy hasn't yet developed. At another international meeting, these countries have been defined in a funny way as 'previously-planned economy countries'. These 'PPEC' countries are in a period when the integration into the global, and especially into the European processes represents one of the main goals. It is the case also in the field of environment. Hungary can be considered as a pioneer in building up environmental contacts with Western European organizations, such as the Council of Europe, OECD, and the European Communities. We have expressed also our readiness to join the planned Environmental Agency and Monitoring Centre of the EC.

While strengthening the environmental contacts with the highly-developed part of Europe, we are well aware of the reality that this process of integration and adjustment to developed environmental standards and policies will take a long time. Therefore the CMEA countries, whose economies are characterized by low energy efficiency, high specific raw-material consumption, an obsolete structure of industry and, consequently, by a highly degraded environment, will need also in the future close environmental contacts. However, the name and structure of cooperation may, and indeed has to change with the necessity of environmental contacts among these countries located in the same geographical region and having similar economic structures.

The point is: that European environmental integration on the one hand, and the Central and Eastern European Cooperation on the other, should be considered and realized as two processes mutually strengthening each other.

In a period of technological revolution and profound changes in the world economy, scientists, environmental organizations and authorities have a difficult and responsible role to play. They have to influence the governmental and sectoral policies so that environmental considerations are enforced. These efforts can only be successful if they are supported by *scientific* results and proposals for solutions. As the environmental *problematique* is highly sophisticated, a cooperation of experts from different fields of science is needed so that scientific arguments may be stronger than economic interests. Therefore, I am very glad that so many fields of science are represented at this Conference, and I look forward with a really great interest to the conclusions of this meeting.

I would mention particularly the field of economics. The impacts of environmental pollution should be quantified in terms of money in order to convince economic decision-makers. The implementation of so-called 'environmental accounting', promoted also by UNEP, would be an urgent and efficient step for a breakthrough of the environmental approach. However, the measurement of certain types of environmental impacts in terms of money is very difficult – if possible at all – and so it is necessary to develop methods of assessment that use terms of 'ecological balance'. This means, if we analyse a production process, that both environmental inputs and outputs can have negative or positive effects on the ecological balance. We should aim at keeping these effects in equilibrium. In this context I would like to maintain that the concept of 'growth', and the category of 'value', should be updated in order to meet the requirements of sustainable development.

22. Looking to the Future

Reid A. Bryson

Professor Emeritus of Meteorology, Geography & Environmental Studies, Senior Scientist, Center for Climatic Research, Institute for Environmental Studies, University of Wisconsin–Madison, Madison, Wisconsin 53706, USA

Introduction

Over the past two decades there have been four *International Conferences on Environmental Future* (ICEFs). Essentially all aspects of the environment have been touched on in these Conferences – starting with the first, held in 1971 in Finland, in which we considered already the world's main environmental problems – including stratospheric ozone depletion, 'greenhouse' warming, deforestation and other devegetation, desertification, salinization, eutrophication, soil destruction and erosion, and above all overpopulation (Polunin, 1972, *see also* pp. 548–53). All these ICEFs have been concerned with the future and with environmental change.

Probably a majority of the papers and addresses given at these conferences have contained what a meteorologist would call a forecast, outlook, or prediction, though in most cases the forecast was implicit and non-quantitative. For example, a speaker might say that the population of a region was increasing rapidly at the same time that the arable land was being lost to erosion. Implicit in the statement was the idea that at some future time (unspecified) there would be inadequate food for the population.

Another example of an implicit forecast would be a statement that a marine wilderness area was needed desperately. This carries the implication that there would result, at some future time, some positive benefit, though neither the nature nor the timing of the benefit was specified.

In none of these conferences was there specific attention paid to the methodology of environmental forecasting, and very little to the construction of the conceptual models which are needed to make a forecast. Even in the case of climatological outlooks and forecasts – and they, in the manner of ecological outlooks, were mostly concerned with changes on the scale of several years to a century – very little was said about predictive models or forecasting methods. While meteorologists have much experience in forecasting for a few days, they have far less experience in forecasts for a month or season, and almost no experience at the decade-to-century scale.

The following paragraphs are intended as a commentary on the general question of models and predictions in the arena of environmental concerns, for its seems to me that the state of environmental questions, and indeed the state of the environment, is such that environmental scholars must soon become more explicit and quantitative. At the same time they must be concerned with credibility, and indeed themselves be fully credible.

Models

A model is nothing more than a formal statement of how the modeller believes that the part of the world of his concern actually works. It may be verbal or mathematical. It may be simple or complex. It may or may not require a large computer to keep track of all the aspects and interactions, whether implicit or explicit. However, as the world – our multivariate, interconnected environment – is very complex, no model of it can be complete.

It should be explicitly stated here that the results devolving from the application of a model *are not data*. They are simply model results, even though there is an unfortunate tendency to refer to such results as though they were equivalent to observed facts or data. The model output comprises data only on the behaviour of the model, and must be compared with actuality in order to test the quality of the model and establish the credibility of the modeller's construct (*i.e.* perception resulting from the orderly arrangement of facts, impressions, etc.).

Fortunately, a carefully constructed simple model can often give useful results. This is particularly true in those cases in which the variable quantities of concern are aggregates of such a nature that variables other than those used in the model 'average out', and general properties of the system are sought (rather than local detail) when a certain type of forecast is desired. Examples will be given in the following.

Forecasting Methods

There are very few truly tested and credible *predictive* models, meaning models or conceptual constructs which state explicitly that a specific event will occur at a certain future time. There are many models that have not been tested to ensure credibility of the outcome.

There are nevertheless a number of forecasting procedures that have been used for various purposes, and methods of testing for accuracy of the model and procedure. It is clear that, unless there is such testing against reality, the forecast remains a mere hypothesis.

Nearly all forecasts are extrapolations of one kind or another. Some of the more commonly used are as follows:

1. Extrapolation of a Linear Trend

This is probably the easiest kind of forecast that can be made, and often gives fairly good results over *short* time- or space-spans. For example, between ages 10 and 13 I grew about 75 millimetres per year. Extrapolating this linear trend to age 14 would have estimated my height within 12 millimetres or so. However, extrapolation to my present age would have predicted a stature of well over 6 metres – mercifully an error of more than 4 metres! For the more general case, linear extrapolation over large spans does not allow for reversal of the trend, even though the fact that complex curves can be approximated by infinitesimal linear segments is the basis of calculus. This is not a trivial matter to note, as statistical fitting of linear trends to data is often used without consideration of the implications of the linearity.

One often reads comments such as 'Global temperature records show a rise of 0.11 degree over the last century' (or some such number with an aura of great precision). This number was usually obtained by linear regression. Unfortunately that value is not representative of other centuries, such as the preceding three or four; nor can one assume that it has any applicability to the coming century. It is a sample of only one century out of the multitude in the past and future. It thus would have an infinite probable error even if the climatic time-series were a *stationary time-series*, which of course it is not. A stationary time-series is essentially one in which the mean and variance are constant throughout the series.

2. *Non-linear Extrapolation*

A common type of non-linear extrapolation is the extension of a rate of change. Population projections are usually based on the extrapolation of a rate such as 2% per year. The result of a simple extrapolation of a constant rate is a logarithmic curve. This extrapolation can go to absurd values even faster than a linear extrapolation (*see* above). To use the example of my own growth-rate, in my early youth the rate used above was about 5% per year. Extrapolation of that rate would have projected my present stature to be about 40 metres! Such a projection is immediately recognized as absurd because all people know that youthful growth-rates diminish and there is a general, limited range of adult statures. The existence of naturally-present limitations is less generally recognized in the case of constant-rate projections of population numbers, food production, energy use, and carbon dioxide content of the atmosphere.

3. *Periodicities*

Regular cycles occur in Nature, often with clearly recognizable physical causes. Examples are the cyclic alternations of day and night, winter and summer, and the ebb and flow of marine tides. Biologists are well aware of adaptation to these periodicities.

Periodic behaviour provides a sound basis for some truly predictive models. Farmers around the world rely on the annual march of the seasons to schedule planting, harvest, etc. The seasons are one of the few 'externalities' that are regularly taken into account in economic models. It is the failure of the seasons to follow a simple annual periodic function that makes the seasons a less-than-perfect forecasting guide and produces most of the climatic problem of the farmer. The growing-season may be wetter or drier, warmer or colder, than expected. Even this normally good forecasting model needs improvement. We need to know why successive growing-seasons differ, and it is an unacceptable research strategy to assume *a priori* that the variation is random. The term 'natural variation' is often used to hide the assumption of randomness, or to avoid the appearance of ignorance.

These are periodicities other than those driven by the rotation of the Earth (the daily cycle), the revolution of the Earth around the Sun (the annual cycle), and that of the Moon around the Earth (the tidal cycle). There are the long, slow periodic changes of the tilt of the Earth's axis, the relative time of the equinoxes, and the eccentricity of the Earth's orbit around the Sun that gives rise to the succession of glacial and interglacial epochs (Milankovitch, 1941; Imbrie *et al.*, 1984; Bryson & Goodman, 1986; and many others). These are of such long periods that very little change is evident from one century to the next. Indeed, 'trends' may be better thought of as segments of very low-frequency signals. There are, however, other, more rapid, periodicities in Nature that may be useful in environmental studies.

Some of the lesser-known periodicities on the time-scale of years and decades are due to the 'Chandler wobble' of the Earth's axis of rotation (Bryson & Starr, 1977) and the interaction of the annual and tidal cycles (Campbell *et al.*, 1983). Though the cycles mentioned above are seen primarily in climatic terms, the pervasive adaptation of life to the climatic environment may result in observable periodicities in biological variables such as animal numbers, and are certainly seen in tree-ring thickness (Bryson & Dutton, 1961).

4. Extrapolated 'Dynamics'

Forecasts are sometimes made by using dynamic models of a system – usually numerical models – for ecosystems and the atmosphere and ocean are all very complex, and the quantitative, symbolic form of mathematics is essential systematically to keep track of all the interactions. Such models are an interlinked set of equations, whether physical, statistical, or a mix of the two. In essence the forecast is based on the assumptions that the model is an adequate statement of how the system actually works, and that it will be equally adequate if some variable is changed. This is not, however, always true or obvious, for a number of reasons.

The assumption that the model is adequate is usually tested by comparing its output with present reality, *i.e.* whether the model output accurately simulates the present. This is a valid test if one remembers that most such models are *equilibrium* models, *i.e.* the assumption is made that the rate of change of the system is zero, and that most such models are 'tuned'. A tuned model is one in which some variable or its behaviour is adjusted to force the model to fit the observed reality more closely than would otherwise be the case. This may or may not be because the 'real' behaviour of the variable is not well-known.

Most dynamic models are also linearized. The reasons for this linearization are straightforward: for short-time steps the linear approximation is nearly valid, and adding non-linearities is very difficult. Consider, however, the implications of this linearization in a *climate* model (for a 30–50 years' period), based on a *weather* model (for the instantaneous state of the atmosphere), that is used to simulate the climate fifty years hence. The equations used, taken from atmospheric physics, are valid at the millisecond time-scale. The linear form is assumed to hold for some short time such as 20 minutes, after which the state of the system is re-evaluated and extrapolated for another 20 minutes. To make a 50-years' forecast, well over a million such time-steps are involved, with thousands of calculations for each time-step. Even if each successive approximation is 99.9% accurate, very large errors may accumulate. There are variations on this theme, however, but to the Author's knowledge none simulates the present other than crudely (*see*, for example, Schlesinger & Zhao, 1989, Figs 9, 12, and 13).

A typical climatic forecast using such a dynamic model starts by running the model with present-day values of the input variables, often including the observed distribution of sea-surface temperatures. The model is run for a large number of simulated days, and the statistics of a select number of days is taken to be a simulation of the present climate. The model is then run for a comparable period with some variable changed as it is expected to change in the future, such as with doubled carbon dioxide content in the atmosphere. The statistics of this model run are then assumed to represent what the climate will be like if and when the amount of carbon dioxide doubles. Several assumptions may be made at this point, and their validity is crucial to the credibility of the forecast.

Because the simulations of the present have large errors, the difference between the modelled present climate and the modelled future climate is assumed to be equal to the actual difference that will occur. Using the observed present and the modelled future produces unbelievable results. Unfortunately there is no evidence that this assumption is correct, and no overwhelming logic that it should be, for most arguments to that effect in turn rest upon another set of assumptions rather than data. Nor can a comparison with the past fifty years or a century be used to validate the assumption, for, as was pointed out above, the probable error of a sample of one is infinite.

A valid test of such a model would be to see whether it can simulate the past sequence of a large number of time-periods of a length comparable with that for which the model is to be used for forecasting. While some, mostly hydrodynamic, General Circulation Models (GCMS) have simulated the past quite well, they have not been able to simulate the fluctuations on the scale of centuries, half-centuries, or decades (Kutzback & Guetter, 1986). Some thermodynamic climate models, on the other hand, have simulated the past centuries-scale temperatures of the hemispheres and of the monsoon rainfall with apparently acceptable accuracy (Bryson, 1988, 1989*a*). *A model that cannot simulate the past can* **not** *be trusted to predict the future.* It follows that we must know and heed the history of the environment.

5. *'Scenarios'*

In lieu of real forecasts, scenarios are often used. They usually are not real forecasts, though they may be if the only unknown is the future behaviour of some particularly important control variable. It is necessary to know whether the forecasting model is correct when the appropriate value of the control variable is known. Generally, scenarios are not very useful because they combine the uncertainty of the forecasting model with the uncertainty of the control variable. They may, however, be reassuring if all the pertinent scenarios are non-threatening; but then one must wonder whether the real scenario has been missed.

6. *Contingency and Other Statistical Forecasts*

Statistical forecasts have the virtue of being based on the past, observed behaviour of a real system, though we should always remember that *Nature itself is the only perfect model of Nature.* Statistical forecasts are usually very conservative, meaning that they do not usually forecast extremes well, even though it is only the extremes that might be important to the user.

Statistical models give some of the best and some of the worst results. It is possible to make a truly predictive statistical model. Considering the antecedent properties of what are thought to be controlling parameters for a variable of interest, and developing a multiple regression or contingency scheme, one may often be able to say that, if certain antecedent values occur, then a certain outcome will follow. For example, Ardis (1961) developed a scheme of threshold value for identifying the air-mass and synoptic conditions that led to thunderstorms the following day. The scheme was right 90% of the time in predicting the occurrence or non-occurrence of thunderstorms, and it seems likely that the technique could probably be adapted to other environmental forecasts.

Experience has shown, in too many cases to list, that statistical models which result from trying all conceivable relationships without a guiding set of physical or biological principles, almost invariably give very poor forecasts. After all, if 100 regressions are calculated between some variable and sets of random variables, one will be a good enough fit to be significant at the 1% level *by chance* on the average. There will, however, be no cause-and-effect meaning.

It seems that one of the problems which plague statistical models and forecasts, arises from failure to consider whether the underlying assumptions in the statistical method which is being used are appropriate to the realm of application. A measure that assumes a monomodal, Gaussian distribution is simply not appropriate to the multimodal, asymmetric distributions that are often found in Nature.

FORECASTING MODELS – SIMPLE AND COMPLEX

A newly-conceived model is an hypothesis. Whether the hypothesis is valid can only be established by multiple testing against reality. Even if these tests are done, the model cannot be regarded as a predictive model unless the tests are of the predictive properties of the model. For example, an ecosystem model may state that the array of life-forms represented in it should be associated with the physical environment in a certain way. The model may be tested against reality in a large number of places, and found to fit the data quite well. It does not follow immediately, however, that the model can be used as a predictive tool. The test of the predictive properties of the model must be made on changes of the physical environment at a place, together with the changes of the biotic array at the same place.

The assumption that the accuracy of a model from place to place can be translated to accuracy from time to time is often made, especially in geology and palynology, even though there may be no data available for the testing of predictions in time. For example, the array of pollen 'rain' at many places in central North America can be related very well to the array of climatic parameters at the same places. It is an *assumption* that this spatial coherence between the two arrays can be used to establish the climates of the past by examining the corresponding pollen array, as did Webb & Bryson (1972). Other data may be used to establish the reasonableness of the assumption, but there is no real test available at present.

Reversing the above example to illustrate a forecasting pitfall appropriate to the discussion of future climates: suppose a similar climatic–biotic relationship were used to say that, in some future climate, a particular biotic array would be found at a different latitude. Unless the model specifically took into account the non-linearity associated with day-length requirements of certain plants, the forecast might fail. For example, when the boreal forest moved northwards in central North America at the close of the last glacial stage, *Fraxinus* did not move with the dominant *Picea* but remained south of a particular latitude.

Let us now consider the often-made statement that environmental impact is proportional to technology *times* dollars. This is a verbal equation, but one that cannot be used, as the dimensions are unspecified. What number and dimensions can one assign to (unspecified) impact or technology? Does the author of the equation really mean dollars, or the equivalent in local currency? Does the term 'environmental impact' mean instantaneous or accumulated? In the latter case the integral should have been specified. It is possible to write the differential equations representing the inchoate thought expressed by the statement above, but, until they are properly specified, the ideas cannot be tested for correctness or used for any practical purpose. (For an attempt at a more precise statement, *see* Bryson, 1989*b* p. 303).

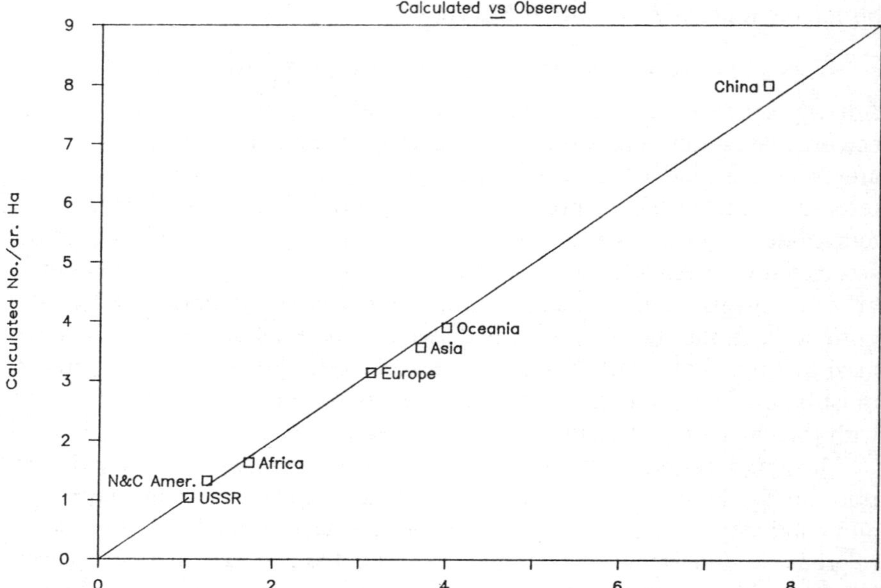

Figure 22.1. People per arable hectare as observed in 1970 compared with the number as calculated in Bryson (1989*b*). The data fit the model, based entirely on climate, very well indeed. This suggests that the population density per arable hectare is quite well adjusted to the mean climate of each region.

Simple environmental models, carefully constructed with due consideration for scale and aggregation, can be shown to fit the data very well and to be of predictive value. Consider Figure 8 in Bryson (1989*b*), which is a plot of actual numbers of persons per arable hectare *versus* the numbers expected on the basis of a regression of the numbers against the rainfall, temperature, and rainfall *times* temperature, at the population centroids of the subcontinental-sized areas used. The covariance of the population per arable hectare observed and the number calculated is very nearly 0.999. This suggests that the areas are very nearly equally saturated with people for the technology and land preparation existing in the year of observation, 1965AD. If this hypothesis is valid, then any rapid departure of the location of a datum towards more people per arable hectare would indicate, *sans* climatic change, population oversaturation of the food-producing resource, outstripping of the slow pace of technology, and production of hunger. That was the hypothesis in the early 1970s when the graph was drawn. Figure 22.1 is the corresponding graph for 1970 and Figure 22.2 is of the observed relationship in 1986. In the interim, the population density per arable hectare increased 64% for Africa, 52% for South America, 41% for South Asia, and less than 25% for the other regions. It is Africa, with the largest percentage increase in people per arable hectare, that had the greatest hunger in the

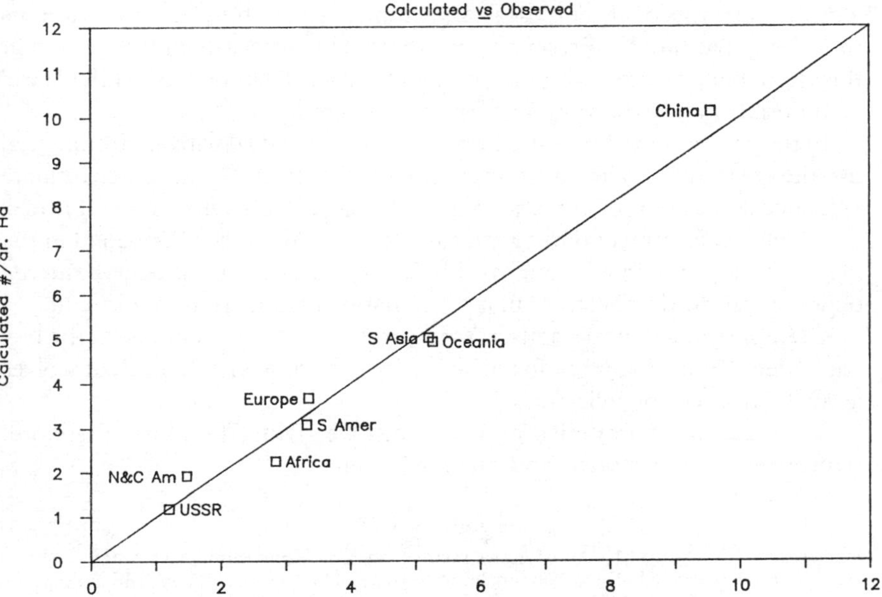

Figure 22.2. People per arable hectare as observed in 1986 compared with the number as calculated in Bryson (1989*b*). The datum for Africa has moved more towards the 'overpopulated' side of the graph than other areas, and Africa has had more famine. It is clear that the population densities are not as neatly adjusted to the climate as in 1970. The larger absolute numbers in 1986 compared with 1970 (*see* Figure 22.1) may reflect either technological change or greater pressure on the land resource.

1970s and 1980s.

Simple forecasting models are cheap to construct and test, though not necessarily easy. Large dynamic models are immeasurably more expensive to construct and run, and have had very little testing in the predictive mode other than for short forward intervals such as hours, days or weeks. Large 'climate' models or GCMs are essentially untested at the year-to-century scale. Knowing that their accuracy is slight at the weeks' scale makes one quite suspicious of their use for forecasts of the next 50 to 100 years.

To the Author's knowledge the predictive properties of large ecosystem and ecocomplex models remain untested, though it is only with extensive testing that forecast schemes and models can gain credibility.

Conclusion and Recapitulation

Environmental science has much to say to a world beset by serious environmental problems. In addition to identifying and describing the problems, much can be said about what will happen if the problems are, or alternatively are not, addressed by whatever means is at the disposal of people. To be useful, however, prescriptions for environmental 'cures' must be shown to be realistic and correct. Because there is a finite amount of wealth in the world, and indeed of land, sea, and human capability, the prescriptions must be quantitative.* Prescriptions are virtually forecasts in the sense that they state that, if nothing is done, certain things will happen, but if the 'cure' is undertaken, a better outcome can be expected.

In the competition for limited world resources to deal with environmental ills, the environmentalist must be credible and correct. We must accumulate evidence that our predictions are reasonably accurate – a not impossible task if we have environmental history available to us. What has happened in the past *is* clearly possible in our world. What has not yet happened *may* be possible, but to demonstrate that it *will* happen is much more difficult.

Credibility is a delicate matter, as illustrated by the ancient tale of the boy who cried 'Wolf!' too often for people to believe him when a pack of wolves actually invaded the village.

We must be quantitative and we must be right. We must, therefore, rigorously test our models and our predictions.

References

Ardis, C. V., Jr (1961). An objective method for forecasting thunderstorms at Truax Field, Madison, Wisconsin. *Bull. Amer. Met. Soc.*, 42(5), pp. 166–74.

Bryson, R. A. (1988). Late Quaternary volcanic modulation of Milankovitch climate forcing. *Theoretical and Applied Climatology*, 39, pp. 115–25.

Bryson, R. A. (1989*a*). Modeling the NW India monsoon for the last 40,000 years. *Climate Dynamics*, 3, pp. 169–77.

Bryson, R. A. (1989*b*). Environmental opportunities and limits for development. *Environmental Conservation*, 16(4), pp. 299–305, 9 figs.

Bryson, R. A. & Dutton, J. A. (1961). Some aspects of the variance spectra of tree rings and varves. *Annals New York Academy of Sciences*, 95, pp. 580–604.

Bryson, R. A. & Goodman, B. M. (1986). Milankovitch and global ice volume simulation. *Theoretical and Applied Climatology*, 37, pp. 22–8.

Bryson, R. A. & Starr, T. B. (1977). Chandler tides in the atmosphere. *Journal of Atmospheric Sciences*, 34(12), pp. 1975–86.

Campbell, W. H., Blechman, J. B. & Bryson, R. A. (1983). Long-period tidal forcing of the Indian monsoon rainfall: An hypothesis. *Journal of Climate and Applied Meteorology*, 22(2), pp. 287–96.

Imbrie, J., Hays, J. D., Martinson, D. G., McIntyre, A., Mix, A C., Morley, J. J., Pisias, N. G., Prell, W. L. & Shackleton, N. J. (1984). The Orbital Theory of Pleistocene Climate: Support from a revised chronology of the marine del^{18}O record. Pp. 269–305 in *Milankovitch and Climate*, Pt. 1. (Eds. A. L. Berger *et al.*). D. Reidel Publishing Co., Dordrecht, The Netherlands: xiii + 510 pp., illustr.

Kutzbach, J. E. & Guetter, P. J. (1986). The influence of changing orbital parameters and surface boundary conditions on climate simulations for the past 18,000 years. *Jour. Atmos. Sci.*, 43 pp. 1726–59.

Milankovitch, M. (1941). Kanon der Erdbestrahlung and seine Anwendung auf das Eiszeiten Problem. *R. Serb. Akad. Spec. Publ.* (Belgrade), 133, pp. 1–613.

*[As an example Professor Bryson wrote us (*in litt.*):
'I once told the late Senator Hubert H. Humphrey that, to feed the annual population-increase of the world by developing newly arable land, would require at least one acre [0.406 ha] per person *times* 80 million new people per year *times* 4,000 US dollars per year to prepare the new land and infrastructure, thus $320,000 million dollars per year *every year*. He asked me to repeat, thought a moment, shook his head, and, grinning, said "You know, Bryson, that is a very large number even to the US Congress!"' Eds.]

(English translation published by Israel Programme for Scientific Translations, Jerusalem, 1969. Available from US Department of Commerce, Washington, DC.)

Polunin, N. (Ed.)(1972). *The Environmental Future: Proceedings of the first International Conference on Environmental Future, held in Finland from 27 June to 3 July 1971*. Macmillan, London & Basingstoke, England, UK, and Barnes & Noble, New York, NY, USA: xiv + 660 pp., illustr.

Schlesinger, M. E. & Zhao, Zong-ci. (1989). Seasonal climate changes induced by doubled CO_2 as simulated by the OSU atmospheric GCM mixed-layer ocean model. *Jour. of Climate,* 2(5), pp. 459–95, illustr.

Webb, Thompson, III & Bryson, R. A. (1972). The Late- and Post-glacial sequence of climatic events in Wisconsin and east-central Minnesota: Quantitative estimates derived from fossil pollen spectra by multivariate statistical analysis. *Quaternary Research,* 2(1), pp. 70–115.

Epilogue: The Budapest Imperative on Surviving With The Biosphere*

1. An international gathering of some 140 scientists and other scholars from more than 30 countries, deeply concerned about the future of the human environment with its natural components, met during 22–27 April 1990 in Budapest at the invitation of the Hungarian Academy of Sciences. The occasion was the fourth in a series of International Conferences on Environmental Future which began in Finland in 1971, and it came to the general conclusion that five leading problems confront the world community, namely:

- the continued rise in human numbers and the ever-increasing consumption of non-renewable and renewable environmental resources;
- the progressive degradation of the environment at local, national, regional, and even global, levels;
- the depressive loss of more and more of the Earth's biological diversity;
- the increasing disparity in wealth among nations, and the crippling poverty in many of the less-developed countries;
- the growing impediments to social advance in many countries – resulting from hunger, disease, ignorance, poverty, and strife.

2. We already have enough evidence to demand remedial action on all the five problems we have identified, although there is a great and continuing need for good environmental research, and for its results to be effectively applied.

3. Political action has lagged too far behind public concern. It *must* catch up, and move ahead. These serious problems will require concerted action among nations. Some progress has been made, as in the mounting effort to

*[Originally promoted through a small *ad hoc* pre-Conference committee consisting of F. Kenneth Hare, Gilbert F. White, and Victor A. Kovda, with powers to co-opt. Drafted early in the Conference by the first-named, amended and generally approved at its Thursday evening session, and subsequently amended in consultation with, particularly, John Burnett, John L. Cloudsley-Thompson, Martin W. Holdgate, Donald F. McMichael, Jean Medawar, Gunavant M. Oza, Nicholas Polunin, Jan W. M. la Rivière, Christopher D. Stone, Francis L. Dale, Gabor Vida, Arthur H. Westing, and David P. S. Wasawo. Passed at the final session of the Conference and subsequently edited as necessary.

The idea of an Imperative originated with Richard G. Miller and follows 'The Reykjavik Imperative on the Environment and Future of Mankind' which was drafted during the 2nd ICEF held in Iceland in 1977 by a committee under the chairmanship of Linus Pauling and published in the proceedings thereof, while further inspiration for the Budapest Imperative was derived from the opening keynote address entitled 'Overview: Our Threatened World' by Martin W. Holdgate and from the Third Baer-Huxley Memorial Lecture. entitled 'Building An Environmental Institutional Framework for the Future'. Eds.]

protect the stratospheric ozone shield. A major United Nations Conference on Environment and Development will be held in Brazil in 1992. But more needs doing now than was the case 20 years ago: Governments have evaded their responsibilities for far too long. Rhetoric has abounded, but relatively little effective action has been taken.

4. Governments and international agencies cannot succeed without the support of the communities which they serve. The experience of the past 20 years demonstrates that Governments need to be moved to action by the demands of individuals and groups within their countries. Action must be based on accurate information and wide understanding.

5. As the new millennium approaches, environmental matters will inevitably move increasingly into the centre of the world's agenda. The welfare of tomorrow's world will depend in large measure on how we deal with the two dominant, interlinked concerns of today: the seemingly inexorable rise in human population, and the poverty and shortages that already threaten two-thirds of the world's population.

6. Without improved economic conditions, education, and health-care, couples are unlikely to have sufficient incentive to limit the numbers of their offspring even if the means to do so are made available to them. Unless women gain access to the same general education and opportunities as men, progress will be further inhibited.

7. Disparities in wealth between nations are exacerbated because poor nations owe vast sums to the rich nations. Finding an equitable way to ease the debt burden would be a first step to reducing these disparities, thereby permitting the attainment of environmentally sound development.

8. The Governments of the world can no longer evade environmental issues. Today we stand in need of a peaceful revolution in which poverty, the rise in human numbers, and environmental degradation, will be brought finally to a halt, and humanity will come to live in enduring harmony with Nature that sustains us all in The Biosphere which constitutes our sole life-support.

THE FOUNDATION FOR ENVIRONMENTAL CONSERVATION
(edited version of the 5th draft, which was passed at the final plenary session of the 4th ICEF, in Budapest, Hungary, on the afternoon of 27 April 1990).

Appendix 1: Our International Conferences on Environmental Future: A Micro-history of the First Twenty Years

NICHOLAS POLUNIN

Foundation for Environmental Conservation, 7 Chemin Taverney, 1218 Grand-Saconnex, Geneva, Switzerland

JYVÄSKYLÄ, FINLAND, 1971

As guest speaker at what was called an 'Environmental Congress', that was held in central Finland in July 1970, I drafted, in consultation with others, a resolution which was passed unanimously at the final plenary session and read as follows:

> 'The Environmental Congress held during 6–11 July 1970 at Jyväskylä, Finland, alarmed at the accelerating changes towards widespread environmental degradation which are threatening the health and very survival of the biosphere and human life as we know them, and realizing that very much more will have to be done to counter these tendencies as human population increases still further, recommends to the appropriate authorities the organization of a high-level International Conference on Environmental Futurology* in which leading specialists of global outlook shall prognosticate what in their expert view is most likely to happen and can be done to avoid further catastrophes to man and nature – the results of their deliberations and discussions to be published in dignified book form in good time for the United Nations Conference on the Human Environment, to be held in Stockholm in June 1972.'†

Through the urging particularly of Professor Kauko Sipponen, at that time Secretary of State in the Prime Minister's Office but subsequently Governor of Central Finland, an invitation was received to hold such a conference in mid-1971 – again in connection with the Jyväskylä Arts Festival, but especially supported by the Government of Finland and the Finnish National Commission for UNESCO. Thus our series of International Conferences on Environmental Future started with the first some 20 years ago and resulted in a large volume (Ed. Polunin, 1972) of which I took copies of the chapters as I edited them over the hill to the United Nations

*[Subsequently pointed out as a 'bad word' by our Conference's Scientific Patron, the late Sir Julian Huxley, and changed to 'Future'.]

†[In the end the book was published only shortly before the Stockholm 'United Nations Conference on the Human Environment', although the chapters were taken to the organizers in Geneva as they were edited.]

office in Geneva where Maurice Strong and his team were preparing for the United Nations Conference on the Human Environment which was to take place in the following year in Stockholm, Sweden. That UN event being planned as a high-level intergovernmental occasion of previously-agreed 'set pieces', our idea was to inculcate the enlightened, duly debated, views of scientific and other specialists having complete freedom of expression.

That this was remembered as successful seemed to be indicated early in 1990 when Maurice Strong telephoned me to say that he had just been appointed by the UN Secretary-General to organize the UN Conference on Environment and Development (UNCED), which is to take place in 1992 in a then undecided location in Brazil.** He subsequently asked for the proceedings volume of the 4th ICEF as soon as available, and so we strove to meet the challenge of repeating this bit of microhistory by providing a breath of reliable fresh-air in good time!

Our 1971 Conference was attended by some 150 invited participants of whom about two-thirds came from other than the host country, and ended with a unanimous vote of the final plenary session to continue with others along similar lines at from two to four years' intervals henceforth if possible, and with the polishing and passage of a Statement that had been drafted by the Resolutions Committee under the Chairmanship of the late Professor Jean Baer. This stated *inter alia:*

'We do not think the collapse of civilization is imminent; there is time to avoid the abyss. This is a solemn call for world-wide action and for collective self-restraint. Now is the time for decisive action for the conservation and amelioration of the human environment. Delayed action may result in irreversible damage and may bring the biosphere to the brink of ecological disaster. In this spirit, we submit the following recommendations to those who will assemble in Stockholm next June, and to the citizens and leaders of the nations of the world who must make many critical decisions.'

One last point about our first Conference. At it were debated, or at least given warnings of, practically all the results of human profligacy which are now threatening our world – stratospheric ozone depletion, the 'greenhouse' effect, deforestation, devegetation and desertification, pollutions of many kinds, nuclear explosions, insufficiency of agricultural land, depletion of raw materials, acidification and salinization, eutrophication, global climatic change, and yet others.

If the general public and their leaders had heeded the large red 'PRIORITY' which the publishers stamped on the dust-jacket of the book of Proceedings, the world would be a far happier and more liveable planet than it is today!

REYKJAVIK, ICELAND, 1977

For this 2nd ICEF we set up an International Steering Committee according to agreed stipulations that it should consist of not fewer than 8 members

**Eventually held from 3–14 June 1992 in Rio de Janeiro.

resident in 6 or more different countries representing at least 4 continents, and electing, at their first meeting, a Chairman from among their number. Details of quorum and procedure were also laid down, and the final clause (No.9) stipulated that 'The Committee for each Conference shall be disbanded at its end but without prejudice as to possible re-election of individuals' on future such occasions. But despite this tightening of discipline and a series of meetings to plan a programme and explore incipient invitations from various countries – three of which were visited before relevant decisions were made – it was not until the summer of 1977 that the 2nd ICEF was actually held, under the subtitle of *Growth Without Ecodisasters*?

By that time the world's human population had increased further and human life had speeded up immensely – especially in its destruction of Nature, so that a leading theme was the need for industrial growth but, at the same time, avoidance of ecodisasters. It was not, however, until early in 1976 that the Government of Iceland decided to invite us to hold the 2nd ICEF in their capital, Reykjavik, in September of the same year, as was confirmed in a letter from their Prime Minister. This proved to be too short a time to organize another truly international conference of the most desirable number of some 140 participants, and so the date was advanced to 5–11 June 1977.

On this occasion, again, the Conference lasted for nearly a week, there were 18 main sessions and corresponding chapters of the Proceedings volume (Ed. Polunin, 1980), participation was entirely by invitation, and the language of the conference was English. The Prime Minister of Iceland, HE Geir Halgrímsson, was Patron of the Conference, and there was a supporting Icelandic National Committee under Professor Gunnar G. Schram and Dr Sturla Fridriksson. The Scientific Patron of the Conference was HE Professor Friedrich T. Wahlen, sometime President of the Swiss Confederation, and the President was Professor Linus Pauling, Nobel Laureate for Chemistry and for Peace. The Sponsors were the Government of Iceland and the Foundation for Environmental Conservation, and there were also donating co-sponsors.

Among cabled or other messages received by the Secretary-General, four were selected and read out by him at the opening of the Conference – from the Scientific Patron of the Conference who, like his predecessor at Jyväskylä, was unable to attend for reasons of health, from the Secretary-General of the United Nations, New York, from Mrs Indira Gandhi, formerly and subsequently Prime Minister of India, and from HE William H. Barton, Ambassador and Permanent Representative of Canada to the United Nations, and President of the UN Security Council, New York.

Besides the 'regular' Proceedings, some memorable social events, and an excursion to Thingvellir, 'the World's Oldest Parliament', two special sessions should be mentioned. The formal one of these was the first Baer–Huxley Memorial Lecture, a series founded by the Secretary-General to commemorate two former associates who had died recently – one while

Chairman of the International Steering Committee of this Conference, and the other its adviser who had been the first Director-General of UNESCO. The lecture was given by Maurice Strong, on 'The International Community and the Environment'. The other, informal, item was a confidential evening debate on the controversial question of 'Is Iceland Polluted?', which ended as a 'drawn match' that so relieved the chief company involved that the next morning the Secretary-General was handed a cheque by that company's President which covered most of the Conference's outstanding debts!

Another outcome of this conference was 'The Reykjavik Imperative on the Environment and Future of Mankind', drafted by the Resolutions Committee consisting of Linus Pauling (Chairman), Donald J. Kuenen, Thomas F. Malone, Letitia E. Obeng, Gunnar G. Schram, and E. Barton Worthington, with powers to co-opt and consult. It adjured the world to look after its limited resources, to stabilize its burgeoning human population, to replace the present-day polluting fuels with renewable sources of energy, to suppress militarism and undue industrial pollution, to encourage environmental impact studies and statements, and to change to a fundamentally different, environmentally aware, approach to economic growth. This New Growth should emphasize quality rather than quantity, reduce the demands on Earth's resources and the concomitant risks to its life-sustaining systems, and so be compatible with our civilization's survival and due conservation of Nature.

EDINBURGH, SCOTLAND, 1987

The 3rd ICEF was a smaller affair of more limited duration than its predecessors, and was arranged at relatively short notice. Under the subtitle of *Maintenance of The Biosphere*, it consisted of five regular sessions and took place in the University of Edinburgh, in honour of its retiring Principal and Vice-Chancellor, who is now Executive Secretary of the World Council For The Biosphere. The five regular sessions were:

I The Global System of Nature and the World Stage of Man
II The Biosphere in Transition
III Global Priorities
IV The Attainable Ideal in Human Ecological Terms
V Practical Targets for Sustainability and Development.

Each of these sessions consisted of two papers (Chapters in the book of Proceedings) by their respective keynote speakers, except that Chapter 10, entitled 'Practical Targets in the Six Inhabited Continents', had special contributions from representatives of each of those six continents (with Australia covering Antarctica). Following each of these sessions there was a 'Commentary' led by three invited Panellists and, thereafter, open discussion as indicated in the book of Proceedings (Eds Polunin & Burnett, 1990).

A sixth and last session was the Second Baer-Huxley Memorial Lecture, prepared by Dr Gro Harlem Brundtland, Prime Minister of Norway, and excellently handled in her unavoidable absence by the Norwegian Ambassador to the United Nations in Geneva, HE Martin Huslid, who added his own Concluding Remarks. The Lecture was entitled 'Our Common Future – A Call for Action', and carried on very well from the report of the World Commission on Environment and Development, of which Dr Brundtland had been Chairman (as she commonly referred to herself, despite being an attractive lady). The 'Brundtland Report' had been entitled *Our Common Future* – hence the title of this outstanding lecture.

In addition, several papers were brought to the Conference by their authors, and were discussed as far as time allowed, four of them being published as Appendices in the Proceedings. These were 'The Economic Policy Framework for Natural Resource Management', by Dr Edward B. Barbier, 'Essential Steps to Protect the Rain-forests from Further Exploitation', by Nicholas Guppy, 'Networking Human Ecology World-wide', by Professor Richard J. Borden, and 'Urban Wastes and Sustainable Development: A Comment on the Brundtland Report', by Dr Christine Furedy.

Altogether the Conference fulfilled its purpose of setting the stage for its successor which was to be, as long intended, the Secretary-General's positively last – at least of primary responsibility.

Budapest, Hungary, 1990

For the 4th ICEF, on *Surviving With The Biosphere*, we had been invited by the Hungarian Academy of Sciences to hold it in Budapest, and with their help in many ways, it was expected to be relatively straightforward to organize. Actually it proved to be notably otherwise, and not at all something which any *individual* should attempt with the world in its present state. To begin with, money proved very difficult to raise and more was long needed to cover our costs; but when we had received firm promises of sufficient for a 'shoestring' operation, and decided to go ahead with this Conference, we found that there were too many events already claiming the services of a large proportion of the most outstanding specialists. (It is gratifying, however, to note that some chosen 'Roving Eminents' and others managed to change their plans to be with us, and that the final total of invited participants returned to our preferred number of about 140.)

In the end we managed to maintain this 4th ICEF at a reasonably high level of papers and debate, all of which proceedings were recorded for transcriptions to be made for reference in editing. Our Patron was HE Professor F. Brunó Straub, until recently President of the Presidential Council of Hungary and a former President of ICSU, while our Co-organizer & Local Agent was Professor Gabór Vida, President of the Hungarian Biological Society and Head of the Department of Genetics in Eötvös Loránd University, Budapest. We also benefited throughout from the interest and unfailing support of Professor István Láng, Secretary-General of the Hungarian Academy of Sciences, who had been a member of Dr Brundtland's World Commission on Environment and Development.

This Conference was *sponsored* by the Hungarian Academy of Sciences and the Foundation for Environmental Conservation, based in Geneva, Switzerland. It was *co-sponsored* by the United Nations Population Fund (UNFPA), the World Conservation Union (IUCN), and the Japan Shipbuilding Industry Foundation (now renamed the Sasakawa Foundation), being further *supported* by the United Nations Environment Programme (UNEP), the Government of Hungary, and the World Wide Fund for Nature (WWF), to all of whom our grateful thanks are due and warmly given.

Finally, it is a matter of deep personal satisfaction to entrust the future of this series of conferences to the Executive Secretary of the World Council For The Biosphere, who had led the (admittedly gratifying to its founder) movement to continue these ICEFs henceforth. A former pupil at Oxford, and practically lifelong friend, whose experience has included the Royal Navy during World War II, the Principalship and Vice-Chancellorship of a major university for rather many years, and the acting Chairmanship of the Nature Conservancy Council through a particularly trying period of its superimposed subdivision, it is a circumstance at once of confident satisfaction and our very good fortune to hand over the leadership of these conferences to Professor Sir John Burnett, after editing together the Proceedings of the 3rd and 4th as cited below.

References

Polunin, N. (Ed.) (1972). *The Environmental Future: Proceedings of the first International Conference on Environmental Future, held in Finland from 27 June to 3 July 1971*. Macmillan, London & Basingstoke, England, UK, and Barnes & Noble, New York, NY, USA: xiv + 660 pp., illustr.

Polunin, N. (Ed.) (1980). *Growth Without Ecodisasters? Proceedings of the Second International Conference on Environmental Future (2nd ICEF), held in Reykjavik, Iceland, 5–11 June 1977*. Macmillan, London & Basingstoke, England, UK, and Halsted Press Division of John Wiley & Sons, New York, NY, USA: xxvi + 675 pp., illustr.

Polunin, N. & Burnett, J. H. (Eds) (1990). *Maintenance of The Biosphere: Proceedings of the Third International Conference on Environmental Future (3rd ICEF)*. [Held in the University of Edinburgh, Scotland, during 24–26 September 1987.] Edinburgh University Press, 22 George Square, Edinburgh EH8 9LF, Scotland, UK: xvi + 228 pp., illustr.

Polunin, N. & Burnett, J. H. (Eds) (in press). *Surviving With The Biosphere: Proceedings of the Fourth International Conference on Environmental Future (4th ICEF)*, held in Budapest, Hungary, 22–27 April 1990. Edinburgh University Press, 22 George Square, Edinburgh EH8 9LF, Scotland, UK: xxii + 572 pp., illust.

Appendix 2: List of Participants* with their Preferred Addresses

Key: K = Keynoter; C = Chairman; P = Panellist; RE = 'Roving Eminent'; M = Moderator; O = Observer; Imp = Budapet Imperative; [...] = contributing in some valued way but not attending

ABRAHAM, Dr Kalman, Director General, Institute for Environmental Management, Alkotmany u. 29, Budapest H-1051, Hungary (O)

ADAM, Professor György, Head of Department, Department of Animal Physiology, Eötvös Lorand Science University, Muzeum krt. 4/a, Budapest H-1088, Hungary (O)

ATANANASOV, Ivan, Director-General, Institute Ocharny Prirodnoi Sredy, Industrialna 7, Sofia, Bulgaria (O)

BAILEY, Dr Robert G., Program Manager, Land Management Planning Systems, USDA Forest Service, Timber & Land Management Planning, 3825 E. Mulberry Street, Fort Collins, Colorado 80524, USA (P)

BATISSE, Dr Michael, Légion d'Honneur, UNESCO, 7 Place de Fontenoy, Paris 75700, France (C & P)

BATISSE, Mme Claude, Friends of the Earth, Villeneuve, c/o Dr Michael Batisse (next above) (O)

BAZZAZ, Professor Fakhri A., H. H. Timken Professor of Science, The Biological Laboratories, Harvard University, 16 Divinity Avenue, Cambridge, Massachusetts 02138, USA (P)

BEREND, Professor T. Ivan, President, Hungarian Academy of Sciences, Roosevelt ter. 9, Budapest, H-1051, Hungary (C)

BERTINE, Dr Kathe, Department of Geology, San Diego State University, San Diego, California 92128, USA (O)

BISIC, Zdravko O., Jugozlav Szocialista Szövetségi Köztársaság Nagykövetsége, Dozsa György u. 92/b, Budapest VI, Hungary (O)

BRYSON, Professor Reid A., Senior Scientist, Center for Climatic Research, Institute for Environmental Studies, University of Wisconsin-Madison, 1225 West Dayton Street, Madison, Wisconsin 53706, USA (C & wind-up)

BURNETT, Professor Sir John, FRSE, 13 Field House Drive, Oxford OX2 7NT, England, UK (K & C)

CLARK, Professor Robert B., Department of Biology, The University, Newcastle upon Tyne NE1 7RU, England, UK(P)

CLOUDSLEY, Mrs Anne, Flat 9, 4 Craven Hill, London W2 3DS, England, UK (O)

CLOUDSLEY-THOMPSON, Professor John L., Department of Biology (Medawar Buildings), University College London, Gower Street, London WC1E 6BT, England, UK (P)

COHEN, Mr Peter, Route 6, Box 6198, Stroudsburg, Pennsylvania 18360, USA (O)

*[Apart from accompanying persons who did not represent any special concern and did not speak but who swelled the numbers to more than 150. Eds.]

CURIKOC, S., Vice-President, Goskompriroda CCCP U1., Nezhdanovoi 11, 103 009 Moscow, USSR (O)

DALE, Ambassador Francis L., President, The Maureen and Mike Mansfield Foundation, Mansfield Library, University of Montana, Missoula, Montana 59812, USA (C & P)

DAOUDY, HE Ambassador Adib, Inspector, Joint Inspection Unit, Office D-512, United Nations, 1211 Geneva 10, Switzerland (C & P)

DAVIS, Mrs Beverley, School of Natural Resourcés, Ohio Sate University, 2021 Coffey Road, Columbus, Ohio 43210, USA (O)

DAVIS, Professor Craig B., School of Natural Resources, Ohio State University, 2021 Coffey Road, Columbus, Ohio 43210, USA (K & P)

DEMETER, Dr András, Scientific Secretry, Biological Section, Hungarian Academy of Sciences, Nádor u. 7, Budapest H-1051, Hungary (O)

DMITRIEVA, Mrs Vera, Executive Director, Vernadskey International Centre for Biosphere Studies, c/o Institutue of Soil Science & Photosynthesis, Akademia Nauk SSSR, Pushchino, Moscow Region 142292, Russia (P)

[EHRLICH, Professor Paul R., Department of Biological Science, Stanford University, Stanford 94305, USA] (K)

[EHRLICH, Dr Anne H., Department of Biological Science, Stanford University, Stanford 94305, USA] (K)

ENYEDI, Professor György, Director-General, Regional Research Centre, Hungarian Academy of Sciences, Kulich Gy. u. 22, Pécs H-7601, Hungary (P)

FEARNSIDE, Professor Philip M., Department of Ecology, INPA, C.P. 478, Manaus-Amazonas 69.011, Brazil (P)

FEHER, Dr Marta, Department of Philosophy, Budapest Technical University, Müegyetem rakpart 3/9, Budapest H-1111, Hungary (P)

FEKETE, Dr Gábor, Head of Department, Institute of Botany and Ecology, Hungarian Academy of Sciences, Alkotmany u. 2–4, Vácratót H-2163, Hungary (O)

FERGUSON, Howard L., Coordinator, Second World Climate Conference, World Meteorological Organization, P. O. Box 2300, 1211 Geneva 2, Switzerland (P)

FORD, Mr Robert, CBE, 1 Latimer Road, Monken Hadley, Barnet, Herts EN5 5NU, England, UK (RE)

FOSBERG, Dr F. Raymond, Botanist Emeritus, National Museum of Natural History, Smithsonian Institution, Washington, DC 20560, USA (P)

FRY, Dr Albert, Director, International Environmental Bureau, 61 Route de Chène, 1208 Geneva, Switzerland. (P)

FUREDY, Dr Christine P., Associate Professor, Faculty of Environmental Studies, York University, 4700 Keele Street, North York, Ontario M3J IP3, Canada (P)

GOLDBERG, Professor Edward D., Mem. Nat. Acad. Sc., Geological Research Division, Scripps Institution of Oceanography, University of California, San Diego, La Jolla, California 92093, USA (K)

*[GULLAND, Dr John A., FRS., Renewable Resources Assessment Group, Centre for Environmental Technology, Imperial College of Science, Technology & Medicine, 8 Prince's Gardens, London SW7 1NA, England] (K)

*[Died 24 June 1990, to the great regret of those who knew him and the loss particularly of the piscatorial fraternity. Eds.]

GYÖRGY, Dr Lajos, Science Adviser, Institute for Experimental Medicine, Szigony u. 43, Budapest H-1083, Hungary (O)

HANKISS, Dr Elemér, Head of Department, Institute for Sociology, Hungarian Academy of Sciences, Uri u. 49, Budapest H-1014, Hungary (O)

HARDI, Dr Peter, Institute for Foreign Affairs, Ministry of Foreign Affairs, Bérc ut 23, Budapest H-1016, Hungary (P)

HARE, Chancellor F. Kenneth, CC, FRSC, 301 Lakeshore Road West, Oakville, Ontario L6K 1G2, Canada (K & Imp)

HARMATHY, Dr Attila, Deputy Director, Institute for Legal and Administrative Sciences, Hungarian Academy of Sciences, Orszaghaz u. 30, Budapest H-1014, Hungary (P)

HOLDGATE, Dr Martin W., CB., Director-General, World Conservation Union (IUCN/UICN), Rue Mauverney 28, 1196 Gland, Switzerland (K & P)

HAMORI, Professor József, Head, Department of Zoology, Janus Pannonius University, Ifjuság utja 6, Pécs H-7604, Hungary (P)

IMRU, HE Mikael, P O Box 2828, Addis Ababa, Ethiopia (K)

IONESCU, Constantin, Vice Minister, Minsterul Apelor Padurilor si Mediului Inconjurator, Str. Negustori 3 sec, 2, 70091 Bucuresti, Romania (O)

JERMY, Professor Tibor, Chairman, Biological Section, Hungarian Academy of Sciences, Nádor u. 7, Budapest H-1051, Hungary (O)

[JIM, Dr Chi Yung, Department of Geography & Geology, University of Hong Kong, Pokfulam Road, Hong Kong]

JUEL-JENSEN, Dr Bent, Medical Officer to the University, Radcliffe Infirmary, Woodstock Road, Oxford OX2 6HE, England (P)

JUHÁSZ-NAGY, Professor Pál, Department of Plant Systematics and Ecology, Eötvös Lorand Science University, Kun B. ter 2, Budapest H-1083, Hungary (K)

KAMINSKI, Bronislaw, Minister, Ochrony srodowiska zasobow naturalnych i lestistwa, U1. Wawelska 52/54, 00-922 Warsaw, Poland

KEFELI, Professor Valentin I., Director, Institute of Soil Science and Photosynthesis, Academy of Sciences of Russia, Pushchino, Moscow Region 142292, Russia (P & C)

KOCH, Professor Sándor, Science Adviser, 2nd Department of Pathology, Semmelweis Medical University, Ullöi ut 93, Budapest H01091, Hungary (P)

*[KOVDA, Professor Viktor A., Institute of Soil Science and Photosynthesis, Academy of Sciences of the USSR, Pushchino, Moscow Region 142292, USSR] ([Imp.])

KUENEN, Professor Donald J., Blauwe Vogelweg 2a, 2333 VK Leiden, The Netherlands (C)

KUNA, Milan, Head of Section, Statni Komise pro Vedeckotechnicky a Investicni Rozvoj, Slezská 9, Praha 2, Czechoslavakia (O)

LÁNG, Prof. Dr István, Secretary-General, Hungarian Academy of Sciences, Roosevelt tér. 9, Budapest H-1051, Hungary (P & C)

LASZLO, Rector Ervin, Villa Franatoni, 56040 Montescudaio, Pisa, Italy (K)

MCCLOSKEY, Mrs Maxine, Washington Representative, Ocean Alliance, 5101 Westbard Avenue, Bethesda, Maryland 20816, USA (P)

*[Died 23 October 1991, to the great loss of his country and soil and environmental scientists throughout the world. Eds.]

McCusker, Dr Alison, Head of Research, International Board for Plant Genetic Resources, c/o of Food and Agriculture Organization of the UN, Via delle Terme di Caracalla, Rome 00100, Italy (P)

McMichael, Dr Donald F., CBE, 244 La Perouse Street, Red Hill, ACT 2603, Australia (P)

McNeely, Jeffrey, Chief Conservation Officer, World Conservation Union (IUCN/UICN), Rue Mauverney 28, 1196 Gland, Switzerland (P)

Marshall, Professor Norman B., FRS, 6 Park Lane, Saffron Walden, Essex CB10 1DA, England, UK (C)

Medawar, Lady (Jean), 25 Downshire Hill, London NW3 1NT, England, UK (P)

Meszaros, Professor Ernö, Director, Central Institute for Atmospheric Physics, National Meteorological Service, Péterhalmi ut 1, Budapest H-1181, Hungary (O)

Miller, Dr Richard G., Director, Foresta Institute for Ocean and Mountain Studies, P. O. Box 41567, Tucson, Arizona 85717, USA (P)

Mische, Dr Gerald F., President, Global Education Associates, Suite 456, 475 Riverside Drive, New York, NY 10115, USA (P)

[Myers, Dr Norman, Consultant in Environment and Development, Upper Meadow, Old Road, Headington, Oxford OX2 8SZ, England] (K)

[Obasanjo, General Olusegun, P O Box 2286, Abeokuta, Ogun State, Nigeria] (K)

[O'riordan, Professor Timothy, School of Environmental Sciences, University of East Anglia, Norwich NR4 7TJ, England, UK] (P)

Oza, Dr Gunavant M. General Secretary of INSONA, Oza Building, Salatwada, Baroda 390 001, India (P)

[Palmberg, Dr Cristel, Forest Resources Development Branch, Forest Resources Division, FAO, Via delle Terme di Caracalla, Rome 00100, Italy]

Papp, Dr Lászió, Senior Scientist, Zoological Department, Hungarian Natural History Museum, Baross u. 13, Budapest H-1088, Hungary (O)

Pechlof, Georgina, Chief of Protocol, Hungarian Academy of Sciences, Roosevelt tér. 9. Budapest H-1051, Hungary (O)

Pellew, Dr Robin A., Director, World Conservation Monitoring Centre, 219c Huntingdon Road, Cambridge CB3 0DL, England, UK (K)

Perczel, Dr György, Deputy Minister, Ministry for Environment and Water Management, Fö utca 44, Budapest H-1011, Hungary (O)

Perez, Juan M. Wong, Head of Department, Comision Nacional de Proteccion del Medio Ambiente (COMARNA), La Habana 4, Cuba (O)

Persányi, Dr Miklós, Senior Adviser, Ministry for Environment and Water Management, PO Box 351, 1394 Budapest, Hungary (RE)

Petts, Mrs Judith I., Centre for Extension Studies, University of Technology, Loughborough, Leicestershire LE11 3TU, England (O)

Petts, Professor Geoffrey E., Department of Geography, University of Technology, Loughborough, Leicestershire LE11 3TU, England (P)

Polunin, Professor Nicholas, President, Foundation for Environmental Conservation, 7 Chemin Taverney, 1218 Grand-Saconnex, Geneva, Switzerland (Organizer & C)

POLUNIN, Dr N. V. C., Research Coordinator, Centre for Tropical Coastal Management Studies, Department of Biology, The University, Newcastle upon Tyne NE1 7RU, England, UK (P)

POORE, Dr M. E. Duncan, Balnacarn, Glenmoriston, Inverness IV3 6YJ, Scotland, UK (K)

PURCELL, Dr Arthur H., Director, Resource Policy Institute, 1745 Selby Avenue No. 11, Los Angeles, California 90024, USA (C)

PUSZTAI, Dr János, Head of Foreign Affairs, Hungarian Academy of Sciences, Roosevelt tér. 9, Budapest H-1051, Hungary (O)

RAMAKRISHNAN, Professor P. A., FNA, School of Environmental Sciences, Jawaharlal Nehru University, New Delhi 110 067, India (K)

RASHID, HE Ambassador Harun ur, Permanent Mission of Bangaladesh to the United Nations, 65 Rue de Lausanne, 1202 Geneva, Switzerland (C)

[RIPLEY, Dr S. Dillon, c/o Embassy of the USA, Shanti Path, Chanakyapuri, New Delhi 110 021, India]

RIVIÈRE, Professor J. W. M. la, International Institute for Hydraulic and Environmental Engineering, Oude Delft 95, P O Box 3015, 2601 DA Delft, The Netherlands (P)

SACKS, Dr Arthur B., Institute for Environmental Studies, University of Wisconsin-Madison, 1225 West Dayton Street, Madison, Wisconsin 53706, USA (P)

SALANKI, Professor János, Director, Lake Balaton Limnological Research Institute, Hungarian Academy of Sciences, Fürdötelepi ut, Tihany H-8237, Hungary (O)

SAMATAR, Professor Said S., Department of History, Rutgers University, Conklin Hall, 175 University Avenue, Newark, New Jersey 07102, USA (K)

SCHRÖDER, Dr Karl-Heinz, Director-General, Ministerium für Umwelt, Naturschutz, Energie und Reaktorsichereit, Schiffbauerdamm 15, Berlin 1040, Germany (O)

[SCHULTES, Professor Richard E., Emeritus Director, Botanical Museum of Harvard University, 26 Oxford Street, Cambridge, Massachusetts 02138, USA] (P)

SHAW, Dr R. Paul, Senior Population and Developemnt Economist, United Nations Population Fund (UNFPA), East 42nd Street, New York, NY 10017, USA (C)

SIMON, Dr David, Centre for Developing Areas Research, Department of Geography, Royal Holloway and Bedford New College, Egham Hill, Egham, Surrey TW20 0EX, England, UK (P)

[SINGH, HE Dr Karan, Mansarovar, 3 Nyanya Marg, Chanakyapuri, New Delhi 110 021, India]

[SMITH, Professor Nigel J. H., Department of Geography, University of Florida, Gainsville, Florida 32611, USA] (P)

SMYTH, Professor John C., President, Scottish Environmental Education Council, University of Stirling, Stirling FK9 4LA, Scotland, UK (K)

SÓMLYODY, Dr László, Director-General, Water Resources Research Centre, Kvassay ut 1, Budapest H-1095, Hungary (P)

STANTON, Professor W. Robert, 73 Main Street, Stanbury, Keighley, West Yorkshire BD22 0HA, England, UK (RE)

[STAPP, Professor William B., Thurnau Professor, Griffith University, Nathan, Queensland 4111, Australia] (K)

STEFANOVITS, Professor Pál, Head, Department of Soil Sciences, Agricultural University, Páter K. u. 1, Gödöllö H-2103, Hungary (O)

STONE, Professor Christopher D., Roy P. Crocker Professor of Law, The Law Centre, University of Southern California, University Park, Los Angeles 90089-0071, USA (K)

STRAUB, HE Professor F. Brunó, Former President of Hungary and of ICSU, Institute of Enzymology, Szeged Biological Centre, Hungarian Academy of Sciences, Karolina ut 29, Budapest H-1113, Hungary (C)

STUMM, Professor Dr Werner, Director, EAWAG, Dübendorf-Zürich 8600, Switzerland (P)

SZABÓ, Dr T. Attila, Head, Botanical Department, Hungarian Natural History Museum, Könyves Kálmán krt. 40, Budapest H-1097, Hungary (P)

SZENTAGOTHAI, Professor János, Former President, Hungarian Academy of Sciences, Roosevelt tér. 9, Budapest H-1051, Hungary (O)

SZIGETI, István, Deputy Secretary of CMEA , Kalinina Prospekt 56, 121 205 Moskva, USSR (O)

THORNDIKE, Mrs Elizabeth, President, Center for Environmental Information Inc., PO Box 245, Fort Myers Beach, Florida 33931, USA (P)

TIGYI, Professor József, Vice-President, Hungarian Academy of Sciences, Roosevelt tér. 9, Budapest H-1051, Hungary (O)

TISDELL, Professor Clement A., Head, Department of Economics, University of Queensland, St Lucia, Queensland 4067, Australia (K)

TOLBA, Dr Mostafa K., Executive Director, United Nations Environment Programme, P O Box 30552, Nairobi, Kenya (C)

TROMPF, Professor Garry W., Head, Department of Comparative Religions, Universtiy of Sydney, Sydney, NSW 2006, Australia (P)

VALLENTYE, Dr John R., Chairman, Canadian Section of Great Lakes Science Advisory Board, 867 Lakeshore Road, P O Box 5050, Burlington, Ontario L7R 4A6, Canada (P)

VARGA, Dr Miklós, Permanent Secretary & Deputy Minister, Ministry for Environment and Water Management, Fö utca 48–50, Budapest H-1011, Hungary (O)

VÁRALLYAY, Dr György, Director, Research Institute for Soil Science and Agricultural Chemistry, Hungarian Academy of Sciences, Hermann Ottó ut 15, Budapest H-1525, Hungary (P)

VASARHELYI, Judit, Independent Ecological Centre, Miklós ter. 1, Budapest H-1035, Hungary (O)

VASARHELYI, Dr Tamás, Deputy Director, Zoological Department, Hungarian Natural History Museum, Baross u. 13, Budapest H-1088, Hungary (O)

VIDA, Pròfessor Gábor, Head, Department of Genetics, Eötvös Lóránd Science University, Muzeum krt. 4/a, Budapest H-1088, Hungary (Co-organizer & K)

WASAWO, Professor David P. S., P O Box 41024, Nairobi, Kenya (Co-C & P)

WESTING, Dr Arthur H., Westing Associates in Environment, Security & Education, RFD1, Box 919, Putney, Vermont 05346, USA (Co-C, K & P)

WESTING, Mrs Carol, Westing Associates in Environment, Security & Education, RFD1, Box 919, Putney, Vermont 05346, USA (O)

[WHITE, Professor Gilbert F., University of Colorado, Campus Box 482, Boulder, Colorado 80302, USA] (Imp)

WORTHINGTON, Dr E. Barton, CBE, Colin Godmans, Furners Green, Near Uckfield, Sussex TN22 3RR, England, UK (M)

ZAKONYI, Dr János, Director-General, Department for International Relations, Ministry for Environment and Water Management, P O Box 351, Budapest H-1394, Hungary (O)

ZÓLYOMI, Academician Bálint, Consultant Scientist, Institute of Botany and Ecology, Hungarian Academy of Sciences, Alkotmany ut 2–4, Vácrátót H-2163, Hungary (O)

ZSOLNAI, Dr László, Associate Professor, Department of Business Economics, Budapest Economics University, Veres Pálne u. 36, Budapest H-1053, Hungary (P)

Appendix 3:
Acronyms in the Text

ACDA	Armament Control and Disarmament Agency (USA)
AGCM	Atmospheric General Circulation Model
AIDS	Auto-Immune Deficiency Syndrome
ANEN	African NGOs Environment Network
APPEN	Asia-Pacific People's Environment Network
BWU	Blue Whale Unit
BP	Before Present
CAP	Common Agricultural Policy (EC)
CCAMLR	Commission for the Conservation of Antarctic Marine Living Resources
CFCS	Chlorofluorocarbons
CGIAR	Consultative Group on International Agricultural Research
CIMMYT	Centro Internacional de Mejoramiento de Maiz y Trigo
CITES	Convention on International Trade in Endangered Species of wild fauna and flora
CMEA	Council for Mutual Economic Assistance
COHMAP	Cooperative Holocene Mapping Project
COMM-89	Commonwealth group of experts (meeting in 1989)
CTA	Technical Centre for Agricultural and Rural Cooperation (Paris)
DNA	Deoxyribose Nucleic Acid
EC	European Community
EEZ	Exclusive Economic Zone
ESA	Endangered Species Act (USA)
FAO	Food and Agricultural Organization of the United Nations
FIO	Forest Industries Organization (India)
FPC(S)	Forest Protection Committee(s) (India)
GATT	General Agreement on Tariffs and Trade
GCM(S)	General Circulation Model(s) (of atmosphere)
GEMS	Global Environment Monitoring System
GESAMP	(Joint) Group of Experts on the Scientific Aspects of Marine Pollution
GDP	Gross Domestic Product
GIS	Geographical Information System
GNP	Gross National Product
GRID	Global Resources Information Database
HIV	Human Immunodeficiency Virus
IAEA	International Atomic Energy Agency
IARC(S)	International Agricultural Research Centre(s)
IBP	International Biological Programme
IBPGR	International Board for Plant Genetic Resources
ICAR	Indian Council for Agricultural Research
ICARDA	International Centre for Agricultural Research in Dry Areas

ICEF	International Conference on Environmental Future
ICOLD	International Commission on Large Dams
ICRISAT	International Crops Research Institute for the Semi-Arid Tropics
ICSU	International Council of Scientific Unions
IGADD	Inter-Governmental Authority on Drought and Development
IGBP	International Geosphere-Biosphere Programme
IIASA	International Institute for Applied Systems Analysis
ILCA	International Livestock Centre for Africa
ILO	International Labour Organization
IMO	International Marine Organization
IPCC	Intergovernmental Panel on Climatic Change
IRPTC	International Register of Potentially Toxic Chemicals
IRRI	International Rice Research Institute
ISEE	International Society for Environmental Education
ITQ	Individual Transferable Quota
ITTO	International Tropical Timber Organization
IUBS	International Union of Biological Sciences
IUCN	The World Conservation Union (International Union for the Conservation of Nature and natural resources)
IWC	International Whaling Commission
IWC	International Wheat Council
LDCS	Least Developed Countries
MAB	Man And the Biosphere programme
MSY	Maximum Sustainable Yield
NAS	National Academy of Sciences (of the USA)
NEPA	National Environment Protection Act (USA)
NG(s)	Non-Governmental Organization(s)
NOAA	National Oceanic and Atmospheric Administration (USA)
NSS	National Security Service (of Somalia)
OECD	Organization for Economic Cooperation and Development
PPEC	Previously Planned Economic Countries ('Eastern bloc')
PRIO	International Peace Research Institute (Oslo)
RFLPS	Restriction Fragment Length Polymorphism(s)
SCOPE	Scientific Committee On Problems of the Environment
SIDA	Swedish International Development Authority
SSDF	Somali Salvation Democratic Front
TBT	Tri-butyltin
UN	United Nations
UNCTAD	United Nations Conference on Trade and Development
UNEP	United Nations Environment Programme
UNESCO	United Nations Educational Scientific and Cultural Organization
UNFPA	United Nations Fund for Population Activities
UNGA	United Nations General Assembly
UNICEF	United Nations International Children's Emergency Fund
WCB	World Council for the Biosphere
WCED	World Commission on Environment and Development
WCMC	World Conservation Monitoring Centre
WCS	World Conservation Strategy (of IUCN)
WMO	World Meteorological Organization
WSF	World Shrimp Farming
WWF	World Wide Fund for nature

Index of Contributors

General Index